The Theory of Critical Phenomena

The Theory of Critical Phenomena
An Introduction to the Renormalization Group

J. J. BINNEY
Merton College and Department of Physics
Oxford University

N. J. DOWRICK and A. J. FISHER
St John's College and Department of Physics
Oxford University

M. E. J. NEWMAN
Laboratory for Atomic and Solid-State Physics
Cornell University

CLARENDON PRESS · OXFORD
1992

Oxford University Press, Walton Street, Oxford OX2 6DP
*Oxford New York Toronto
Delhi Bombay Calcutta Madras Karachi
Petaling Jaya Singapore Hong Kong Tokyo
Nairobi Dar es Salaam Cape Town
Melbourne Auckland
and associated companies in
Berlin Ibadan*

Oxford is a trade mark of Oxford University Press

*Published in the United States
by Oxford University Press, New York*

© *J. J. Binney, N. J. Dowrick, A. J. Fisher, and M. E. J. Newman 1992*

*All rights reserved. No part of this publication may be reproduced,
stored in a retrieval system, or transmitted, in any form or by any means,
electronic, mechanical, photocopying, recording, or otherwise, without
the prior permission of Oxford University Press*

*This book is sold subject to the condition that it shall not, by way
of trade or otherwise, be lent, re-sold, hired out, or otherwise circulated
without the publisher's prior consent in any form of binding or cover
other than that in which it is published and without a similar condiiton
including this condition being imposed on the subsequent purchaser*

A Catalogue record for this book is available from the British Library

*Library of Congress Cataloging in Publication Data
(data available)
ISBN 0–19–851394–1 (h/b)
ISBN 0–19–851393–3 (p/b)*

*Typeset by the authors using TeX
Printed in Great Britain by
Bookcraft (Bath) Ltd, Midsomer Norton*

Preface

There are now quite a number of books available on the theory of critical phenomena. Does the world really need another? We believe this book does fill a gap in the literature for several reasons.

First, it combines discussions of exact solutions, numerical simulations, real-space renormalization and field-theoretic methods, in a way which we hope illuminates the similarities and differences, and the strengths and weaknesses, of all these approaches to the study of continuous phase transitions.

Second, we have tried hard to make the book accessible to students with a good undergraduate background in physics but no knowledge of quantum field theory. Thus, we have taken pains to exclude as much jargon as possible, and to define clearly those technical terms that we do require. We have also tried to make the book as self-contained as possible by covering in boxes and appendices the technical details and mathematical techniques which are necessary for the solution of some of the more difficult problems in the field, but with which the reader may be unfamiliar.

Finally, at the end of each chapter we give several problems, which we hope will help readers to become familiar with the concepts and techniques introduced in that chapter. Complete solutions to all the problems are given at the back of the book.

While writing this book we have drawn freely on many sources, but especially on the books by Daniel Amit (1989), Shang-Keng Ma (1976), Giorgio Parisi (1988), Eugene Stanley (1971), and Jean Zinn-Justin (1989). In addition we are indebted to many people for helpful comments and suggestions, and for interesting conversations and seminars on the subject of critical phenomena, but we would in particular like to thank Eytan Domany for the outstanding series of lectures he gave at Oxford in the summer of '91, and Robert Phillips for his careful reading of the manuscript and his many helpful criticisms. Thanks are due to Lawrence Harwood and the Dyson Perrins Laboratory for making Figure 1.8 possible.

Oxford JJB NJD
January 1992 AJF MEJN

Contents

1 Introduction   1
- **1.1 Continuous phase transitions and critical points**   2
  - 1.1.1 Divergences and critical exponents   5
  - 1.1.2 Fluctuations and critical opalescence   8
- **1.2 The order parameter**   9
  - 1.2.1 Liquid–gas transition   11
  - 1.2.2 Binary fluids   11
  - 1.2.3 Ferromagnetic/paramagnetic transition   12
  - 1.2.4 Anti-ferromagnetic/paramagnetic transition   13
  - 1.2.5 Helium I/helium II transition   13
  - 1.2.6 Conductor/superconductor transitions   14
  - 1.2.7 Helium three   15
- **1.3 Correlation functions**   16
- **1.4 Universality**   21
- **1.5 Thermodynamic potentials**   21
  - 1.5.1 The Widom and Kadanoff scaling hypotheses   27
- **1.6 Why study phase transitions?**   30
- **Problems**   31

2 Statistical mechanics   33
- **2.1 Thermodynamic quantities**   35
- **2.2 Fluctuations and correlation functions**   40
- **2.3 Metastability and spontaneous symmetry breaking**   47
  - 2.3.1 Metastability   47
  - 2.3.2 Spontaneous symmetry breaking   48
- **Problems**   52

3 Models — 54

3.1 Description of models — 55
- 3.1.1 The Ising model — 55
- 3.1.2 The lattice gas — 56
- 3.1.3 β-brass — 57
- 3.1.4 The XY and Heisenberg models — 57
- 3.1.5 Potts model — 58
- 3.1.6 Gaussian and spherical models — 58
- 3.1.7 Percolation model — 60

3.2 Transfer matrices and the Ising ring — 61
- 3.2.1 Solution of the Ising ring — 62
- 3.2.2 Correlation functions — 64

3.3 The partition function of the spherical model — 66

3.4 High-temperature expansions and the Ising model — 72
- 3.4.1 High-temperature expansions — 72
- 3.4.2 The partition function of the Ising model — 73
- 3.4.3 The correlation functions of the Ising model — 78
- 3.4.4 Numerical evaluation of high-temperature expansions — 80

Problems — 82

4 Numerical simulations — 84

4.1 Direct evaluation of thermal averages — 85

4.2 Sampling configurations — 87
- 4.2.1 Importance sampling — 89
- 4.2.2 General structure of numerical algorithms — 91

4.3 Monte Carlo methods — 92
- 4.3.1 The Metropolis algorithm — 94

4.4 Molecular dynamics — 95
- 4.4.1 Ergodicity and integrability — 97
- 4.4.2 From microcanonical to canonical averages — 98

4.5 Langevin equations — 100
- 4.5.1 Comparison of the Langevin and molecular-dynamics methods — 103

4.6 Independence of configurations — 103
- 4.6.1 Correlations along the path — 104
- 4.6.2 Critical slowing down — 104
- 4.6.3 The Swendsen–Wang algorithm — 106
- 4.6.4 The Wolff algorithm — 108

4.7 Calculation of critical exponents from simulations — 111

Problems — 111

Contents

5 Real-space renormalization ... 113
   5.1 Renormalizing the lattice ... 114
   5.2 Block variables ... 115
   5.3 The renormalization of the Hamiltonian ... 117
      5.3.1 Fixed points ... 120
      5.3.2 The calculation of ν ... 124
   5.4 The renormalization of B, M, χ and G_c ... 127
      5.4.1 The value of ω ... 128
      5.4.2 Non-zero external field ... 129
      5.4.3 The renormalization of M, χ and G_c ... 131
      5.4.4 Critical exponents for the renormalized model ... 132
   5.5 The critical exponents for $T = T_c$ ... 133
      5.5.1 The exponent η ... 133
      5.5.2 The exponent δ ... 134
   5.6 The critical exponents for $T \neq T_c$ ... 135
      5.6.1 The exponent β ... 136
      5.6.2 The exponent γ ... 137
      5.6.3 The exponent α ... 137
   5.7 The scaling laws ... 140
   5.8 Bond percolation in two dimensions ... 143
   5.9 The Ising model ... 147
   5.10 Monte Carlo renormalization ... 153
   Problems ... 156

6 Mean-field theory ... 158
   6.1 Mean-field theory of the Ising model ... 159
   6.2 Mean-field theory of percolation ... 161
   6.3 Mean-field theory of the non-ideal gas ... 162
   6.4 A variational derivation of mean-field theory ... 164
   6.5 Correlation functions in mean-field theory ... 168
   6.6 Infinite-range interactions ... 171
   6.7 Critical exponents in mean-field theory ... 173
      6.7.1 Calculating η from $\widetilde{G}_c^{(2)}(\mathbf{k})$ ... 175
   6.8 What is missing from mean-field theory? ... 176
   Problems ... 177

7 The Landau–Ginzburg model ... 178
   7.1 Formulation of the Landau–Ginzburg model ... 178
   7.2 Landau theory ... 183
   Problems ... 187

8 Diagrammatic perturbation theory — 188

- 8.1 The Gaussian partition function — 189
  - 8.1.1 Correlation functions in the Gaussian model — 192
- 8.2 The partition function for the full Landau–Ginzburg model — 195
  - 8.2.1 The Feynman rules — 195
  - 8.2.2 The symmetry factor — 199
- 8.3 The Helmholtz free energy of the Landau–Ginzburg model — 203
  - 8.3.1 Feynman rules in wavevector space — 204
  - 8.3.2 Vertex functions — 211
- 8.4 The Gibbs free energy of the Landau–Ginzburg model — 215
  - 8.4.1 The rules for finding $\Gamma[\widetilde{\varphi}]$ — 216
  - 8.4.2 The loop expansion — 221
  - 8.4.3 The one-loop Gibbs free energy — 223
- Problems — 226

9 Renormalization — 228

- 9.1 Mass renormalization — 230
- 9.2 Field renormalization — 235
- 9.3 Renormalizing the coupling constant — 238
- 9.4 Renormalization at higher orders — 241
- 9.5 More on field renormalization — 245
- 9.6 The Ginzburg criterion — 246
- Problems — 248

10 The calculation of critical exponents for $T \geq T_c$ — 249

- 10.1 Ultraviolet and infrared divergences — 249
- 10.2 The calculation of γ — 254
  - 10.2.1 $d = 4$ and above — 255
  - 10.2.2 Below four dimensions — 256
- 10.3 The calculation of η — 261
  - 10.3.1 $d = 4$ and above — 263
  - 10.3.2 Below four dimensions — 264

Contents   xi

    10.4 The ϵ-expansion   268
        10.4.1 Dimensional regularization   270
        10.4.2 Calculating γ by dimensional regularization   271
        10.4.3 Calculating η by dimensional regularization   273
        10.4.4 Feynman parameters   273
        10.4.5 The calculation of η again   275
        10.4.6 Calculation of η by the ϵ-expansion   278
    Problems   281

11 The renormalization group   282
    11.1 The renormalization group at $T = T_c$ 283
    11.2 The exponents η and δ   293
        11.2.1 The exponent η 293
        11.2.2 The exponent δ 294
    11.3 The calculation of β and γ_1   296
        11.3.1 The calculation of γ_1 to order ϵ^2 296
    Problems   298

12 The renormalization group at $T \neq T_c$ 299
    12.1 Expansion about the critical temperature   300
        12.1.1 Functional Taylor expansions   301
        12.1.2 Diagrammatic representation of the ϕ^2 correlation   301
               functions
        12.1.3 Wavevector space   302
        12.1.4 Vertex functions   304
        12.1.5 Renormalization   305
        12.1.6 Expanding the renormalized vertex functions   308
        12.1.7 The validity of the expansion   309
    12.2 The renormalization group equations   310
        12.2.1 The exponent ν 311
        12.2.2 The exponent γ 312
        12.2.3 The exponent α 312
    12.3 The renormalization group below T_c   314
        12.3.1 The exponent β 315
    12.4 Calculating γ_2 to one loop   316
    Problems   318

13 The lower critical dimension   320
    13.1 Order below T_c   321
        13.1.1 The case $D = 1$ 321
        13.1.2 Systems with more than one component   322
        13.1.3 Goldstone modes   324

13.2 The non-linear σ-model		325
13.2.1 The two-point vertex function		329
13.2.2 The renormalization group equation		333
13.3 The Kosterlitz–Thouless transition		339
13.3.1 The two-dimensional Coulomb gas		342
13.3.2 General Remarks		350
Problems		352

14 Universality 353

14.1 Perturbing the Gaussian Hamiltonian		355
14.1.1 The applicability of these results		361
14.2 Perturbing the Landau–Ginzburg Hamiltonian		361
14.2.1 The case of three dimensions		362
14.2.2 The case of two dimensions		366
14.3 Relevance and renormalizability		369
Problems		373

Appendices

A	The magnetic scattering of neutrons	375
B	The natural variables for thermodynamic potentials	378
C	Magnetic energy	381
D	Connected correlation functions and $\log Z[J]$	384
E	The Gibbs free energy	386
F	Discrete Fourier transforms	390
G	The method of steepest descent	393
H	Counting closed loops on a square lattice	394
I	Einstein's fluctuation theory	398
J	The Gaussian transformation	399
K	The Landau–Ginzburg model and the Ising model	400
L	Functional differentiation and integration	404
M	The Feynman rules for the vertex functions	417
N	Feynman rules for generalized Landau–Ginzburg models	421

Answers 429

References 448

Index 453

1
Introduction

The phase transitions of H_2O are matters of everyday experience: the boiling of water in a kettle; the formation of frost on clear winter nights; the melting of ice-cubes in a fizzy drink. Water turns into vapour, vapour into ice, ice into water. Life without the endless movement of H_2O from one phase to another would be a sorry thing. Furthermore, since the earliest times the phase changes of substances other than H_2O have been of the greatest technological (and therefore cultural and political) significance: the founding of metals to make swords, ploughshares and printer's type; the evaporation and condensation of ammonia in refrigerators; the evaporation of petroleum distillates in carburettors—to name but a few that were important by the end of the nineteenth century. Phase transitions are central to life on Earth and understanding them is one of the prime tasks of condensed-matter physicists. This book seeks to contribute to that understanding by discussing a small but particularly interesting subset of all phase transitions, the 'continuous' ones.

Under ordinary circumstances the phase transitions of H_2O, or the solidification of molten metal, are 'first-order' phase transitions rather than continuous ones. First-order phase transitions are those that involve latent heat: when a material makes a first-order transition from a high-temperature phase to a low-temperature phase, a non-zero quantity of heat, the latent heat, is given out as the material cools through an infinitesimally small temperature range around the transition temperature T_t. This emission of heat at the transition tells us that the structure of the material is being radically

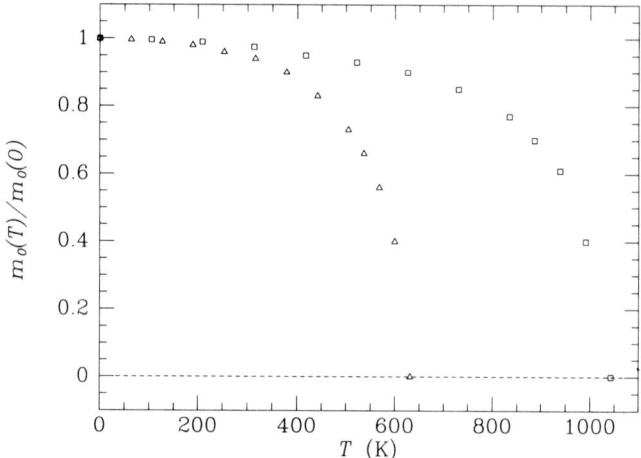

Figure 1.1 The magnetization $m_0(T)$ of iron (squares) and nickel (triangles) in the absence of a magnetic field. Notice that for this continuous phase change dm_0/dT has an infinite discontinuity at the Curie point. This fact makes this a second-order phase change in Ehrenfest's classification—see Box 1.3.

reordered at T_t. For example, the latent heat $L \simeq 334\,\mathrm{J\,g^{-1}}$ of the water–ice transition is the energy released when H_2O molecules neatly pack themselves into a face-centred cubic lattice, rather than wandering around in a disorganized series of constantly dissolving huddles. Above the freezing point of water, there is no crystal lattice. Below the freezing point, the lattice is well defined even if not free of imperfections ('defects'). The transition from disordered water to ordered ice is an all-or-nothing affair. Either the lattice is there and the vast majority of H_2O molecules are comparatively tightly bound, or there is no lattice and the molecules are not optimally packed. As water freezes the transition is made, and energy released as latent heat.

1.1 Continuous phase transitions and critical points

The paradigm of a continuous phase transition is the conversion at the **Curie temperature** $T_c = 1043\,\mathrm{K}$ of iron from paramagnetic to ferromagnetic form. At $T > T_c$ iron, like copper or zinc, is paramagnetic. That is, the material is not magnetized in the absence of an applied magnetic field, and if a weak field **B** is applied, the material's magnetic moment per unit volume[1] **m** is proportional to the applied field: $\mathbf{m} \simeq \mu\mathbf{B}$ with μ a positive constant. In

---

[1] In this book we shall use capital letters for extensive quantities such as dipole moment or entropy when referring to total dipole moment, entropy or whatever, and shall use small letters when referring to the dipole moment per unit volume, entropy per unit volume, etc.

1.1 Continuous phase transitions and critical points

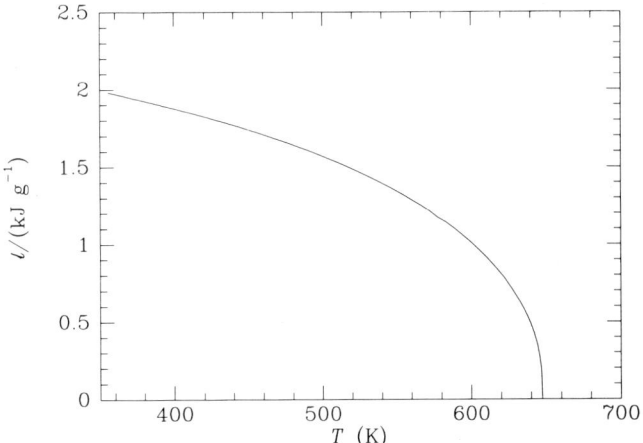

Figure 1.2 The latent heat of water as a function of temperature.

the ferromagnetic state ($T < T_c$), the material is magnetized even when no field is applied, and when an external field **B** is applied this magnetization swings almost instantaneously to align with **B**.[2] Consequently, **m** is no longer linearly related to **B**. As Figure 1.1 shows, the magnitude of the magnetization $\mathbf{m}_0(T)$ at $\mathbf{B} = 0$ vanishes as one approaches T_c from below. Therefore, as one heats a sample of iron through T_c in zero applied field nothing very dramatic happens at T_c. The iron's magnetization steadily decreases as T_c is approached, vanishing entirely at T_c and for all higher temperatures. What changes discontinuously at T_c is the *rate* of change of $\mathbf{m}_0$ rather than $\mathbf{m}_0$ itself. This is the essence of a continuous phase change: the properties of the system do not change discontinuously at T_c, but at least one of their rates of change does. By contrast, when water freezes, there is an abrupt change in the properties of the system and not merely in their rates of change, for example the density ρ and specific heat c_v.

Although the phase changes of H_2O are normally first-order, there is a special circumstance under which the water–steam transition is continuous. One can conveniently track this transition through a sequence of steadily increasing temperatures and pressures by heating a quantity of H_2O inside a stout vessel. As the vessel is heated, water steadily converts to steam, raising the density of the steam in the vessel. Meanwhile the water expands slightly, so the density difference between the water and steam in the vessel becomes steadily smaller. Since the latent heat absorbed when water vaporizes reflects the different inter-molecular binding energies of H_2O molecules in steam and

---

[2] We idealize slightly; in reality when $B = 0$ the direction of magnetization is constant only within small, but macroscopic, 'domains', and a non-negligible magnetic field may be required to coerce **m** in different domains into alignment with **B**.

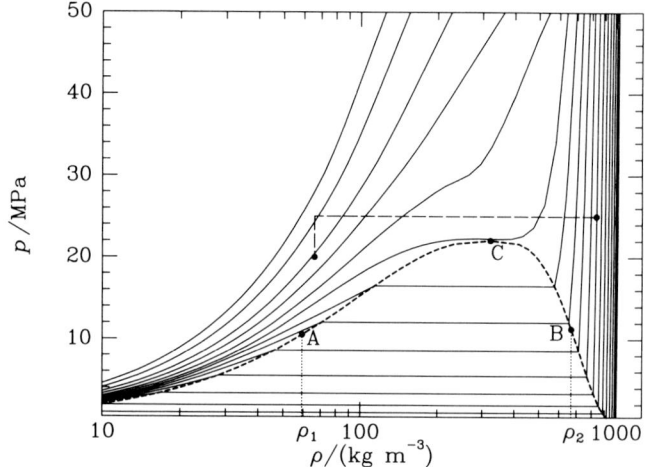

Figure 1.3 The isotherms of water. Each point may be thought of as giving the pressure and volume $= 1/\rho$ of a vessel containing unit mass of H_2O. Below the dashed curve the vessel contains a mixture of water and steam. To the right of the dashed curve and below the critical isotherm (that which passes through C) the vessel contains only water. Above the critical isotherm and/or to the left of the dashed curve the vessel contains only steam. If one heats unit mass of cold water in a vessel of volume $v = 1/\rho_1$, the last water evaporates at point A. If one heats unit mass of cold water in a smaller vessel, volume $v = 1/\rho_2$, thermal expansion of the water steadily diminishes the volume occupied by steam, until at point B the last steam is squeezed out of existence. If the vessel has the special volume $v_c = 3.1\,\text{cm}^3\,\text{g}^{-1}$, the last steam disappears at just the temperature T_c at which the distinction between water and steam ceases to be meaningful. The dash-dot lines show how water may be converted to steam without boiling it.

water, this diminishes as the temperature rises. If the volume of the vessel is just right, the latent heat eventually vanishes altogether—see Figure 1.2. At this point the density and temperature of the H_2O in the vessel are $(\rho_c = 0.323\,\text{g}\,\text{cm}^{-3}, T_c = 647\,\text{K})$. These values define the **critical point** of the water–steam transition. At this particular density, ρ_c, the evaporation of water is a continuous phase change.

At $T > T_c$ water and steam cease to be distinct entities. In particular, one can convert a quantity of water to steam without boiling it as follows. First heat the water to $T > T_c$ under a pressure that exceeds the pressure exerted by steam (or, equivalently, water) at (ρ_c, T_c). Then reduce the pressure to any required value by isothermal expansion. The dashed-dot curve in Figure 1.3 illustrates this process.

First-order phase changes are generally defined to be those that involve a non-zero latent heat, all other phase changes being deemed continuous.[3] Most first-order phase changes reduce to a continuous phase change at one

---

[3] In §1.5 we shall relate this definition to the 'thermodynamic potentials' f and g.

1.1 Continuous phase transitions and critical points

special point in the way that the water–steam transition does. But by no means all continuous phase changes arise in this way; as we shall see in §1.2, many important systems, like iron, undergo a continuous phase change that is in no way associated with a first-order one.

1.1.1 Divergences and critical exponents

The vanishing of the latent heat at the critical point does not ensure that the specific heat of a sample which is undergoing a continuous phase change is a smooth function of temperature, or even finite. In fact the specific heat c often diverges in the neighbourhood of T_c as $c \sim |T - T_c|^{-\alpha}$, with $\alpha > 0$. The number α is called a 'critical exponent' and is one of a set of such exponents which describe the singular behaviour of interesting quantities at a continuous phase transition.[4]

By the definition of a continuous phase change the total heat absorbed $l = \int_{T_c-}^{T_c+} c(T)\,dT$ on passing through the transition must vanish, so $\alpha < 1$. However, in practice α is invariably significantly smaller than unity and the divergence of c with T is slow. In fact, in some systems c grows only as $\log(T_c/|T - T_c|)$ when T tends to T_c. Since

$$\log(1/x) = \lim_{\alpha \to 0+} \frac{1}{\alpha}(x^{-\alpha} - 1), \tag{1.1}$$

we may consider such cases of logarithmically diverging c as cases in which α vanishes.

In yet other systems c has a finite peak as T passes through T_c. Then c can usefully be fitted to the formula

$$c(T) \simeq \text{constant} \times \frac{1}{|\alpha|}\left(1 - x^{|\alpha|}\right) \quad \text{where} \quad x \equiv \frac{|T - T_c|}{T_c}. \tag{1.2}$$

Comparison of this formula with (1.1) suggests that we regard such cases of finite peaks in c as cases in which α is less than zero. Then all three cases—power-law divergence, logarithmic divergence and a finite peak—can be embraced in the single formula

$$c(T) \sim \frac{1}{\alpha}\left(\left|\frac{T - T_c}{T_c}\right|^{-\alpha} - 1\right). \tag{1.3}$$

Notice that though c remains finite for $\alpha < 0$, dc/dT need not. In fact, α can be conveniently defined as follows. Let $d^n c/dT^n$ be the lowest derivative of c which diverges as a power of $|T - T_c|$ in the limit $T \to T_c$, and let it diverge as $|T - T_c|^{-k}$. Then $\alpha = k - n$.

---

[4] By 'singular' behaviour we mean non-analytic dependence on the independent variable, in this case T—see §1.5.

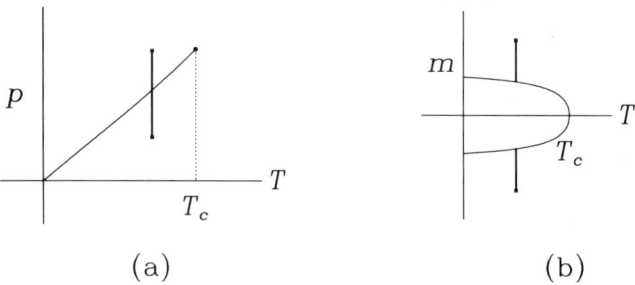

Figure 1.4 (a) The (T, p) plane for a quantity of H_2O. Below the curve all the H_2O is liquid, while above this curve it has all vaporized; water and steam coexist along the curve. As the system moves upwards along the vertical line, latent heat of condensation is emitted where the line crosses the curve. (b) The corresponding diagram for a ferromagnet. The heavy vertical line shows the trajectory of the system as the applied field changes slowly from a negative to a positive value. As B passes through zero, m changes discontinuously between energetically equivalent values. No latent heat is given out.

The specific heat c_B of iron has a finite peak at the Curie temperature. Thus for iron $\alpha \simeq -0.03$ is less than zero. In the case of H_2O, by contrast, the specific heat c_v diverges as T approaches T_c from above, so for H_2O, $\alpha > 0$.

Whenever c is infinite at T_c, it is not immediately obvious that c should diverge as the same power of $|T - T_c|$ as T_c is approached from below as above. That is, one might anticipate that

$$c \sim \begin{cases} (T - T_c)^{-\alpha} & \text{for } T > T_c \\ (T_c - T)^{-\alpha'} & \text{for } T < T_c \end{cases} \quad \text{with} \quad \alpha' \neq \alpha. \tag{1.4}$$

However, theory (see §5.6.3) and the (rather meagre) experimental evidence (e.g., Greer and Moldover 1981, Kumar et al. 1983) suggest that $\alpha = \alpha'$ whenever the distinction between α and α' can be meaningfully drawn.

In the case of H_2O, α' cannot be defined since the water–steam transition is first-order at $T < T_c$. Figure 1.4 illustrates this state of affairs. The heavy vertical line in (a) shows the trajectory of a quantity of H_2O condensed from steam into water by compression in a cylinder at a fixed temperature $T < T_c$. As the H_2O crosses the sloping coexistence curve, it condenses and latent heat is emitted. Since this heat represents the change in the system's internal energy from one side of the coexistence curve to the other, it is equal to the heat $\int c_v(T) dT$ that would have been released had the H_2O been condensed by isobaric cooling across the curve from right to left rather than by isothermal compression from below to above. Thus below T_c the specific heat of H_2O diverges as strongly as a Dirac δ-function, rather than having an integrable singularity characterizable by α.

The heavy vertical lines in Figure 1.4(b) show the trajectory of iron coerced at $T < T_c$ from negative to positive magnetization by an increasing

1.1 Continuous phase transitions and critical points

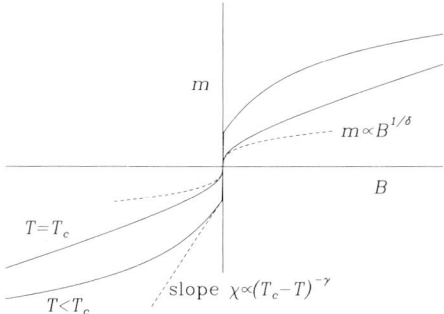

Figure 1.5 Definition of the critical exponents γ and δ. Near T_c the susceptibility in the limit of small B is proportional to $(T_c - T)^{-\gamma}$. At T_c itself $m \propto B^{1/\delta}$.

magnetic field. In this case the abrupt change from $m = -m_0(T)$ to $m = m_0(T)$ is between physically equivalent configurations, so no latent heat is involved. Thus below T_c the specific heat of iron is in no way singular, and it is meaningful to ask how it peaks as T tends to T_c from below.

Several thermodynamic quantities other than the specific heat c usually diverge at a continuous phase change. For example, the isothermal compressibility κ_T of a vessel containing H_2O at the critical density diverges near T_c as $\kappa_T \sim (T - T_c)^{-\gamma}$, where $\gamma \simeq 1.2$. Below T_c water and steam have different densities, ρ_w and ρ_s, and the isothermal compressibility of a vessel containing both fluids is infinite even through a finite change in the volume, since water can always be converted into steam to ensure that the pressure is equal to the saturated vapour pressure, no matter what the volume (so long as some water is left). At T_c the difference $\Delta\rho \equiv \rho_w - \rho_s$ vanishes and the isothermal compressibility diverges only at the single density ρ_c at which water is liable to condense out. Figure 1.3 illustrates this situation.

Notice the parallel between the behaviour for H_2O of c_v and κ_T near the critical point: in much the same way that a divergence in c_v at the critical point is the precursor of a non-zero difference l in the heat contents (effectively the specific entropies) of water and steam below T_c, so a divergence of κ_T at the critical point is the precursor of a non-zero difference in the specific volumes of water and steam below T_c.

The analogue for iron of the divergence in the compressibility of H_2O near (ρ_c, T_c) is a divergence in the susceptibility: $\chi_T \equiv (\partial m/\partial B)_T \sim (T - T_c)^{-\gamma}$ with $\gamma \simeq 1.3$. Below T_c the analogy between compressibility of a water–steam mixture and the susceptibility of iron is imperfect: while the pressure of a water–steam mixture is entirely independent of volume, below the Curie temperature the magnetization is of the form $m(T, B) \simeq m_0(T) + \chi_T B$, where $\chi_T \sim (T_c - T)^{-\gamma'}$. Although the experimental evidence on this point is by no means clear-cut (Rastelli and Realto 1969, Rocker et al. 1971), the exponent γ' here is thought to equal that which governs the divergence of χ_T as T_c is approached from above (see §5.6.2).

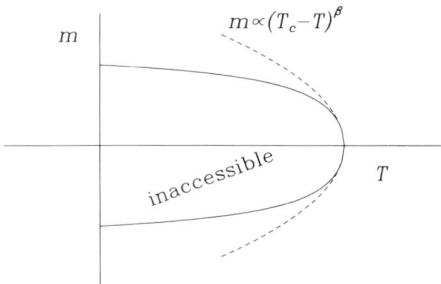

Figure 1.6 Definition of the critical exponent β. In the absence of an externally applied field the magnetization tends to zero as $(T_c - T)^\beta$ when T_c is approached from below. More highly magnetized states can be produced by applying an external field, but states inside the curve $m_0(T)$ are physically inaccessible.

In a magnetic system m_0 tends to zero as $m_0 \sim (T_c - T)^\beta$, where, for example, $\beta \simeq 0.35$ for iron. Correspondingly, as T approaches T_c from below, the densities of coexisting liquid and vapour approach each other such that $\rho_{\text{liquid}} - \rho_{\text{gas}} \sim (T_c - T)^\beta$, where, for example, $\beta \simeq 0.35$ for H_2O.

At T_c itself, m becomes proportional to a power of B—specifically $m \sim B^{1/\delta}$, where $\delta \simeq 4.3$ for iron; thus at T_c, m responds sensitively and highly non-linearly to small fields B. Correspondingly, on the critical isotherm the density of a fluid varies with pressure as $\rho \sim (p - p_c)^{1/\delta}$, where p_c is the mean pressure in a vessel containing the fluid at the critical temperature and density. For example, $\delta \simeq 4$ for H_2O.

These definitions of the critical exponents α, β, γ and δ are summarized in Table 1.1 below.

1.1.2 Fluctuations and critical opalescence

The large values attained by κ_T and χ_T just above T_c give rise to large fluctuations in the density or magnetization of the material. In the case of a gas the physical origin of these fluctuations is easy to understand qualitatively: the pressure which an ordinary gas exerts on any small surface fluctuates constantly due to the atomic nature of matter. Hence each small parcel of gas can be thought of as subject to a constantly fluctuating confining pressure from the gas enveloping it. If the compressibility of the gas is unusually large, as it is near (ρ_c, T_c), the volume of the parcel fluctuates by a correspondingly large amount in response to the fluctuating bounding pressure.

The larger the parcel one considers, the smaller will be the fluctuations in the mean pressure exerted on its bounding surface by the gas enveloping it. So the magnitude of fluctuations in the mean density of a parcel decreases as the parcel's size increases. In a near-ideal gas such as ordinary air, density fluctuations are very small indeed for parcels comparable in size to the wavelength $\lambda \approx 0.5\,\mu\text{m}$ of visible light. However, the fluctuations in the refractive index of air to which density fluctuations give rise do scatter light when the air is illuminated. Such scattering of sunlight by the Earth's atmosphere is only significant when one considers light paths several kilometres in length,

1.2 The order parameter

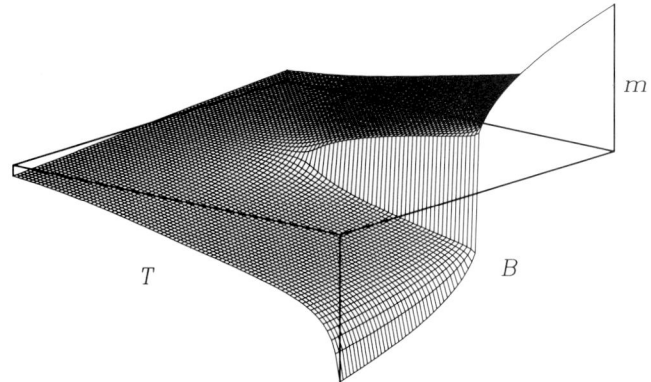

Figure 1.7 The relationship between B, m and T for a typical magnetic material. The vertical face in the $m(T, B)$ surface narrows to a point at $T = T_c$. Figure 1.5 shows two sections of this surface in planes of constant T.

and then the effect is much more pronounced for blue than for red light—this is the origin of blue skies and red sunsets. Near a critical point, most gases are quite dense and therefore have rather large refractive indices, so fluctuations on scales $\approx 0.5\,\mu$m can be large enough to be readily observed in the laboratory through the phenomenon of **critical opalescence**: the blurring of images seen through a small volume of gas similar to that caused by a heat haze—see Figure 1.8.

Since the compressibility of a vessel containing a water–steam mixture (which of course must be at $T < T_c$) is infinite, it is natural to ask whether large fluctuations are to be expected in the volume of such a vessel. The answer is 'no' because below T_c non-zero latent heat has to be absorbed or emitted when H_2O passes between water and steam. Consequently, a sudden upward fluctuation in the pressure on the vessel leads to the release of latent heat by freshly condensed vapour, which tends to restore equilibrium by heating the vessel's contents. At the critical point the latent heat vanishes and this stabilizing process is ineffective.

1.2 The order parameter

The first step towards a quantitative theory of a continuous phase transition is to identify a quantity, called the **order parameter**, whose 'thermal average' vanishes on one side of the transition (almost invariably the high-temperature side) and moves away from zero on the other side. We shall reserve the symbol ϕ for order parameters.

We define ϕ to be a quantity which fluctuates in time and space in order to be able to interpret phenomena such as critical opalescence in terms

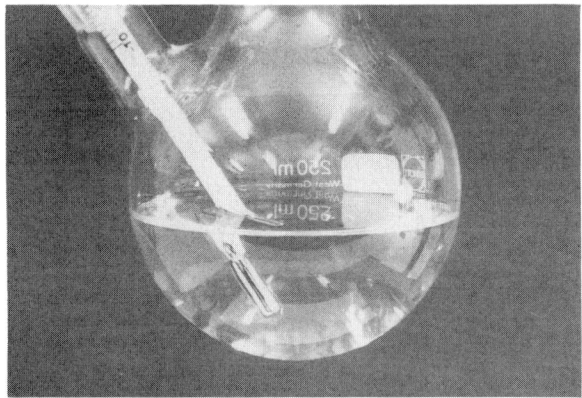

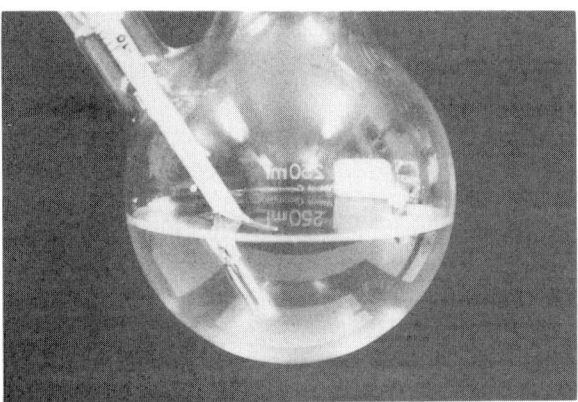

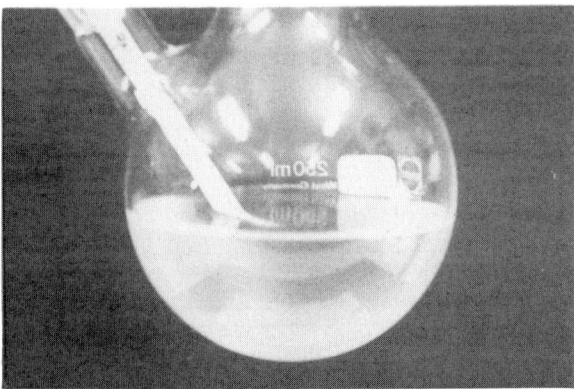

Figure 1.8 Critical opalescence in a mixture of methanol and n-hexane. Top: $T > T_c$; bottom: $T \simeq T_c$. In the middle frame T is very slightly larger than T_c.

1.2 The order parameter

of fluctuations in $\phi(\mathbf{x},t)$. However, in many applications we are only interested in the **thermal average** of ϕ: this is ϕ's value when averaged over a long period of equilibrium at constant temperature. In particular, the critical exponents α, β, γ and δ introduced above describe the variation with temperature, magnetic field, pressure, etc. of the thermal average of ϕ.

There is no general scheme for defining order parameters; one has to consider each new physical system afresh. We can best indicate what sort of thing an order parameter is by giving a number of examples, starting with the phase transitions with which we are already familiar.

1.2.1 Liquid–gas transition

We have seen that as T_c is approached from below, the mean densities of the liquid and vapour phases converge on a common value. Thus a suitable order parameter for the liquid–gas transition is

$$\phi(\mathbf{x}) \equiv \rho(\mathbf{x}) - \rho_{\text{gas}}(\mathbf{x}). \tag{1.5}$$

Here the density $\rho(\mathbf{x})$ is a fluctuating quantity rather than a thermal average; it should be considered to be an average density in a volume centred on $\mathbf{x}$ that is small compared with the dimensions of the apparatus, but large enough to contain a great many molecules. By contrast ρ_{gas} is the density of gas at the temperature and time-averaged pressure prevailing at $\mathbf{x}$, i.e., the density associated with the appropriate point on the left-hand branch of the dashed curve in Figure 1.3.[5] This order 'parameter' is a scalar field $\phi(\mathbf{x})$ rather than a simple parameter as its outmoded name suggests.

Above T_c, ϕ fluctuates in a small range centred on zero, as it also does below T_c wherever there is vapour. But where there is liquid, ϕ fluctuates around a positive number. As T_c is approached from below, the mean of these fluctuations diminishes towards zero, and their amplitude simultaneously increases.

1.2.2 Binary fluids

At room temperature methanol (CH_3OH), which, like H_2O, has distinctly polar molecules, cannot be mixed in arbitrary proportions with n-hexane (C_6H_{14}), a non-polar substance. For example, if equal numbers of CH_3OH and C_6H_{14} molecules are shaken up in a bottle and then allowed to settle at room temperature, two phases will form separated by a meniscus; above the meniscus the fluid will consist of hexane in which a little methanol has been dissolved, while the fluid below will consist of methanol with traces of dissolved hexane. If the experiment is repeated with methanol and hexane that have been heated to a temperature in excess of 42.4 °C, no meniscus will form; above 42.4 °C methanol can be mixed with hexane in any proportion to form a fluid of uniform composition.

---

[5] In the presence of gravity ρ_{gas} varies from place to place.

Now imagine cooling a uniform methanol–hexane mixture extremely slowly. At some temperature below 43 °C two distinct phases will be seen to form. If the initial mixture consists primarily of hexane, a meniscus will form at the bottom of the container and move gradually upwards as more and more methanol-with-dissolved-hexane separates out. Conversely, if the initial mixture is predominantly methanol, the meniscus will form at the top of the container and move downwards. In either case the new phase is completely distinct from the bulk of the fluid from the moment it appears, just as water droplets in a cooling flask of H_2O at $T < 647$ K are entirely distinct from the surrounding steam. But if the numbers of methanol and hexane molecules in the bottle are in the ratio $0.435:0.665$, a meniscus will not form immediately. Instead, as the temperature approaches 42.4 °C, large fluctuations in the fluid's refractive index will make the contents of the bottle cloudy—see Figure 1.8. In this critical mixture the transition from a single phase to two distinct phases is continuous.

Below T_c the order parameter of the system may be taken to be the difference $\phi(\mathbf{x}) \equiv X'(\mathbf{x}) - X''$ between the local molar density of one constituent (for example, methanol) and its thermally averaged value in one of the two phases at coexistence. For definiteness we assume that X'' is the smaller of the two possible values at any given temperature. Above T_c there is only one phase and ϕ fluctuates around zero. Below T_c there can be two phases, one richer in methanol than the other. In one phase ϕ fluctuates around zero, and in the other it fluctuates around a strictly positive value. As T_c is approached from below, the range of overall compositions compatible with the coexistence of these phases narrows towards the critical value methanol:hexane $= 0.435:0.665$, and, if the phases do coexist, ϕ everywhere fluctuates with larger and larger amplitude. Meanwhile the positive mean taken by ϕ in one phase steadily diminishes, vanishing at T_c itself.

1.2.3 Ferromagnetic/paramagnetic transition

Since in zero magnetic field the magnetization $\mathbf{m}$ changes from a non-zero value at $T < T_c$ to zero at T_c, it is a suitable order parameter for the ferromagnetic/paramagnetic transition. However, we shall reserve the symbol $\mathbf{m}$ for the thermally averaged magnetization and the order parameter should be a fluctuating quantity rather than a thermal average. Therefore we define the order parameter $\boldsymbol{\phi}(\mathbf{x})$ to be the instantaneous mean magnetization in a small volume around $\mathbf{x}$. Thus the order parameter of a magnetic system is a vector field $\boldsymbol{\phi}(\mathbf{x})$ rather than a simple parameter.

It is often convenient to work with an order parameter which is well defined even on atomic scales. So when we consider a magnetic substance consisting of N spins distributed on a lattice, we shall usually adopt for the order parameter the values $\mathbf{s}_i$ of the spins on the lattice sites; the order parameter is then only defined on lattice sites rather than being a continuous function of the position vector $\mathbf{x}$. The macroscopic magnetization $\boldsymbol{\phi}$ is related

1.2 The order parameter

to $\mathbf{s}_i$ by

$$\boldsymbol{\phi}(\mathbf{x}) = \frac{\mu}{\delta V} \sum_{i \in \delta V} \mathbf{s}_i, \qquad (1.6)$$

where μ is a suitable constant and the sum is over all sites lying in a small volume δV centred on $\mathbf{x}$.

1.2.4 Anti-ferromagnetic/paramagnetic transition

In a ferromagnetic material it is energetically favourable for all spins to align. In an anti-ferromagnetic material, by contrast, the energetically favoured configuration is one in which neighbouring spins are anti-aligned. Often the optimal configuration is that in which each spin is anti-aligned with all its nearest neighbours. It is easy to see that such a configuration is possible on a cubic lattice. But on certain other lattices, for example a two-dimensional triangular lattice, it is impossible to arrange the spins so that each is anti-parallel to all its nearest neighbours. The system then does not have a simple, periodic lowest-energy state and is said to be **frustrated**. We shall not consider frustrated systems in this book.

Even at low temperatures an anti-ferromagnet has no net magnetic moment. So what is the order parameter of an anti-ferromagnet? We divide the sites into two classes A and B such that in the lowest-energy configuration the spins on all sites in class A are parallel to one another and anti-parallel to all spins on sites in class B. We then define the order parameter on site i to be

$$\boldsymbol{\phi}_i = \begin{cases} \mathbf{s}_i & \text{for } i \in \text{A}, \\ -\mathbf{s}_i & \text{for } i \in \text{B}. \end{cases} \qquad (1.7)$$

Thus defined, the order parameter at $T = 0$ is everywhere equal to the spins on sites in class A, and has vanishing thermal average for $T > T_c$. In general $\boldsymbol{\phi}$ is a fluctuating, position-dependent vector field on a lattice. By averaging it locally after the fashion of (1.6), it may be converted into a smoothly varying function of position.

1.2.5 Helium I/helium II transition

Above about 2 K liquid ^{4}He is a normal viscous liquid, called **He I**. But when ^{4}He is cooled through about 2 K (the exact temperature depends on the pressure), a 'superfluid' component starts to form in the liquid; that is, below $\approx$ 2 K the liquid appears to contain a component that flows without viscosity. This zero-viscosity component coexists with ordinary, viscous ^{4}He rather as water coexists with steam.[6] The lower the temperature, the greater the proportion of ^{4}He that is in the superfluid 'condensate'. ^{4}He that contains some of the superfluid condensate is called **He II**. The transition from He I

---

[6] The analogy is inexact—in particular, no latent heat is involved in the passage of He atoms in and out of the condensate.

to He II is continuous over the whole range of pressures within which the transition can take place.

The superfluid condensate in He II is made up of ^{4}He atoms with momenta so small that their de Broglie wavelengths are macroscopic in size. Above T_c there are only a few such atoms, but as T falls below T_c, macroscopically many atoms accumulate in states with large de Broglie wavelengths and one says that a 'Bose condensate' has formed—see e.g. Reichl (1980) for details. A Bose condensate has peculiar properties because its (delocalized) constituent particles move coherently and are insensible of the small-scale structures which dominate the scattering of normal, thermally excited helium atoms.

The natural order parameter of this system is the complex scalar field $\psi(\mathbf{x})$ whose value at $\mathbf{x}$ is the quantum amplitude to find a particle of the Bose condensate at $\mathbf{x}$. Thus the order parameter of the He I/He II transition is a complex scalar function of position.[7]

1.2.6 Conductor/superconductor transitions

In 1911 Kamerlingh Onnes (1853–1926) discovered that mercury lost all trace of electrical resistivity when cooled through about 4.2 K. Many other materials have subsequently been found to become such **superconductors** at low temperatures, and in recent years this phenomenon has become of considerable technological importance.

According to the standard 'BCS' theory of superconductivity (Bardeen *et al.* 1957) the phenomenon is closely related to superfluidity in ^{4}He: at low temperatures electrons with oppositely aligned spins form **Cooper pairs** by exchanging phonons. Each Cooper pair is a charged, spin-zero particle. Since Cooper pairs are bound by phonons, which by electronic standards are slow-moving beasts, they form with near-zero momentum. Consequently, as soon as significant numbers of Cooper pairs are present, they constitute a Bose condensate, which is in some ways analogous to He II. Since this Bose condensate is charged and can flow without viscosity, all trace of electrical resistance vanishes when the condensate forms. In the absence of an externally applied magnetic field the phase change from the resistive to the superconducting state is continuous.

By analogy with ^{4}He we adopt as order parameter the field $\psi(\mathbf{x})$ whose value at $\mathbf{x}$ is the quantum amplitude to find a Cooper pair at $\mathbf{x}$.[8] Thus the order parameter of the superconductor transition is a complex scalar field.

Imposition of an external magnetic field complicates discussion of the superconductor transition. Many superconductors are not so much materials

---

[7] Since the absolute phase of a quantum amplitude is irrelevant, in simple cases one can choose $\psi(\mathbf{x})$ so that it is real. However, a set of amplitudes that describes a non-vanishing probability current $\psi^*\nabla\psi - \psi\nabla\psi^*$ must be non-trivially complex.

[8] An alternative order parameter is the so-called **gap parameter** whose modulus is the difference in the energy per electron of the Bose condensate and the Fermi energy.

1.2 The order parameter

with vanishing resistivity as perfect diamagnets: such **type I superconductors** rigorously exclude magnetic flux.[9] In fact if a block of type I material is cooled to the point at which it becomes a superconductor while immersed in a magnetic field, magnetic flux is not trapped within the block as it would be if the transition were merely to a state of zero resistivity, but driven out of the block: as the material becomes superconducting, currents spontaneously arise in its surface to drive out the field. If the strength of the external field is now increased, the field will eventually force its way back into the block, which ceases to be a superconductor as the field re-enters. When material passes in this way between the superconducting and normal states in the presence of an external field, the transition is first-order rather than continuous.

Superconducting materials other than pure metals are generally capable of remaining in the superconducting state after a magnetic field has penetrated them. Materials with this capacity are called **type II superconductors**. At very low temperatures even a type II superconductor excludes an applied magnetic field, but as the material is heated in a field through a critical temperature T_{c_1}, the field begins to penetrate the bulk of the material without destroying the Cooper pairs. As the temperature is raised further, the magnitude of the field in the superconductor rises smoothly until at a second critical temperature, T_{c_2}, the discontinuity in B at the material's surface vanishes and the material ceases to be superconducting. In type II superconductors the phase transition is continuous even in non-zero applied field.

The problem of measuring reliable critical exponents for standard superconductors is still unsolved because it is thought that these can be measured only extremely close to T_c. That is, relations such as that between the mean magnitude $|\psi|$ of the order parameter below T_c and $(T_c - T)$ are expected to settle to asymptotic forms of the type $|\psi| \sim (T_c - T)^{-\beta}$ only when $(T_c - T)$ is significantly smaller that $1\,\mu\mathrm{K}$. It is currently not possible to control the temperature with enough precision to probe this region. It may, however, prove possible to measure the critical exponents of certain high-T_c superconductors shortly.

1.2.7 Helium three

Below about $4\,\mathrm{K}$ $^3\mathrm{He}$ can be liquefied by applying a suitable pressure. Since $^3\mathrm{He}$ atoms are spin-half fermions they cannot form a Bose condensate similar to that responsible for the superfluid phase of $^4\mathrm{He}$, but below $2.7\,\mathrm{mK}$ $^3\mathrm{He}$ can become a superfluid by a process which is reminiscent of superconductivity: $^3\mathrm{He}$ atoms stick together in pairs to form bosons, which form a Bose condensate. The atoms are held together by the attractive part of the van der Waals force. They eliminate the danger of penetrating the repulsive

---

[9] Actually B tends to zero with distance x into the sample as $B \sim e^{-x/\lambda}$, where the **penetration depth** λ is large just below T_c.

core of the potential by rotating around one another. Hence the constituent particles of superfluid ^{3}He have considerable intrinsic angular momentum—^{3}He atoms orbit around their common centre of mass such that they have orbital angular momentum $l = 1$ and, moreover, combine with their spins parallel, to give total spin angular momentum $S = 1$.[10] Thus depending on the mutual orientation of the spin and orbital angular momenta, the total angular momentum of a ^{3}He boson can be $j = 0, 1$ or 2. Since there are $2j + 1$ states associated with each j-value, ^{3}He has no less than nine possible angular momentum configurations. Each such configuration can form a Bose condensate characterized by a complex-valued wavefunction, so the order parameter of ^{3}He is a field of nine complex, or eighteen real, numbers.

Given that the order parameter is such a complicated thing, it is not surprising that there are several phases of superfluid ^{3}He. In zero magnetic field there are just two, A and B. But other phases appear if a magnetic field is imposed or the dewar is set rotating—see for example Salomaa et al. (1987). The transition in zero magnetic field from normal to superfluid ^{3}He appears to be continuous, but several of the other transitions are thought to be first-order.

1.3 Correlation functions

Much of our knowledge of continuous phase transitions derives from experiments in which particles are scattered by a nearly critical system. The scattered particles may be photons, whether visible or x-ray, or they may be phonons, electrons or neutrons. Neutrons are especially valuable probes since (i) they are relatively penetrating, enabling one to ignore multiple scattering at least in a first approximation, and (ii) they are sensitive to the orientation of atomic spins, so they are good probes of magnetic materials. In Appendix A we develop the theory of scattering of polarized neutrons by a lattice of spins and show that studies of such scattering lead rather directly to the **two-point correlation function** of the spins, namely

$$G^{(2)}(\mathbf{i},\mathbf{j}) \equiv \langle \mathbf{s}_i \cdot \mathbf{s}_j \rangle, \tag{1.8}$$

where $\mathbf{i}$ and $\mathbf{j}$ are the position vectors of sites i and j, respectively, and angle brackets indicate thermal averaging. In a great many situations the system is translationally invariant, either because it forms a crystal lattice or because it is highly disordered; $G^{(2)}$ then depends only on the difference vector $\mathbf{i}-\mathbf{j}$. If the system is also isotropic, $G^{(2)}$ becomes a function $G^{(2)}(|\mathbf{i}-\mathbf{j}|)$ of the distance between spins i and j. Of course a lattice cannot be strictly isotropic. But if the three lattice directions are equivalent, it appears isotropic when probed at much lower resolution than that required to resolve

---

[10] $S = 1$ because once they have decided to be antisymmetric in real space, these fermions are obliged to be symmetric in spin space.

1.3 Correlation functions

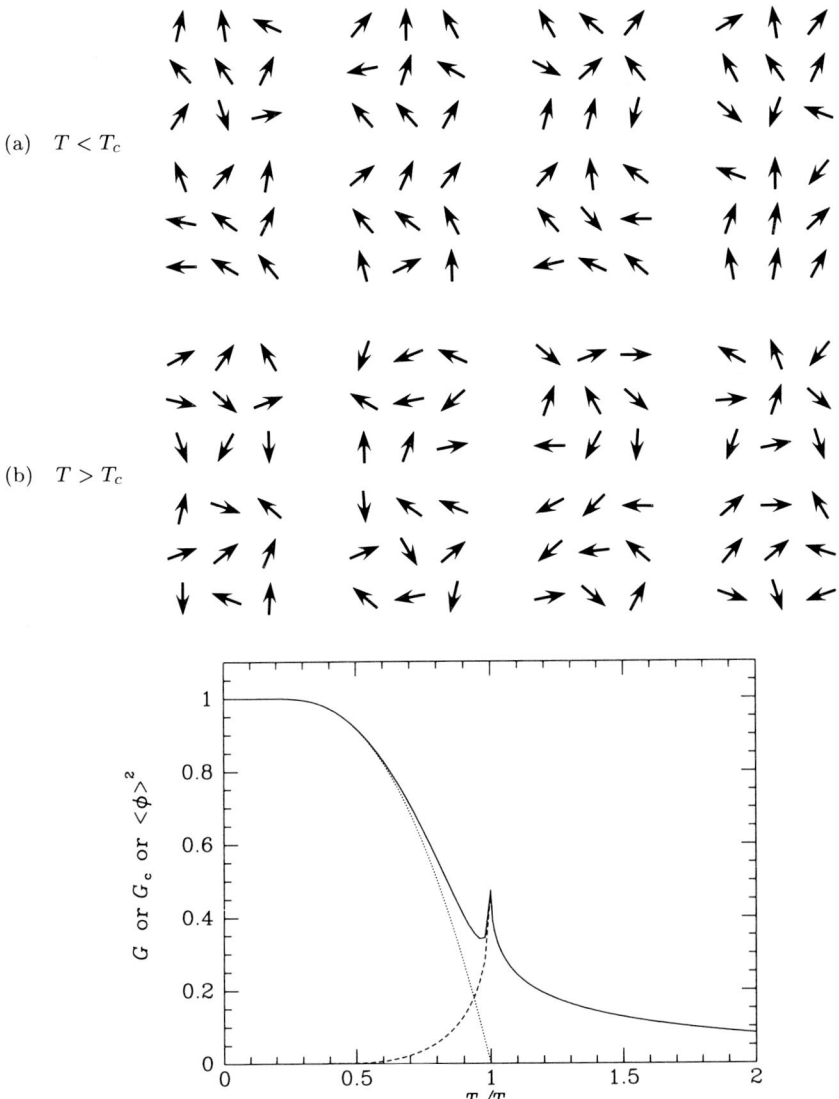

Figure 1.9 The distinction between the ordinary two-point correlation function $G^{(2)}$ and the connected two-point corrrelation function $G_c^{(2)}$. (a) shows eight snapshots of a system of spins below T_c. Although there is a good deal of randomness, on the average the spins point upwards. (b) shows the system above T_c. Now there is no overall alignment, but neighbouring spins still tend to point in the same direction. Whereas $G_c^{(2)}$ is comparable in both (a) and (b), $G^{(2)}$ is greater in (a) because it includes the contribution to the correlation from the overall alignment. In the graph the full curve shows the variation with temperature of $G^{(2)}$ at a particular separation, while the dashed curve shows that of $G_c^{(2)}$. The dotted curve shows the square of the thermally averaged order parameter, so that the full curve is the sum of the dotted and dashed curves.

individual unit cells, and is commonly called an 'isotropic lattice'. Since the theory of critical phenomena is concerned with macroscopic phenomena and we are primarily interested in isotropic lattices, we shall generally assume that $G^{(2)}$ is a function of $|\mathbf{i} - \mathbf{j}|$ only.

Studies of photon scattering by nearly critical liquids and phonon scattering by superfluids yield the two-point correlation functions of these systems. In general we define[11]

$$G^{(2)}(r) \equiv \langle \boldsymbol{\phi}(0) \cdot \boldsymbol{\phi}(\mathbf{r}) \rangle, \tag{1.9}$$

where $\boldsymbol{\phi}$ is the order parameter of the system in question and the dot represents any suitable scalar product. Well below T_c, $G^{(2)}$ becomes large for all values of its argument. This leads us to define the **connected two-point correlation function** $G_c^{(2)}$:

$$G_c^{(2)}(r) \equiv \langle \boldsymbol{\phi}(0) \cdot \boldsymbol{\phi}(\mathbf{r}) \rangle - |\langle \boldsymbol{\phi} \rangle|^2. \tag{1.10}$$

Above T_c, $\langle \boldsymbol{\phi} \rangle = 0$, so $G_c^{(2)}$ is identical to $G^{(2)}$ and both measure the degree of coordination of the order parameter at different points. Below T_c, $G_c^{(2)}(r)$ differs from $G^{(2)}(r)$ in discounting the general alignment of the order parameter at different points, so that only *fluctuations* in the order parameter contribute to $G_c^{(2)}$. (See §2.2 for more on connected correlation functions.)

Scattering experiments show that for $r \neq 0$, $G_c^{(2)}(r)$ is small at both large and small T/T_c. Furthermore, when $T = T_c$ one finds that the asymptotic form of $G_c^{(2)}$ for r large compared with inter-molecular distances is

$$G_c^{(2)}(r) \sim \frac{1}{r^{d-2+\eta}}, \qquad (r \text{ large and } T = T_c). \tag{1.11}$$

Here d is the dimensionality of the system, and η is a further critical exponent. Typically, $0 \leq \eta \leq 0.1$. Away from T_c, $G_c^{(2)}$ cannot be represented by a simple power law. In fact, for small $|T - T_c|/T_c$ one has, approximately,

$$G_c^{(2)}(r) \sim \frac{e^{-r/\xi}}{r^{d-2+\eta}} \qquad (r \text{ large and } |T - T_c|/T_c \ll 1). \tag{1.12}$$

The characteristic length ξ is called the **correlation length**. According to (1.12) the order parameter fluctuates in blocks of all sizes up to size ξ, but fluctuations that are significantly larger are exceedingly rare. As T_c is approached from either above or below, ξ grows without limit. In fact, one finds empirically that

$$\xi \sim |T - T_c|^{-\nu} \qquad (|T - T_c|/T_c \ll 1), \tag{1.13}$$

---

[11] The general spin–spin correlation function, $G^{(2)}_{\mu\nu}(r) \equiv \langle \phi_\mu(0)\phi_\nu(\mathbf{r}) \rangle$, is a tensor-valued object. However, in this book we shall only use the linear combination of these that is the average of $\boldsymbol{\phi}(0) \cdot \boldsymbol{\phi}(\mathbf{r})$.

1.3 Correlation functions

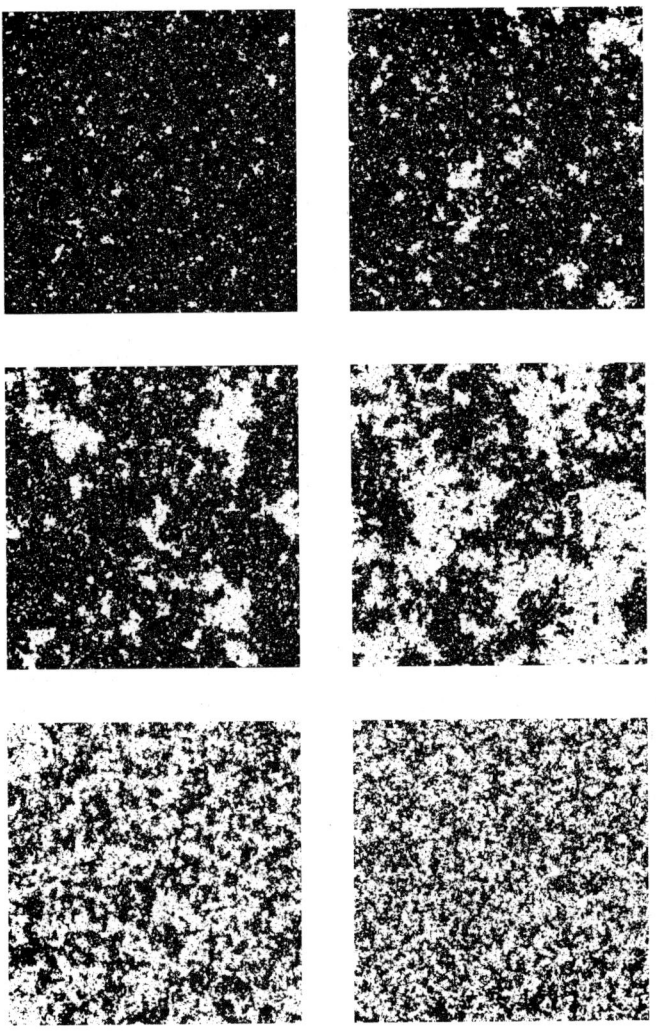

Figure 1.10 As the temperature of the two-dimensional Ising model (see §3.1.1) is increased from near zero (top left), the typical size of the white regions, in which the order parameter differs from the background value, steadily increases until at a critical temperature T_c the white and black regions on average occupy equal areas (middle right panel). As the temperature is increased still further, the black and white regions become smaller and more numerous, generating a pepper-and-salt effect (lower right). The correlation length ξ offers a measure of the size of a typical white region. At the critical temperature $\xi = \infty$ because the biggest white region contains an infinite number of lattice points. From top left to bottom right $T/T_c = 0.97, 0.99, 1, 1.01, 1.06, 1.15$. Each square is 512 spins on a side and periodic boundary conditions have been imposed. The Wolff algorithm (see §4.6.4) was used to relax the system to thermal equilibrium.

Table 1.1. Definitions of critical exponents

Exponent	Definition
α	$c_B \sim \alpha^{-1}\big((|T-T_c|/T_c)^{-\alpha} - 1\big),\ T \to T_c,\ B=0$
β	$m \sim (T_c - T)^\beta,\ T \to T_c$ from below, $B=0$
γ	$\chi_T \sim |T - T_c|^{-\gamma},\ T \to T_c,\ B=0$
δ	$m \sim B^{1/\delta},\ B \to 0,\ T = T_c$
η	$G^{(2)}(r) \sim 1/r^{d-2+\eta},\ T=T_c,\ B=0$
ν	$\xi \sim |T - T_c|^{-\nu},\ B=0$

Box 1.1: Why are power laws scale-free?

Since only dimensionless numbers can be raised to arbitrary powers in a meaningful way, it is natural to ask in what sense a power law is more scale-free than, say, an exponential; both $f_1 = (r/r_0)^\eta$ and $f_2 = \exp(r/r_0)$ involve a scale length r_0. The difference between these two laws is the following. If one measures f_2 for values of r in the range $(0.5 r_0, 2 r_0)$, say, the ratio of largest to smallest value will be $e^{1.5}$, while the corresponding ratio for measurements made over two octaves in r centred on $10 r_0$ will be e^{15}, and for two octaves in r centred on $100 r_0$ the ratio will be e^{150}. Thus graphs of $f_2(r)$ over the three pairs of octaves are not scale models of one another: they cannot be superimposed by linear changes in the scales of the two axes. By contrast, when $f_1 \sim (r/r_0)^\eta$ is measured over pairs of octaves centred on r_0, $10 r_0$ and $100 r_0$, the ratio of largest to smallest values measured is always $4^{|\eta|}$, so the three graphs of $f_1(r)$ can be superimposed by simple changes of scale. In this sense a phenomenon obeying a power law looks the same no matter on what scale one probes it.

---

where ν (which is typically found to be approximately $\frac{2}{3}$) is our sixth and final critical exponent.

At the critical point large-scale fluctuations, which coordinate the values of the order parameter at widely separated points, are not exponentially rare as they are above or below T_c. Much of the interest in critical phenomena arises from this fact, which implies that at a critical point the short-range inter-molecular forces contrive to generate large dynamical structures. Moreover, whereas the underlying inter-molecular forces have a well-defined length scale, the structures to which they give rise do not—hence the power law in (1.11). Figure 1.10 illustrates this state of affairs.

1.4 Universality

Table 1.2 gives numerical values for the critical exponents of several systems. Surprisingly, the critical exponents of such very different transitions as the liquid–gas transition of xenon and the separation of a mixture of two organic chemicals are equal to within the (admittedly sometimes large) experimental errors. This phenomenon, whereby dissimilar systems exhibit the same critical exponents, is called **universality**.

One can assign each system to a **universality class** in such a way that any two systems in the same universality class have the same dimensionality d (almost always 3) and order parameters of the same dimensionality D. Experimental results such as those listed in Table 1.2 suggest that all systems in the same universality class have the same critical exponents.

Figure 1.11 demonstrates an even more striking congruence in the properties of apparently diverse systems than that evidenced by systems in the same universality class having the same critical exponents; the liquid–gas coexistence curves of many fluids can be superimposed by simple scalings of the variables. The data for different fluids are said to 'collapse' onto a common curve. Magnetic systems also exhibit data collapse of the type illustrated by Figure 1.11. One of the chief goals of the theory of critical phenomena is to explain how systems that must have very different microphysics contrive to exhibit data collapse and yield the same critical exponents. For there is a paradox here: on the one hand it is clear that inter-atomic forces are responsible for the very existence of a phase transition; on the other hand the details of these forces cannot play any rôle in determining the critical exponents, since these stay the same when the atoms, and therefore the forces, change.

1.5 Thermodynamic potentials

The standard theory of thermodynamics tells us that knowledge of a thermodynamic potential as a function of its natural variables[12] completely specifies the thermodynamics of a system. Several potentials have this property of encoding complete thermodynamic information, but in this book we shall be concerned with just two of them, the Helmholtz free energy F and the Gibbs free energy G. F's natural variables are the temperature (or inverse temperature $\beta \equiv 1/k_B T$) and whatever macroscopic parameters determine the system's energy levels; for example, in the case of a simple fluid we have

---

[12] The natural variables of the internal energy U are S and V since when we choose these as the independent variables $dU = TdS - pdV$ satisfies two conditions: (i) it splits naturally into heat and work, and (ii) the partial derivatives $\partial U/\partial V$ etc. are physically significant quantities—see Appendix B for a discussion of this point. The other thermodynamic potentials are obtained from U by Legendre transformations, and inherit their natural variables from U. For example, $F = F(T,V)$ because $dF = d(U - TS) = -SdT - pdV$.

Table 1.2. Values of critical exponents

	Xe	Binary fluid	β-brass	^{4}He	Fe	Ni
D	1	1	1	2	3	3
α	< 0.2	$0.113\pm.005$	$0.05\pm.06$	$-0.014\pm.016$	$-0.03\pm.12$	$0.04\pm.12$
β	$0.35\pm.015$	$0.322\pm.002$	$0.305\pm.005$	$0.34\pm.01$	$0.37\pm.01$	$0.358\pm.003$
γ	$1.3^{+.1}_{-.2}$	$1.239\pm.002$	$1.25\pm.02$	$1.33\pm.03$	$1.33\pm.015$	$1.33\pm.02$
δ	$4.2^{+.6}_{-.3}$	$4.85\pm.03$		$3.95\pm.15$	$4.3\pm.1$	$4.29\pm.05$
η	$0.1\pm.1$	$0.017\pm.015$	$0.08\pm.07$	$0.021\pm.05$	$0.07\pm.04$	$.041\pm.01$
ν	≈ 0.57	$0.625\pm.006$	$0.65\pm.02$	$.672\pm.001$	$0.69\pm.02$	$0.64\pm.1$

NOTES: From data published in Ahlers (1980), Anders and Stierstadt (1981), Bally et al. (1968), Chang et al. (1979), Cohen and Carver (1977), Collins (1969), Heller (1967), Hiroyoshi (1980), Kobeissi (1981), Kumar et al. (1983), Rocker et al. (1971), Rowlinson and Winton (1982), Soeffge (1980), Suter and Hohenemser (1978), Vygovskiy and Yergin (1972). α and β for a binary fluid are for methanol–hexane, while γ, η and ν are for a mixture of trimethylpentane and nitroethane. The data for helium relate to the He I– He II transition. δ for a binary fluid, α for β-brass, and both γ and η for helium have been calculated from the other exponents using formulae from §1.5.1.

$F(T,V)$, where V is the system's volume, while for a simple magnetic system we have $F(T,B)$. The Gibbs free energy is the Legendre transform of F with respect to its second variable. (Legendre transforms are described in Box 1.2.) Thus we have

$$G(T,p) \equiv F + pV \quad \text{and} \quad G(T,M) \equiv F + BM \qquad (1.14)$$

for fluid and magnetic systems, respectively.

F and G are both **extensive variables**, that is they are proportional to the mass of the system. It is often convenient to work with the free energies per unit mass ('specific' free energies), which we shall denote by $f(T,B)$ and $g(T,m)$, respectively.

f and g are thermodynamic 'potentials' in the sense that other interesting quantities can be expressed as derivatives of them—see Table 1.3. Since the physics of a system can be encapsulated in the functional forms of f and g, it is interesting to investigate the implications of the singular behaviour observed near critical points for the forms of the functions $f(T,B)$ and $g(T,m)$.

For any system, the natural variables of one of F and G are T and an intensive variable, while the natural variables of the other are temperature and an extensive variable. Let X denote the free energy whose natural variables are both intensive. Then the condition for two phases to be in thermodynamic equilibrium with each other is that the specific free energy x have the same value in the two phases. For example, a fluid is in equilibrium with its vapour when $g(T,p)$ is the same in fluid and vapour, while the ferromagnetic and paramagnetic phases of a magnetic material are in equilibrium when

1.5 Thermodynamic potentials

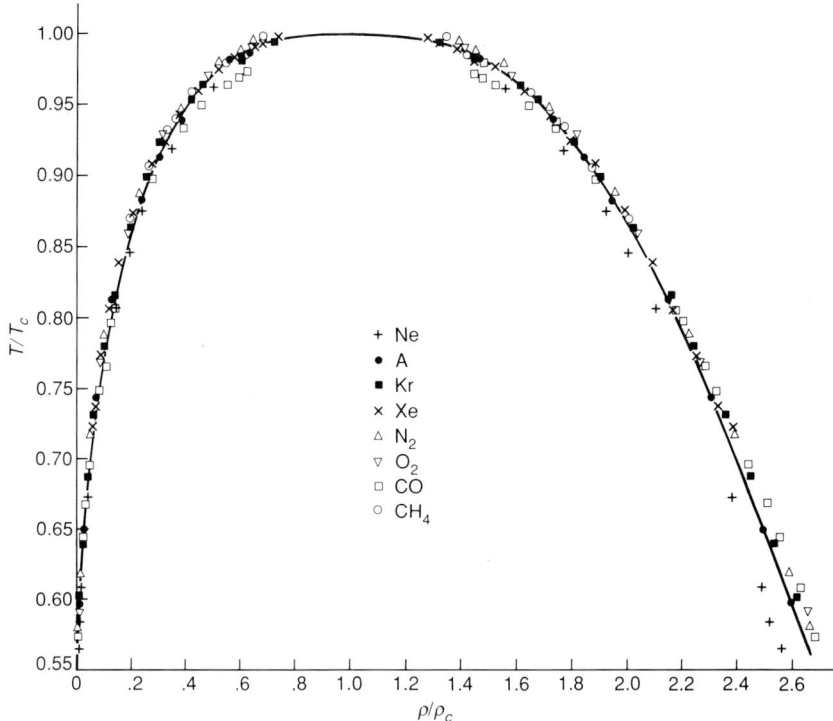

Figure 1.11 The liquid–gas coexistence curves of many fluids can be virtually superimposed when the temperature and density are scaled by their corresponding critical values T_c and ρ_c. Plots of this kind are manifestations of the 'law of corresponding states'—see, e.g., Stanley (1971) for details. This plot from Guggenheim (1945) played an important rôle in the development of the theory of critical phenomena by showing that the data cannot be fitted with a quadratic curve as van der Waals theory implies (see §6.3), but require the cubic shown here, which corresponds to the fact that $\beta \simeq 1/3$.

$f(T, B)$ is the same in both phases.[13] Hence if we represent $f(T, B)$ by the distance of a surface above the (T, B) plane, the surface will be free of vertical cliffs. Since $g = f + Bm$, and m changes continuously or abruptly between phases according as the transition is continuous or first-order, when $g(T, m)$ is represented as a surface there will be cliffs only at points corresponding to first-order phase changes with $B \neq 0$.

If subscripts $\pm$ denote values on the high- and low-temperature sides of

---

[13] The per-particle value of the free energy whose natural variables are intensive is called the 'chemical potential' μ, so one may equivalently require that two phases in equilibrium have equal chemical potentials.

Box 1.2: Legendre transforms

Let $g(x)$ be a **convex function**, that is, a function such that $g''(x) > 0$. Then the **Legendre transform** $\bar{g}(p)$ of g is defined by

$$\bar{g}(p) \equiv xp - g(x) \quad \text{where } x(p) \text{ is implicitly defined as the root for given } p \text{ of} \quad p = \frac{\partial g}{\partial x}. \quad (1)$$

The convexity of g guarantees that the equation defining $x(p)$ can be solved for any p that lies between the maximum and minimum gradients of g. Thus $\bar{g}(p)$ is well defined. It is straightforward to show that Legendre transforms are invertible. In fact a Legendre transform is its own inverse: $\bar{\bar{g}}(x) = g(x)$ (see Problem 1.6).

It is often helpful to consider the function $\mathcal{G}(x, p) \equiv xp - g(x)$ of two independent variables (x, p). Graphically, $\mathcal{G}(x, p)$ is the vertical displacement at ordinate x between the straight line $y = px$ and the upward curving graph of $g(x)$:

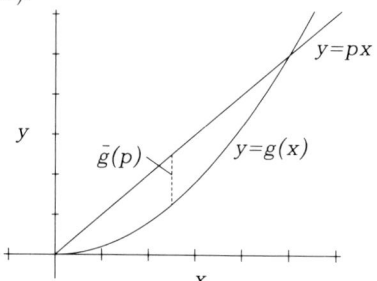

The Legendre transform $\bar{g}(p)$ is the value of $\mathcal{G}$ at the point $x(p)$ at which the curve runs parallel to the line. Since

$$\frac{\partial \mathcal{G}}{\partial x} = p - \frac{\partial g}{\partial x}, \quad (2)$$

$x(p)$ is the value of x which extremizes $\mathcal{G}$ for given p, as is already evident from the figure.

a phase transition respectively, then the latent heat l per unit mass is

$$l = T(s_+ - s_-) = T\left[\left(\frac{\partial g_-}{\partial T}\right)_m - \left(\frac{\partial g_+}{\partial T}\right)_m\right]$$
$$= T\left[\left(\frac{\partial f_-}{\partial T}\right)_B - \left(\frac{\partial f_+}{\partial T}\right)_B\right]. \quad (1.15)$$

At a first-order phase change $l > 0$, so the surfaces $g(T, p)$ and $f(T, B)$ have discontinuous first derivatives—indeed, it was this fact which caused Ehrenfest to give first-order transitions their name (see Box 1.3). At a continuous transition the free-energy surfaces have continuous first derivatives.

1.5 Thermodynamic potentials

Table 1.3. Use of the thermodynamic potentials

Variable	Fluid		Magnetic system	
s	$-\left(\dfrac{\partial g}{\partial T}\right)_p = -\left(\dfrac{\partial f}{\partial T}\right)_v$		$-\left(\dfrac{\partial g}{\partial T}\right)_m = -\left(\dfrac{\partial f}{\partial T}\right)_B$	
p or m	$-\left(\dfrac{\partial f}{\partial v}\right)_T$		$-\left(\dfrac{\partial f}{\partial B}\right)_T$	
v or B	$\left(\dfrac{\partial g}{\partial p}\right)_T$		$\left(\dfrac{\partial g}{\partial m}\right)_T$	
u	$f - T\left(\dfrac{\partial f}{\partial T}\right)_v = \left(\dfrac{\partial \beta f}{\partial \beta}\right)_v$		$f - T\left(\dfrac{\partial f}{\partial T}\right)_B = \left(\dfrac{\partial \beta f}{\partial \beta}\right)_B$	
κ_T or χ_T	$-\dfrac{1}{v}\left(\dfrac{\partial^2 g}{\partial p^2}\right)_T$		$-\left(\dfrac{\partial^2 f}{\partial B^2}\right)_T$	
c_v or c_B	$-T\left(\dfrac{\partial^2 f}{\partial T^2}\right)_v$		$-\left(\dfrac{\partial^2 f}{\partial T^2}\right)_B$	

NOTES: Most relations follow immediately from $df = -sdT - mdB$, $dg = -Tds + Bdm$, or the corresponding expressions for a fluid.

But they may nevertheless have infinite second-order derivatives—we have seen that both $c_B = -(\partial^2 f/\partial T^2)_B$ and $\chi_T = -(\partial^2 f/\partial B^2)_T$ are liable to diverge near T_c. However, these divergences are too weak to integrate up to discontinuities in the first-order derivatives of f.

If f were an analytic function of its arguments, all its derivatives would be continuous and the critical exponents α, β, γ, and δ would all vanish. In fact, if we express f as a sum of an analytic or **regular** part and a non-analytic or **singular** part, the critical exponents can be deduced from a knowledge of the singular part alone (see Problem 1.2). We shall occasionally find it useful to ignore a regular contribution to f on the grounds that it cannot affect the values taken by critical exponents.

It is possible to define a potential f_3 which is in some ways an extension of f and is a potential in a deeper sense than f. The **extended free energy** f_3 is a function $f_3(T, B, m)$ not of two but of three variables; it is defined such that the probability dP of m lying in the interval $(m + dm, dm)$ is proportional to $\exp[-\beta F_3(T, B, m)]dm$. Away from a critical point fluctuations are small, and this probability distribution is sharply peaked about $m(T, B) = -\partial f/\partial B$, where f is the conventional Helmholtz free energy. Near a critical point f_3 has a much less well-defined minimum at the thermally averaged value of m, with the result that significant fluctuations about this minimum are liable to occur.

In Appendix E we show that above T_c we may take f_3 to be

$$f_3(T, B, m) = g(T, m) - Bm \quad (T > T_c). \tag{1.16}$$

> **Box 1.3: Ehrenfest's classification of phase transitions**
>
> P. Ehrenfest (1880–1933) proposed a classification of phase changes based on the smoothness of the **chemical potential** μ, which is defined to be the value per particle of the thermodynamic potential whose natural variables are intensive. μ is continuous across a phase change. Since the entropy per particle is $s = -\partial \mu/\partial T$, equation (1.15) tells us that a transition involves non-zero latent heat, and thus is first order, if and only if $\partial \mu/\partial T$ is discontinuous at the transition. Ehrenfest's scheme was to classify a phase change as n^{th}-order if the lowest-order derivative of μ to be discontinuous across the transition is an nth derivative.
>
> This scheme fell into some disfavour when it was realized that derivatives of μ can diverge as a transition is approached without actually being discontinuous *at* the transition. For example, in §3.4.2 we shall see that the specific heat $c_B = -\beta^2(\partial^2 \beta\mu/\partial \beta^2)$ of the Ising model of a two-dimensional ferromagnet diverges logarithmically as T_c is approached. The specific heat is not discontinuous, so according to Ehrenfest the transition is third-order. On the other hand the internal energy of such a ferromagnet changes arbitrarily rapidly near T_c, much as if it had a non-zero latent heat (see Figure 3.5).
>
> It is now more usual to classify transitions as first order if they have non-zero latent heat, and to lump all other transitions together as 'continuous' ones.

That is, for $T > T_c$, f_3 is $-\mathcal{G}$ where $\mathcal{G}(T, B, m) \equiv Bm - g(T, m)$ is the function involved in the Legendre transformation between f and g. From Box 1.2 it follows that f_3 is minimized with respect to m at the equilibrium value $m_0(T, B)$; Appendix E demonstrates that the probability of m lying in the interval $(m + dm, dm)$ is proportional to $\exp(-\beta F_3)$. Below T_c (1.16) is not true, and f_3 is no longer simply related to a thermodynamic function. Statistical mechanics enables us to calculate f_3 nevertheless.

Since the most probable value of m for given (B, T) occurs at the minimum of f_3, the latter is a potential in the sense that whenever the order parameter is disturbed, it rolls 'downhill' to the bottom of the 'potential' f_3.

As one changes T and/or B in the neighbourhood of a phase change, the function $f_3(m)$ evolves in a way that is qualitatively different in a first-order transition and a continuous transition. In the interests of definiteness, imagine that we are changing T through the transition temperature T_t at fixed B. For $T \gg T_t$, f_3 is minimized by $m = 0$, so a plot of $f_3(m)$ at fixed (T, B) looks something like the bottom curve in Figure 1.12(a). As the system is cooled, f_3 increases and distorts in such a way that for $T \ll T_t$ the minimum of f_3 occurs for $m \neq 0$. The top curve in Figure 1.12(a) illustrates

1.5 Thermodynamic potentials

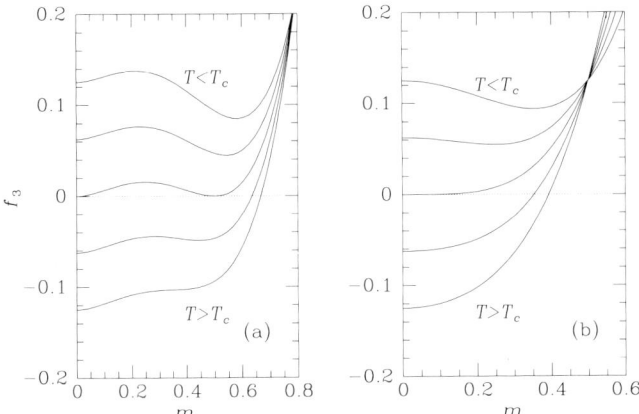

Figure 1.12 The equilibrium value of the order parameter m is that which minimizes the extended free energy f_3 at fixed (T, B). (a) In a first-order phase change the global minimum of f_3 migrates away from $m = 0$ as the temperature is lowered through T_t because a new minimum forms quite independently of the old. (b) In a continuous phase change the plot of f_3 versus m first becomes very flat around $m = 0$ and then develops a minimum at $m \neq 0$ to which the order parameter can migrate without slipping over a hill.

this state of affairs.

There are two ways in which the minimum at $m \neq 0$ can appear. If the transition is first-order, the new minimum starts out quite independent of the old minimum $m = 0$ (see Figure 1.12(a)). In this case the order parameter suddenly jumps to the new minimum when the latter becomes the global minimum of f_3. By contrast, in the case of a continuous phase transition the new minimum does not appear until the old minimum has ceased to be a minimum, as shown in Figure 1.12(b). A hill never separates the old and the new minima, so as the temperature is lowered, the order parameter can remain at all times at the lowest possible value of f_3, moving continuously, if perhaps rather rapidly at $T = T_t$, from $m = 0$ to $m \neq 0$. When $T = T_t$ the plot of f_3 is very flat near $m = 0$ so that small fluctuations in f_3 correspond to large fluctuations in m. This is the cause of critical opalescence.

In the case of a first-order transition, the plot of $f_3(m)$ is never flat so the fluctuations in m are never particularly large. As a result there are no critical phenomena. What can happen, however, is that the order parameter can become trapped in a minimum that is a local, but not a global, minimum, and then suddenly move across to the true global minimum. This is what happens when supercooled steam in a cloud chamber suddenly forms water droplets as a charged particle flies through.

1.5.1 The Widom and Kadanoff scaling hypotheses

The critical exponents α, β, γ, δ, μ, ν turn out not to be independent of one another. Historically, it was first shown by Rushbrooke (1963), Griffiths

(1965), Josephson (1967) and Fisher (1969) that basic thermodynamics together with a few reasonable assumptions oblige the six exponents to satisfy four inequalities. Gradually, experimental evidence accumulated that these inequalities were, in fact, equalities, and in 1965 Widom showed that two of them would indeed be equalities if the Helmholtz free energy were not any old function of two variables, but could be approximated by a function ψ of one variable. For a magnetic system Widom suggested that near T_c, f can be approximated by

$$f(T, B) = t^{1/y}\psi(B/t^{x/y}) \quad \text{where} \quad t \equiv \frac{|T - T_c|}{T_c}. \tag{1.17}$$

A corresponding expression would hold for any other system with two degrees of freedom. Equation (1.17) is called the **Widom scaling hypothesis**. One of the triumphs of the theory of critical phenomena is to derive this hypothesis from first principles—see §§5.7, 12.2 and 12.3. Here we take the hypothesis as given and derive from it two relations between the four exponents α, β, γ and δ—these are the critical exponents which are unconnected to the correlation length.

We first use (1.17) to calculate the zero-field magnetization near T_c:

$$m\big|_{B=0} = -\left(\frac{\partial f}{\partial B}\right)_T = -(t^{1/y}/t^{x/y})\psi'(0). \tag{1.18}$$

Comparing this with the definition of the exponent β, $m \sim t^\beta$, we conclude that

$$\beta = \frac{1 - x}{y}. \tag{1.19}$$

Similarly, equation (1.17) predicts that near T_c the zero-field susceptibility χ varies as

$$\chi\big|_{B=0} = -\left(\frac{\partial^2 f}{\partial B^2}\right)_T = t^{(1-2x)/y}\psi''(0). \tag{1.20}$$

Comparing this with the definition $\chi \sim t^{-\gamma}$ of γ, we conclude that

$$\gamma = \frac{2x - 1}{y}. \tag{1.21}$$

The calculation of the zero-field specific heat c_B predicted by (1.17) is slightly more tedious. It involves two derivatives with respect to t, but fortunately we can neglect all derivatives of ψ since they are multiplied by at least one power of B. We find

$$\begin{aligned} c_B &= -T\left(\frac{\partial^2 f}{\partial T^2}\right)_B \\ &\simeq -\frac{1}{T_c}\frac{\partial^2}{\partial t^2}\left[t^{1/y}\psi(B/t^{x/y})\right] \\ &= -\frac{\psi(0)}{T_c}\frac{1}{y}\left(\frac{1}{y} - 1\right)t^{(1/y - 2)}. \end{aligned} \tag{1.22}$$

1.5 Thermodynamic potentials

Comparing this with the definition $c_B \sim t^{-\alpha}$ of α, we have

$$\alpha = 2 - \frac{1}{y}. \tag{1.23}$$

Finally we use equation (1.17) to calculate the relationship between B and m at $T = T_c$. In this connection it is convenient to re-express (1.17) in the form

$$f(T, B) = B^{1/x}\widetilde{\psi}(B/t^{x/y}) \quad \text{where} \quad \widetilde{\psi}(z) \equiv z^{-1/x}\psi(z). \tag{1.24}$$

Differentiating this form of (1.17) at $t = 0$ we find[14]

$$m\Big|_{T=T_c} = -\left(\frac{\partial f}{\partial B}\right)_T = \frac{\widetilde{\psi}(\infty)}{x} B^{1/x-1}. \tag{1.25}$$

Comparing this with the definition $m \sim B^{1/\delta}$ of δ we have

$$\delta = \frac{x}{1-x}. \tag{1.26}$$

We can eliminate x and y between the four relations (1.19), (1.21), (1.23) and (1.26) to obtain two relations connecting the exponents α, β, γ and δ:

$$\alpha + 2\beta + \gamma = \frac{(2y-1) + 2(1-x) + (2x-1)}{y} = 2,$$
$$\alpha + \beta(\delta+1) = \frac{(2y-1)}{y} + \frac{1}{y} = 2. \tag{1.27}$$

These equations are known, respectively, as Rushbrooke's and Griffiths' laws in honour of those who demonstrated that the left sides of (1.27) must be greater than or equal to the right sides.

As Kadanoff (1966) first pointed out, two further relations that are satisfied by experimentally determined critical exponents can be understood by hypothesizing that the two-point correlation function is of the form

$$G_c^{(2)}(r, t) = \frac{\psi(r^d t^{2-\alpha})}{r^{d-2+\eta}}, \tag{1.28}$$

where t is defined by (1.17) as before, ψ is some function of one variable, and d is the dimensionality of the system. Comparing (1.28) with (1.12) we see that $\psi(z) \sim e^{-z}$. From (1.28) it is easy to show that (see Problem 1.3)

$$(2 - \eta)\nu = \gamma$$
$$\nu d = 2 - \alpha. \tag{1.29}$$

These relations are called Fisher's law and Josephson's law, respectively, having been first derived as inequalities by Fisher (1969) and Josephson (1967).

---

[14] We assume that $\widetilde{\psi}(z)$ tends to its limit as $z \to \infty$ sufficiently rapidly that any terms involving $\widetilde{\psi}'(\infty)$ can be neglected.

1.6 Why study phase transitions?

Now that we have reviewed the phenomenology of continuous phase transitions and presented some of the field's key results, we must address the question 'is it worth going on with this book—who cares about critical phenomena anyway?'

To the question 'why study critical phenomena?' it is perfectly legitimate to reply in the spirit of Leigh Mallory, who when asked why he had wanted to climb Mount Everest replied 'because it was there'. Indeed much current research in low-temperature physics centres on studies of continuous phase transitions in systems as diverse as semiconductors and liquid ^{3}He. But there are much larger reasons for studying phase changes, and especially continuous ones. First phase changes are immensely influential in every corner of the Universe—indeed it is widely argued that the very existence of the observable Universe is attributable to a phase change in the state of some pre-existing vacuum, and that the disposition of matter in and around galaxies should be understood in terms of fluctuations associated with some such transition (see for example Kolb and Turner 1989). A prerequisite for an attack on these fundamental questions is the theory of a fluctuating thermal medium. The study of critical phenomena develops just such a theory.

What makes continuous phase changes especially interesting is the scale-freedom of the fluctuations at T_c. Not only is the creation of long-range structure by short-range inter-molecular forces intriguing, but any example of scale-freedom is worthy of close examination since this phenomenon occurs in several physical systems that are inadequately understood—the clustering of galaxies (e.g., Peebles 1980), the distribution of earthquakes (e.g., Carlson and Langer 1989, Bak *et al.* 1988), turbulence in fluids and plasmas (e.g., Mandlebrot 1974), polymers (de Gennes 1972), snow flakes (Ball *et al.* 1989, Meakin and Tolman 1989)—to name but a few. In each case there is a wide range of scales over which some phenomenon varies as a power law of the scale, presumably because there is a gross mismatch between the largest and smallest scales in the problem. For example, when air is stirred by heat from the Sun, convective cells kilometres in size are set into motion. Ultimately the kinetic energy of these cells must be thermalized on the scale of a fraction of a millimetre at which the small viscosity of air first becomes effective. Over most of the intervening decades between a kilometre and a millimetre, fluctuations in the velocity and density of air conform to power-law distributions. Many of the standard techniques for handling phenomena of this kind were developed for the theory of critical phenomena.

A further motivation for studying the statistical field theory of critical phenomena is for the light it throws on quantum field theory. The connections here are both theoretical and practical. On a practical level, much exploration of gauge field theories has to be carried out numerically by representing the field, in principle a continuum concept, on a discrete lattice of events. An elementary particle is represented by a structure of a certain physical size on the lattice. As the lattice is refined this structure should retain its physical size by covering more and more lattice sites. Hence, as the discrete model approaches the continuum limit of real quantum fields, the particle must be represented by correlations on the lattice of longer and longer range, and the field theory that gives rise to these correlations must be approaching what in statistical mechanics we would call a critical point.

The goal of quantum field theory is to explain the rich diversity of experimental particle physics in terms of a theory in which a very small number of fundamental fields propagate under the influence of a highly symmetric Hamiltonian. Since the experimentally studied world does not display the high degree of symmetry required of this fundamental Hamiltonian, the solutions to the equations of motion to which the latter gives rise must be less symmetric than it is itself. We shall see in §2.3 that at a phase transition the order parameter field loses symmetry, with the result that at high temperatures the state of a thermal system generally displays the full symmetry of its governing Hamiltonian, but at low temperatures the system is much less symmetric. Thus phase transitions enable one to study in directly measurable cases a process which is thought to play a crucial rôle in structuring physics as we know it.

Statistical field theory also helps illuminate the renormalization of a quantum field theory. The statistical field theory of Chapter 7 onwards is a phenomenological treatment of known microphysics. Quantum field theories are likewise thought to be merely phenomenological in nature, but one does not know the relevant microphysics *a priori*; the relevant scale is likely to be something like the Planck length $(G\hbar/c^3)^{1/2} \approx 10^{-35}$ m. Because this scale is so much smaller than the scale of practicable experiments, one hopes to be able to model the latter with a theory in which no very small scale appears. 'Renormalization', which is the subject of Chapter 9, is the process by which one ensures that one's theory is indeed scale-free in this sense. In the theory of critical phenomena we similarly appeal to renormalization to ensure that as T_c is approached the underlying atomic scale disappears from view just as the Planck scale plays no rôle in conventional gauge field theories. A thorough understanding of the rôle of renormalization in the theory of critical phenomena helps one to understand what is happening in the more fundamental and mysterious field of high-energy physics.

Problems

1.1 Adapt Table 1.1 to the case of the liquid–gas transition.

1.2 Given that the free energy $f = f_a + f_s$ is a sum of a part f_a analytic in its arguments and a non-analytic part f_s, explain why the critical exponents α, β, γ and δ can be deduced from knowledge of f_s alone.

1.3 Derive Josephson's law from equation (1.28). The 'linear response theorem' (see §2.2) states that the susceptibility χ is related to the connected two-point correlation function by $\chi = \int G_c^{(2)}(r)\mathrm{d}^3\mathbf{r}$. Derive Fisher's law from this result and equation (1.28).

1.4 The transverse elastic displacement $\boldsymbol{\xi}(\mathbf{x})$ in a three-dimensional cube with periodic boundary conditions satisfies $\ddot{\boldsymbol{\xi}}/c^2 - \nabla^2\boldsymbol{\xi} = 0$. Writing a typical normal mode's displacement as $\boldsymbol{\xi} = \boldsymbol{\Xi}_\mathbf{k} e^{i(\mathbf{k}\cdot\mathbf{x}-\omega_\mathbf{k}t)}$, where $\boldsymbol{\Xi}_\mathbf{k}$ has a random phase, show that the thermal average of $\Xi_\mathbf{k}^2$ is $\langle\Xi_\mathbf{k}^2\rangle = k_BT/(\rho\omega_\mathbf{k}^2 L^3)$, where L is the cube's linear dimension and ρ is its mass density. By assuming that the phases of the $\boldsymbol{\Xi}_\mathbf{k}$ are uncorrelated, show that

$$\langle\boldsymbol{\xi}(\mathbf{x})\cdot\boldsymbol{\xi}(\mathbf{x}')\rangle \simeq \frac{2k_BT}{(2\pi)^2\rho c^2 r}I(r),$$

where $r \equiv |\mathbf{x}-\mathbf{x}'|$ and

$$I(r) \equiv \int_{2\pi r/L}^{\infty} \mathrm{d}x\,\frac{\sin x}{x}.$$

Explain physically why this integral is dominated by contributions from $x < 1$. What value of the critical exponent η is predicted by this model?

1.5 Show that the model of the last problem, adapted to the case of sound waves in gas filling a three-dimensional cube with periodic boundary conditions, predicts that the critical exponent η associated with the density–density correlation function is $\eta = 2$.

1.6 $\bar{g}(p)$ is the Legendre transform of the convex function $g(x)$. Show that the Legendre transform is its own inverse: $\bar{\bar{g}}(x) = g(x)$.

1.7 Use thermodynamics to prove Rushbrooke's law as an inequality. [Hint: use Table 1.3 to express the condition for f to have an extremum (rather than a saddle point) at $(T,B) = (T_c, 0)$.]

1.8 Use thermodynamics to prove Griffiths' law as an inequality. [Hint: Do the previous problem first. Then replot the central portion of Figure 1.5 on an enlarged scale.]

2
Statistical mechanics

Statistical mechanics is the art of predicting the behaviour of a system with a large number of degrees of freedom, given the laws governing its microscopic behaviour. It divides naturally into two parts. The first is the more fundamental. It tries to answer general questions concerning the behaviour of systems with many degrees of freedom, such as how order can arise from the apparent chaos of the microscopic behaviour, and how time-symmetric microscopic laws can give rise to the time asymmetry observed in the macroscopic world. The second part of statistical mechanics is concerned with using the results of this analysis to determine the behaviour of specific systems. This is what we will be doing in this book. Statistical mechanics will be used as a tool to study critical phenomena in particular systems.

We will only be concerned with systems in thermal equilibrium. For such systems there is a very simple relationship between microscopic properties and macroscopic behaviour. Let us label the different possible microscopic configurations—**microstates**—of a system by α. The set of microstates may be either continuous or discrete, depending on the system. A classical system has a continuous set of possible microstates. For example, the microstates of a classical system consisting of a single particle in a box are labelled by the position and momentum of the particle. A quantum system (in a finite volume) may have a discrete set of possible microstates—the eigenstates of the Hamiltonian, for example.[1] The system may also be completely fictitious.

---

[1] One might object that even a system whose Hamiltonian has discrete eigenstates can

It is very rare to be able to model a real system exactly, and so it is quite usual to construct models which only vaguely resemble real systems but which, one hopes, capture their essence. These idealizations may be either continuous or discrete; examples of each type will be found throughout the book.

Let E_α be the energy of the microstate labelled by α, and let us suppose that the system is in thermal equilibrium with a heat-bath at temperature T. Then:

> The probability p_α of the system being found in the microstate α is proportional to $e^{-E_\alpha/k_B T}$.

In this expression k_B is **Boltzmann's constant**, $1.38 \times 10^{-23}\,\mathrm{J\,K^{-1}}$. It is usual to write $1/k_B T$ as β, and we shall do so in this book.[2] The quantity $e^{-\beta E_\alpha}$ is known as the **Boltzmann factor**. This fundamental result of statistical mechanics will be the starting point of our analysis. From it will follow all the time-independent properties of a system in thermal equilibrium.

Since the system must always be in *some* state, the total of the p_α should be one, and the normalized probability p_α is therefore

$$p_\alpha = \frac{1}{Z} e^{-\beta E_\alpha}, \tag{2.1}$$

where the normalizing factor Z is

$$Z = \sum_\alpha e^{-\beta E_\alpha}. \tag{2.2}$$

This is the **Gibbs probability distribution**. From it we can calculate the thermal average $\langle X \rangle$ of any property X of the system, provided that:
- We know what the possible microstates α of the system are;
- We know what the energy E_α of each microstate is;
- We know the value X_α taken by X in each microstate α;
- We can perform the sum

$$\begin{aligned}\langle X \rangle &= \sum_\alpha p_\alpha X_\alpha \\ &= \frac{1}{Z} \sum_\alpha X_\alpha e^{-\beta E_\alpha}.\end{aligned} \tag{2.3}$$

The fourth of these in particular can be very hard if not impossible to do exactly, but this is a problem of technique and not of principle. The Gibbs

---

be in a continuum of possible states, since a state is specified by giving the amplitude for the system to be in each one of a complete set of basis states and these amplitudes are (continuous) complex variables. However, in statistical mechanics one calculates averages, and taking the average of the expectation value of some operator over *all* possible states gives the same result as averaging over the members of a complete set of basis states only.

[2] There should be no danger of confusing this β with the critical exponent of Chapter 1. They represent quite different quantities, and appear in different contexts.

2.1 Thermodynamic quantities

distribution provides a remarkably simple and universal connection between the microscopic laws governing a system and its behaviour when in thermal equilibrium.

The quantity Z defined in (2.2) is called the **partition function**. It is a function of temperature. Less obviously, it is also a function of the parameters which determine the energy E_α of each microstate. We shall refer to these parameters as **constraints**. For example, if our system consists of molecules of gas inside a rigid container, this can be represented by a potential $U(\mathbf{x})$ which is zero inside the container but infinite outside it. In this case the 'constraint' $U(\mathbf{x})$ does literally constrain the system. Or if a system of spins $\mathbf{s}_i$ is put in an external magnetic field $\mathbf{B}$ there will be an addition to the energy $-\sum_i \mathbf{s}_i \cdot \mathbf{B}$, reflecting this. In this case the external field $\mathbf{B}$ is the constraint. We shall denote the set of constraints applied to a general system by $\{V\}$. It is by varying the constraints that mechanical work is done on the system. Far from Z being just a normalizing constant, we shall see that all the properties of the system can be obtained from the functional dependence of Z on the temperature and the constraints.

There remains the question of what the averages we calculate with the Gibbs distribution mean. There are two possibilities. We can imagine a very large number of copies of a system, and ask what the distribution of the microstates of these copies is at one particular time. Or we can imagine carrying out a series of measurements on one particular system at different times, and asking how these measurements are distributed. Usually it is correct to assume these two distributions to be equal to each other and to the Gibbs distribution, but problems can arise if it takes a long time for the system to travel between different parts of the space of microstates available to it. It may be that the microstates of a single system follow the Gibbs distribution over a sufficiently long time, but over the time of observation they do not. We will return to this matter in §2.3.

2.1 Thermodynamic quantities

The mean energy of a system in thermal equilibrium is, from equation (2.3),

$$U \equiv \langle E \rangle = \frac{1}{Z} \sum_\alpha E_\alpha \, e^{-\beta E_\alpha}. \tag{2.4}$$

This is the sum of the energy of each microstate over all microstates α, weighted by the probability that the system will be in that microstate. From the definition (2.2) of the partition function Z it follows that

$$U = -\left(\frac{\partial \log Z}{\partial \beta}\right)_{\{V\}}. \tag{2.5}$$

So knowing the partition function for a system as a function of β allows us to find the mean energy. Differentiating U with respect to temperature while holding the constraints fixed gives the heat capacity at constant $\{V\}$, which we write as $C_{\{V\}}$:

$$C_{\{V\}} \equiv \left(\frac{\partial U}{\partial T}\right)_{\{V\}}$$
$$= k_B \beta^2 \left(\frac{\partial^2 \log Z}{\partial \beta^2}\right)_{\{V\}}. \qquad (2.6)$$

In a change in which the constraints $\{V\}$ do not vary, no mechanical work is done on the system. So any change in the energy of the system must be due entirely to a flow of heat. Therefore

$$C_{\{V\}} \, dT = T \, dS, \qquad (2.7)$$

where S is the entropy as defined in classical thermodynamics. Hence

$$C_{\{V\}} = -\beta \left(\frac{\partial S}{\partial \beta}\right)_{\{V\}}. \qquad (2.8)$$

Substituting from (2.6) for $C_{\{V\}}$ and integrating with respect to β, this yields the following expression for the system's entropy:

$$S = -k_B \beta \left(\frac{\partial \log Z}{\partial \beta}\right)_{\{V\}} + k_B \log Z + \Phi(\{V\}), \qquad (2.9)$$

where $\Phi(\{V\})$ is a function of the constraints alone. The third law of thermodynamics requires the entropy of the system to tend to a constant value as $T \to 0$, independently of the values of the constraints. Hence it follows that $\Phi(\{V\}) = 0$ (see Problem 2.3) and the entropy is

$$S = -k_B \beta \left(\frac{\partial \log Z}{\partial \beta}\right)_{\{V\}} + k_B \log Z. \qquad (2.10)$$

This equation has an interesting interpretation. We know (and shall prove in the next section) that the fluctuations δU in the energy of a macroscopic system are usually small. So we can approximate the partition function of such a system by neglecting the microstates whose energy lies outside the range $U \pm \delta U$, since the system is very unlikely to be found in any of these microstates. This gives

$$Z \approx \Omega e^{-\beta U}, \qquad (2.11)$$

where Ω is the number of microstates of the system in the range $U \pm \delta U$. Using (2.11) in the formula (2.10) for the entropy, and bearing in mind equation (2.5) for U, we find

$$S = k_B \log \Omega. \qquad (2.12)$$

2.1 Thermodynamic quantities

So the classical entropy is a measure of the number of microstates accessible to the system.

Equations (2.5) and (2.10) allow us to find the Helmholtz free energy

$$F \equiv U - TS. \tag{2.13}$$

It has a particularly simple form. Combining equations (2.5) and (2.10) gives

$$F = -\frac{1}{\beta} \log Z, \quad \text{or} \quad Z = e^{-\beta F}. \tag{2.14}$$

Z is a function of T and $\{V\}$, which are the 'natural' variables for the Helmholtz free energy. All the usual results of classical thermodynamics can now be applied to (2.14). In particular the 'pressure' p_i conjugate to the constraint V_i is

$$\begin{aligned} p_i &\equiv -\left(\frac{\partial F}{\partial V_i}\right)_{\beta, V_j \neq i} \\ &= \frac{1}{\beta}\left(\frac{\partial \log Z}{\partial V_i}\right)_{\beta, V_j \neq i}. \end{aligned} \tag{2.15}$$

These equations are the **equations of state** for the system. There is one such equation for each constraint.[3] Knowing the p_i allows us to find the Gibbs free energies:

$$\begin{aligned} G &\equiv F + \sum_i P_i V_i \\ &= -\frac{1}{\beta} \sum_i \frac{\partial}{\partial V_i}(V_i \log Z). \end{aligned} \tag{2.16}$$

A system with N constraints has $2^N - 1$ possible Gibbs free energies, which differ in the choice of terms included in the sum of equation (2.16).

So knowing the partition function Z as a function of the temperature and the constraints allows us to find all the quantities which are of interest in classical thermodynamics. In turn, Z can be found from a knowledge of the energies of all the system's microstates. It is conventional to regard this information as being conveyed by a function $H(\alpha, \{V\})$; thus we write

$$E_\alpha = H(\alpha, \text{constraints}). \tag{2.17}$$

---

[3] Expanding the formula (2.15) for the pressure gives

$$p_i = \frac{1}{Z} \sum_\alpha -\left(\frac{\partial E_\alpha}{\partial V_i}\right)_{V_j \neq i} e^{-\beta E_\alpha}.$$

In words, the pressure associated with the constraint V_i is minus the rate at which the energy of each microstate varies as the constraint changes, averaged over all the microstates.

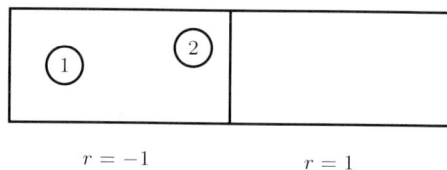

Figure 2.1 A very simple model, the statistical mechanics of which is considered in the text. The two particles can each be in one of the two boxes. The energy of the system depends on which boxes the particles are in in a manner discussed in the text.

H is known as the **Hamiltonian**.

Since classical thermodynamics assumes no particular model for the system under consideration, it provides no link between microscopic and macroscopic physics. But it leads to results of great generality which, when combined with some minimal experimental input such as an equation of state, yield powerful predictions. In contrast, statistical mechanics allows us to start with the Hamiltonian of the system and get everything from that, provided that we can calculate Z.

As an example of the above techniques, consider the system illustrated in Figure 2.1. This is a very simple system. It can be thought of as two distinguishable particles, each of which can be in either of two boxes. The system has $2 \times 2 = 4$ microstates. The energy of each particle is $2A$ more in the right box than in the left box. There is also a 'binding energy' term, which lowers the energy of the system by κ if the two particles are in the same box. We specify the state of the system with two variables, r_1 and r_2. r_1 takes the value -1 or $+1$, according to whether the first particle is in the left box or the right box; r_2 describes the position of the second particle in the same way. The Hamiltonian for the system is

$$H(r_1, r_2) = (r_1 + r_2)A - \kappa \delta_{r_1, r_2}. \tag{2.18}$$

Suppose that the binding energy κ is much greater than the difference A between the energies of the two boxes, and that both A and κ are positive. At low temperatures both particles are almost certain to be found in the left-hand box. As the temperature rises the particles will spend more and more time in the right-hand box, and the mean energy of the system will rise to reflect this fact. They will, however, be in the same box as each other for most of the time. As the temperature continues to rise there will come a point when the particles become 'unbound' and start to spend time in different boxes. More energy will be absorbed at this point. As T becomes very large the energy of the system will become constant (unlike a physical system, our model has a maximum energy). The particles will then move independently of each other.

2.1 Thermodynamic quantities

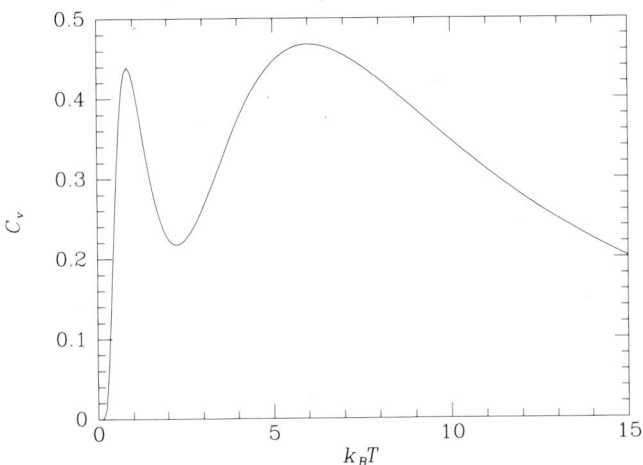

Figure 2.2 Heat capacity as a function of temperature T for the model defined by (2.18) with $A = 0.5$ and $\kappa = 15$.

How do we check this? We shall see in the next section how to find out in which boxes the particles are spending their time, but we already know how to calculate the temperature dependence of the system's heat capacity. Our guess is that the system should absorb heat around two temperatures—the first when the particles start to spend time in the right-hand box, and the second when the particles become unbound. As the temperature goes to infinity the heat capacity should go to zero. The first step in calculating the heat capacity C is to find the partition function Z:

$$Z = \underbrace{e^{-\beta(-2A-\kappa)}}_{r_1=-1;\, r_2=-1} + \underbrace{1}_{r_1=1;\, r_2=-1} + \underbrace{1}_{r_1=-1;\, r_2=1} + \underbrace{e^{-\beta(2A-\kappa)}}_{r_1=1;\, r_2=1} \quad (2.19)$$

$$= 2\left[e^{\beta\kappa}\cosh(2\beta A) + 1\right].$$

Now we can find U, using equation (2.5). The result is

$$U = -\frac{e^{\beta\kappa}\left[\kappa\cosh(2\beta A) + 2A\sinh(2\beta A)\right]}{e^{\beta\kappa}\cosh(2\beta A) + 1}. \quad (2.20)$$

The heat capacity C can be found by differentiating this with respect to temperature, and is plotted as a function of temperature in Figure 2.2 for $A = 0.5$ and $\kappa = 15$. Two peaks in the heat capacity are clearly visible. We would guess that the first marks the temperature at which the particles start to spend time in the right-hand box, and the second marks the temperature at which the particles become unbound. But simply knowing the classical thermodynamic potentials does not allow us to check this. In the next section we shall see how that may be done.

The algebra in this example was rather heavy (if you carried out the differentiation of U to obtain C), and yet this system has only four microstates. Fortunately the difficulty does not necessarily increase as the number of microstates, and as we shall see in the next chapter there are several non-trivial models whose partition functions can be found exactly.

2.2 Fluctuations and correlation functions

Statistical mechanics can also be used to study quantities about which classical thermodynamics has nothing to say. For example, consider the fluctuations that occur in the energy of a system at given values of temperature and constraints. The energy will not always have its equilibrium value, and it would be interesting to know over what sort of range the energy fluctuates. The thermal average of $(E - \langle E \rangle)^2$ tells us this. Now

$$\langle (E - \langle E \rangle)^2 \rangle = \langle E^2 \rangle - \langle E \rangle^2. \tag{2.21}$$

We know how to find $\langle E \rangle \equiv U$ from the partition function (equation (2.5)), but how do we find $\langle E^2 \rangle$? One look at the definition (2.2) of Z suggests the answer: differentiate Z twice with respect to β to bring down two factors of E_α, and then divide by Z to normalize the probabilities:

$$\begin{aligned}
\langle E^2 \rangle &= \frac{1}{Z} \sum_\alpha E_\alpha^2 \, e^{-\beta E_\alpha} \\
&= \frac{1}{Z} \left(\frac{\partial^2 Z}{\partial \beta^2} \right)_{\{V\}}.
\end{aligned} \tag{2.22}$$

Combining this with the expression (2.5) for U gives

$$\begin{aligned}
\langle (E - \langle E \rangle)^2 \rangle \equiv (\delta U)^2 &= \frac{1}{Z} \left(\frac{\partial^2 Z}{\partial \beta^2} \right)_{\{V\}} - \left(\frac{1}{Z} \left(\frac{\partial Z}{\partial \beta} \right)_{\{V\}} \right)^2 \\
&= \left(\frac{\partial^2 \log Z}{\partial \beta^2} \right)_{\{V\}} \\
&= \frac{C_{\{V\}}}{k_B \beta^2}.
\end{aligned} \tag{2.23}$$

The first line follows from differentiating the definition (2.2) of Z, the second line is an identity, and the third line follows from (2.6). The RMS fluctuation δU in the energy of the system is therefore

$$\delta U = T(k_B C_{\{V\}})^{1/2}. \tag{2.24}$$

2.2 Correlation functions

This is a completely general result, and does not depend on any particular properties of the system being considered. For 1 kg of water at room temperature δU is about 4.2×10^{-8} J, which is small—the amount of energy needed to change the water's temperature by 1 K is 10^{11} times this. Note that since the heat capacity $C_{\{V\}}$ grows linearly with the size of the system, the fractional energy fluctuations $\delta U/U$ fall as the square root of the system size. They therefore become negligible in the limit that the size goes to infinity. Because of this, this limit is called the **thermodynamic limit**. The exception to this is when the heat capacity of the system diverges, as it does at a critical point. Then the fluctuations do not go away as the system becomes larger, but are present on all scales.

It was surprisingly easy to find something as interesting as the fluctuations in the energy. The technique of differentiating the partition function to find thermal averages is a very powerful one. If the partition function is known as a function of the appropriate constraints, the thermal average of any quantity can be found. If the required constraint term is not already present in the Hamiltonian, it can be added in.

Suppose that we want to know the thermal average of a quantity X. One approach would be to work out the sum (2.3) directly. But another way is to add to the Hamiltonian a term

$$\Delta H = -XY, \tag{2.25}$$

where Y is some number, if such a term is not already present. The partition function can be worked out as a function of Y, which plays the rôle of a constraint. This done, $\langle X \rangle$ can be found by differentiating $\log Z$ with respect to Y, and then putting Y equal to zero:

$$\begin{aligned} \frac{1}{\beta} \frac{\partial \log Z}{\partial Y} \bigg|_{Y=0} &= \frac{1}{\beta Z} \frac{\partial}{\partial Y} \bigg|_{Y=0} \sum_\alpha e^{-\beta(E_\alpha - X_\alpha Y)} \\ &= \frac{1}{Z} \sum_\alpha X_\alpha e^{-\beta E_\alpha} \\ &= \langle X \rangle. \end{aligned} \tag{2.26}$$

In terms of the Helmholtz free energy F,

$$\langle X \rangle = -\frac{\partial F}{\partial Y}\bigg|_{Y=0}. \tag{2.27}$$

Differentiating again with respect to Y gives the fluctuations in X:

$$\langle X^2 \rangle - \langle X \rangle^2 = \frac{1}{\beta} \frac{\partial \langle X \rangle}{\partial Y}\bigg|_{Y=0} \equiv \frac{\chi}{\beta}, \tag{2.28}$$

where χ is a generalized susceptibility. This relation between the fluctuations in X and the linear response of $\langle X \rangle$ to Y is known as the **linear response theorem**.

The quantities looked at so far relate to the whole of the system being studied—its total energy, its entropy, and so on. But knowing the probability distribution for the microstates of the system we are studying allows us to calculate thermal averages that relate not just to the system as a whole, but to parts of it. For example, instead of calculating the total magnetization of a system of spins, we can see how the magnetization varies from place to place, and how the spins at different places are correlated with each other. These questions are beyond the scope of classical thermodynamics, since they require a model of the microscopic structure of the system.

Let us consider a system whose microstates can be specified by giving the values of N variables $s_1, \ldots, s_N$. We shall refer to these variables as 'spins' to make the discussion less abstract, but what follows is completely general. Suppose we wish to know the 'magnetization' M, the thermal average of the total spin $S = \sum_i s_i$. We add a term $\Delta H = (-1/\beta) J S$ to the Hamiltonian, and calculate the partition function Z as a function of J.[4] Then

$$M = \left. \frac{\partial \log Z}{\partial J} \right|_{J=0}. \tag{2.29}$$

However, if the field J is allowed to be different at each spin site, the thermal averages of individual spins can be extracted from Z. Instead of $-JS/\beta$, let us add the term

$$\Delta H' = -\frac{1}{\beta} \sum_i J_i s_i \tag{2.30}$$

to the Hamiltonian. Then knowing Z as a function of the J_i allows us to calculate

$$\langle s_i \rangle = \frac{1}{Z} \frac{\partial Z}{\partial J_i};$$

and also $\quad \langle s_i s_j \rangle = \frac{1}{Z} \frac{\partial^2 Z}{\partial J_i \partial J_j};\tag{2.31}$

$$\langle s_i s_j s_k \rangle = \frac{1}{Z} \frac{\partial^3 Z}{\partial J_i \partial J_j \partial J_k};$$

and so on. Quantities such as $\langle s_i s_j \rangle$ are called **correlation functions**. The thermal average of $s_{i_1} \ldots s_{i_n}$ (n spins in total) is called the **n-point correlation function** of the spins, and is written

$$G^{(n)}(i_1, \ldots, i_n) \equiv \langle s_{i_1} \ldots s_{i_n} \rangle$$
$$= \frac{1}{Z} \frac{\partial^n Z}{\partial J_{i_1} \ldots \partial J_{i_n}}. \tag{2.32}$$

---

[4] Note that if the s_i really are spins, then J is related to the magnetic field B by $J = -\beta B$. The factor of $-\beta$ simplifies the following equations.

2.2 Correlation functions

G is a function of the locations $(i_1, \ldots, i_n)$ of the spins being averaged.

What do correlation functions tell us about a system? If the spins are non-interacting then they give us no more information than could be obtained by considering each spin individually. In such a case averages of products of different spins factorize into the product of the individual averages:

$$\langle s_i s_j \rangle = \langle s_i \rangle \langle s_j \rangle \qquad (i \neq j; \text{ system of non-interacting spins only!}). \quad (2.33)$$

However, if the spins interact with each other, correlation functions start to become interesting. They tell us how 'correlated' different parts of the system are. Recall the 'two particles in two boxes' example of the last section. We guessed that the second peak in the heat capacity marked when the two particles began to spend a significant amount of time in separate boxes. The correlation function of their two positions $\langle r_1 r_2 \rangle$ gives us this information. If r_1 is nearly always equal to r_2, this correlation function is close to one; if r_1 and r_2 vary essentially independently of each other, it is close to zero; if they spend most of their time in different boxes, it is close to -1. The correlation function gives us information that we couldn't get from the thermal averages of r_1 and r_2 separately. This is true of correlation functions in general. When different parts of a system interact they fluctuate together, and correlation functions tell us about this behaviour.

However, the correlation functions $G^{(n)}$ introduced above are not the best way of describing the behaviour of a system. To illustrate this, suppose that our system of N spins represents a ferromagnet in no external field, above the Curie point. There is therefore no net magnetization of the system, and the thermal average of each spin is zero: $\langle s_i \rangle = 0$. However, the two-point correlation function $G^{(2)}(i,j) = \langle s_i s_j \rangle$ does not vanish, but tells us how correlated different spins are with each other. Since this is a ferromagnet we expect that spins which are close together will tend to point in the same direction as each other on average, the correlation decreasing as we consider spins that are further apart. So $G^{(2)}(i,j)$ will be positive for small values of the site separation $|\mathbf{r}_i - \mathbf{r}_j|$, going to zero as this becomes large. Now suppose that a magnetic field $\mathbf{B}$ is applied to the magnet. Each spin will tend to line up with it, and so $\langle s_i s_j \rangle$ will no longer go to zero at large distances. Even if the spins do not interact with each other at all, so that equation (2.33) is true, $G^{(2)}$ is non-zero since $\langle s_i \rangle$ and $\langle s_j \rangle$ are themselves non-zero. It would be nice to have some measure of correlation between the spins s_i and s_j that measures only the part of the correlation that is due to interactions. Such a quantity is the two-point **connected correlation function** $G_c^{(2)}(i,j)$, defined as

$$G_c^{(2)}(i,j) \equiv \langle s_i s_j \rangle - \langle s_i \rangle \langle s_j \rangle. \quad (2.34)$$

The contribution to $\langle s_i s_j \rangle$ that comes from each spin separately has been subtracted off; what is left is due to the two spins together. The connected correlation function always vanishes for a system of non-interacting spins (if $i \neq j$), even if a magnetic field is present. Another way of writing $G_c^{(2)}(i,j)$ is

$$G_c^{(2)}(i,j) = \Big\langle (s_i - \langle s_i \rangle) \times (s_j - \langle s_j \rangle) \Big\rangle. \tag{2.35}$$

This shows clearly that $G_c^{(2)}$ measures the correlation between the fluctuations at the two sites i and j, which vanishes if there is no interaction between the sites. Note that

$$G_c^{(2)}(i,j) = \frac{\partial^2 \log Z}{\partial J_i \partial J_j}, \tag{2.36}$$

as can easily be checked.

Higher-order correlation functions ($G^{(n)}$ for $n > 2$) are also useful, but again they carry superfluous information. In the magnetic example above it is clear that $G^{(3)}(i,j,k)$ will be non-zero in the presence of a magnetic field in just the same way that $G^{(2)}(i,j)$ was. Just as $G_c^{(2)}$ extracted from $G^{(2)}$ the information that was not already present in $G^{(1)}$, so higher-order connected correlation functions may be defined that contain only the new information concerning n spins. For example, the connected three-point correlation function $G_c^{(3)}$ is defined by

$$G^{(3)}(i,j,k) = G^{(1)}(i)G^{(1)}(j)G^{(1)}(k) + G_c^{(2)}(i,j)G^{(1)}(k) \tag{2.37}$$
$$+ G_c^{(2)}(j,k)G^{(1)}(i) + G_c^{(2)}(k,i)G^{(1)}(j) + G_c^{(3)}(i,j,k).$$

$G_c^{(3)}$ is the contribution to $G^{(3)}$ 'left over' when all other contributions have been subtracted off. The first term on the right is the contribution to $G^{(3)}$ that comes from the average of each of the three spins considered separately; the next three terms are the contributions coming from a 'connected' pair of spins and the third spin. What is left must be a property of all three spins together, and is called $G_c^{(3)}$. It represents the difference between the actual value of $G^{(3)}$ and our 'best guess' at it, based on our knowledge of lower-order correlation functions. The same goes for higher-order correlation functions. $G^{(n)}$ is rather like a list of all possible menus that can be prepared with n ingredients, while $G_c^{(n)}$ is a like a list of individual dishes that require n ingredients. If we know the individual dishes (the $G_c^{(n)}$) we can work out the possible menus (the $G^{(n)}$), and knowledge of the individual dishes is a much more compact way of storing the information.

The definitions for higher-order connected correlation functions are similar to that for $G^{(3)}$ above, and therefore even longer (a definition of $G_c^{(n)}$ is given in Appendix D). It would be hard work to calculate the $G_c^{(n)}$ by first finding the $G^{(n)}$ and then using formulae such as (2.37). But there is a beautiful result which states that the connected correlation functions can be

2.2 Correlation functions

obtained by differentiating $\log Z$:

$$G_c^{(n)}(i_1,\ldots,i_n) = \frac{\partial}{\partial J_{i_1}} \cdots \frac{\partial}{\partial J_{i_n}} \log Z \qquad (2.38)$$

$$\equiv \left\langle s_{i_1} \ldots s_{i_n} \right\rangle_c$$

(compare (2.32)). We have noted in equation (2.36) that this is true for $G_c^{(2)}$, and it can be checked for the three-point function above as well. It is proved in general in Appendix D. Since the connected correlation functions are much more useful than the ordinary ones, this is a very helpful result. Looked at in another way, it gives the connected correlation functions an additional physical significance; $-G_c^{(n)}/\beta$ is a coefficient in the Taylor series expansion of the Helmholtz free energy as a function of the J_i.

Let us return to the simple model of two particles in two boxes which we considered in the previous section. There we calculated its energy and plotted its heat capacity. Now that we know about correlation functions we can work out what the individual particles are doing by working out the partition function for the modified Hamiltonian

$$H' = A_1 r_1 + A_2 r_2 - \kappa \delta_{r_1, r_2}. \qquad (2.39)$$

Having different coefficients for r_1 and r_2 allows us to bring down r_1 or r_2 separately, by differentiating with respect to either A_1 or A_2. This is quite analogous to the introduction of the site-dependent field J_i in equation (2.30) earlier in this section. After the differentiation A_1 and A_2 can be set equal to A to restore the original model. The partition function Z' for the model described by (2.39) is

$$Z'(\beta, A_1, A_2) = 2 \left[e^{\beta \kappa} \cosh \beta(A_1 + A_2) + \cosh \beta(A_1 - A_2) \right]. \qquad (2.40)$$

It reduces to the expression (2.19) for Z when $A_1 = A_2 = A$. From it we can find $\langle r_1 \rangle$ (which by symmetry is equal to $\langle r_2 \rangle$):

$$\langle r_1 \rangle = -\frac{1}{\beta} \left(\frac{\partial \log Z'}{\partial A_1} \right)_{A_2} \bigg|_{A_1 = A_2 = A} \qquad (2.41)$$

$$= -\frac{e^{\beta \kappa} \sinh(2\beta A)}{e^{\beta \kappa} \cosh(2A\beta) + 1}.$$

This is plotted in Figure 2.3. As expected, at low temperatures the particle spends most of its time in the left-hand box, while at temperatures much greater than A it is equally likely to be in either. But we don't have complete information about the system until we have found out about the correlation between the positions of the two particles. The thermal average of $r_1 r_2$ tells

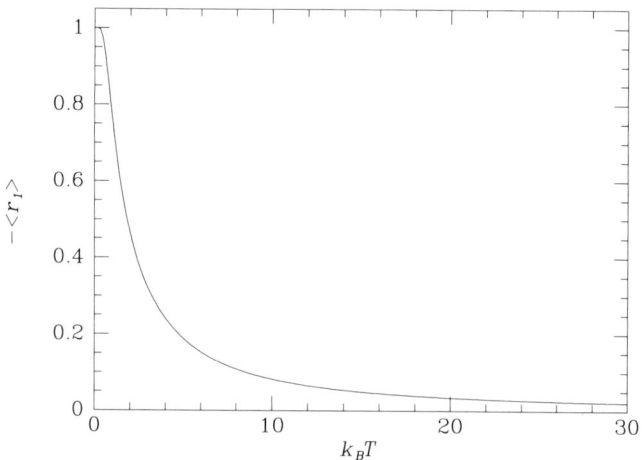

Figure 2.3 The mean position of one particle $\langle r_1 \rangle$ as a function of temperature T for the simple two-particle model considered in the text, with $A = 0.5$ and $\kappa = 15$.

us this. This product is $+1$ when the particles are in the same box, and -1 when they are in different boxes. Differentiating Z' gives us

$$\langle r_1 r_2 \rangle = \frac{1}{Z'} \frac{1}{\beta^2} \frac{\partial^2 Z'}{\partial A_1 \partial A_2}\bigg|_{A_1 = A_2 = A}$$
$$= \frac{e^{\beta \kappa} \cosh(2\beta A) - 1}{e^{\beta \kappa} \cosh(2\beta A) + 1}. \tag{2.42}$$

This is plotted as the full curve in Figure 2.4. As expected it is 1 at low temperatures, going to zero at high temperatures.

However, (2.42) shows that $\langle r_1 r_2 \rangle$ would behave in this general way even if the binding energy κ vanished. The reason for this is easy to understand. At very low temperatures both particles will be in the left-hand box simply because they don't have enough energy to reach the right-hand box, whether or not there is any binding energy. Just as in the case of the spins, the connected correlation function

$$G_c^{(2)}(1,2) = \langle r_1 r_2 \rangle - \langle r_1 \rangle \langle r_2 \rangle \tag{2.43}$$

extracts the effect that genuinely results from interaction between the particles. Using equations (2.41) and (2.42) $G_c^{(2)}$ is

$$G_c^{(2)}(1,2) = \frac{e^{2\beta \kappa} - 1}{(e^{\beta \kappa} \cosh(2\beta A) + 1)^2}. \tag{2.44}$$

This can also be calculated by differentiating $\log Z'$, as in equation (2.36). $G_c^{(2)}(1,2)$ is plotted as the dashed curve in Figure 2.4. It rises from zero at low

2.3 Metastability and symmetry breaking

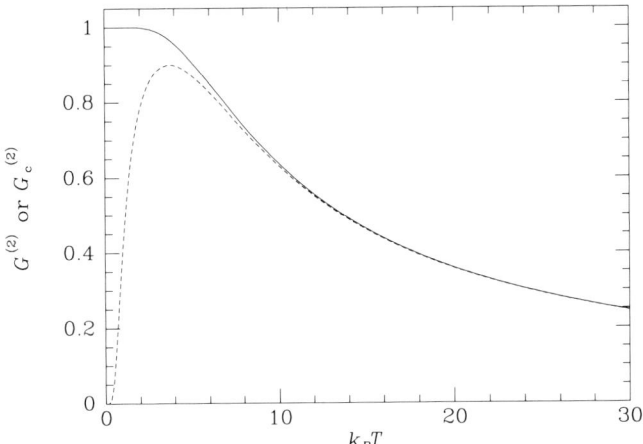

Figure 2.4 The two-point correlation function $\langle r_1 r_2 \rangle$ (full curve) and the connected two-point correlation function $\langle r_1 r_2 \rangle - \langle r_1 \rangle \langle r_2 \rangle$ (dashed curve) as functions of temperature T for the two-particle model.

temperature, and slowly becomes small again at high temperatures, showing that the binding term is most important at intermediate temperatures. This is reasonable; at high enough temperature we would expect any binding between the particles to be broken.

2.3 Metastability and spontaneous symmetry breaking

2.3.1 Metastability

The methods described so far in this chapter give a unique value for the thermal average of any property of a system in thermal equilibrium. But there are cases in which the condition of thermal equilibrium alone does not fix the properties of a system. Consider a mixture of two moles of hydrogen and one mole of oxygen at room temperature. This appears to have two equilibrium states—one in which the hydrogen and oxygen exist as separate gases, and one in which they have reacted to form two moles of water. Is the theory of thermal equilibrium presented in this chapter incomplete?

What we have not yet considered is the period over which we must average our observations if they are to follow the Gibbs distribution. A mixture of hydrogen and oxygen at room temperature will eventually react to form water but this takes a very long time, because the mean kinetic energy of the molecules is not enough to allow a pair of them to overcome the potential energy barrier keeping them apart. When calculating the partition function for this system, equation (2.2) tells us to sum over *all* of its possible microstates, including both those containing the gas mixture and those containing water. But unless we wait an extremely long time for the system to pass back and forth between these two sets of states this does not correspond to experiment. To calculate the properties of the mixture of hydrogen and oxygen observed in the laboratory we must artificially restrict the sum over microstates to only those states containing the gas mixture, since the states containing water will not be reached during the experiment. We have had to augment statistical mechanics with some knowledge of the dynamics of our system. The hydrogen–oxygen mixture is said to be **metastable**.

One might expect that when studying water, we should exclude for the same reason the states in which the system takes the form of a mixture of hydrogen and oxygen. While this is true in principle, the situation is rather different in practice. The Helmholtz free energy of two moles of water is 4.62×10^5 J less than that of the mixture of gases, at 298 K and 1 atm. Using the second of equations (2.14) we find that

$$Z_{\text{water}} = e^{1.1 \times 10^{26}} Z_{\text{mixture}}. \tag{2.45}$$

The contribution to Z from the states involving the mixture of gases is extremely small, and it does not matter whether or not we include it when studying water. In contrast, it is vital that the states involving water are excluded if we are to learn anything concerning the properties of the gas mixture.

From the above it would appear that the most stable state for a system at given temperature and subject to given constraints is the one for which the Helmholtz free energy is the lowest. This is correct, the result being familiar from classical thermodynamics. We can use this to decide which of two possible configurations of a system is the more probable, by seeing which has the lowest Helmholtz free energy.

2.3.2 Spontaneous symmetry breaking

Consider a ferromagnet whose Hamiltonian is spherically symmetric. (Real ferromagnets are not spherically symmetric, but never mind.) Above its Curie point it has no permanent magnetic moment, but when cooled to below its Curie point it becomes spontaneously magnetized in a particular direction. Because of its spherically symmetric Hamiltonian its energy is independent of this direction but nevertheless a particular direction is chosen, determined by random fluctuations in the spins as the magnet is being cooled. The state of the magnet does not show the symmetry of its Hamiltonian. When an exact symmetry of the laws governing a system is not manifest in the state of the system the symmetry is said to be **spontaneously broken**. Since the symmetry of the laws is not actually broken it would perhaps be better described as 'hidden', but the term 'spontaneously broken symmetry' has stuck.

2.3 Metastability and symmetry breaking

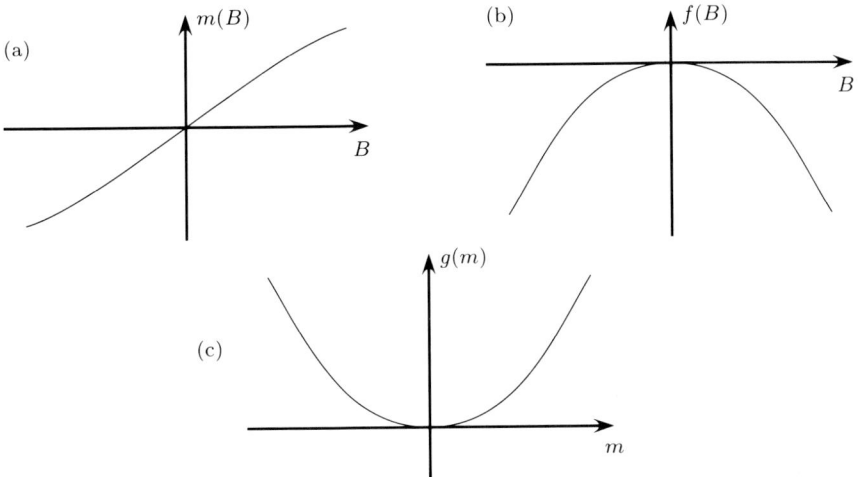

Figure 2.5 $m(B)$, $f(B)$, and $g(m)$ for a magnetic system with unbroken symmetry. (a) shows $m \equiv \langle S \rangle / N$ as a function of magnetic field B. m vanishes when B does, and is proportional to B when B is small. (b) shows f, the Helmholtz free energy per spin, as a function of B. Because of the form assumed for $m(B)$ in the first graph, F depends quadratically on B for small B. (c) shows g, the Gibbs free energy per spin, as a function of m. This is again a parabola.

However, the magnetization as calculated by averaging over all the microstates of the system vanishes even below the Curie point, since the total spin of the magnet can point in any direction with equal probability. The reason for this discrepancy between statistical mechanics and experiment is the same as for metastable systems: the time needed for the total spin of a magnet to pass through all of its possible values is in practice very long compared with the time over which the system is observed. In calculating the partition function by summing over all possible microstates of the spin system we are ignoring the possibility that the spin may not rotate significantly during the experiment. For the spin to change direction it is no good if just a few of the dipoles in the magnet rotate to a new direction; if the majority have not rotated, these few get dragged back again. If the total spin is to change direction significantly it is necessary for a large fraction of the spins to rotate together. The probability of this happening decreases rapidly as the sample gets larger. In a way the situation is a bit like that in a metastable system. The difference between this situation and metastability is that here all directions of the spin are equivalent, whereas of the two 'equilibrium' states of the hydrogen–oxygen mixture only one is a true equilibrium.

Let us consider a magnetic system of N spins $(s_1, \ldots, s_n)$ (scalars, for simplicity), whose Hamiltonian is symmetric when all of the s_i change sign,

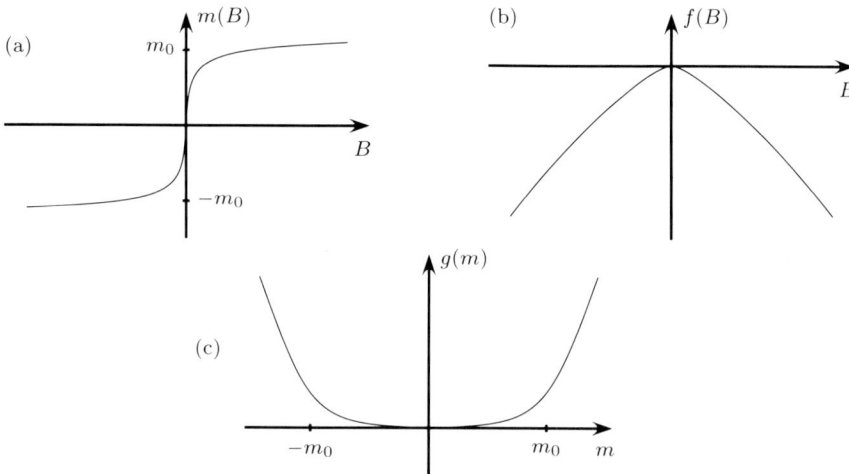

Figure 2.6 $m(B)$, $f(B)$, and $g(m)$ for a magnetic system with spontaneously broken symmetry. (a) shows $m(B)$. Since the symmetry is broken, m responds strongly to a small applied field as the spins 'swing around' to follow B. In the limit $N \to \infty$ its gradient would be infinite at the origin. (b) shows $f(B)$. In the limit $N \to \infty$ its gradient would be discontinuous at the origin; for a finite system the graph turns around sharply but continuously. (c) shows $g(m)$. From graph (a) we see that there is a range of values of m between $\pm m_0$ for which B is very small. Since the gradient of $g(m)$ is proportional to B, $g(m)$ must be almost flat between $\pm m_0$. In the limit $N \to \infty$ its gradient would vanish.

except for a term $-B \sum_i s_i$ representing the interaction of the system with a magnetic field. We shall write the total spin as S. Figure 2.5 shows the behaviour of the magnetization per spin $m \equiv \langle S \rangle / N$, along with the Helmholtz and Gibbs free energies per spin, when the symmetry is not spontaneously broken. m is proportional to B for small B, and the forms of $f(B, \beta)$ and $g(m, \beta)$ follow from this and the relations

$$m = -\left(\frac{\partial f}{\partial B}\right)_\beta \quad \text{and} \quad B = \left(\frac{\partial g}{\partial m}\right)_\beta. \qquad (2.46)$$

The corresponding functions when the symmetry of the system is spontaneously broken are shown in Figure 2.6. In this situation the thermal average of $|S|$ is a large quantity $M_0 \equiv N m_0$ proportional to N. The probability of S taking the value $+M_0$ is $e^{2\beta M_0 B}$ times the probability of its taking the value $-M_0$. If $B = 0$ these two probabilities are equal and the magnetization vanishes, but a value of B of order N^{-1} is enough to make m of order m_0. Thus the derivative $(\partial m / \partial B)$ is $O(N)$ at $B = 0$, whereas it is $O(1)$ if the symmetry is unbroken. This very rapid change in m shows up in the free energies too. Using equations (2.46) we see that the gradient of $f(B, \beta)$ changes very

2.3 Metastability and symmetry breaking

rapidly near $B = 0$ over a range in B of order N^{-1}, and the Gibbs free energy becomes almost flat between $\pm m_0$, its gradient being $O(N^{-1})$. Incidentally, the relation

$$\frac{\partial^2 g(m,\beta)}{\partial m^2} = -\left(\frac{\partial^2 f(B,\beta)}{\partial B^2}\right)^{-1} \qquad (2.47)$$

implies that g is a convex function of m, because

$$\frac{\partial^2 f}{\partial B^2} = -\frac{\beta}{N}\langle(S - \langle S\rangle)^2\rangle. \qquad (2.48)$$

The right-hand side of (2.48) is always non-positive, so $(\partial^2 g/\partial m^2)$ is always non-negative. Therefore $g(m, \beta)$ always curves upwards; it can never have a 'hump'.

For a system of fixed size, deciding whether or not the symmetry is spontaneously broken is a little arbitrary. The corresponding graphs in Figures 2.5 and 2.6 can be continuously deformed into each other, and there is no clear point where (for example) the gradient of m at $B = 0$ stops being $O(1)$ and starts being $O(N)$. Only for in the thermodynamic limit is the distinction clear-cut. For an infinite system with unbroken symmetry $\lim_{B \to 0\pm} m = 0$, the gradient of f is continuous at $B = 0$ and the gradient of g vanishes only at $m = 0$. For an infinite system with broken symmetry $\lim_{B \to 0\pm} m = m_0$, the gradient of f is discontinuous at $B = 0$ and g has a flat portion between $\pm m_0$. Any of the three functions m, f, and g may therefore be used to test for spontaneous symmetry breaking.

For the systems we study later in the book the Gibbs free energy provides the most convenient test. At the critical temperature T_c the gradient of f becomes discontinuous at $B = 0$, so $(\partial^2 f/\partial B^2)$ diverges. Equation (2.47) then implies that $(\partial^2 g/\partial m^2)$ vanishes at $m = 0$. So knowing $g(m, \beta)$ allows us to locate T_c as the highest temperature at which $(\partial^2 g/\partial m^2)$ vanishes for $m = 0$. It also allows us to find the spontaneous magnetization. This is the maximum value that m can take when $B = 0$. Now the second of equations (2.46) tells us that the magnetic field vanishes all along the flat portion of $g(m, \beta)$. The spontaneous magnetization is therefore the value of m at the edge of the flat portion of $g(m, \beta)$.

Instead of calculating $\langle S \rangle$, it is possible to calculate the probability distribution for S and to use this to investigate spontaneous symmetry breaking. In a system with unbroken symmetry we expect this distribution to be peaked around $S = 0$, whereas if the symmetry is broken with $\langle |S| \rangle = M_0$, we expect peaks around $S = \pm M_0$ and a minimum around $S = 0$. It is shown in Appendix E that in the thermodynamic limit the probability of S lying between M and $M + dM$ is

$$\rho(M, \beta)\,dM \propto e^{-\beta G(M,\beta)}\,dM, \qquad (2.49)$$

where $G(M, \beta)$ is the Gibbs free energy of the whole system. This result is valid as long as $(\partial^2 G/\partial M^2)$ is non-zero, so it does not apply to the 'flat' part of $G(M,\beta)$. In such a case $\rho(M,\beta)$ can still be calculated independently of G; we shall see an example of this in Chapter 8.

Problems

2.1 "If a system is in thermal equilibrium at temperature T, the probability of its having energy E is directly proportional to $e^{-E/k_B T}$." True or false?

2.2 The Hamiltonian of a (classical) gas of N interacting particles, each of mass m, is

$$H = \sum_{i=1}^{N} \frac{p_i^2}{2m} + U(\mathbf{q}_1, \ldots, \mathbf{q}_N). \tag{2.50}$$

Show that the partition function is the product of two terms, one depending only on the kinetic term and one only on the potential term. Find an expression for the mean kinetic energy of each particle as a function of temperature.

Some marbles of mass $\sim 10\,\text{g}$ are being shaken about in a box, with a mean speed of $\sim 1\,\text{m/s}$. Find the temperature associated with the motion of the marbles. Why doesn't the box catch fire?

2.3 At sufficiently low temperatures, the partition function of a system with discrete microstates can be written

$$Z \approx n_0 e^{-\beta E_0} + n_1 e^{-\beta E_1}, \tag{2.51}$$

where E_0 and E_1 are the two lowest energies of the system, and n_0 and n_1 are the number of microstates with each energy. Use this expression for Z to calculate the entropy S using (2.9) in the limit $T \to 0$, and so show that the third law of thermodynamics requires $\Phi(\{V\}) = 0$.

2.4 Consider a model involving two distinguishable particles, each of which can be in any one of n boxes. When the two particles are in the same box, the energy of the system is lowered by κ. The Hamiltonian for this system is

$$H(r_1, r_2) = -\kappa \delta_{r_1, r_2}, \tag{2.52}$$

where r_1 and r_2 are integers running from 1 to n labelling the positions of the two particles. Show that the probability of finding the two particles in the same box is

$$p = \frac{1}{1 + (n-1)e^{-\beta\kappa}}. \tag{2.53}$$

The system is thermally isolated, and then n is increased from 2 to 6 without any work being done. What is the new temperature of the system?

2.5 A single spin one-half particle in a magnetic field B has Hamiltonian

$$H = -\mu B s, \tag{2.54}$$

where s can take the values ± 1. Calculate the partition function for N such spins, and use one of the tests mentioned in §2.3.2 to show that the symmetry of the system is not spontaneously broken. What is the probability of the total spin being zero in a magnetic field B?

Problems

2.6 Consider a system of N interacting two-state spins, with interactions so strong that the only significant terms in the partition function are those with all the spins pointing in the same direction. Calculate the partition function for this system, and use one of the tests mentioned in §2.3.2 to show that its symmetry is spontaneously broken.

2.7 The probability of the total spin S of a magnetic system lying between M and $M + \mathrm{d}M$ is

$$\rho(M)\mathrm{d}M \propto e^{-\beta G(M)}\mathrm{d}M, \tag{2.55}$$

where $G(M)$ is the Gibbs free energy of the system (equation (2.49)). This result, derived rigorously in Appendix E, can also be obtained as follows. Convince yourself that the required probability is

$$\rho(M)\mathrm{d}M = \frac{1}{Z}\sum_\alpha e^{-\beta E_\alpha}, \tag{2.56}$$

where the sum is over only those states α for which S lies in the required range. Imagine applying a magnetic field B to the system of such a strength that $\langle S \rangle = M$. By observing that the fluctuations in the value of S for a macroscopic system are extremely small, derive the expression (2.55) for $\rho(M)$.

3
Models

In this chapter we introduce a number of models that have played important rôles in the development of the theory of phase transitions. To avoid disappointment later it should be emphasized here that these are all toy models designed to be tractable rather than realistic. In fact some are frankly unphysical and are of interest mainly for the light they throw on other, more realistic models. But among the systems we shall play with are ones that seem to capture the essence of what happens when a real system undergoes a continuous phase transition.

Many important results relating to phase transitions have been derived by studying particular models. In view of this it is a good idea to familiarize ourselves with some of the most widely discussed models, and to derive some results for a subset of them. We shall solve a few models exactly and present numerical results for some others.

We shall be exclusively concerned with models of things that happens on a lattice of N sites. This lattice may be one-, two- or three-dimensional—the letter d will be used to denote this dimensionality—and we shall assume that the sample is cubic, having $L \equiv N^{1/d}$ sites on a side. The model's order parameter s (see §1.2) is defined at each lattice point. The order parameter may be a 'scalar' in that its value at each lattice point is a single real number, or it may take complex values, or vectors, or even tensors for its values. We shall denote by D the real-dimensionality of s's values: $D = 1$ implies that s is a scalar, $D = 2$ implies that s is two-vector or takes complex values, and so forth.

3.1 Description of models

3.1.1 The Ising model

Far and away the most influential model of a system capable of a phase transition is the **Ising model**. This was invented by W. Lenz (1888–1957) in 1920 as a simple model of a ferromagnet, though we shall see that it can be interpreted as a model of other systems too. It was first solved by E. Ising in 1925, who treated the case $d = 1$. In 1944 Onsager (1903–1976) solved the model for $d = 2$ in the absence of an externally applied magnetic field and showed that the model's critical exponents were quite different from those predicted by Landau theory (see §7.2), which had until then been thought correct. Despite decades of intensive effort, we still have no exact solution for the $d = 3$ model or for the $d = 2$ model in a non-zero magnetic field.

The system that has kept some of the brightest physicists busy for nearly three-quarters of a century is easily described. The lattice is usually assumed to be cubic, and associated with each point of the lattice is a number s that is either 1 or -1. The system's Hamiltonian is

$$H = \tfrac{1}{2} \sum_{ij} \mathcal{J}_{ij} s_i s_j - B \sum_i s_i, \qquad (3.1)$$

where B is an externally imposed field, the subscripts label lattice sites, and $\mathcal{J}_{ij}$ is defined such that

$$\mathcal{J}_{ij} = \begin{cases} \mathcal{J}, & i \text{ and } j \text{ neighbouring sites,} \\ 0, & \text{otherwise.} \end{cases} \qquad (3.2)$$

The model's partition function can now be written

$$Z_{\text{Ising}} = \sum_{\{s_i\}} \exp\left[\beta\left(B \sum_i s_i - \tfrac{1}{2} \sum_{ij} \mathcal{J}_{ij} s_i s_j\right)\right], \qquad (3.3)$$

where $\{s_i\}$ indicates that the sum should be extended over all possible assignments of ± 1 to lattice sites. The physical picture is of an array of magnets that are obliged to be either parallel or anti-parallel to a uniform magnetic field **B**.[1]

If in equation (3.2) we set $\mathcal{J} < 0$, neighbouring spins try to align parallel to one another and parallel to **B**—the model is a ferromagnet. If we set $\mathcal{J} > 0$, neighbouring spins try to align anti-parallel to one another and (3.3) becomes the partition function of an anti-ferromagnet (see §1.2.3). One may show (see Problem 3.1) that the thermodynamic properties of an anti-ferromagnetic system are identical with those of the corresponding ferromagnetic system.[2]

---

[1] It is tempting to replace the magnets by spin-half dipoles and to associate $\hbar s_i/2$ with the eigenvalues of the spin-half operator **S**. Unfortunately, the interaction Hamiltonian of spin-half dipoles is not diagonal in the representation in which $\mathbf{B} \cdot \mathbf{S}$ is, so its average has contributions from states in which the value of $\mathbf{B} \cdot \mathbf{S}$ is not well defined.

[2] We assume that the system is not frustrated—see §1.2.3.

3.1.2 The lattice gas

The Ising model turns out to be mathematically equivalent to the following highly stylized model of a non-ideal gas.

We divide the d-dimensional space occupied by the gas up into cells of just the same size as an individual molecule. Each molecule is obliged to occupy a single cell, and no cell may contain more than one molecule. Since the gas is non-ideal, molecules attract each other and the energy of the gas is lower when molecules are in adjacent cells than when each lives in glorious isolation. We model this state of affairs by changing the energy of the gas by $4\mathcal{J} < 0$ for every pair of molecules in adjacent cells. Let e_i be zero if the cell i is vacant and one otherwise. Then inter-molecular attraction changes the system's energy by

$$2 \sum_{ij} \mathcal{J}_{ij} e_i e_j, \qquad (3.4)$$

where $\mathcal{J}_{ij}$ is defined by equation (3.2). With these definitions, the grand partition function of the gas is

$$\mathcal{Z} = \sum_{\{e_i\}} \exp\left(\beta\mu \sum_i e_i - 2\beta \sum_{ij} \mathcal{J}_{ij} e_i e_j\right), \qquad (3.5)$$

where μ is the chemical potential of the gas and $\{e_i\}$ indicates that the sum should be extended over all possible assignments of zeros and ones to the cells.

The grand partition function (3.5) of the lattice gas is fundamentally identical with the partition function (3.3) of the Ising model. We show this by eliminating the e_i in favour of the variables s_i defined by

$$s_i \equiv 2e_i - 1. \qquad (3.6)$$

In terms of the s_i equation (3.5) reads

$$\mathcal{Z} = \sum_{\{s_i\}} \exp\left[\beta\left((\tfrac{1}{2}\mu - z\mathcal{J})\sum_i s_i - \tfrac{1}{2}\sum_{ij} \mathcal{J}_{ij} s_i s_j\right) + \beta N(\tfrac{1}{2}\mu - z\mathcal{J})\right], \qquad (3.7)$$

where N is the total number of cells and we have assumed that each cell has z nearest neighbours. (z is known as the **coordination number**. On a d-dimensional cubic lattice $z = 2d$.) Apart from the multiplicative factor $\exp\left[\beta N(\tfrac{1}{2}\mu - z\mathcal{J})\right]$, (3.7) is identical with equation (3.3) for the partition function of the Ising model when $B = \tfrac{1}{2}\mu - z\mathcal{J}$.

3.1 Description of models

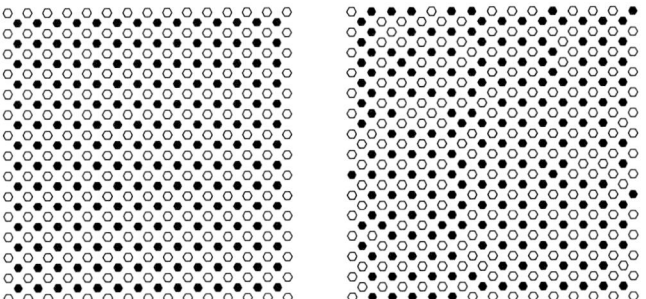

Figure 3.1 Projection of a slice through the lattice of β-brass, well below (left) and near (right) the critical temperature.

3.1.3 β-brass

β-brass is an alloy consisting of equal numbers of copper and zinc atoms. At $T = 0$ the alloy consists of two interpenetrating cubic lattices, one of copper and one of zinc atoms, arranged such that each copper atom is surrounded by eight zinc atoms, and *vice versa* for each zinc atom. As the temperature is raised, more and more copper atoms stray onto the zinc sub-lattice and *vice versa*, until at 739 K the division into two distinct sub-lattices breaks down altogether; above 739 K both sub-lattices contain equal numbers of each kind of atom (see Figure 3.1). This system can be represented by the Ising model as follows.

At low temperatures the system is ordered because it is energetically preferable for unlike atoms to be nearest neighbours rather than like atoms. Suppose the system's energy is lowered by an amount $\mathcal{J}$ for every bond between unlike atoms on adjacent sites, and raised by $\mathcal{J}'$ for every bond between like atoms on adjacent sites. By a suitable choice of the arbitrary zero point of the energy scale we can ensure that $\mathcal{J} = \mathcal{J}'$.

Now we set the order parameter on the ith site s_i to $+1$ if the site is occupied by a copper atom and to -1 if it is occupied by a zinc atom. Then the system's partition function becomes

$$Z = \sum_{\{s_i\}} \exp\left(-\tfrac{1}{2}\beta \sum_{ij} \mathcal{J}_{ij} s_i s_j\right), \tag{3.8}$$

which is identical with equation (3.3) for the Ising model's partition function in the case $B = 0$.

3.1.4 The XY and Heisenberg models

The magnetic dipoles of the Ising model can only point in two directions—'up' or 'down'. It turns out that qualitatively different phase transitions can occur in systems of spins that have greater flexibility of orientation.

The spins of the **XY model** are each capable of pointing in any direction within some given plane—each spin is characterized by a (classical) two-component vector **s**. The **Heisenberg model** consists of an array of D-dimensional[3] classical spins **s**. The XY model may be considered to be the special case $D = 2$ of the Heisenberg model. In these models neighbouring spins are assumed to have interaction energy $\mathcal{J}\mathbf{s}_i \cdot \mathbf{s}_j$, so that the system's partition function becomes

$$Z_{\text{Heisen}} = \sum_{\{s_i\}} \exp\left[\beta\left(\mathbf{B} \cdot \sum_i \mathbf{s}_i + \tfrac{1}{2}\sum_{ij} \mathcal{J}_{ij}\mathbf{s}_i \cdot \mathbf{s}_j\right)\right]. \qquad (3.9)$$

Since two-dimensional vectors may be represented by complex numbers, the XY model's order parameter **s** may be replaced by a complex order parameter ψ, the interaction energy being written $\tfrac{1}{2}\mathcal{J}(\psi_i^*\psi_j + \psi_i\psi_j^*)$.

3.1.5 Potts model

In the Potts model the order parameter can take one of several possible values. The interaction energy between neighbouring sites is zero if the corresponding values of the order parameter are different and $\mathcal{J}$ if they happen to be the same. Thus, as in the Heisenberg model, the order parameter has an incentive to take the same value on neighbouring sites, but, by contrast with the Heisenberg model, there is no energetic reward for its taking 'similar' but not identical values.

3.1.6 Gaussian and spherical models

In the wake of Onsager's epochal solution of the two-dimensional Ising model, M. Kac invented two distinctly unphysical models which he hoped would illuminate Onsager's highly involved solution. The first of these, the 'Gaussian model', is incapable of undergoing a phase transition in any number of dimensions. It is nonetheless widely discussed because it forms the starting point for perturbative calculations such as those discussed in Chapter 8 onwards. The second of Kac's inventions, the 'spherical model', has the distinction of being one of the very few exactly soluble models for a three-dimensional lattice ($d = 3$)—Berlin and Kac (1952) solved it for $d = 1$ to 3. Moreover, in 1969 Stanley showed that its partition function is the same as that of the Heisenberg model in the limit $D \to \infty$.

The idea behind both these models is to make the Ising model more tractable by allowing the spin to take any real value, ($-\infty < s < \infty$), rather than just the values ± 1. In Appendix K we show how an Ising system can be mapped onto such a model exactly, but here we follow Kac in arbitrarily

---

[3] Actually, Heisenberg considered only the case $D = 3$, and that for quantized spin, so what we here call the Heisenberg model would be more precisely named the 'generalized classical Heisenberg model'.

3.1 Description of models

replacing each spin variable with a continuous variable. In the **Gaussian model** each such variable is associated with a self-energy $\frac{1}{2}s_i^2/\beta$. The Boltzmann factor due to this self-energy causes an isolated spin s to be normally distributed with unit variance, $\langle s^2 \rangle = 1$, just as $\langle s^2 \rangle = 1$ in the Ising model. The system's Hamiltonian is

$$H = \frac{1}{2\beta}\sum_i s_i^2 + \tfrac{1}{2}\sum_{ij} \mathcal{J}_{ij} s_i s_j - B \sum_i s_i, \qquad (3.10)$$

where $\mathcal{J}$ is defined by (3.2) with $\mathcal{J} < 0$. At high temperatures (large β^{-1}), the lowest value of this Hamiltonian is achieved when $s_i = 0$, and the average of the order parameter $\langle s \rangle$ vanishes. However, below a certain critical temperature the terms in H proportional to $\mathcal{J}$ overwhelm the positive contributions from the first sum, and H/N (where N is the number of lattice sites) can be made as negative as we please by allowing the s_i to grow without bound. Thus at the critical temperature the spins don't just align, they explode in magnitude, making the model silly.

The spherical model puts a stop to this nonsense by replacing the Gaussian model's self-energy term $\tfrac{1}{2}\beta^{-1}\sum s_i^2$ by the condition

$$\sum_i s_i^2 = N. \qquad (3.11)$$

This condition is also satisfied by the spins of the Ising model. The partition function of the spherical model is

$$Z = \int_{\sum_i s_i^2 = N} ds_1 \ldots ds_N \, \exp\left[\beta\left(B \sum_i s_i - \tfrac{1}{2}\sum_{ij} \mathcal{J}_{ij} s_i s_j\right)\right]. \qquad (3.12)$$

With the condition (3.11) in place, a few of the s_i can take on macroscopic values ($\sim \sqrt{N}$), but they cannot all do so, and H/N is guaranteed finite for all β. It is easy to show that the overwhelming majority of configurations satisfying (3.11) are such that no s_i is of macroscopic size—see Problem 3.5.

A geometrical picture may help to clarify the relation between the Gaussian, spherical and Ising models. The numbers s_i may be thought of as the components of an N-dimensional vector. Then each configuration of the system may be represented by a point in an N-dimensional space. The partition function of the Gaussian model is obtained by integrating a Boltzmann factor all through this space. That of the spherical model is obtained by integrating a slightly simpler Boltzmann factor over a spherical shell of radius $\sqrt{N}$. The partition function of the Ising model is obtained by adding the values taken by the same, simple Boltzmann factor at the corners of the unit hypercube. At these corners the hypercube touches the spherical model's sphere, so the spherical model's partition function includes that of the Ising model.

3.1.7 Percolation model

Classical statistical mechanics can sometimes seem a dreary matter of doing sums—it is usually easy enough to write down the partition function, the only difficulty is adding it up. So it is good to know that there are interesting physical problems that don't centre on a partition function and yet can be solved by the rescaling techniques that are the subject of this book. Here is such a problem. We shall solve it approximately by two different methods in §§5.8 and 6.2.

The system consists of a lattice, between the sites of which bridges can be thrown. We call any set of sites that are connected one to another by bridges, a cluster. Let the probability that there is a bridge between any two sites be p. Then the problem is to understand how the size of a typical cluster depends on p.

When p is sufficiently large there will be a cluster that extends from one side of the system all the way to the other. When p is very small there will not be. As we increase p from zero there must be some critical value p_c at which it becomes possible to march over bridges from one side of the system to the other. In §5.8 we shall show that $p_c = \frac{1}{2}$ for a square two-dimensional lattice.

One way to measure the 'typical cluster size' is to define the 'correlation length' ξ as follows. Let $P(r)$ be the probability that it is possible to march over bridges between two randomly chosen sites distance r apart. Then for $p < p_c$, P becomes small at large r as $P \propto e^{-r/\xi}$, where $\xi(p)$ is the correlation length. In §5.8 we explore the dependence of ξ on p for small $|p - p_c|$.

This **percolation model** has several interesting applications. For example, the lattice sites might represent small pockets of porous rock in an otherwise impervious stratum, while the 'bonds' represent capillaries joining these pockets. Then if $p > p_c$, even a very thick slab of the rock will allow a fluid such as oil to escape through it from a subterranean reservoir, while if $p < p_c$, a sufficiently thick layer of the rock will seal in underlying oil.

Alternatively, the lattice sites might represent regions of high electrical conductivity in an otherwise insulating medium, while the 'bonds' represent conducting junctions between these regions.[4] Then the whole system is insulating or not according as $p < p_c$.

The kind of percolation we have described is often referred to as 'bond percolation' to distinguish it from **site percolation**. A system is said to exhibit site percolation if the element that is sometimes there and sometimes not is the connection between the bonds that reach a site, rather than the bonds themselves. Thus in site percolation every site is joined to every neighbour by a bond, but the bonds feeding into a given site are only connected to one another with probability p.

---

[4] Such a matrix of conducting zones arises when a conductor is immersed in a very strong magnetic field.

3.2 Transfer matrices and the Ising ring

Suppose we have a chain of N dipoles, s_i, that interact only with their neighbours on each side. To keep everything symmetrical, we assume that the chain is wrapped round on itself to form a ring. So dipole 0 interacts with dipoles 1 and $N-1$ and we consider that $s_N \equiv s_0$. Let $H_0(s_i, s_{i+1})$ be the interaction energy between dipoles i and $i+1$. (For example, in the Ising model $H_0 = \mathcal{J} s_i s_{i+1}$.) Then $H = \sum_i H_0(s_i, s_{i+1})$, and the partition function of the system is

$$Z = \sum_{\{s_i\}} \exp\left[-\beta \sum_{i=0}^{N-1} H_0(s_i, s_{i+1})\right]$$
$$= \sum_{\{s_i\}} \prod_{i=0}^{N-1} \exp\left[-\beta H_0(s_i, s_{i+1})\right]. \tag{3.13}$$

We can express this rather complex expression in a compact form by defining the **transfer matrix**

$$T_{\mu\nu} \equiv \exp\left[-\beta H_0(s^\mu, s^\nu)\right]. \tag{3.14}$$

Here, by s^μ we mean the μth possible value of s. For example, in the case of the Ising model each s_i has two possible values, $s^\pm = \pm 1$, so **T** is a 2×2 matrix. Since the interaction Hamiltonian H_0 is guaranteed a symmetrical function of its arguments, **T** is a real, symmetrical matrix all of whose elements are greater than zero. We can now write Z in terms of **T** as

$$Z = \text{Tr}\, \mathbf{T}^N. \tag{3.15}$$

In this expression each of the matrix products involved in raising **T** to the Nth power introduces one of the sums in (3.13) over the possible values of a spin.

It is possible to generalize the definition (3.15) of the transfer matrix to systems whose energy depends on an external field in addition to the interaction Hamiltonian H_0. If the energy of a single dipole s_i in a field B is $V(s_i)$, then we define **T** as

$$T_{\mu\nu} = \exp\left\{-\beta H_0(s^\mu, s^\nu) - \tfrac{1}{2}\beta[V(s^\mu) + V(s^\nu)]\right\}. \tag{3.16}$$

Defined like this **T** is still symmetrical in its indices. It is straightforward to check that we still have $Z = \text{Tr}\, \mathbf{T}^N$.

3.2.1 Solution of the Ising ring

We now use transfer matrices to solve the problem set to Ising by his supervisor Lenz: find the partition function of the $d = 1$ Ising model. We consider first the case $B = 0$. In equation (3.2) we set

$$\mathcal{J} = -\epsilon \qquad (\epsilon > 0). \tag{3.17}$$

Then equations (3.1) and (3.14) give the transfer matrix as

$$\mathbf{T} \equiv \begin{pmatrix} e^{\beta\epsilon} & e^{-\beta\epsilon} \\ e^{-\beta\epsilon} & e^{\beta\epsilon} \end{pmatrix}. \tag{3.18}$$

In the limit of large N the trace of $\mathbf{T}^N$ is easily calculated. The trace is invariant under orthogonal transformations of $\mathbf{T}$, and since $\mathbf{T}$ is symmetric and positive it has real, positive eigenvalues and can be diagonalized by such a transformation. In the frame in which $\mathbf{T}$ is diagonal, $\mathbf{T}^N$ is also diagonal, consisting of $\mathbf{T}$'s eigenvalues to the power N. Thus if we denote $\mathbf{T}$'s eigenvalues in order of decreasing magnitude by λ_0, λ_1, then

$$\lim_{N\to\infty} \frac{1}{N} \log\left(\operatorname{Tr} \mathbf{T}^N\right) = \lim_{N\to\infty} \frac{1}{N} \log\left\{\lambda_0^N \left[1 + (\lambda_1/\lambda_0)^N\right]\right\} \tag{3.19}$$
$$= \log \lambda_0,$$

where λ_0 is $\mathbf{T}$'s larger eigenvalue, which is easily shown to be $2\cosh(\beta\epsilon)$. Thus the ring's partition function and Helmholtz free energy per site are

$$Z = 2^N \cosh^N(\beta\epsilon) \quad \text{and} \quad f = -\frac{1}{\beta} \log[2\cosh(\beta\epsilon)]. \tag{3.20}$$

The internal energy per site, $\mathrm{d}(\beta f)/\mathrm{d}\beta$, is

$$u = \epsilon \tanh(\beta\epsilon), \tag{3.21}$$

which is a perfectly smooth function of β. So this system has no specific heat anomaly such as that which usually occurs at the Curie point of a ferromagnetic material.

By equation (3.16) the transfer matrix of the Ising ring in a constant field is

$$\mathbf{T} = \begin{pmatrix} e^{\beta(\epsilon+B)} & e^{-\beta\epsilon} \\ e^{-\beta\epsilon} & e^{\beta(\epsilon-B)} \end{pmatrix}, \tag{3.22}$$

which has larger eigenvalue

$$\lambda_0 = e^{\beta\epsilon}\left[\cosh\beta B + \sqrt{\cosh^2 \beta B - (1 - e^{-4\beta\epsilon})}\right]. \tag{3.23}$$

3.2 Transfer matrices and the Ising ring

Thus in the presence of a magnetic field the free energy per site is

$$f = -\epsilon - \frac{1}{\beta} \log \left[\cosh \beta B + \sqrt{\cosh^2 \beta B - (1 - e^{-4\beta\epsilon})} \right]. \quad (3.24)$$

Differentiating with respect to B we obtain the average of s:

$$\langle s \rangle = -\left(\frac{\partial f}{\partial B}\right)_T$$
$$= \frac{\sinh \beta B}{\sqrt{\cosh^2 \beta B - (1 - e^{-4\beta\epsilon})}}. \quad (3.25)$$

No matter what the value of β, $\langle s \rangle \to 0$ as $B \to 0$. Thus the system never becomes a ferromagnet. Evidently this simply solved system does not exhibit a phase change. Imagine poor Ising's disappointment on deriving this result![5]

It is not difficult to understand why the Ising ring never makes the transition to a ferromagnetic state. Suppose that it did and that we examined the system at a temperature well below T_c, when the great majority of the spins would be aligned. We consider how the entropy of the Universe would change if the spins in some section of the ring decided to flip over. The ring's energy would usually go up by 4ϵ, so the entropy of the rest of the Universe, whence the energy came, would drop by $4\epsilon/T$. But the beginning and end of the flipped section can be placed in N^2 different places. So the entropy of the ring goes up by $2k_B \log N$. Hence for sufficiently large N, flipping a section of the ring, even a large section, will always produce a net entropy gain no matter how low the ring's temperature.

We shall see that the $d = 2$ Ising model does spontaneously magnetize. It is instructive to see whether an extension of the argument we have just given enables us to anticipate this result. The energy required to flip a section of the Ising ring is independent of the length of the section. In higher dimensions the 'surface energy' of a flipped block of spins increases with the number of spins in the block. The minimum energy for a block containing a given number M of spins is achieved when the block is a d-sphere: if a is the lattice spacing and r the radius of the block, it is for $d = 2$

$$2\epsilon \times \frac{2\pi r}{a} = 4\epsilon\sqrt{\pi M}.$$

For $d = 3$ the corresponding number is $2\epsilon(36\pi M^2)^{1/3}$—the energetic cost of flipping a block rises more sharply with the block's size the bigger d is. The gain in the system's entropy is harder to calculate. The centre of the block can be located at any of N sites, which gives us a straight entropy gain

---

[5] He never published another paper.

$k_B \log N$. If the block had to be spherical, we could immediately conclude that flipping blocks with M comparable with N must lead to a net entropy decrease since the entropy cost of the energy, which is $\propto \beta M^p$ with $p = \frac{1}{2}$ or $\frac{2}{3}$, can be offset by an entropy gain $\propto \log N$ only for $M \ll N$. However, the block doesn't need to be spherical, so we should consider the number of ways n_{wrap} in which a boundary of given area can be wrapped around the centre. This is a hard problem, but one can see that n_{wrap} must increase very rapidly with M if $\log(n_{\text{wrap}})$ is going to overwhelm M^p and prevent the system settling to an ordered state at sufficiently low temperature. Hence it is not *a priori* unlikely that the $d = 2$ Ising model can spontaneously magnetize. On the other hand it is not *a priori* clear that it *will* magnetize either: we shall see that the spherical model, which shares many features with the Ising model, first undergoes a phase change at $d = 3$.

3.2.2 Correlation functions

We can use transfer matrices to calculate correlations as well as the partition function and the thermodynamic variables that follow from it. Specifically, suppose we wish to calculate the average of $f(s_0)f(s_n)$, where $f(s)$ is some function of the value of an individual spin. Then we form the diagonal matrix with elements

$$F_{\mu\nu} = f(s^\mu)\delta_{\mu\nu} \quad \text{(no summation on } \mu\text{)}, \tag{3.26}$$

where as before s^μ denotes the μth value of s_i. Considering for simplicity the case of no external field, we have

$$\begin{aligned}(\mathbf{F} \cdot \mathbf{T})_{\mu\nu} &= \sum_\lambda f(s^\mu)\delta_{\mu\lambda} \exp[-\beta H_0(s^\lambda, s^\nu)] \\ &= f(s^\mu) \exp[-\beta H_0(s^\mu, s^\nu)].\end{aligned} \tag{3.27}$$

So the overall effect of inserting $\mathbf{F}$ before any spin's copy of the transfer matrix is to weight the usual sum by f evaluated on that spin. Thus to calculate the correlation $\langle f(s_0)f(s_n) \rangle$, all we need to do is to evaluate

$$\langle f(s_0)f(s_n) \rangle = \frac{1}{Z} \operatorname{Tr}\left(\mathbf{F} \cdot \mathbf{T}^n \cdot \mathbf{F} \cdot \mathbf{T}^{N-n}\right). \tag{3.28}$$

Similar expressions apply for higher-order correlations.

In the thermodynamic limit $N \to \infty$ the expression (3.28) is readily evaluated in the frame of $\mathbf{T}$'s eigenvectors. In that frame $\mathbf{F}$ is no longer diagonal, but $\mathbf{T}$ is, so

$$T^n_{\mu\nu} = \lambda^n_\mu \delta_{\mu\nu} \quad \text{(no summation on } \mu\text{)}; \tag{3.29}$$

3.2 Transfer matrices and the Ising ring

where the λ_μ are **T**'s eigenvalues; we may assume them ordered such that $\lambda_0 > \lambda_1 > \ldots$. Then

$$\langle f(s_0)f(s_n)\rangle = \frac{1}{\lambda_0^N}\sum_{\mu,\nu,\kappa,\sigma} F_{\mu\nu}\lambda_\nu^n \delta_{\nu\kappa} F_{\kappa\sigma}\lambda_\sigma^{N-n}\delta_{\sigma\mu}. \tag{3.30}$$

Bearing in mind that **F** is symmetric, this becomes in the limit $N \to \infty$

$$\begin{aligned}\langle f(s_0)f(s_n)\rangle &= \sum_\nu F_{0\nu}^2 \left(\frac{\lambda_\nu}{\lambda_0}\right)^n \\ &= F_{00}^2 + F_{01}^2\left(\frac{\lambda_1}{\lambda_0}\right)^n + \cdots.\end{aligned} \tag{3.31}$$

By a similar argument we can show that $\langle f(s_0)\rangle = F_{00}$, so we can rewrite (3.31)

$$\begin{aligned}\langle f(s_0)f(s_n)\rangle_c &\equiv \langle f(s_0)f(s_n)\rangle - \langle f(s_0)\rangle\langle f(s_n)\rangle \\ &= F_{01}^2\left(\frac{\lambda_1}{\lambda_0}\right)^n + \cdots.\end{aligned} \tag{3.32}$$

Comparing this with equation (1.11) we have

$$G_c^{(2)} = F_{01}^2 e^{-na/\xi}, \tag{3.33}$$

where

$$\xi = \frac{a}{\log(\lambda_0/\lambda_1)} \tag{3.34}$$

is the correlation length. For the Ising ring $\xi = 1/\log[\coth(\beta\epsilon)] \simeq \frac{1}{2}e^{2\beta\epsilon}$ for $\beta\epsilon \gg 1$.

Equation (3.34) shows that so long as **T**'s largest two eigenvalues are positive and distinct (i.e. so long as λ_0 is non-degenerate), the correlation length is finite and there is no long-range order. Actually it is not hard to prove that the largest eigenvalue of any symmetric matrix **T** which has strictly positive elements must be non-degenerate—see Box 3.1. Hence no system that admits a transfer matrix of strictly positive elements can exhibit long-range order. The Ising ring is just an instance of this general rule.

> **Box 3.1: Non-degeneracy of T's largest eigenvalue**
>
> Let λ_0 be the largest eigenvalue of **T** and **u** be a corresponding, normalized eigenvector. Then by Rayleigh's theorem **u** maximizes $\mathbf{u} \cdot \mathbf{T} \cdot \mathbf{u}$ subject to $|\mathbf{u}|^2 = 1$. Since all of **T**'s elements are greater than zero, if any component of **u** were negative, the double sum $\mathbf{u} \cdot \mathbf{T} \cdot \mathbf{u}$ would contain negative terms as well as positive ones. But then we could increase $\mathbf{u} \cdot \mathbf{T} \cdot \mathbf{u}$ without altering $|\mathbf{u}|^2$ by changing the sign of **u**'s negative components. So all components of **u** must be non-negative. Moreover, $\mathbf{T} \cdot \mathbf{u} = \lambda_0 \mathbf{u}$, and every element of **T** exceeds zero, so no element of **u** can vanish. This guarantees the uniqueness of **u** because two vectors cannot be mutually orthogonal if they both have strictly positive elements.

3.3 The partition function of the spherical model

Before we tackle the two-dimensional Ising model, which is much harder to solve than the Ising ring, we solve Berlin and Kac's spherical model for arbitrary dimensionality d. This calculation will introduce us to some of the devices which we shall employ in Chapters 7–9 and is instructive as an example of how a discontinuous phenomenon can be described by analytic functions. For simplicity we set the external field B to zero.

We first replace the restriction $\sum_i s_i^2 = N$ on the spherical model's spins with a δ-function:

$$Z = \int \mathrm{d}s_1 \ldots \mathrm{d}s_N \exp\left(-\tfrac{1}{2}\beta \sum_{ij} \mathcal{J}_{ij} s_i s_j\right) \delta\left(\sum_i s_i^2 - N\right)$$

$$= \int \mathrm{d}s_1 \ldots \mathrm{d}s_N \exp\left(-\tfrac{1}{2}\beta \sum_{ij} \mathcal{J}_{ij} s_i s_j\right) \frac{1}{2\pi \mathrm{i}} \int_{-\mathrm{i}\infty}^{\mathrm{i}\infty} \mathrm{d}p' \exp\left[p'\left(N - \sum_i s_i^2\right)\right]. \tag{3.35}$$

Our next goal is to make the integration over the dummy variable p' the last rather than the first integral. As things stand we cannot do this because the integrals over the s_i are not absolutely convergent.[6] However, for any α we have

$$-\tfrac{1}{2}\beta \sum_{ij} \mathcal{J}_{ij} s_i s_j = N\alpha - N\alpha - \tfrac{1}{2}\beta \sum_{ij} \mathcal{J}_{ij} s_i s_j$$

$$= N\alpha - \alpha \sum_i s_i^2 - \tfrac{1}{2}\beta \sum_{ij} \mathcal{J}_{ij} s_i s_j. \tag{3.36}$$

---

[6] $\int_0^\infty \mathrm{d}x\, f(x)$ is said to be **absolutely convergent** if $\lim_{X \to \infty} \int_0^X \mathrm{d}x\, |f(x)|$ exists. Similarly, $\int_{x,y \geq 0} \mathrm{d}x \mathrm{d}y\, f(x,y)$ is absolutely convergent if $\lim_{X \to \infty} \lim_{Y \to \infty} \int_0^X \mathrm{d}x \int_0^Y \mathrm{d}y\, |f(x,y)|$ exists. The value of a double integral is guaranteed independent of the order of integration only if the integral is absolutely convergent.

3.3 Partition function of the spherical model

For sufficiently large α the quadratic form in s_i on the last line of (3.36) will be negative definite and thus the s_i integrations absolutely convergent. Interchanging the p' and s integrations now yields

$$Z = \frac{e^{N\alpha}}{2\pi i} \int_{\alpha-i\infty}^{\alpha+i\infty} dp\, e^{pN} \int ds_1 \ldots ds_N \exp\left[-\sum_{ij}\left(p\delta_{ij} + \tfrac{1}{2}\beta \mathcal{J}_{ij}\right)s_i s_j\right], \tag{3.37}$$

where $p \equiv p' + \alpha$. We next seek a unitary linear transformation from the variables s_i to new variables of integration in which the matrix $\mathcal{J}$ is diagonal. If we can find such a transformation, the inner N integrals in (3.37) will decompose into the product of N integrals of the type

$$\int_{-\infty}^{\infty} dx\, e^{-cx^2} = \sqrt{\pi/c}. \tag{3.38}$$

The numbers c of our integrals will be $p + \tfrac{1}{2}\beta\lambda_q$, where λ_q is the qth diagonal element of $\mathcal{J}$ in the new coordinate system. Hence if such a transformation can be found, Z will be of the form

$$Z = \frac{\pi^{N/2} e^{N\alpha}}{2\pi i} \int_{\alpha-i\infty}^{\alpha+i\infty} dp\, e^{pN} \Big/ \prod_q (p + \tfrac{1}{2}\beta\lambda_q)^{1/2}. \tag{3.39}$$

So we may reduce the original N integrals to one if we can diagonalize $\mathcal{J}$. Fortunately this is easily done.

The key point is that we can arrange for the components of the interaction matrix $\mathcal{J}$ of *any* translationally invariant system to depend only on the difference of their indices—i.e., $\mathcal{J}_{ij}$ can be expressed in the form $\mathcal{J}_{ij} = \mathcal{J}(i-j)$. To see this, recall that each index labels a lattice site. So far we have not chosen a scheme for enumerating sites. On a square, two-dimensional lattice L sites on a side, we make the site labels i and j into two-dimensional vectors $\mathbf{i} = (i_1, i_2)$ and $\mathbf{j} = (j_1, j_2)$, whose integer components i_1 etc. tell us to which row and column the site belongs. Then it is obvious that $\mathcal{J}_{ij}$ depends only on the difference vector $(\mathbf{i} - \mathbf{j})$. The product $\sum_j \mathcal{J}_{ij} s_j$ now becomes a convolution

$$\sum_j \mathcal{J}_{ij} s_j = \sum_{\mathbf{j}} \mathcal{J}(\mathbf{i}-\mathbf{j}) s_{\mathbf{j}}, \tag{3.40}$$

which we evaluate with the aid of the discrete Fourier convolution theorem (see Appendix F). In terms of the transforms $\tilde{\mathcal{J}}$ and $\tilde{s}$

$$\left.\begin{array}{l} \tilde{\mathcal{J}}_{\mathbf{q}} \equiv \sum_{\mathbf{j}} \mathcal{J}_{\mathbf{j}} e^{-2\pi i(\mathbf{j}\cdot\mathbf{q})/L} \\[4pt] \tilde{s}_{\mathbf{q}} \equiv \sum_{\mathbf{j}} s_{\mathbf{j}} e^{-2\pi i(\mathbf{j}\cdot\mathbf{q})/L} \end{array}\right\} \quad (q_l = 0, \ldots, L-1), \tag{3.41}$$

we have
$$\sum_j \mathcal{J}(\mathbf{i}-\mathbf{j})s_\mathbf{j} = \frac{1}{N}\sum_\mathbf{q} \tilde{\mathcal{J}}_\mathbf{q}\tilde{s}_\mathbf{q} e^{2\pi i(\mathbf{i}\cdot\mathbf{q})/L}, \quad (3.42)$$

where $L \equiv N^{1/d}$ and $\mathbf{q}$ is a d-dimensional vector with integer components. Multiplying (3.42) by $s_\mathbf{i}$ and summing over $\mathbf{i}$ we find

$$\sum_{\mathbf{ij}} \mathcal{J}(\mathbf{i}-\mathbf{j})s_\mathbf{i}s_\mathbf{j} = \frac{1}{N}\sum_\mathbf{q} \tilde{\mathcal{J}}_\mathbf{q}|\tilde{s}_\mathbf{q}|^2. \quad (3.43)$$

Thus the quadratic form $\sum_{\mathbf{ij}}\mathcal{J}_{\mathbf{ij}}s_\mathbf{i}s_\mathbf{j}$ is diagonalized by coordinates proportional to the $\tilde{s}_\mathbf{q}$. By Parseval's theorem, equation (F.14), the transformation from the coordinate set $s_\mathbf{i}$ to the set $\tilde{s}_\mathbf{q}/\sqrt{N}$ is unitary.[7] So $\lambda_\mathbf{q} = \tilde{\mathcal{J}}_\mathbf{q}$ and the partition function (3.39) is

$$Z = \frac{\pi^{N/2}e^{N\alpha}}{2\pi i}\int_{\alpha-i\infty}^{\alpha+i\infty} dp \, \exp\left[pN - \tfrac{1}{2}\sum_\mathbf{q} \log\left(p + \tfrac{1}{2}\beta\tilde{\mathcal{J}}_\mathbf{q}\right)\right]. \quad (3.44)$$

To proceed further we need to know the precise form of $\tilde{\mathcal{J}}_\mathbf{q}$. When $\mathcal{J}$ is defined by equation (3.2) with $\mathcal{J} = -\epsilon < 0$, $\mathcal{J}_{\mathbf{ij}} = \mathcal{J}(\mathbf{i}-\mathbf{j})$ is non-zero only when

$$\mathbf{i}-\mathbf{j} = \begin{cases} (\pm 1, 0, 0, \ldots) \\ (0, \pm 1, 0, \ldots) \\ (0, 0, \pm 1, \ldots) \\ \ldots \end{cases} \quad (3.45)$$

depending on the value of d. Taking the discrete Fourier transform we find

$$\begin{aligned}\tilde{\mathcal{J}}_\mathbf{q} &= \sum_\mathbf{j} \mathcal{J}(\mathbf{j})e^{-2\pi i \mathbf{j}\cdot\mathbf{q}/L} \\ &= -2\epsilon \sum_{l=1}^d \cos(2\pi q_l/L).\end{aligned} \quad (3.46)$$

Inserting this result into the logarithm of (3.44), we have

$$\sum_\mathbf{q} \log\left(p + \tfrac{1}{2}\beta\tilde{\mathcal{J}}_\mathbf{q}\right) = N\log(\beta\epsilon) + \sum_\mathbf{q} \log\left[\zeta - \sum_{l=1}^d \cos(2\pi q_l/L)\right], \quad (3.47)$$

where
$$\zeta \equiv \frac{p}{\beta\epsilon}. \quad (3.48)$$

---

[7] Since the $s_\mathbf{j}$ are real, $\tilde{s}_{\mathbf{q}'} = \tilde{s}_\mathbf{q}^*$, where $q'_l \equiv L - q_l$, so only half of the N complex variables $\tilde{s}_\mathbf{q}$ are independent. The real and imaginary parts of the $s_\mathbf{q}$ with $q_1 \geq 0$, say, constitute a suitable set of N real variables.

3.3 Partition function of the spherical model

Since $L = N^{1/d}$ is large, the sum over **q** in equation (3.47) can be approximated by a d-dimensional integral:

$$\sum_{\mathbf{q}} \log\left[\zeta - \sum_{l=1}^{d} \cos(2\pi q_l/L)\right] \simeq \frac{N}{(2\pi)^d} \int_0^{2\pi} d\omega_1 \ldots d\omega_d \log\left(\zeta - \sum_{k=1}^{d} \cos \omega_k\right)$$
$$\equiv N\phi(\zeta), \qquad (3.49)$$

where the angles ω_l are defined to be

$$\omega_l \equiv 2\pi q_l/L. \qquad (3.50)$$

As $L \to \infty$ the increment in ω_l when q_l is increased by one becomes smaller and smaller, making the approximation in equation (3.49) better and better.

The partition function (3.44) can now be written

$$Z = (\beta\epsilon)^{1-N/2} \frac{\pi^{N/2} e^{N\alpha}}{2\pi i} \int_{\alpha'-i\infty}^{\alpha'+i\infty} d\zeta \, e^{g(\zeta)}, \qquad (3.51)$$

where

$$g(\zeta) \equiv N[\beta\epsilon\zeta - \tfrac{1}{2}\phi(\zeta)], \qquad (3.52)$$

and α' is any sufficiently large real number.

The integral over ζ can be evaluated by the method of 'steepest descent' explained in Appendix G. (Since $g \propto N$, the approximation involved in this evaluation becomes ever more accurate as $N \to \infty$.) Thus

$$Z \simeq (\beta\epsilon)^{1-N/2} \pi^{N/2} e^{N\alpha} \frac{e^{g(\zeta_s)}}{\sqrt{2\pi g''(\zeta_s)}}, \qquad (3.53)$$

where ζ_s is the value of ζ at which the exponent $g(\zeta)$ peaks. From (3.49) and (3.52) this is the solution of

$$2\beta\epsilon = \frac{1}{(2\pi)^d} \int_0^{2\pi} \ldots \int_0^{2\pi} \frac{d\omega_1 \ldots d\omega_d}{\zeta_s - \sum_k \cos \omega_k}. \qquad (3.54)$$

As the temperature is lowered, the left side of (3.54) increases and ζ_s decreases to enable the integral to keep in step. It can be shown that for $d \leq 2$ the integral can be made to take any value, no matter how large, without decreasing ζ_s past the value $\zeta_s = d$ at which the integrand becomes singular. Consequently, for $d \leq 2$, $\zeta_s(\beta)$ is a smooth function defined for all β, and $Z(\zeta_s)$ becomes a smooth function of β also. So the spherical model does not undergo a phase change for $d \leq 2$.

By contrast, when $d = 3$ the right-hand side of (3.54) takes a finite value $0.50546\ldots$ even when $\zeta_s = 3$ and the integrand is at one point singular.

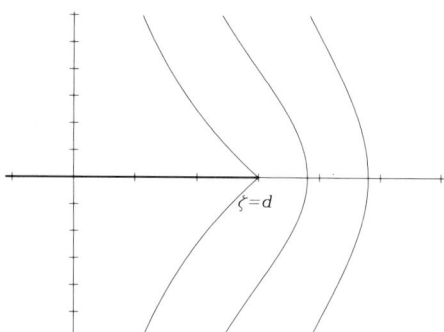

Figure 3.2 Steepest descent paths for the evaluation of the integral of equation (3.51). On account of the logarithm in (3.49), g is non-analytic at $\zeta = d$ (here 3) and we first make g analytic by excising this point from the complex plane by a cut along the real axis from $-\infty$. For small β, g peaks at a point $\zeta_s > d$ on the real axis, and in the limit $N \to \infty$ the integral can be evaluated from the behaviour of g near ζ_s. When $d \geq 3$, for sufficiently large β g peaks at $\zeta = d$ and the contour of evaluation swings hard around the excised point.

Consequently, ζ_s is a smooth function of β only for $\beta < 0.25272/\epsilon$. For larger values of β another strategy has to be devised for the evaluation of the integral, and the change in strategy is reflected in discontinuities in the gradient of Z at $\beta = 0.25272/\epsilon$. Figure 3.2 illustrates this state of affairs. Thus the spherical model exhibits a phase change only for $d \geq 3$. We now concentrate on the case $d = 3$.

Taking the logarithm of Z and proceeding to the limit $N \to \infty$ we obtain an expression for the free energy per site:

$$\begin{aligned}
\beta f &= \tfrac{1}{2}\log(\beta\epsilon) - g(\zeta_s)/N - \tfrac{1}{2}\log\pi - \alpha \\
&= \tfrac{1}{2}\log(\beta\epsilon) - \beta\epsilon\zeta_s \\
&\quad + \tfrac{1}{2}\frac{1}{(2\pi)^d}\int_0^{2\pi} d\omega_1\ldots d\omega_d \log\left(\zeta_s - \sum_{k=1}^{d}\cos\omega_k\right) - \tfrac{1}{2}\log\pi - \alpha.
\end{aligned} \quad (3.55)$$

The internal energy and specific heat are readily calculated from (3.55) (see Box 3.2 and Figure 3.3). There is no specific heat anomaly. The first discontinuity occurs in dc/dT so the critical exponent α (see §1.1.1) takes a negative value, $\alpha = -1$, when $d = 3$. We can also show that the susceptibility $\chi = \beta\langle\sum_i s_i^2\rangle$ diverges at the critical temperature such that the associated critical exponent $\gamma = 2$ when $d = 3$—see Box 3.3.

To obtain the critical exponent δ we need an expression for f in the presence of an external field B. This is readily obtained (see Problem 3.6) and is

$$\beta f = \tfrac{1}{2}\log(\beta\epsilon) - g(\zeta_s)/N - \frac{(\tfrac{1}{2}\beta B)^2}{\beta\epsilon(\zeta_s - d)} - \tfrac{1}{2}\log\pi - \alpha, \quad (3.56)$$

3.3 Partition function of the spherical model

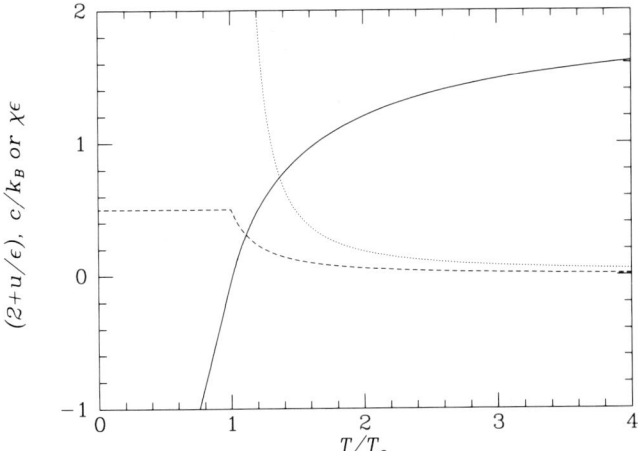

Figure 3.3 The behaviour of the internal energy (full curve), specific heat (dashed curve) and susceptibility (dotted curve) of the $d=3$ spherical model.

where ζ_s is to be determined from

$$2\beta\epsilon = \frac{1}{(2\pi)^d} \int_0^{2\pi} \cdots \int_0^{2\pi} \frac{d\omega_1 \ldots d\omega_d}{\zeta_s - \sum_k \cos\omega_k} + \frac{2(\frac{1}{2}\beta B)^2}{\beta\epsilon(\zeta_s - d)^2}. \tag{3.57}$$

From equation (3.57) one may show that on the critical isotherm, $(\zeta_s - 3) \propto B^{4/5}$ for small B, and when this result is used in (3.56) one finds that $\delta = 5$ (see Problem 3.9).

Equation (3.56) can also be used to find the critical exponent β, which describes the variation of the mean magnetization $\langle s \rangle = -(\partial f/\partial B)_{B=0}$ with temperature just below T_c. Differentiating (3.56) we have

$$\langle s_i \rangle = \frac{B}{2\epsilon(\zeta_s - d)}. \tag{3.58}$$

In this equation $\zeta_s - d$ is a function of B through (3.57). The integral in (3.57) depends only weakly on ζ_s for $T \simeq T_c$ (see Problem 3.8), so we may approximate the first term on the right-hand side of (3.57) by its value at the critical point, namely $2\beta_c\epsilon$. The equation can then be written $B/[2\epsilon(\zeta_s - d)] = \sqrt{1 - \beta_c/\beta}$. Substituting this into (3.58) we conclude that $\langle s_i \rangle = \sqrt{1 - \beta_c/\beta}$ and thus that $\beta = \frac{1}{2}$.

To calculate the remaining critical exponents η and ν we need the correlation function $\langle s_0 s_n \rangle$—they cannot be evaluated from a knowledge of Z alone. However, the calculation can be done (Berlin and Kac 1952), and for $d = 3$ one finds $\eta = 0$ and $\nu = 1$.

These results are summarized in Table 3.1 below. Notice that the exponents satisfy the scaling laws of §1.5.1.

> **Box 3.2: Internal energy of the $d = 3$ spherical model**
>
> The calculation of the internal energy per site u involves taking a derivative of (3.55) with respect to β. ζ_s is a function of β, but fortunately ζ_s is by construction a stationary point of $g(\beta, \zeta)$ when ζ is varied at fixed β. Therefore
>
> $$ u = \frac{\mathrm{d}\beta f}{\mathrm{d}\beta} = \left(\frac{\partial \beta f}{\partial \beta}\right)_{\zeta_s} + \frac{\mathrm{d}\zeta_s}{\mathrm{d}\beta}\left(\frac{\partial \beta f}{\partial \zeta_s}\right)_\beta \qquad (1) $$
> $$ = \frac{1}{2\beta} - \epsilon\zeta_s. $$
>
> Differentiating again we obtain the specific heat per site
>
> $$ c = -\beta^2 \frac{\mathrm{d}u}{\mathrm{d}\beta} \qquad (2) $$
> $$ = \tfrac{1}{2} + \beta^2 \epsilon \frac{\mathrm{d}\zeta_s}{\mathrm{d}\beta}. $$
>
> $(\mathrm{d}\zeta_s/\mathrm{d}\beta)$ proves to vanish at $\beta_c = 0.25272/\epsilon$ (see Problem 3.8), so there is no specific heat anomaly. Consequently, the critical exponent α of §1.1.1 is negative: $\alpha = -1$ for $d = 3$. $(\mathrm{d}c/\mathrm{d}\beta)$ does change discontinuously at $\beta = \beta_c$, however—see Figure 3.3.

3.4 High-temperature expansions and the Ising model

Since Onsager's classic paper of 1944, the partition function of the two-dimensional Ising model has been evaluated by a number of different techniques—see for example Landau and Lifshitz 1969 and Polyakov 1987. The flavour of all these calculations is very different from that of the preceding evaluation of the partition function for the spherical model. We follow Kac and Ward (1952) and Feynman (1972) in evaluating the model's 'high-temperature expansion'.[8]

3.4.1 High-temperature expansions

Valuable results for many different models have been obtained by high-temperature expansion of the partition function. For most models the expansion can be carried through only approximately. In the case of the Ising model, however, it is possible to sum the expansion to all orders, and thus obtain exact results, when $d = 1$ or $d = 2$.

The idea behind a high-temperature expansion is to expand the partition function in powers of some parameter that vanishes in the limit $\beta \to 0$. In

---

[8] See Burgoyne (1963) for an account of the method's history.

3.4 High-temperature expansions and the Ising model

Box 3.3: The susceptibility of the $d = 3$ spherical model

Above the critical temperature $\langle s_i \rangle = 0$, so by equation (2.28)

$$\chi = \beta \sum_i \langle s_0 s_i \rangle = \frac{\beta}{N} \sum_{ij} \langle s_i s_j \rangle.$$

Now $\tilde{s}_0 = \sum_i s_i$, so $\langle \tilde{s}_0^2 \rangle / N = N^{-1} \sum_{ij} \langle s_i s_j \rangle = \chi/\beta$. We can calculate $Z \langle \tilde{s}_0^2 \rangle / N$ by multiplying the integrand of equation (3.37) by $\tilde{s}_0^2/N$. We evaluate this integral as before by transforming to the variables $\tilde{s}_\mathbf{q}/\sqrt{N}$ defined by equation (3.41). Since these variables reduce the exponent of (3.37) to a sum of squares, it follows from the relation $\int x^2 e^{-cx^2} dx / \int e^{-cx^2} dx = 1/2c$ that multiplying the integrand of (3.37) by $\tilde{s}_0^2/N$ modifies equations (3.39), (3.44) and (3.51) by dividing each integrand by $2(p + \frac{1}{2}\beta\lambda_0) = \beta(2\epsilon\zeta + \lambda_0)$. When we evaluate the modified form of (3.51) by steepest descent the answer is just the right-hand side of (3.53) divided by a factor of $\beta(\epsilon\zeta_s + \lambda_0)$. Dividing through by the original right-hand side of (3.53), we have finally

$$\frac{1}{N}\langle \tilde{s}_0^2 \rangle = \frac{1}{\beta(2\epsilon\zeta_s + \lambda_0)}. \tag{1}$$

But from (3.46) we have for $d=3$, $\lambda_0 = \tilde{\mathcal{J}}_0 = -6\epsilon$. Thus

$$\chi = \frac{\beta}{N}\langle \tilde{s}_0^2 \rangle = \frac{1}{2\epsilon(\zeta_s - 3)}, \tag{2}$$

which diverges as $\zeta_s \to 3$ (see Figure 3.3). Using the result $\zeta_s - 3 \sim (\beta_c - \beta)^2$ of Problem 3.8, we conclude that the critical exponent $\gamma = 2$.

---

the most straightforward examples the expansion parameter is just βH:

$$Z = \sum_{\{s_i\}} e^{-\beta H(\{s_i\})} = \sum_{\{s_i\}} \sum_{n=0}^{\infty} \frac{[-\beta H(\{s_i\})]^n}{n!}$$

$$= \sum_{n=0}^{\infty} \frac{(-1)^n}{n!} \sum_{\{s_i\}} [\beta H(\{s\})]^n. \tag{3.59}$$

Since the series for e^{-x} converges for any x, we should be able to truncate the sum over n and thus get a good approximation to Z by adding only a finite number of terms.

3.4.2 The partition function of the Ising model

Let us now apply the high-temperature expansion to the Ising model. By

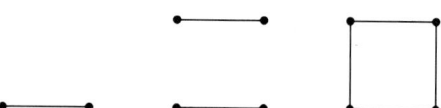

Figure 3.4 Terms in equation (3.64) of the form $s_i s_j$ are represented by drawing a line between the neighbouring sites i and j. A term $s_i s_j s_k s_l$ is represented by lines between the elements of both the pairs contributing to the term. The first terms to make a non-zero net contribution to Z are ones due to octets that are represented by squares.

(3.3) we can write

$$Z_{\text{Ising}} = \sum_{\{s_i\}} \prod_{\text{pairs (i,j)}} e^{\beta \epsilon s_i s_j}, \tag{3.60}$$

where the product is over nearest-neighbour pairs only and $\epsilon = -\mathcal{J}$. Since $s_i s_j = \pm 1$ and

$$e^{\pm A} = \cosh A \pm \sinh A = \cosh A (1 \pm \tanh A), \tag{3.61}$$

we can write

$$\begin{aligned} Z_{\text{Ising}} &= \sum_{\{s_i\}} \prod_{\text{pairs (i,j)}} \cosh(\beta\epsilon)[1 + s_i s_j \tanh(\beta\epsilon)] \\ &= \cosh^{Nz/2}(\beta\epsilon) \sum_{\{s_i\}} \prod_{\text{pairs (i,j)}} (1 + s_i s_j v), \end{aligned} \tag{3.62}$$

where

$$v \equiv \tanh(\beta\epsilon). \tag{3.63}$$

Here $Nz/2$ is the number of nearest-neighbour pairs and we have introduced our expansion parameter v, which is small at high temperatures. We now expand the product in (3.62) in powers of v. It is a vast polynomial, a piece of which looks like this

$$\prod_{\text{pairs (i,j)}} (1 + v s_i s_j) = \ldots (1 + v s_i s_j)(1 + v s_i s_{j'}) \ldots,$$

where sites **j** and **j'** are nearest neighbours of site **i**. So on multiplying everything out we get

$$\prod_{\text{pairs (i,j)}} (1 + v s_i s_j) = 1 + v \sum_{\text{pairs (i,j)}} s_i s_j + v^2 \sum_{\substack{\text{pairs (i,j)} \\ \text{pairs (k,l)}}} s_i s_j s_k s_l + \cdots. \tag{3.64}$$

A pictorial representation will help keep track of all these terms. Each product $s_i s_j$ in (3.64), which involves neighbouring spins, we represent by drawing a line between the two spins on the lattice (see Figure 3.4). So the first sum in (3.64) would be represented by joining every point to all its neighbours. The four spins in each term of the second sum consist of two pairs of nearest-neighbour spins. So each quartet can be represented by two lines, each of which joins nearest neighbours. And so on up the hierarchy.

3.4 High-temperature expansions and the Ising model

Now a term whose graphical representation involves some spin, say s_j, with an odd number of lines leading to it, will not contribute to Z when we sum over all possible values for the spins. To see this, imagine you have already summed over the values of any other spins involved in such a product, reserving to the last summation over the values ± 1 of s_j. Since the number of lines leading to it is odd, your sum is proportional to an odd power of s_j, and the final summation will produce zero.

Hence the only terms in (3.64) which contribute to Z are those in whose graphical representation each spin is touched by an even number (including zero) of lines. In other words, the only contributing terms are those represented by one or more loops on the lattice. Clearly one cannot make a loop with only one line, so the terms in (3.64) proportional to v do not contribute to Z. Furthermore, lines are not allowed to double back on themselves (each pair occurs once only in the left side of (3.64)), so there are no loops involving just two lines, and the terms in (3.64) proportional to v^2 also make no net contribution to Z. If a term represented by l lines does make a net contribution, this will be $2^N v^l$ since the product of spins will be equal to $+1$ for each of the 2^N possible values of the set of spins. Since each spin occurs z times in the left side of (3.64), at most $z/2$ lines can pass through any spin.

So far our argument has assumed nothing about the dimensionality d. If we apply our apparatus to the Ising ring, we see that the only contributing loops are the null loop (no lines)—which gives us the leading 1 in equation (3.64)—and the loop that goes all the way round the ring (N lines). Since $v < 1$, in the limit $N \to \infty$ we have $v^N \to 0$, so only the null loop contributes and the partition function is $Z = 2^N \cosh^N(\beta \epsilon)$ in agreement with (3.20).

In two dimensions the first non-trivial contribution to the partition function comes from loops made up of four lines and is

$$\cosh^{Nz/2}(\beta\epsilon) 2^N g(4) v^4, \tag{3.65}$$

where $g(l)$ is the number of loops on the lattice that can be made with l lines. Generalizing this result to other terms, we have for the Ising model on a lattice of any dimension that

$$Z_{\text{Ising}} = \cosh^{Nz/2}(\beta\epsilon) 2^N \sum_{l=0}^{\infty} g(l) v^l, \tag{3.66}$$

where it is understood that $g(0) \equiv 1$ and $g(l) = 0$ for $l > \frac{1}{2}zN$. Thus the high-temperature expansion reduces Ising's problem to a matter of counting loops on a lattice. Nobody has yet solved this problem for a three-dimensional lattice, but it has been solved for the square two-dimensional lattice.

The basic scheme for counting loops is the following. We introduce a matrix $\mathbf{M}$ that has non-zero $M_{\mathbf{ij}}$ if it is possible to join site $\mathbf{i}$ to site $\mathbf{j}$ by a single line; **M** is a very sparse matrix. Squaring **M** we obtain a matrix which has a non-zero element $(\mathbf{M}^2)_{\mathbf{ik}}$ if it is possible to join sites **i** and **k** with two lines. The summation inherent in matrix multiplication corresponds to summing over all the sites at which we could have stopped between taking a step from **i** and making the final step to **k**, so $(\mathbf{M}^2)_{\mathbf{ik}}$ is proportional to the number of ways of getting from **i** to **k**. Taking further products we obtain the matrix $\mathbf{M}^l$ that tells us how many ways there are of joining any two sites with l lines. The numbers we require, the $g(l)$, can then be calculated from the diagonal elements of $\mathbf{M}^l$, although doing so unfortunately obliges us to master some irritating difficulties introduced by self-intersecting loops. Since the machinery used to solve this problem is rather specialized, we have relegated it to Appendix H. There we show that (3.66) can be written

$$Z_{\text{Ising}} = \cosh^{2N}(\beta\epsilon) 2^N$$
$$\times \prod_q \left\{ (1+v^2)^2 - 2v(1-v^2)[\cos(2\pi q_1/L) + \cos(2\pi q_2/L)] \right\}^{1/2}$$
$$= 2^N \prod_q \left\{ \cosh^2(2\beta\epsilon) - \sinh(2\beta\epsilon)[\cos(2\pi q_1/L) + \cos(2\pi q_2/L)] \right\}^{1/2}.$$
(3.67)

Taking the logarithm of Z, dividing by N, and replacing the resulting sum over q by a double integral as in equation (3.49), we obtain an expression for the free energy per site as $N \to \infty$:[9]

$$\beta f = -\log 2 - \tfrac{1}{2} \frac{1}{(2\pi)^2} \int_0^{2\pi} d\omega_1 d\omega_2 \\ \times \log\left[\cosh^2(2\beta\epsilon) - \sinh(2\beta\epsilon)(\cos\omega_1 + \cos\omega_2) \right].$$
(3.68)

Differentiating (3.68) with respect to β we find

$$u = -\frac{\epsilon \coth(2\beta\epsilon)}{(2\pi)^2} \int_0^{2\pi} \frac{d\omega_1 d\omega_2 [2\sinh^2(2\beta\epsilon) - \sinh(2\beta\epsilon)(\cos\omega_1 + \cos\omega_2)]}{\cosh^2(2\beta\epsilon) - \sinh(2\beta\epsilon)(\cos\omega_1 + \cos\omega_2)}$$
$$= -\epsilon \coth(2\beta\epsilon)\left(1 - \frac{1}{(2\pi)^2} \int_0^{2\pi} \frac{d\omega_1 d\omega_2 [\cosh^2(2\beta\epsilon) - 2\sinh^2(2\beta\epsilon)]}{\cosh^2(2\beta\epsilon) - \sinh(2\beta\epsilon)(\cos\omega_1 + \cos\omega_2)} \right)$$
$$= -\epsilon \coth(2\beta\epsilon)\left\{ 1 - [1 - 2\tanh^2(2\beta\epsilon)]\frac{2}{\pi}K(k) \right\}.$$
(3.69)

Here

$$k \equiv \frac{2\sinh 2\beta\epsilon}{\cosh^2 2\beta\epsilon},$$
(3.70)

---

[9] Berlin and Kac (1952) were struck by the resemblance of equation (3.68) to equation (3.55) for the free energy of the spherical model. This prompted them to conjecture the form of the partition function of the $d=3$ Ising model on the basis of their result for the $d=3$ spherical model. Sadly, this conjecture is now known to be false.

3.4 High-temperature expansions and the Ising model

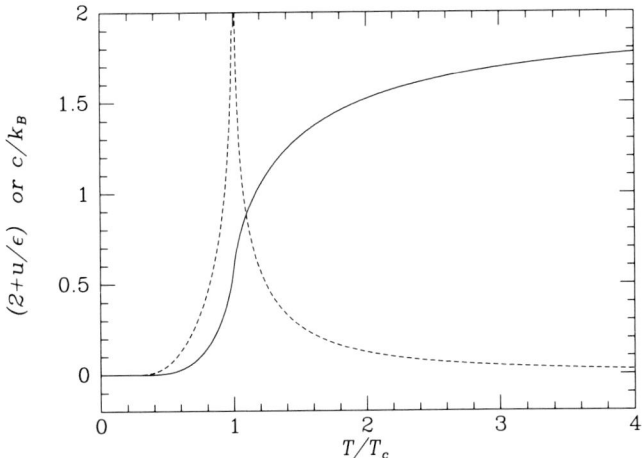

Figure 3.5 The internal energy u (full curve) and specific heat (dashed curve) c of the Ising model.

K is the complete elliptic integral of the first kind,

$$K(k) \equiv \int_0^{\pi/2} \frac{\mathrm{d}\phi}{\sqrt{1 - k^2 \sin^2 \phi}}, \qquad (3.71)$$

and we have used the result

$$\frac{1}{(2\pi)^2} \int_0^{2\pi} \frac{\mathrm{d}\omega_1 \mathrm{d}\omega_2}{1 - \zeta(\cos\omega_1 + \cos\omega_2)} = \frac{2}{\pi} K(2\zeta). \qquad (3.72)$$

Differentiating again we find that the specific heat is given by

$$\begin{aligned} c = & \frac{4k_B}{\pi} (\beta\epsilon \coth 2\beta\epsilon)^2 \\ & \times \left\{ K(k) - E(k) - \operatorname{sech}^2 2\beta\epsilon \left[\frac{\pi}{2} - (1 - 2\tanh^2 2\beta\epsilon) K(k) \right] \right\} \end{aligned} \qquad (3.73)$$

where E is the complete elliptic integral of the second kind:

$$E(k) \equiv \int_0^{\pi/2} \mathrm{d}\phi \sqrt{1 - k^2 \sin^2 \phi}. \qquad (3.74)$$

These results are plotted in Figure 3.5. The internal energy varies from $u = -2\epsilon$ at $T \ll T_c$ when all spins are aligned, to zero at $T \gg T_c$ when as many bonds are anti-aligned as aligned. It is straightforward to show that for $T \gg T_c$, $u \sim -\beta$.

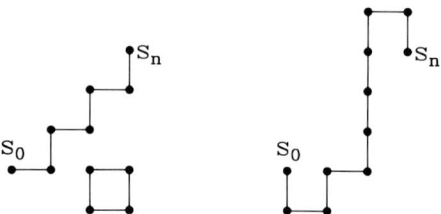

Figure 3.6 Two paths with $l = 10$ that contribute to $\langle s_0 s_{\mathbf{n}} \rangle$.

$K(k)$ diverges logarithmically near

$$k = 1 \quad \Rightarrow \quad \sinh(2\beta\epsilon) = 1 \quad \Rightarrow \quad \beta = \beta_c \equiv \frac{1}{2\epsilon} \log(1 + \sqrt{2}) \simeq 0.4407/\epsilon. \tag{3.75}$$

The divergence of K does not cause u to grow since in the right-hand side of (3.69) K is multiplied by a factor that vanishes as $\beta \to \beta_c$. But c does diverge logarithmically near β_c, leading to the value $\alpha = 0$ for the critical exponent defined in §1.1.1 to characterize specific heat anomalies.

3.4.3 The correlation functions of the Ising model

The technique used to evaluate Z_{Ising} in the last section can be extended to the calculation of the two-point correlation function $\langle s_0 s_{\mathbf{n}} \rangle$. In the notation of §3.4.2 we have

$$\begin{aligned}\langle s_0 s_{\mathbf{n}} \rangle &= \frac{1}{Z} \sum_{\{s_i\}} \prod_{\text{pairs (i,j)}} s_0 s_{\mathbf{n}} e^{\beta \epsilon s_i s_j} \\ &= \frac{1}{Z} \cosh^{Nz/2}(\beta\epsilon) \sum_{\{s_i\}} \prod_{\text{pairs (i,j)}} s_0 s_{\mathbf{n}} (1 + s_i s_j v). \end{aligned} \tag{3.76}$$

As in §3.4.2 we argue that when the product of factors $(1 + s_i s_j v)$ is multiplied out, the only terms which do not cancel when we sum over all nearest-neighbour pairs $(\mathbf{i}, \mathbf{j})$, are those in which each spin occurs an even number of times. Since s_0 and $s_{\mathbf{n}}$ inevitably occur at least once, the contributing terms are represented diagrammatically by paths which run from s_0 to $s_{\mathbf{n}}$, possibly accompanied by one or more additional closed paths—see Figure 3.6. If a path represented by l lines does make a net contribution, this will be $2^N v^l$. Thus

$$\langle s_0 s_{\mathbf{n}} \rangle = \frac{1}{Z} \cosh^{Nz/2}(\beta\epsilon) 2^N \sum_{l} g(l, 2) v^l, \tag{3.77}$$

where $g(l, 2)$ is the number of ways of joining $s_{\mathbf{n}}$ to s_0 with l lines, not excluding the possibility that some of the lines form a closed loop.

For example, in the case of the Ising ring there are only two contributing paths from s_0 to s_n, one with n lines and one with $N - n$ lines. In the limit

3.4 High-temperature expansions and the Ising model

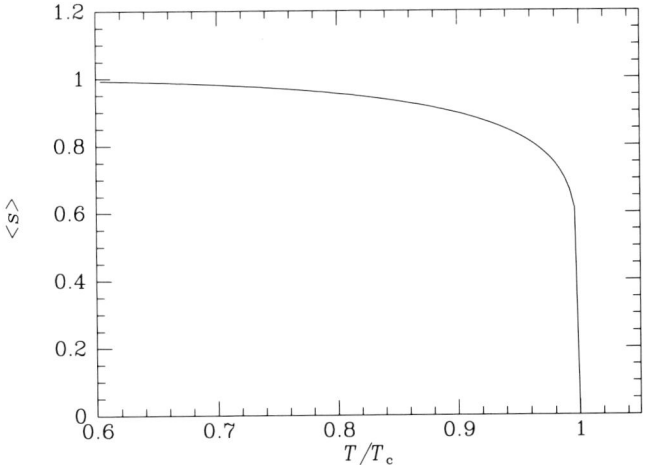

Figure 3.7 The zero-field magnetization of the $d = 2$ Ising model from equation (3.79).

$N \to \infty$, the first dominates the sum and with equation (3.20) we deduce that

$$\langle s_0 s_n \rangle = v^n = \tanh^n(\beta\epsilon) \tag{3.78}$$

in agreement with the result one obtains from transfer matrices (see Problem 3.2).

In the case of a two-dimensional lattice, no elegant expression is available for $\langle s_0 s_\mathbf{n} \rangle$ with $|\mathbf{n}|$ finite, although Kaufman and Onsager (1949) and Montroll et al. (1963) were able to express $\langle s_0 s_\mathbf{n} \rangle$ in terms of $n \times n$ Toeplitz determinants.[10] However, an extremely simple result is available for the limit $n \to \infty$: the zero-field magnetization at $T < T_c$ can be obtained by calculating $\langle s \rangle = \lim_{n \to \infty} \langle s_0 s_n \rangle^{1/2}$, and in 1948 Onsager gave this as

$$\langle s \rangle = \left[1 - \mathrm{cosech}^2(2\beta\epsilon)\right]^{1/8}. \tag{3.79}$$

This is plotted in Figure 3.7. As β approaches its critical value $\beta_c = \sinh^{-1}(1)/(2\epsilon)$, the quantity in square brackets vanishes as $\beta_c - \beta$, and $\langle s \rangle$ vanishes as $\langle s \rangle \sim (\beta - \beta_c)^{1/8}$. Hence the critical exponent $\beta = \frac{1}{8}$. Curiously Onsager never published a derivation of equation (3.79), and only Onsager knew its origin until Yang published a derivation in 1952.

---

[10] A **Toeplitz determinant** is one in which the kth column is obtained by shifting the $(k-1)^{\mathrm{th}}$ column down one place, scrapping the bottom element, and inserting a new element at the top.

3.4.4 Numerical evaluation of high-temperature expansions

Two problems which at first sight seem to require only small extensions of the calculations of the last two sub-sections have obstinately resisted attack for forty years: the analytic evaluation of the partition function of the Ising model for either $B \neq 0$ or $d > 2$. Nonetheless, for practical purposes we have complete knowledge of the properties of the Ising model, in and out of a field for dimension $d \leq 3$, by virtue of numerical evaluations of the model's high- and low-temperature expansions. Here we describe what is involved in evaluating the Ising model's high-temperature expansion in non-zero external field, and discuss techniques for extracting critical exponents from an approximate numerical summation of the expansion. Similar techniques permit one to determine critical exponents from low-temperature expansions—see Sykes et al. (1965) for details.

The Ising model's partition function in the presence of a magnetic field B is given by equation (3.3). In the notation of §3.4.2 this can be written

$$Z_{\text{Ising}} = \sum_{\{s_i\}} \prod_{\substack{\text{pairs} \\ (i,j)}} \cosh(\beta\epsilon)[1 + s_i s_j \tanh(\beta\epsilon)] \prod_{\mathbf{k}} \cosh(\beta B)[1 - s_\mathbf{k} \tanh(\beta B)]$$

$$= \cosh^{Nz/2}(\beta\epsilon) \cosh^N(\beta B) \sum_{\{s_i\}} \prod_{\substack{\text{pairs} \\ (i,j)}} (1 + s_i s_j v) \prod_{\mathbf{k}} (1 - s_\mathbf{k} u),$$

(3.80)

where

$$u \equiv \tanh(\beta B). \qquad (3.81)$$

The products of round brackets in (3.80) can again be multiplied out to give a vast sum of terms of the form $v^l u^m s_i \times \cdots \times s_\mathbf{k}$, and we argue that the only terms to contribute to the sum over sets of spins are those in which each spin occurs an even number of times. In particular, the total number of spins in a contributing term will be even, so m must be even.

Diagrammatically, we represent each factor u by a circle around the corresponding spin $s_\mathbf{k}$; from the structure of equation (3.80) it follows that no spin can have more than one circle. So the diagram representing any contributing term will have 0, 2, 4,... encircled spins. From the requirement that every spin occur an even number of times it follows that the encircled spins must be joined to one another in pairs by lines representing factors of v—see Figure 3.8. Thus

$$Z_{\text{Ising}} = \cosh^{Nz/2}(\beta\epsilon) \cosh^N(\beta B) 2^N \sum_{l,m} g(l,m) v^l u^m, \qquad (3.82)$$

where $g(l, m)$ is the number of arrangements of m points and l lines, in which the points are all joined in pairs. (In effect, equation (3.82) is an extension of equation (3.77) for the Ising model's two-point correlation function.)

3.4 High-temperature expansions and the Ising model

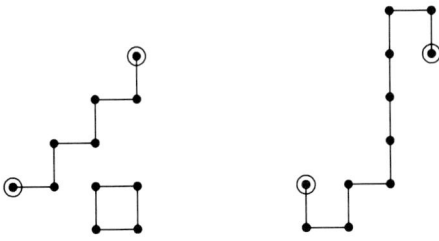

Figure 3.8 Two paths with $l = 10$, $m = 2$ that contribute to $Z_{\text{Ising}}(T, B)$.

While nobody has found a general analytic formula for $g(l,m)$, cunning algorithms have been devised for the numerical evaluation of $g(l,m)$ for any specified (l,m). Thus it has been possible to obtain as many as 70 terms in the series (3.82). We now discuss the extraction of critical exponents from such series.

Most critical exponents can be derived from logarithmic derivatives of the quantities Z and $\langle s_0 s_{\mathbf{n}} \rangle$, for which we have obtained high-temperature expansions. For example, denoting $t \equiv T_c - T$ we have

$$\beta = -\lim_{t \to 0} t \frac{\partial \log m}{\partial t}$$

$$= \lim_{t \to 0} t \frac{\partial}{\partial t} \left[\log \left(\frac{k_B T}{Z} \frac{\partial Z}{\partial B} \right) \right]. \tag{3.83}$$

β is non-zero notwithstanding the factor t that comes after the limit operator, because the derivative that follows has a simple pole at $t = 0$. This pole derives from precisely the non-analyticity of Z near T_c which is the essential ingredient of a phase transition—see §4.1. So high-temperature expansions such as (3.82) will fail to converge at the critical value $v = v_c$, just where one most needs them.[11]

Given that we cannot deduce quantities of interest by brute-force summation, we adopt the following strategy. We use our high-temperature expansion to obtain a power series for whatever derivative $D(v)$ we require for a given critical exponent. $D(v)$ will have some sort of singularity at $v = v_c$, but as the example of equation (3.83) suggests, this will often be just a pole. Now we look for an analytic function whose power-series expansion starts with the terms we have evaluated for D, and has a similar pole. Then we identify D with this function. If we are lucky the identification will be exact—this is effectively what happens in Appendix H on the counting of closed loops on a square lattice, where the series for $\log(1-x)$ is identified. More often the identification is not exact, but experience shows that it generally provides a reliable guide to the behaviour of D near v_c.

---

[11] In the case of the $d = 2$ Ising model, $v_c = 0.414\ldots$, and given the rapid increase with l in the number of paths $g(l,m)$, one cannot expect the series (3.82) to be convergent for such a large value of v.

Table 3.1. Critical exponents

	Variables related	Mean field	Ising $d=2$	Ising $d=3$	Heisenberg $d=3, D=3$	Spherical $d=3, D=\infty$
α	(C,T)	0 (disctv)	0 (log)	$0.119\pm.006$	$-0.08\pm.04$	-1
β	(M,T)	$1/2$	$1/8$	$0.326\pm.004$	$0.38\pm.03$	$1/2$
γ	(χ,T)	1	$7/4$	$1.239\pm.003$	$1.38\pm.02$	2
δ	(B,M)	3	15	$4.80\pm.05$	$4.63\pm.29$	5
η	$(G_c^{(2)},R)$	0	$1/4$	$0.024\pm.007$	$0.07\pm.06$	0
ν	(ξ,T)	$1/2$	1	$0.627\pm.002$	$0.715\pm.02$	1

NOTES: β, γ and ν for the $d=3$ Ising model are from Ferrenberg and Landau (1991). α, δ and η have been calculated from these values using scaling relations from §1.5.1. α, β, γ and ν for the Heisenberg model are from high-T expansions, and δ and η have been deduced from them assuming scaling relations.

The method of **Padé approximants** provides a powerful general technique for fitting an analytic function to the first n terms of a power series. One assumes that one's series for $D(v)$ is the expansion of a ratio

$$\frac{\mathcal{M}(v)}{\mathcal{N}(v)} \equiv \frac{m_0 + m_1 v + \cdots m_M v^M}{1 + n_1 v + \cdots n_N v^N} \tag{3.84}$$

of two polynomials, of order M and N. It is a simple matter to choose the coefficients m_i and n_i such that the first $M+N+1$ coefficients in the Maclaurin expansion of $\mathcal{M}/\mathcal{N}$ agree with the corresponding coefficients in $D(v)$—see, for example, Cabannes (1975). The power of the Padé method arises from the ability of the zeros of the denominator $\mathcal{N}$ to endow $\mathcal{M}/\mathcal{N}$ with singularities, including the one we know D must possess at v_c.

Many of the critical exponents listed in Table 3.1 have been obtained by the technique just described. For more detail see, for example, Gaunt and Guttmann (1974).

Problems

3.1 The Hamiltonian of the Ising anti-ferromagnet is given by equations (3.1) and (3.2) with $B=0$ and $J>0$. Show that if the lattice is square there is a ferromagnetic system whose states can be put into one-to-one correspondence with the states of the antiferromagnet, such that corresponding states have the same energy.

3.2 Use transfer matrices to show that the correlation function of the Ising ring is $\langle s_0 s_n \rangle = \tanh^n(\beta \epsilon)$.

3.3 Without using the linear response theorem show that the susceptibility of the Ising ring is

$$\chi = \beta \cosh(\beta B) \left[\cosh^2(\beta B) - (1 - e^{-4\beta\epsilon}) \right]^{-3/2} e^{-4\beta\epsilon}.$$

Use the result of the previous problem to verify the linear response theorem for this system for the case $B = 0$.

3.4 Let Ω_n denote the area of the unit n-sphere $\sum_{i=0}^{n} x_i^2 = 1$. By expressing the integral $\int e^{-x^2} d^{n+1}x = \pi^{(n+1)/2}$ in spherical polar coordinates, show that

$$\Omega_n = \frac{2\pi^{(n+1)/2}}{[\frac{1}{2}(n-1)]!}.$$

Verify that this formula correctly gives the areas of the 1- and 2-spheres, and gives $\Omega_3 = 2\pi^2$.

3.5 Use the expression for Ω_n derived in Problem 3.4 show that at very high temperatures the expectation of the pth power of a single spin of the d-dimensional spherical model tends to

$$\langle s_i^p \rangle \simeq \begin{cases} \frac{2^{(p+1)/2}}{\sqrt{\pi}} [\frac{1}{2}(p-1)]! & \text{for even } p \text{ and } N \gg p, \\ 0 & \text{for odd } p. \end{cases}$$

3.6 Exploit the fact that $\tilde{s}_0 = \sum_i s_i$ to show that the partition function of the spherical model in the presence of an external field B is given by equations (3.56) and (3.57).

3.7 Show that the susceptibility χ obtained by differentiating equation (3.58) for the magnetization of the $d = 3$ spherical model agrees with the value derived in Box 3.3 from the linear response theorem.

3.8 Prove that near β_c we have for the $d = 3$ spherical model $(\zeta_s - 3) \sim (\beta_c - \beta)^2$.

3.9 Use equation (3.56) to show that for $d = 3$ the spherical model has critical exponent $\delta = 5$. [Hint do Problem 3.8 first.]

3.10 Show that at high temperatures the internal energy of any system whose energy eigenvalues are bounded above varies as $u \sim -\beta$.

4
Numerical simulations

We have now seen how powerful the tools of statistical mechanics can be in the study of complex systems in thermal equilibrium and how they can be used to determine exactly the behaviour of some simplified, but nevertheless interesting, models of phase transitions. However, these exact solutions are sometimes lengthy and difficult and there is no guarantee that any particular model will yield exact results. For example, we were able to solve the two-dimensional, but not the three-dimensional, Ising model, and that only in zero magnetic field. So, it is natural to consider whether we can use a computer to simulate these systems. In this chapter we shall see that this is indeed possible; a great deal of our understanding of critical phenomena is based wholly or partly on the results of such numerical work. However, as we shall also see, such simulations are not entirely straightforward and a number of tricks are needed to obtain accurate results. This is partly a consequence of the very physics that makes the phase transitions interesting to start with. One should probably think of numerical simulation as a whole subject, almost an art, of its own; we shall have time only to scratch at the surface and discuss a few of the most important ideas.

This chapter describes computer *simulations*; these are essentially computer experiments. One sets up in the computer a system described by the same Hamiltonian as one's physical model, and tries to see how it behaves in thermal equilibrium. There are, however, also other ways of using computers to tackle these problems. We caught a flavour of these in §3.4.4, where we saw that high-temperature expansions that cannot be summed exactly can

4.1 Direct evaluation of thermal averages

nevertheless be analyzed numerically to high order and critical properties extracted. The treatment of such expansions by computer is beyond our scope in this chapter.

4.1 Direct evaluation of thermal averages

As we saw in Chapter 2, almost all the questions one might ask about a physical system in thermal equilibrium can be answered by evaluating the thermal expectation of some quantity X which depends on the system's configuration:

$$\langle X \rangle_p = \sum_{\alpha=1}^{A} X_\alpha p_\alpha. \tag{4.1}$$

Here $\langle X \rangle_p$ means the average of the quantity X with respect to the distribution p, α labels configurations of the system, A is the total number of such configurations and p_α is the Gibbs probability for the occurrence of state α in thermal equilibrium:

$$p_\alpha = \frac{e^{-\beta E_\alpha}}{Z}, \quad \text{where} \quad Z \equiv \sum_\alpha e^{-\beta E_\alpha}. \tag{4.2}$$

In Chapters 2 and 3 we made use of a standard analytical technique to evaluate quantities like (4.1), which involves differentiating the partition function with respect to the variable thermodynamically conjugate to X. However, this approach is very costly numerically since it requires us to evaluate the partition function for a whole range of values of the conjugate variable, and then perform an inaccurate numerical differentiation. So, how else can we evaluate (4.1)? One method which suggests itself is simply to evaluate the sum in (4.1) exactly for a system of modest size.

Consider, for example, a two-dimensional Ising model on a square lattice. We might start by solving a 3×3 system, containing nine sites in all. Since each spin can take one of two configurations, there are $A = 2^9 = 512$ possible configurations for the system. It only takes a few floating-point operations to evaluate the energy associated with each configuration, so the sum over all 512 states can be done in a fraction of a second, even on a very modest microcomputer. Figure 4.1 shows the results of such calculations of the internal energy and specific heat of this system as a function of temperature. The exact results for the corresponding quantities in an infinite system are plotted for comparison; we see that the agreement is good at high and low temperatures but that the phase transition is broadened over a large temperature range. This makes it very difficult to extract, for example, critical exponents from our results.

Suppose we try to improve things by calculating the properties of a somewhat larger system, say a 10×10 lattice with 100 sites. Now the number of

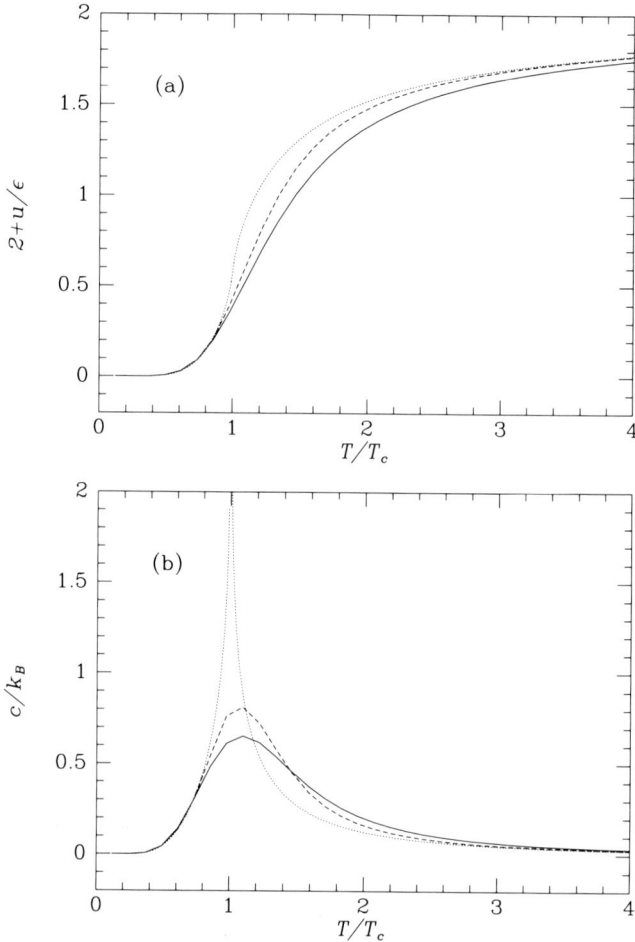

Figure 4.1 Internal energy (a) and specific heat (b) as a function of temperature for 3×3 (full curve), 4×4 (dashed curve), and $\infty \times \infty$ (dotted curve) Ising systems. The finite lattices satisfy periodic boundary conditions. Below T_c good results are obtained even on a very small lattice because then the only significant contributors to the partition function are states in which flipped spins are isolated objects. Above $2T_c$ clusters of correlated spins are small enough to fit well on a 4×4 lattice, though they fit less well onto the 3×3 lattice. On any finite lattice both c_B and the slope of $u(T)$ is finite at $T = T_c$.

possible states is $A = 2^{100} \approx 1.3 \times 10^{30}$, and even with full-time access to a (1992) state-of-the-art supercomputer capable of 10^9 floating point operations per second, evaluating the energy of all these states would take $\gtrsim 10^{21}$ seconds, or $\gtrsim 1000$ times the age of the Universe.

At the root of our difficulties is the fact that the number A of configurations available to a system grows exponentially with its size.[1] Any calculation which involves including all the configurations will therefore very quickly become unmanageable. We can get around this difficulty only if we can develop methods which estimate thermodynamic properties by sampling a small subset of all configurations. How should we do this?

4.2 Sampling configurations

One possible strategy for sampling the configurations of a system is to pick states completely at random, i.e., so that the probability of choosing each configuration is the same. For the Ising model, we could arrange this by sweeping through the system's spins and using a random number generator to decide whether the each spin is to point up or down. Suppose we use an average of the quantity $e^{-\beta E}$ to estimate the partition function, and an average of the quantity $Xe^{-\beta E}/Z$ to estimate the thermal average of X. Let the number of states used in the average be B. This amounts to choosing for the quantity X the **estimator**

$$X_B = \frac{\frac{1}{B}\sum_{k=1}^{B} X_{\alpha_k} e^{-\beta E_{\alpha_k}}}{\frac{1}{B}\sum_{l=1}^{B_1} e^{-\beta E_{\alpha_l}}}. \tag{4.3}$$

Let us first calculate the thermal average of the denominator, given that each of the A states of the system may be chosen at each step with probability $1/A$. We find

$$\frac{1}{B}\sum_{l=1}^{B}\sum_{\alpha_l=1}^{A} \frac{1}{A} e^{-\beta E_{\alpha_l}} = \frac{Z}{A}. \tag{4.4}$$

Similarly, for the numerator, we obtain

$$\frac{1}{B}\sum_{k=1}^{B}\sum_{\alpha_k=1}^{A} \frac{1}{A} X_{\alpha_k} e^{-\beta E_{\alpha_k}} = \frac{Z}{A}\langle X\rangle_p. \tag{4.5}$$

The expected value of our estimator is then the ratio of these two results, namely

$$\langle X_B \rangle = \langle X \rangle_p. \tag{4.6}$$

This shows that we have devised an **unbiased estimator** for the quantity X; its mean value, averaged over all possible choices of our B representative states, is equal to the thermal average $\langle X \rangle_p$.

Unfortunately, there is a serious problem with this approach. This is illustrated by Figure 4.2, which shows the results of two such calculations for

---

[1] This is necessarily so in order that the entropy shall be an extensive quantity.

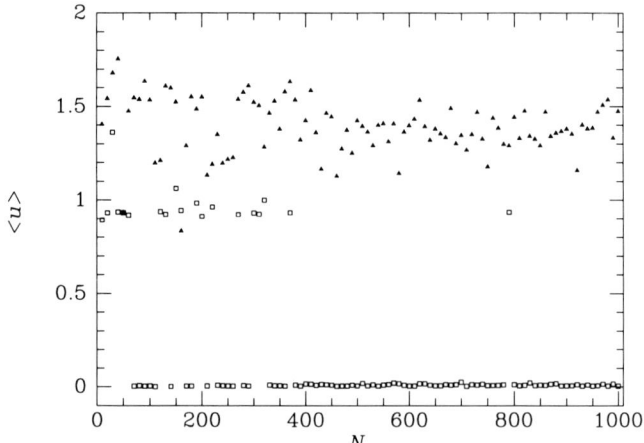

Figure 4.2 Estimates of the internal energy obtained by randomly sampling N configurations for $T = \frac{1}{2}T_c$ (open squares) and $T = 2T_c$ (filled triangles). At $T = \frac{1}{2}T_c$ the estimated value of u is very poor unless one's sample happens to include the ground state.

the 3×3 Ising system. We see that while at high temperatures the statistical estimate converges quickly to the exact value, at low temperatures there are much larger fluctuations in the results. To understand this we need to study the variance of our estimator X_B.

The variance of a sum of independently chosen random variables is equal to the sum of the individual variances; therefore the variance of the numerator in (4.3) is $\frac{1}{B}\text{var}(Xe^{-\beta E})$. Similarly, the variance of the denominator is $\frac{1}{B}\text{var}(e^{-\beta E})$. Both these expressions have the same structure: a factor of $\frac{1}{B}$ multiplied by a factor which is independent of B. The factor of $\frac{1}{B}$ guarantees that, by making the number of samples sufficiently large, we can reduce the variance to any desired level. Notice, however, that this decrease in statistical error happens rather slowly with B—the RMS fluctuation in the estimator decreases only as $B^{-1/2}$. We shall therefore be in serious trouble if the other factor, the one independent of B, in our expression for the variance ever becomes large. Unfortunately, this is almost always the case at low temperatures where, irrespective of the how X_α behaves, the Boltzmann factor $e^{-\beta E_\alpha}$ fluctuates wildly between different configurations. This is illustrated, for the case of the sampling of the internal energy of a 3×3 Ising model, in Figure 4.3; the variance of the quantity $Ee^{-\beta E}$ becomes very much larger than the physical, thermal fluctuations of the internal energy at low temperatures.

At the heart of our difficulties is the fact that we are including with equal weight in our sampling procedure states which make utterly negligible contributions to the answers we are trying to calculate. For example, just 7 of the 512 states provide 99.9% of the contribution to the partition function

4.2 Sampling configurations

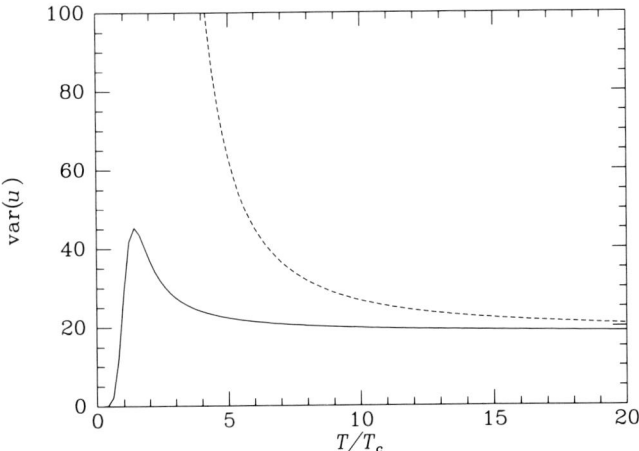

Figure 4.3 The variance of the internal energy in the Gibbs distribution (full curve) and the uniform distribution (dashed curve) for a 3 × 3 Ising lattice as a function of temperature.

of the 3 × 3 Ising system at $\beta\epsilon = 1.1$ (we set $\epsilon = -\mathcal{J}$ as in Chapter 3) and only 2 states are needed for the same proportion of the partition function at $\beta\epsilon = 2.2$.

4.2.1 Importance sampling

We can alleviate the problems just discussed by **importance sampling**. This consists in arranging the way we sample our configurations so that we spend as much time as possible looking at those which make a large contribution to the quantities we are trying to calculate, and do not waste our computer budget on the others.

This much is common sense. Some mathematical care is needed, however, to ensure that we do the best possible job and that the estimators we use to calculate physical quantities remain unbiased. We introduce a positive, normalized probability distribution ρ_α for the configurations of the system:

$$\rho_\alpha \geq 0 \;\forall\; \alpha \quad \text{and} \quad \sum_{\alpha=1}^{A} \rho_\alpha = 1, \tag{4.7}$$

and let

$$\xi_\alpha \equiv \frac{p_\alpha}{\rho_\alpha} X_\alpha. \tag{4.8}$$

That is to say, we choose an arbitrary distribution ρ, and then choose ξ so that the product $\rho_\alpha \xi_\alpha$ is equal to the contribution $p_\alpha X_\alpha$ of state α to the sum in (4.1). Then the expectation of ξ in the probability distribution ρ is

$$\langle \xi \rangle_\rho = \sum_{\alpha=1}^{A} \xi_\alpha \rho_\alpha = \sum_{\alpha=1}^{A} X_\alpha p_\alpha = \langle X \rangle_p. \tag{4.9}$$

So, estimating $\langle \xi \rangle_\rho$ (the average of ξ in our artificial distribution ρ) is equivalent to estimating $\langle X \rangle_p$ (the desired thermal average of X). An unbiased estimator for $\langle X \rangle$ is therefore

$$X_B^{(\rho)} = \frac{1}{B} \sum_{k=1}^{B} \xi_{\alpha_k}, \tag{4.10}$$

where the configurations α_k are to be selected according to the probability distribution ρ.

Why is this useful? This becomes apparent when we calculate the variance of the new estimator, $\text{var}(X_B^{(\rho)}) = \text{var}(\xi)/B$. But

$$\text{var}(\xi) = \langle (\xi - \langle \xi \rangle_\rho)^2 \rangle_\rho = \sum_{\alpha=1}^{A} (\xi_\alpha - \langle X \rangle_p)^2 \rho_\alpha. \tag{4.11}$$

So we can reduce $\text{var}(\xi)$ and thus $\text{var}(X_B^{(\rho)})$ by choosing the ρ_α such that $\xi_\alpha \simeq \langle X \rangle_p$ for all α. In other words, we reduce the variance by isolating as much as possible of the configuration dependence of the product $X_\alpha p_\alpha$ in the probability distribution ρ and making the quantity ξ as nearly as possible constant; we sample the available states using a distribution ρ which mirrors as closely as possible the importance of the states in contributing to the thermal average we are trying to calculate. The art of efficient importance sampling involves making the best possible choice for ρ.

In an ideal world, we could set $\xi_\alpha = \langle X \rangle_p$ independent of α. However, this involves prior knowledge of $\langle X \rangle_p$, which is what we set out to determine. There is also another difficulty: if ρ is to be interpreted as a probability distribution, the elements ρ_α must all be non-negative. If the signs of X_α vary from configuration to configuration, we must absorb any such sign change into the value of ξ_α and we certainly cannot hope to keep it constant.

For these reasons, the usual strategy is to choose the distribution ρ not so that it follows the variation with α of the product $X_\alpha p_\alpha$, but so that it imitates the Gibbs probability p_α. This strategy has several advantages. First, we do not have to adapt our algorithm for selecting configurations when we change the property we are calculating, since our choice of ρ is independent of the quantity X. For example, we can use the same fundamental algorithm for the correlation functions as for the internal energy. Second,

4.2 Sampling configurations

p is guaranteed to be everywhere positive and we require this property also for ρ. Third, most of the variation in the magnitude of the product Xp is contained in the p factor anyway. For example, if we take X to be the energy of the 3×3 Ising model at $\beta\epsilon = 2$, the possible values of X_α range from -36 to $+36$ in steps of 4; the values of p_α, however, vary by the enormous factor of e^{72} between the lowest- and highest-energy states. Therefore, unless the value of X_α is exactly zero, by far the most important factor determining the magnitude of the contribution of the state to $\langle X \rangle_p$ is the Gibbs probability p_α.

Remember, however, that we still cannot calculate (4.2) for p_α directly, since we do not know the value of Z. In order to avoid carrying out a second normalizing calculation for the partition function, as we did in equation (4.3), we shall describe methods of sampling the Gibbs distribution without being able to evaluate explicitly any of the probabilities involved. These methods will produce for us a sequence of states which can be shown, with a greater or lesser degree of rigour, to be selected according to the Gibbs distribution. We shall devote most of the rest of the chapter to studying some of these algorithms.

It is a great gain to be able to sample the Gibbs distribution without needing to evaluate the partition function. The price we pay for it is that the successive states in the sequence which we produce are not independent. That is to say, the relative probabilities of picking different states at step k in the procedure depend on what the state was at step $k-1$.

4.2.2 General structure of numerical algorithms

Box 4.1 shows a **structure chart** which lists the essential steps involved in evaluating the thermal average of X by the technique introduced in §4.2.1. We have left two important gaps in the structure: one, at position C, is where the crucial algorithm for producing the sequence of states for the system should go. Much of the rest of the chapter is devoted to specifying appropriate contents for this gap. The second gap, at position D, is for tests of the statistical independence of two configurations. This is necessary because our methods for generating sequences of configurations do not guarantee that successive states are drawn independently from the Gibbs distribution. We shall defer this problem until §4.6.

We shall study three sampling algorithms suitable for step C of Box 4.1: Monte Carlo, molecular dynamics, and integration of a Langevin equation. Not all are applicable to all the models of Chapter 3; in particular only the first, the Monte Carlo method, is suitable for models such as the Ising model which involve a set of variables that are permitted to take only discrete values.

Box 4.1: General structure chart

Main program	Input data Produce initial configuration: A Evolve state of system: B Average the stored values of quantity X Output results
A. For each degree of freedom	Obtain random number r Assign spin initially on basis of r
B. Repeat	Update configuration: C Test for independence: D If independent: E
C. Evolution step	Depends on chosen method See Boxes 4.2, 4.3 and 4.4
D. Test for independence	See §4.6
E. Sample X	Evaluate X in chosen configuration Store result

4.3 Monte Carlo methods

The **Monte Carlo method** for sampling the Gibbs distribution involves an element of chance, hence its name. Random (or, more strictly, pseudo-random) numbers are used to select the configurations of the system. To appreciate how it works, we first need to cover some of the general theory of 'Markov processes'.

A **Markov process** is a rule for randomly generating a new configuration of a system from the present one. The important thing about this rule is that it should depend only on the present state of the system, and should not require knowledge of any previous one. We can express these rules in the form of a set of probabilities; for each possible pair of states α and α', there is an associated probability, $P(\alpha \to \alpha')$, that if the system is now in state α, then it will be in state α' at the next step. These probabilities satisfy a sum rule which expresses the fact that at each step the system must go somewhere:

$$\sum_{\alpha'} P(\alpha \to \alpha') = 1. \tag{4.12}$$

We are interested in producing a **Markov chain**—a sequence of states generated by a Markov process—in which the frequency of occurrence of each state α is proportional to the associated Gibbs probability p_α. In order to do this, we shall find that we need to place two conditions on the transition probabilities $P(\alpha \to \alpha')$.

4.3 Monte Carlo methods

(i) From a given starting point, it must be possible to evolve the system to any of its other configurations by applying the evolution rule a sufficiently large number of times. This is often called the **accessibility assumption**.

(ii) The transition probabilities must satisfy the **microreversibility** or **detailed balance** condition:

$$p_\alpha P(\alpha \to \alpha') = p_{\alpha'} P(\alpha' \to \alpha). \tag{4.13}$$

We can find sets of transition probabilities which obey these rules for any Gibbs distribution p_α: we could take $P(\alpha \to \alpha') \propto e^{\beta(E_\alpha - E'_\alpha)/2}$, for example. However, most of the sets of transition probabilities we might write down do not satisfy these conditions and, in consequence, they are not useful for modelling states drawn from the Gibbs distribution.

Suppose that we have chosen P so that (i) and (ii) above are satisfied. We use this to show two things. First, we prove that if each configuration α appears at step n of a Markov chain with probability $W(\alpha, n)$ equal to its Gibbs probability,

$$W(\alpha, n) = p_\alpha, \tag{4.14}$$

then it also appears with this probability at step $n+1$. This follows because the distribution at step $n+1$ will be

$$\begin{aligned} W(\alpha, n+1) &= \sum_{\alpha'} p_{\alpha'} P(\alpha' \to \alpha) \\ &= p_\alpha \sum_{\alpha'} P(\alpha \to \alpha'), \end{aligned} \tag{4.15}$$

where we have used the microreversibility condition. However, the normalization condition (4.12) now tells us that

$$W(\alpha, n+1) = p_\alpha, \tag{4.16}$$

as was to be shown.

Second, we show that the deviation between the actual probabilities of occurrence and the Gibbs distribution decreases as we progress along the Markov chain. We define a 'difference' D_n at step n between the actual probability distribution of the states and the Gibbs distribution:

$$D_n \equiv \sum_\alpha |W(\alpha, n) - p_\alpha|. \tag{4.17}$$

Then

$$\begin{aligned}
D_{n+1} &= \sum_\alpha |W(\alpha, n+1) - p_\alpha| \\
&= \sum_\alpha \left|\sum_{\alpha'} W(\alpha', n) P(\alpha' \to \alpha) - p_\alpha\right| \\
&= \sum_\alpha \left|\sum_{\alpha'} [W(\alpha', n) P(\alpha' \to \alpha) - p_\alpha P(\alpha \to \alpha')]\right| \\
&= \sum_\alpha \left|\sum_{\alpha'} [W(\alpha', n) - p_{\alpha'}] P(\alpha' \to \alpha)\right|,
\end{aligned} \quad (4.18)$$

where we have used the sum rule (4.12) to go from the second line to the third and the microreversibility condition to go from the third line to the fourth. Since P, being a probability, must be everywhere positive, the triangle inequality applied to (4.18) gives

$$\begin{aligned}
D_{n+1} &\leq \sum_{\alpha\alpha'} |W(\alpha', n) - p_{\alpha'}| P(\alpha' \to \alpha) \\
&= \sum_{\alpha'} |W(\alpha', n) - p_{\alpha'}| \\
&= D_n,
\end{aligned} \quad (4.19)$$

where we have again used (4.12) to get from the first line to the second. In other words, the deviation D_n from the Gibbs distribution decreases steadily along a Markov chain.

4.3.1 The Metropolis algorithm

The most important and most frequently used algorithm for the Markov process is one invented by Metropolis *et al.* in 1953. The change in the energy of the system consequent upon the change from configuration α to α' is calculated. If the energy change is negative, then the new configuration is automatically accepted; if, however, it is positive, the new configuration is accepted with probability $e^{-\beta(E_{\alpha'} - E_\alpha)}$. In other words,

$$P(\alpha \to \alpha') = \begin{cases} A^{-1} & \text{if } E_{\alpha'} < E_\alpha \\ A^{-1} e^{-\beta(E_{\alpha'} - E_\alpha)} & \text{if } E_{\alpha'} > E_\alpha \end{cases} \quad (4.20)$$

for those states α' which may be reached from α, and zero for all other $\alpha' \neq \alpha$. Here A is a normalization constant, chosen to ensure that equation (4.12) is satisfied. The lack of symmetry in this algorithm is at first surprising, but it is straightforward to show that the crucial microreversibility assumption is satisfied. The accessibility criterion is satisfied if new states are chosen in

4.4 Molecular dynamics

Box 4.2: Structure chart for the Metropolis method

C. For each variable s_i
$\begin{cases} \text{Generate new configuration ('flip spin')} \\ \text{Calculate energy change } E_{\alpha'} - E_\alpha \\ \text{Calculate } \mathcal{P} = \min\left(1, e^{-\beta(E_{\alpha'} - E_\alpha)}\right) \\ \text{Find random variable } r \text{ uniform on } [0,1] \\ \text{Accept move if } r \leq \mathcal{P} \end{cases}$

such a way that any new configuration α' can in principle be obtained from α in a finite number of steps. An alternative to the Metropolis algorithm is discussed in Problem 4.1.

The practical implementation of the Metropolis algorithm is extremely straightforward and this is one of the main reasons for its great success. We show a structure chart for this procedure in Box 4.2; we have used the word 'spin' as shorthand for 'degree of freedom', without implying any restriction to the case of magnetic systems.

First, a new configuration α' of the system is generated from the current one α by some method. For example, in a simulation of the Ising model, one might try reversing the direction of a single spin. This spin can be selected at random, or each of the spins in the sample may be reversed in turn. Either of these procedures will ensure that the accessibility criterion is satisfied. Then the energies of the new and old configurations are compared; this is usually easy, since the energy change on reversing a spin involves only the values of a few neighbouring spins. Finally a pseudo-random number generated by the computer is used to accept or reject the move with probabilities given by (4.20).

4.4 Molecular dynamics

The method known as **molecular dynamics** is an alternative to the Monte Carlo methods we have just been discussing. It involves numerical integration of Newton's equations of motion for a classical dynamical system. It is important to realize that for us this dynamics is artificial; we are concerned solely in this book with thermodynamic properties and as a result none of our models contain any time dependence. We introduce the dynamics only as a convenient device for the evaluation of thermal averages. To avoid confusion, however, it is worth noting that molecular dynamics is also applied extensively to systems that *do* have genuine time dependence.

It is straightforward to create a dynamical system from one of the model Hamiltonians of Chapter 3, provided only that the variables describing the system are free to take continuously varying values. So, for example, we might imagine using the molecular-dynamics technique to study the Heisenberg model, the spherical model, or the Gaussian model (although we shall see later that there are other difficulties in this last case) but not the Ising model. We create our dynamical system by taking the Hamiltonian $H(\{s_i\})$ as the potential energy. We also give the system a kinetic energy

$$K = \sum_i \tfrac{1}{2} \dot{s}_i^2. \qquad (4.21)$$

This expression involves assigning to each of the system's N degrees of freedom s_i a unit 'mass', which is completely artificial and need have nothing to do with the actual physical mass of any of the particles in the system. The equation of motion of the system then becomes

$$\ddot{s}_i = -\frac{\partial H(\{s_k\})}{\partial s_i}. \qquad (4.22)$$

We now have a set of N coupled second-order differential equations. Unless we happen to be extremely lucky, we shall not be able to solve these equations analytically; indeed, we shall see shortly that it is extremely important, if the method is to work at all, that we should *not* be able to solve them analytically. Instead, let us integrate equation (4.22) numerically. There are a number of possible ways of doing this; we shall illustrate the method using a rather unsophisticated technique, known to physicists as the Verlet algorithm and to numerical analysts as the 'explicit central difference scheme'. This employs the discretized form of the second derivative

$$\ddot{s}_i \approx \frac{s_i(t+\Delta t) - 2s_i(t) + s_i(t-\Delta t)}{(\Delta t)^2}, \qquad (4.23)$$

where Δt is a suitably small time interval. If we take (4.23) to be an exact equality, equation (4.22) becomes

$$\begin{aligned} s_i(t+\Delta t) &= 2s_i(t) - s_i(t-\Delta t) - (\Delta t)^2 \frac{\partial H(\{s_k\})}{\partial s_i} \\ &= s_i(t) + \Delta t\, v_i(t) - (\Delta t)^2 \frac{\partial H(\{s\})}{\partial s_i}, \end{aligned} \qquad (4.24)$$

where

$$v_i(t) \equiv \frac{s_i(t) - s_i(t-\Delta t)}{\Delta t} \qquad (4.25)$$

is the discretized 'velocity' of the ith variable.

4.4 Molecular dynamics

4.4.1 Ergodicity and integrability

Given an initial value and 'velocity' for each of the variables s_i, (4.24) enables us to compute the evolution of the system as a function of 'time'. What will this have to do with the thermal average of its behaviour? The answer comes in several stages, by which we can show (subject to some rather dubious assumptions) that the time average of a quantity calculated in molecular dynamics is equal to the average taken in the canonical ensemble for the original Hamiltonian, i.e., *without* the extra kinetic energy term.

The first (and shakiest) step involves the **ergodic hypothesis**, which may be stated as follows.

The evolution of a complex classical dynamical system takes it, with equal probability, through all states which are accessible from the starting point subject to the constraint of energy conservation.

In other words, according to this hypothesis, the time-evolution of a complex classical system provides, given long enough, a fair sample of all the states which are available to it given its initial energy.

Given the importance that this hypothesis has for the molecular dynamics method, it is very disappointing to discover that it is often not correct! One very simple counter-example is provided by the dynamics of two identical coupled harmonic oscillators for which the equations of motion are:

$$\ddot{s}_1 + \omega_0^2 s_1 = -k(s_1 - s_2),$$
$$\ddot{s}_2 + \omega_0^2 s_2 = -k(s_2 - s_1).$$
(4.26)

The system has two normal modes, one in which the variables s_1 and s_2 oscillate in phase with each other and one in which they oscillate in antiphase. If the system starts moving in one of the normal modes, the energy is never transferred into the other mode. There is no possibility of the system exploring the whole range of states which are available to it with the appropriate energy. We can prove that the ergodic hypothesis fails for this system because it is **integrable**: quantities exist in addition to the total energy which are conserved during its motion and these conservation laws prevent it exploring all the states with a given energy. The reason for our warning earlier that molecular dynamics is no use for equations of motion which can be solved exactly is now clear; exactly soluble equations of motion always have additional constants of the motion associated with them. The existence of even one additional constant of the motion is sufficient to prevent the system exploring all the states of a given energy and to nullify the ergodic hypothesis.[2]

---

[2] Ergodicity is a much stronger requirement than that the system should not be completely integrable, since it requires that *no* constants of the motion besides the total energy should exist. In fact, both completely integrable systems (where the number of constants of the motion is equal to the number of independent coordinates) and completely ergodic systems (where there are no constants of the motion besides the energy) are believed to be extremely special cases among general dynamical systems.

Exactly integrable equations of motion can occur for systems much more complicated than this double-pendulum example. Consider, for example, the Gaussian model, which we met in §3.1.6. In zero external field its Hamiltonian is

$$H = \frac{1}{2\beta} \sum_i s_i^2 + \tfrac{1}{2} \sum_{ij} \mathcal{J}_{ij} s_i s_j. \tag{4.27}$$

Now, by an analysis almost identical to that in §3.3, we see that if the system is translationally invariant this Hamiltonian can be diagonalized by making a discrete Fourier transformation. We find

$$H = \frac{1}{2N} \sum_{\mathbf{q}} \tilde{\mathcal{J}}_{\mathbf{q}} |\tilde{s}_{\mathbf{q}}|^2, \tag{4.28}$$

where, by analogy with (3.46),

$$\tilde{\mathcal{J}}_{\mathbf{q}} = \frac{1}{2\beta} - 2\epsilon \sum_{l=1}^{d} \cos(2\pi q_l / L). \tag{4.29}$$

Moreover, Parseval's theorem allows us to write the artificial 'kinetic energy' term (4.21) as

$$K = \frac{1}{N} \sum_{\mathbf{q}} |\dot{\tilde{s}}_{\mathbf{q}}|^2. \tag{4.30}$$

Adding (4.30) to (4.28), we see that the total 'energy' of the dynamical system corresponding to the Gaussian model is that of N d-dimensional uncoupled harmonic oscillators. The energy of each oscillator is separately constant, and once again there is no hope that our system will explore all the configurations with a given energy and thereby satisfy the ergodic hypothesis. It might appear that this result is a consequence of our assuming translational invariance for the kinetic energy as well as for the potential energy, but this is not in fact the case; whatever the fictitious masses we assign to the spins, the dynamical system we produce is a harmonic one and has a set of normal modes which cannot exchange energy. The only thing which is special about the case of translational symmetry is that the normal modes are the discrete Fourier transforms of the original variables.

4.4.2 From microcanonical to canonical averages

Let us suppose, however, that the dynamical system which we have invented *is* at least approximately ergodic. Then performing a time average of some property X of the system during its dynamical evolution is equivalent to averaging X over all states with a given energy. This is the average within the **microcanonical ensemble**, which is the set of states of the system with a given energy. However, using the results of §2.2 we can see that this

4.4 Molecular dynamics

average is usually very nearly equal to the usual thermal average, which includes *all* states but weights them with $e^{-\beta H}$; this approximate equality follows from the fact that the RMS variation in the energy H increases only as the square root of the system's volume V, while the internal energy $U = \langle H \rangle$ increases as V (see equation (2.23)). Hence in the thermodynamic limit of infinite V the only states which contribute to thermal averages are those for which $H = U$. So unless we are very close to T_c, where the heat capacity C_v in equation (2.23) is very large, the microcanonical average of X that we obtain as the time average of X within our dynamical system provides a valid estimate of the thermal average of X in that system.

Remember, however, that the dynamical system is completely artificial—its kinetic energy term has nothing to do with reality. How is the thermal average of X in this system related to the average we want, namely the thermal average at inverse temperature β of the *original* system? Fortunately, these two averages are equal, as may be seen by recalling that the energy of the dynamical system separates into a kinetic part, which depends only on the momenta $p_i \equiv \dot{s}_i$, and a potential part which depends only on the variables s_i:

$$E(\{p_k\}, \{s_k\}) = K(\{p_k\}) + H(\{s_k\}). \quad (4.31)$$

Therefore, for any quantity X which depends only on the values of $\{s_k\}$, we have for the thermal average of X in the dynamical system

$$\langle X \rangle_{\text{Dynamical}} = \frac{\int (\prod_i dp_i) \int (\prod_j ds_j) X(\{s_k\}) \exp[-\beta K(\{p_k\}) - \beta H(\{s_k\})]}{\int (\prod_i dp_i) \int (\prod_j ds_j) \exp[-\beta K(\{p_k\}) - \beta H(\{s_k\})]}$$

$$= \frac{\int (\prod_i ds_i) X(\{s_k\}) \exp[-\beta H(\{s_k\})]}{\int (\prod_i ds_i) \exp[-\beta H(\{s_k\})]}. \quad (4.32)$$

But the second line on the right-hand side is just the required thermal average in the original, static system.

We still need to decide to what temperature our molecular-dynamics simulation corresponds. One way to do this is by invoking the equipartition theorem of classical statistical mechanics, since the kinetic energy (4.21) of our fictitious system is quadratic in the momenta. One can therefore perform a simulation and afterwards use the average value of the artificial kinetic energy to measure the temperature:

$$\langle K \rangle = \tfrac{1}{2} N k_B T, \quad (4.33)$$

where as usual N is the number of variables in the system. Recently, more sophisticated methods have been developed (see Nose (1984) and Hoover (1985)) which involve exerting on the system a fictitious friction force, proportional to the value of a new dynamical variable. The equations of motion of this variable can be adjusted to keep the average kinetic energy of the system at any desired value.

> **Box 4.3: Structure chart for molecular dynamics**
>
> C. For each variable s_i $\begin{cases} \text{Calculate 'force' } -\partial H/\partial s_i \\ \text{Find } s_i(t + \Delta t) \text{ from equation (4.24)} \\ \text{Calculate } v_i(t + \Delta t) \text{ from equation (4.25)} \\ \text{Replace } s_i(t) \text{ by } s_i(t + \Delta t) \end{cases}$

A structure chart of the molecular dynamics method is shown in Box 4.3. At each time-step, the 'forces' on each variable s_i are calculated and used to evolve the system according to equation (4.24) or one of its more sophisticated relatives. Thermal averages of the thermodynamic quantities are estimated as 'time' averages over a sufficiently large number of steps.

Because the chain of reasoning connecting molecular-dynamics simulations to the canonical ensemble is flawed, care is necessary when performing these calculations. At the very least, calculations should be repeated with several different sets of initial conditions, preferably chosen at random to prevent the system becoming trapped in a state of high symmetry. More sophisticated methods have also been developed which combine the deterministic evolution given by (4.24) with a random impulsive 'force'—see Andersen (1980). These are very similar in spirit to the integration of the Langevin equation, which we consider next.

4.5 Langevin equations

The main drawback of the molecular-dynamics method is anxiety that the deterministic evolution of the equations of motion of the system will not always sample properly the states available to it. To get round these problems, it is helpful to introduce an element of randomness into the calculation, for example through the initial conditions. Our final method involves, like the Monte Carlo method, randomness at every step in the evolution.

This method is the integration of a so-called **Langevin equation**, an equation of motion which is first-order rather than second-order:

$$\dot{s}_i = -\Gamma \frac{\partial H(\{s\})}{\partial s_i} + f_i(t). \qquad (4.34)$$

Here $f_i(t)$ is a vector of random variables. In the absence of this noise term, the variables s_i will relax towards the values that minimize H (i.e., to the configuration of the system at zero temperature), at a rate determined by the second derivatives of H at the position of the minimum and the value of the constant Γ. We might hope that the random variables appearing in (4.34) will frustrate this relaxation process and thereby mimic the effect of finite temperature. We shall see that this is indeed what happens.

4.5 Langevin equations

In order to avoid mathematical problems with defining the distribution of the noise, we shall start from the discretized version of (4.34):

$$s_i(t+\Delta t) = s_i(t) - \Gamma \frac{\partial H(\{s\})}{\partial s_i} \Delta t + f_i(t)\Delta t. \qquad (4.35)$$

Consider the case in which the random variables f_i take the form of 'Gaussian white noise'. This means that at each time-step the components f_i are drawn independently from a Gaussian probability distribution of mean zero and variance $\sigma^2/\Delta t$. (The factor Δt is necessary to make the results independent of the time-step at a later stage in the calculation.) What can we say about the behaviour of the system governed by this equation? Since there is an element of randomness in its evolution, we cannot hope to make any definite predictions about it. We shall therefore focus on the probability density $W(\{s_i\}, t)$, which is the probability per unit volume of configuration space that the variables $\{s_i\}$ take given values at time t.

Let us suppose for a moment that we know this distribution at some time t. We decompose the change Δs_i in s_i during the next step into deterministic and random parts $\Delta s_i^{(D)}$ and $\Delta s_i^{(R)}$ respectively:

$$\Delta s_i = \Delta s_i^{(D)} + \Delta s_i^{(R)}, \quad \text{where} \quad \begin{cases} \Delta s_i^{(D)} \equiv -\Gamma \frac{\partial H}{\partial s_i} \Delta t \\ \Delta s_i^{(R)} \equiv f_i \Delta t. \end{cases} \qquad (4.36)$$

The deterministic term in (4.36) causes systems to flow through configuration space at velocity $-\Gamma \partial H/\partial s_i$. This produces a corresponding 'current' of probability density

$$j_i = -\Gamma W \frac{\partial H}{\partial s_i}. \qquad (4.37)$$

From the conservation condition for probabilities, it follows that the deterministic part of the rate of change of probability density W is

$$\begin{aligned} \frac{\partial W^{(D)}}{\partial t} &= -\sum_i \frac{\partial j_i}{\partial s_i} \\ &= \Gamma \sum_i \frac{\partial}{\partial s_i}\left[W \frac{\partial H}{\partial s_i}\right]. \end{aligned} \qquad (4.38)$$

The second part of the change in the probability density comes from the random term. In order to find it, we must sum over all the possible values of the random forces f_i at time t:

$$\Delta W^{(R)} = \left[\mathcal{A}\int_{-\infty}^{\infty}(\prod_i \mathrm{d}f_i)\exp\left(-\Delta t \sum_i f_i^2/2\sigma^2\right)W(\mathbf{s}-\Delta \mathbf{s}^{(R)}, t)\right] - W(\mathbf{s}, t),$$

$$(4.39)$$

where $\mathcal{A} \equiv (2\pi\sigma^2/\Delta t)^{-N/2}$ is a normalization factor. However, we can expand $W(\mathbf{s} - \Delta\mathbf{s}^{(\mathrm{R})}, t)$ in the form

$$W(\mathbf{s}-\Delta\mathbf{s}^{(\mathrm{R})},t) \approx W(\mathbf{s},t) - \sum_i \Delta s_i^{(\mathrm{R})} \frac{\partial W}{\partial s_i} + \frac{1}{2}\sum_{ij} \Delta s_i^{(\mathrm{R})} \Delta s_j^{(\mathrm{R})} \frac{\partial^2 W}{\partial s_i \partial s_j}. \quad (4.40)$$

As the time-step Δt becomes small, the neglect of higher terms in (4.40) becomes a more and more accurate approximation. Only terms which contain an even number of powers of each random component f_i will survive the integration in (4.39), so we can write

$$\Delta W^{(\mathrm{R})} \approx \tfrac{1}{2}(\Delta t)^2 \mathcal{A} \int_{-\infty}^{\infty} \left(\prod_i \mathrm{d}f_i\right) \exp\left(-\Delta t \sum_i f_i^2/2\sigma^2\right) \sum_i f_i^2 \frac{\partial^2 W}{\partial s_i^2}. \quad (4.41)$$

Now we perform the integral over the f_i to obtain:

$$\Delta W^{(\mathrm{R})} = \frac{\sigma^2}{2} \sum_i \frac{\partial^2 W}{\partial s_i^2} \Delta t. \quad (4.42)$$

Letting $\Delta t \to 0$ and adding the random and deterministic parts of the motion together, we obtain the **Fokker–Planck equation** (also in this context sometimes called the Smoluchowski equation)

$$\frac{\partial W}{\partial t} = \sum_i \frac{\partial}{\partial s_i}\left(\Gamma W \frac{\partial H}{\partial s_i} + \tfrac{1}{2}\sigma^2 \frac{\partial W}{\partial s_i}\right). \quad (4.43)$$

This is a partial differential equation governing the evolution of the probability distribution. Notice that if $\Gamma = 0$, i.e., in the absence of any 'driving force' towards lower energies, it reduces to the diffusion equation.

Now suppose that the distribution W is the Gibbs distribution

$$W(\{s_i\},t) \propto \exp\left[-\beta H(\{s_i\})\right]. \quad (4.44)$$

Substituting this into the right-hand side of (4.43), we obtain

$$\frac{\partial W}{\partial t} = \sum_i \frac{\partial}{\partial s_i}\left((\Gamma - \tfrac{1}{2}\beta\sigma^2) W \frac{\partial H}{\partial s_i}\right), \quad (4.45)$$

the right-hand side of which vanishes if we choose $2\Gamma/\sigma^2 = \beta$. In other words, the parameters σ and Γ can be chosen so that the steady-state probability distribution for variables satisfying the Langevin equation is the Gibbs distribution for any desired temperature. The proof that the distribution always tends to this equilibrium distribution with time is left to Problems 4.2 and 4.3, where it is also shown that the Langevin dynamics is a Markov process.

Box 4.4: Structure chart for the Langevin method

C. For each variable s_i
$\begin{cases} \text{Calculate force } -\Gamma\partial H/\partial s_i \\ \text{Find random variable } f_i \\ \text{Find } s_i(t+\Delta t) \text{ from equation (4.35)} \\ \text{Replace } s_i(t) \text{ by } s_i(t+\Delta t) \end{cases}$

4.5.1 Comparison of the Langevin and molecular-dynamics methods

In order to clarify the relationship between the Langevin and molecular dynamics methods, it is convenient to rewrite equation (4.35) with $\Gamma = \Delta t$. We then obtain

$$s_i(t+\Delta t) = s_i(t) + \Delta t f_i(t) - (\Delta t)^2 \frac{\partial H(\{s_k\})}{\partial s_i}. \tag{4.46}$$

Now let us compare this with equation (4.24), the corresponding expression for evolution through a time-step in the molecular-dynamics method. We see that the two results differ only in that $v_i(t)$, the discretized 'velocity' of the ith variable, which appears in (4.24), is replaced in (4.46) by the random variable f_i. In other words, the Langevin method is equivalent to the molecular-dynamics method but with the velocity randomized at each time-step. On account of this difference, the Langevin method is more likely to break away from a local region of the configuration space in which the molecular-dynamics method would spend a large amount of time; as a result, we can loosely say that the Langevin method is more efficient at sampling the global, and the molecular-dynamics method the local, configuration space of the system. In some applications it is advantageous to alternate the two, performing a molecular-dynamics step or a Langevin step randomly at each time; see Kogut (1986) for a discussion in the context of simulations of quantum field theories. These ideas are similar to those we met in the context of overcoming non-ergodicity in molecular dynamics in §4.4.2.

We show a structure chart for the integration of the Langevin equation in Box 4.4.

4.6 Independence of configurations

We now know several different ways to produce a sequence of states in such a way that, over a long enough period of time, each state appears with a probability equal to the corresponding probability in the equilibrium Gibbs distribution. The remaining problem is to interpret the phrase 'a long enough period of time'.

It is clear that with any of our evolution algorithms, the probability of finding a particular state at the $(n+1)^{\text{th}}$ step is strongly dependent on the state of the system at the nth step. For example, with the Metropolis algorithm only N states can be reached by flipping a single spin. It would therefore be pointless to evaluate the quantity X in two configurations separated only by a single spin-flip. The values would not be statistically independent. Similar arguments apply for the molecular-dynamics and Langevin methods. Common sense tells us that we should wait some number of steps for the system to lose its 'memory' of the previous value of the quantity. But how long should we wait?

4.6.1 Correlations along the path

One way of deciding when it is time to evaluate X again is to calculate the autocorrelation function of X along our path through configuration space. This is

$$C(k) \equiv \frac{\langle X_{\alpha_j} X_{\alpha_{j+k}} \rangle - \langle X_{\alpha_j} \rangle^2}{\langle X_{\alpha_j}^2 \rangle - \langle X_{\alpha_j} \rangle^2}, \tag{4.47}$$

where X_{α_j} is the value of X in the configuration α_j selected at step j and the averages are taken over a particular path of the system through configuration space. C has been normalized so that it is unity at $k = 0$, where the values of X are completely correlated, and decays to zero when k is large enough for them to have become completely uncorrelated. We can evaluate $C(k)$ for the first few steps of our system's evolution, find the value of k necessary for its value to fall, say, to 0.05, and then sample the quantity X every k steps subsequently on the grounds that the values will be 'almost' independent. This procedure is summarized in Box 4.5.

4.6.2 Critical slowing down

Figure 4.4 shows the autocorrelation function of the internal energy of a Monte Carlo simulation of the Ising model along a path through configuration space generated by the Metropolis algorithm for three different values of β. It is clear that as we approach the critical temperature, the characteristic 'time' (more strictly number of steps) τ for the autocorrelation to disappear gets longer and longer.

This phenomenon is known as **critical slowing down**. A full discussion of critical slowing down is beyond the scope of this book, but it is easy to understand the physics behind it. As the transition is approached, the correlation length becomes longer and the system contains larger and larger 'islands' in which all the spins are aligned. In order to produce a statistically independent configuration, it is necessary that we stand a reasonable chance of turning over *all* the spins in such an island; this by itself will make rather a small difference to the energy. However, to do this by the Metropolis algorithm involves turning the spins over one at a time. Since the neighbouring

4.6 Independence of configurations

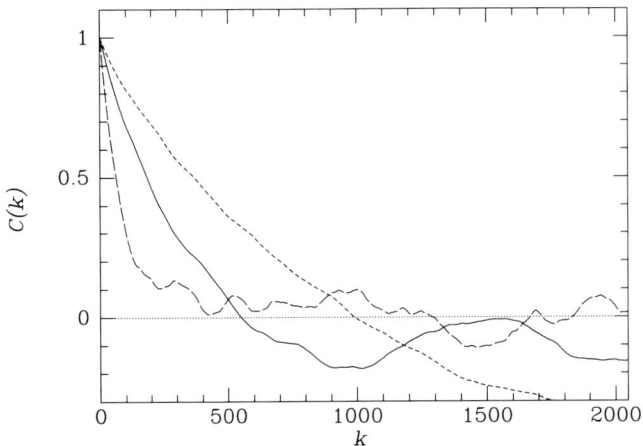

Figure 4.4 The autocorrelation function C of the internal energy in a simulation of a 15×15 Ising model at $T = 0.7T_c$ (full curve), $T = T_c$ (short dashed curve), and $T = 3T_c$ (long dashed curve).

Box 4.5: Structure chart for decision whether configurations are independent

D. At start of calculation
- Evaluate X at each step
- Calculate autocorrelation function (4.47)
- Wait until it has decayed to a small value
- Store the number of steps K needed

Subsequently
- Test whether K steps elapsed
- If so, evaluate X again

---

spins are aligned, the intermediate states will have a substantially increased energy and any move to them is unlikely to be accepted.

Empirically, it is found that the decorrelation time τ is related to the correlation length[3] ξ by a relation of the form

$$\tau = \xi^z, \qquad (4.48)$$

where z is known as the 'dynamical critical exponent'. For the Ising model, z is found to be approximately two for the Metropolis algorithm and for any other Markov process in which the spins are flipped one at a time. This is a statement of what is often called **dynamical universality**, and it has very

---
[3] For a finite system, it is to be understood that ξ cannot exceed the system size L.

serious consequences for our ability to perform simulations for large systems in the interesting regime close to the critical point.

Clearly what is needed to overcome this disastrous increase in τ is a method which allows us to flip many spins simultaneously. Considerable ingenuity is necessary to devise a method which permits this and still satisfies the crucial requirements of accessibility and microreversibility. We shall describe two such methods.

4.6.3 The Swendsen–Wang algorithm

The first successful algorithm of this kind was introduced by Swendsen and Wang (1987). The essential idea is to let the system itself decide the shape of the 'clusters' of spins which are to be flipped simultaneously.

We will describe the algorithm as it is applied to the Ising model. Suppose we select two neighbouring sites, l and m, and define a new Hamiltonian equal to the old one minus the contribution corresponding to the bond between them:

$$H_{lm} = \sum_{(i,j) \neq (l,m)} \mathcal{J}_{ij} s_i s_j.$$

Next, we calculate the partition functions corresponding to this Hamiltonian when the spins s_l and s_m are constrained to be the same, or to be different:

$$\begin{aligned}Z_{lm}^{\text{Same}} &\equiv \sum_{\{s_i\}} \exp(-\beta H_{lm}) \delta_{s_l, s_m}, \\ Z_{lm}^{\text{Diff}} &\equiv \sum_{\{s_i\}} \exp(-\beta H_{lm}) (1 - \delta_{s_l, s_m}),\end{aligned} \quad (4.49)$$

and also in the case in which s_l and s_m are allowed to vary independently:

$$\begin{aligned}Z_{lm}^{\text{Ind}} &\equiv \sum_{\{s_i\}} \exp(-\beta H_{lm}) \\ &= Z_{lm}^{\text{Same}} + Z_{lm}^{\text{Diff}}.\end{aligned} \quad (4.50)$$

Then we can reconstruct the full partition function for the original system by taking

$$\begin{aligned}Z &= e^{-2\beta \mathcal{J}} Z_{lm}^{\text{Same}} + e^{2\beta \mathcal{J}} Z_{lm}^{\text{Diff}} \\ &= e^{-2\beta \mathcal{J}} \big[(1 - e^{4\beta \mathcal{J}}) Z_{lm}^{\text{Same}} + e^{4\beta \mathcal{J}} Z_{lm}^{\text{Ind}} \big].\end{aligned} \quad (4.51)$$

The last line has a very simple interpretation: apart from a physically irrelevant constant factor $e^{-2\beta \mathcal{J}}$, it is the partition function for a system in which spins l and m are bonded together (constraining the spins to be the same), with probability $1 - e^{4\beta \mathcal{J}}$ and allowed to vary freely without any energy penalty with probability $e^{4\beta \mathcal{J}}$. Notice that for a ferromagnetic Ising model, with $\mathcal{J} < 0$, both these probabilities lie between 0 and 1 as required;

4.6 Independence of configurations

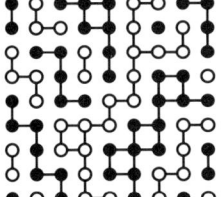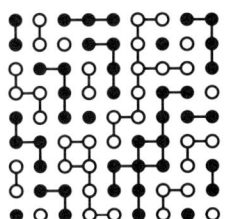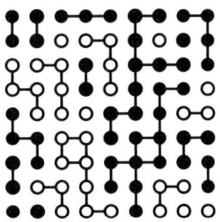

Figure 4.5 The sequence of steps involved in the Swendsen–Wang algorithm for the Ising model: (i) formation of bonds between neighbouring spins; (ii) random deletion of bonds; (iii) random reorientation of the resulting clusters of spins; (iv) restoring the underlying lattice.

for an anti-ferromagnetic model with $\mathcal{J} > 0$, simply change the signs of $\mathcal{J}$ in (4.51) and replace Z_{lm}^{Same} by Z_{lm}^{Diff} and *vice versa*. By repeating this process on all the different links in the lattice we transform the problem from its specification in terms of spins to one in terms of the links between spins—in fact, to a percolation problem.

This parallel inspired Swendsen and Wang to suggest an evolution algorithm which passes from the spin picture to the bond picture and back again:

(i) Take a spin configuration and form a network of bonds by joining together all aligned neighbours.
(ii) Keep these bonds with probability $1 - e^{4\beta\mathcal{J}}$, or delete them with probability $e^{4\beta\mathcal{J}}$. This step will cut most large clusters of spins into two or more smaller clusters.
(iii) Treat the spin orientation of the new, smaller, clusters as the basic units and orient them randomly, i.e., up or down with equal probability.
(iv) Finally reconstruct the original lattice of spins from the reoriented clusters.

This whole procedure is taken as a single Monte Carlo evolution step for the system. It is illustrated in Figure 4.5.

It is easy to see that this procedure satisfies the accessibility criterion, since it is possible that any bond between two like spins will be deleted and therefore any state can be reached from any other in a single step. The following argument shows that it also satisfies the microreversibility condition. Consider a pair of configurations α and α' which are transformed into each other in a single Swendsen–Wang step via a particular pair of bond patterns as shown in Figure 4.5. However, there is also a route from configuration α' back to configuration α which passes through exactly the same bond pattern after the deletion step (ii). Suppose the forward route involves the deletion of n_1 bonds and the backward route the deletion of n_2 bonds; then the ratio of the probabilities for the forward and backward routes is $e^{4(n_1-n_2)\beta\mathcal{J}}$. However, $n_1 - n_2$ is exactly the number of neighbouring

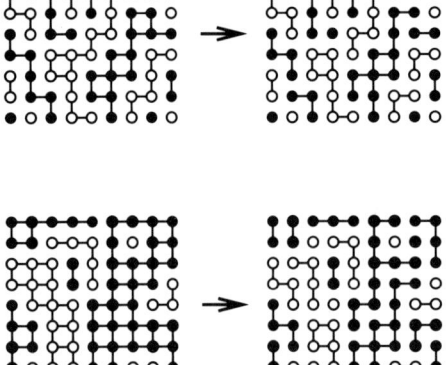

Figure 4.6 The same intermediate configuration of bonds is produced in the Swendsen–Wang procedure by deleting n_1 bonds from configuration α, or n_2 bonds from configuration α'.

Box 4.6: Structure chart for the Swendsen–Wang algorithm

C. At each step
$\Big\{$
Draw bonds between aligned neighbouring spins
Delete bonds at random with probability $e^{4\beta\mathcal{J}}$
Treat spins of resulting smaller clusters as single units
Orient these spins randomly

---

pairs of spins aligned in configuration α minus the corresponding quantity in configuration α' (see Figure 4.6), so the energy difference between the configurations is

$$E_\alpha - E_{\alpha'} = 4\mathcal{J}(n_1 - n_2). \tag{4.52}$$

The ratio of the forward and backward transition probabilities via this particular route is therefore $\exp[-\beta(E_{\alpha'} - E_\alpha)]$, just as required by microreversibility. Furthermore, all the possible forward and backward paths can be put into correspondence in this way, so the total transition probabilities also satisfy the microreversibility condition.

Box 4.6 summarizes the Swendsen–Wang algorithm. Numerical tests show that the relaxation times for the method satisfy equation (4.48) with an exponent z of approximately 0.35—a great improvement over the $z = 2$ characteristic of the conventional Metropolis algorithm.

4.6.4 The Wolff algorithm

Wolff (1989) has described an algorithm, similar in spirit to that of Swendsen and Wang, which eliminates the problem of critical slowing down completely.

4.6 Independence of configurations

Put another way, this algorithm obeys equation (4.48), but with $z = 0$: there is *no* dependence of the number of steps required to remove the autocorrelation of X on the system size or the correlation length. We describe it first of all for an Ising system.

The evolution step of the Wolff algorithm involves building up a cluster of spins which are then flipped *en bloc*. The cluster is built up as follows.

(i) Randomly choose a site i on the lattice.

(ii) Visit all the neighbouring sites j and add site j to the cluster containing i with probability

$$P_{\text{Add}}(s_i, s_j) \equiv 1 - e^{\min(0, 4\beta \mathcal{J} s_i s_j)}. \tag{4.53}$$

For a ferromagnetic model ($\mathcal{J} < 0$), P_{Add} is zero if the spins are antiparallel and $1 - e^{4\beta \mathcal{J}}$ if they are parallel.

(iii) Repeat step (ii) with i set equal to the index j of each of the new sites which have joined the cluster.

(iv) Repeat (iii) until no more new sites are added.

(v) Flip the spins on all the sites in the cluster thus produced.

This procedure satisfies the accessibility criterion, because there is always a non-zero probability that the cluster will contain only a single spin and flipping a sequence of single spins allows one to reach any other configuration. It also satisfies the microreversibility criterion: consider two spin configurations α and α', in which the spins take values $\{s_i\}$ and $\{s_i'\}$ and which can be transformed into each other by reversing a single cluster of spins, which we shall call C. Then the ratio of the transition probabilities in opposite directions is

$$\begin{aligned}\frac{P(\alpha \to \alpha')}{P(\alpha' \to \alpha)} &= \prod_{(i \in C, j \notin C)} \frac{1 - P_{\text{Add}}(s_i, s_j)}{1 - P_{\text{Add}}(s_i', s_j')} \\ &= \exp\Big[4\beta \mathcal{J} \sum_{(i \in C, j \notin C)} s_i s_j \Big] \\ &= \exp[-\beta(E_{\alpha'} - E_\alpha)]. \end{aligned} \tag{4.54}$$

In (4.54), we have used the invariance of the Ising Hamiltonian under the reversal of both of a pair of interacting spins to cancel the probabilities for adding the internal spins to the cluster for the two configurations; the probabilities for failing to add the spins at the edge of the cluster, however, do not cancel. They give a factor of $e^{4\beta \mathcal{J}}$ for each pair (i, j) of aligned spins on the boundary, and a factor of $e^{-4\beta \mathcal{J}}$ for each anti-aligned pair. The product of all these produces exactly the factor required by the detailed balance condition.

Wolff also pointed out that his algorithm (and, indeed, that of Swendsen and Wang) can be applied to the generalized Heisenberg model, whose spins

> **Box 4.7: Structure chart for the Wolff algorithm**
>
> C. At each step
> - Select random starting spin $\mathbf{s}_i$
> - Select random unit vector $\hat{\mathbf{n}}$
> - Add neighbouring spins $\mathbf{s}_j$ with probability (4.53)
> - Continue until no more spins added
> - Transform all cluster spins using (4.55)

$\mathbf{s}_i$ can point in any direction rather than just up or down, by the following extension of the idea of flipping an Ising spin.

(i) Choose a random n-component unit vector $\hat{\mathbf{n}}$.

(ii) Reflect the spin in the plane perpendicular to $\hat{\mathbf{n}}$:

$$\mathbf{s} \rightarrow R_{\hat{\mathbf{n}}}(\mathbf{s}) \equiv \mathbf{s} - 2(\mathbf{s}\cdot\hat{\mathbf{n}})\hat{\mathbf{n}}. \tag{4.55}$$

Notice that the interaction energy between two spins in the Heisenberg model is invariant if such a reflection is performed on both spins, just as it is in the Ising model if both spins are reversed. The generalization of equation (4.53), giving the probability of adding spin j to the cluster containing its neighbour i, is now

$$P_{\text{Add}}(\mathbf{s}_i,\mathbf{s}_j) \equiv 1 - \exp\big\{\min\big[0, 4\beta\mathcal{J}(\hat{\mathbf{n}}\cdot\mathbf{s}_i)(\hat{\mathbf{n}}\cdot\mathbf{s}_j)\big]\big\}. \tag{4.56}$$

In effect, one treats the component of $\mathbf{s}$ parallel to $\hat{\mathbf{n}}$ like an Ising spin.

There is a close relationship between the Wolff and Swendsen–Wang methods. The probability (4.53) for joining a new spin to the cluster is the same as the probability for keeping a bond in the Swendsen–Wang algorithm, so the clusters of spins 'grown' by the two methods will be on average the same. However, whereas in the Swendsen–Wang procedure *all* the clusters are subject to the possibility of a spin-flip at each step, in the Wolff method it is only the cluster containing the starting spin. This means that larger clusters are subjected to such flips more often in the Wolff procedure; this greater investment of effort in the larger clusters makes the algorithm more effective. A structure chart for the general Heisenberg version of the Wolff method is given in Box 4.7.

Despite this success, a word of caution is in order. The number of steps, as defined by Box 4.7, to achieve statistically independent results is indeed independent of the temperature. However, this is unfortunately not true of the computer time used by the method: as we approach the critical temperature, large clusters are grown more and more often and the number of operations required per step grows.

4.7 Calculation of critical exponents from simulations

These simulation algorithms work well for a wide variety of systems. By using the Swendsen–Wang or Wolff methods, calculations can even be done on systems with large correlation lengths in such a way that the number of Monte Carlo steps necessary to obtain statistically independent results does not become too large. However, this is not of itself enough to guarantee that interesting quantities characterizing critical behaviour, such as the critical exponents, can be directly extracted from the simulations. In order to calculate, for example, the magnetization exponent β, it would be necessary to perform simulations at a number of different temperatures and plot a curve of magnetization against temperature from which the exponent could be extracted. However such an undertaking is fraught with difficulty: we know that the physical fluctuations in the magnetization become very large near T_c, and these fluctuations cannot be entirely suppressed by clever importance sampling. We should thus have to run our simulation for an inordinately long time in order to obtain good averages. In the finite system that is simulated, there is the additional difficulty that there is not a sharp transition between zero and non-zero magnetizations or a sharp peak in the specific heat, and therefore T_c cannot be located precisely. Finally, all the noise that has accumulated along the way can make the task of extracting a critical exponent from the gradient of a log-log plot or whatever all but impossible.

Do not despair! The expertise we have developed for sampling thermal averages will not be wasted. It is simply not helpful, if one is trying to calculate critical exponents, to do it by computing directly the behaviour of the system's thermodynamic variables and correlation functions near the critical point. Instead, we shall see in §5.10 that there are other quantities, which do not diverge or fluctuate strongly at the critical point, from which the critical exponents can also be calculated. It is to these that our sampling expertise should be applied. First, however, we need to study the powerful techniques of real-space renormalization.

Problems

4.1 Consider the following alternative to the usual Metropolis algorithm in Monte Carlo calculations:

$$P(\alpha \to \alpha') \propto \frac{\exp[-\frac{1}{2}\beta(E_{\alpha'} - E_\alpha)]}{2\cosh[\frac{1}{2}\beta(E_{\alpha'} - E_\alpha)]}.$$

Show that it satisfies the microreversibility and accessibility criteria. What are its limiting forms as $E_{\alpha'} - E_\alpha$ becomes (a) large and positive; (b) large and negative? Why do you think that the Metropolis algorithm is usually preferred in practical Monte Carlo calculations?

4.2 Consider the differential operator on the right-hand side of the Fokker–Planck equation (4.43):

$$\hat{O}_{\rm FP} = \sum_i \frac{\partial}{\partial s_i}\left[\Gamma\frac{\partial H}{\partial s_i} + \frac{\sigma^2}{2}\frac{\partial}{\partial s_i}\right].$$

Show that this operator is not self-adjoint when it acts on functions which vanish at infinity, but can be brought into self-adjoint form by the transformation

$$\hat{O}'_{\rm FP} \equiv e^{\beta H/2}\hat{O}_{\rm FP}e^{-\beta H/2}$$
$$= -\frac{\sigma^2}{2}\left[-\frac{\partial}{\partial s_i} + \tfrac{1}{2}\beta\frac{\partial H}{\partial s_i}\right]\left[\frac{\partial}{\partial s_i} + \tfrac{1}{2}\beta\frac{\partial H}{\partial s_i}\right],$$

with $\beta = 2\Gamma/\sigma^2$. Associated with this is a transformation of the probability density $W(\{s_i\})$ to

$$f(\{s_i\}) \equiv e^{\beta H(\{s_i\})/2}W(\{s_i\}).$$

Argue that $\hat{O}'_{\rm FP}$ is a negative semi-definite operator and therefore that its zero eigenvalue, corresponding to the eigenfunction

$$f_0 \propto \exp(-\tfrac{1}{2}\beta H),$$

is the largest. Use an expansion of the time-dependent solutions of the Fokker–Planck equation in terms of this and the other eigenfunctions to show that the probability distribution relaxes towards its equilibrium form at large times.

4.3 Show that the time-evolution governed by the discretized Langevin equation (4.35) can also be regarded as a Markov process in which the transition probability between states in which the spins take the values $\{s\}$ and $\{s + \Delta s\}$ is given by

$$P(\{s_i\} \to \{s_i + \Delta s_i\}) \propto \exp\left[-\frac{(\Delta t)^2}{2\sigma^2}\sum_i\left(\frac{\Delta s_i}{\Delta t} + \Gamma\frac{\partial H}{\partial s_i}\right)^2\right].$$

Show that, in the limit $\Delta t \to 0$, these transition probabilities satisfy the accessibility and microreversibility criteria. Hence complete a second proof that the probability distribution for the Langevin equation relaxes to the Gibbs distribution.

4.4 Show that the generalization of the Wolff algorithm to the n-component Heisenberg model satisfies the accessibility and microreversibility criteria.

5
Real-space renormalization

Until the early 1970s all calculations of critical exponents were made either by exactly solving a model for its thermodynamic properties and then examining its behaviour in the critical regime (as with the Ising model in one and two dimensions) or by direct numerical simulations (as with the three-dimensional Ising model) or by extrapolation from approximate solutions which become invalid in the vicinity of a phase transition (high-temperature expansions, for example). However, in a now-famous paper published in 1966, Kadanoff presented arguments which, he maintained, would allow one to simplify calculations in the critical regime to the point at which critical exponents could be extracted, without ever working out the partition function for the problem. However, though Kadanoff's ideas showed great physical intuition about the processes giving rise to critical phenomena, they lacked the mathematical precision which might have given one faith in his results. A more quantitative realization of the argument was given by Wilson and others, who introduced the so-called **renormalization group** techniques (Wilson and Kogut 1974). In their present state, these techniques divide roughly into two classes. There are those which were developed by pursuing the analogy between statistical mechanics and quantum field theory, which we will refer to as 'field-theoretical' or occasionally 'k-space' techniques, and then there are the **real-space renormalization** techniques, which are simpler and closer in spirit to the original ideas of Kadanoff. In the later chapters of this book we will discuss field-theoretical methods extensively (Chapter 7 onwards); this chapter deals with the real-space methods. The term 'real-space' in this

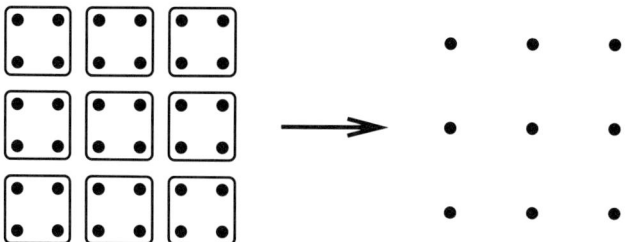

Figure 5.1 Renormalization of a square lattice. The linear dimensions of the lattice on the right must be shrunk by a factor of $b = 2$ to render it similar to the original one. The final lattice therefore has fewer sites than the original by a factor of $b^2 = 4$.

context refers to the fact that these techniques involve quantities dependent on position coordinates in ordinary space. The field-theoretical techniques, conversely, are simplest when the equations are written in terms of spatially Fourier-transformed quantities, hence the name 'k-space' techniques.

5.1 Renormalizing the lattice

Real-space renormalization techniques are applicable only to models based on a lattice. And more than this, the lattice must be regular in a very special kind of way; it must have a 'discrete scaling symmetry'. To understand what this means consider taking a lattice and **blocking** it. This means dividing the sites of the lattice into groups or **blocks**, and then replacing each block by just one single site, which may be at the position occupied by one of the sites in that block, or at some other position within the area covered by the block. The lattice has a **discrete scaling symmetry** if we can block it in this way so as to produce a lattice exactly like the one we started with, except for an increase in the lattice parameter $a \to a' \equiv ba$. The process of **renormalizing** the lattice is then completed by reducing all dimensions in the new lattice by a factor of b so that we end up with exactly the same lattice that we started with.

Actually, one thing does change when we renormalize our lattice. If we group sites into blocks containing p sites on average, then the renormalized lattice will contain fewer sites than the original by a factor of p. This will be important when we come to define site-averaged quantities, such as susceptibility, on our renormalized lattice. Since we have scaled our whole lattice down by a factor of b, its volume must have shrunk by a factor of b^d, where d is the number of spatial dimensions. Clearly then, if the sites in the renormalized lattice are arranged in exactly the same way as those in the original lattice, their number must have been reduced by a factor of $p = b^d$.

The most common lattice displaying a discrete scaling symmetry is the square lattice. Figure 5.1 illustrates the renormalization of the square lattice

5.2 Block variables

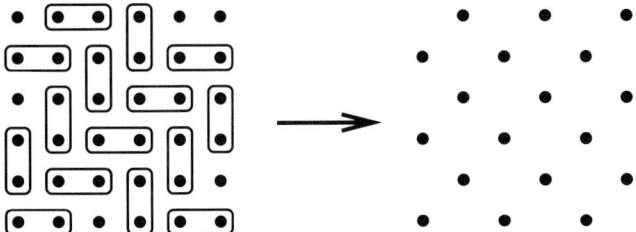

Figure 5.2 An alternative blocking scheme for renormalizing the square lattice. Here all length scales must be divided by a factor of $b = \sqrt{2}$. The final lattice will therefore have fewer sites than the original by a factor of $b^2 = 2$.

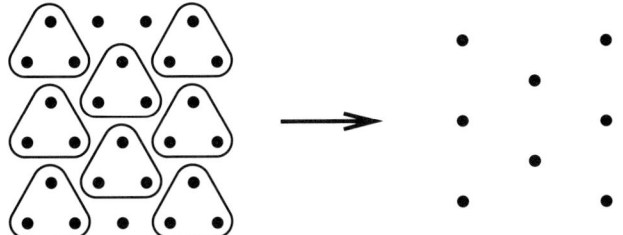

Figure 5.3 Renormalization of a triangular lattice. The linear dimensions of the lattice on the right must be shrunk by a factor of $b = \sqrt{3}$ to render it similar to the original one. The final lattice therefore has fewer sites than the original by a factor of $b^2 = 3$.

in two dimensions. In this case $b = 2$ and the lattice is left with a quarter the number of sites it started with. Clearly this transformation may also be performed on a rectangular lattice. Actually this is not the only way to block a square lattice. Figure 5.2 illustrates an alternative method. In this case $b = \sqrt{2}$ and the number of sites is halved. Figure 5.3 shows how one would renormalize a triangular lattice. Here $b = \sqrt{3}$ and the renormalized lattice has a third as many sites as the original one.

5.2 Block variables

Now let us suppose we have a model in which we have variables, which we will refer to as 'spins', defined on each site on our lattice. For simplicity, let us confine our discussion to models in which only a single, scalar spin s_i is defined on each site i. Now let us renormalize our lattice. For each block k, we define a new variable $\sigma_k^{(1)}$, which is some function f of the spins in that block. The variables $\sigma_k^{(1)}$ we call **block variables**. If $\mathcal{S}_k$ denotes the set of sites in the kth block, then we write

$$\sigma_k^{(1)} = f(\{s_i\}), \qquad \text{where } i \in \mathcal{S}_k. \tag{5.1}$$

With the block variables defined in this way, we now replace each block with a single site and shrink our lattice down to the size of the original. We arrive at a model just like the one we started with, with the same lattice and with a single variable at each site. The technique of real-space renormalization involves the successive application of this kind of transformation. If we denote the block variables after the nth renormalization by $\sigma_i^{(n)}$, then by analogy with (5.1)

$$\sigma_k^{(n+1)} = f(\{\sigma_i^{(n)}\}), \qquad \text{where } i \in \mathcal{S}_k. \tag{5.2}$$

The recursive nature of this definition—the fact that to define the variables $\sigma_i^{(n+1)}$ we only need the variables $\sigma_i^{(n)}$, and not the variables $s_i \equiv \sigma_i^{(0)}$—is crucial to the renormalization approach.

As an example, a simple definition of the block variables is the linear definition

$$\sigma_k^{(n+1)} = A^{(n)} \sum_{i \in \mathcal{S}_k} \sigma_i^{(n)}, \tag{5.3}$$

where $A^{(n)}$ is a (re)normalization constant. We will make considerable use of this particular definition of the block variables later in this chapter. When we come to tackle the Ising model at the end of this chapter, however, we will use a different scheme in which the value of the block variable is taken to be the mode, or commonest value of the spins in the block. (For blocks made up of an even number of Ising spins, exactly half of which are up and half down, there is no commonest value. In this case we choose the block spin to be either up or down at random, with equal probability.) For want of a better name, let us refer to this scheme as the **majority rule**. Another common way of defining $\sigma_k^{(n+1)}$ is to set it equal to the value of one of the variables $\sigma_i^{(n)}$ in the block:

$$\sigma_k^{(n+1)} = \sigma_i^{(n)}, \qquad \text{where } i \in \mathcal{S}_k. \tag{5.4}$$

This definition of the $\sigma_k^{(n)}$ goes under the name of **decimation**. The particular spin $\sigma_i^{(n)}$ which we use might be chosen at random from those in the block k, or more commonly according to some rule such as always choosing the top left spin in a square block. We must be careful how we employ the decimation technique however. In many cases it gives rather poor results for the critical exponents by comparison with, for example, the majority rule. (See Swendsen (1979b) for a detailed comparison of the two schemes for the Ising model.)

There are many other ways of defining the block variables. However, our choice of definition is not completely free. First, the definition must be a 'sensible' one. For example, if we choose the function $f(\{s_i^{(n)}\})$ to be equal to 1, independent of all the variables $\{s_i^{(n)}\}$, then all information about the real physical system will be lost after the first stage of renormalization. Clearly, this would not be a sensible definition. However, there are more subtle things that can go wrong with a particular definition. We will leave the discussion of these problems until we have defined a few more of our terms (§5.7).

However, there is one very important constraint which it is worth pointing out at this stage. We must ensure that the set of states available to each of the new variables is the same as that available to the original ones. This means that, whilst the additive definition (5.3) would be fine for the Gaussian model (see §3.1.6) it would be unacceptable for, say, the Ising model, whose Hamiltonian is given in equation (3.1). To understand this, consider applying (5.3) to the Ising chain. To simplify things let us set the external field B to zero. We can block this model by grouping the spins in the chain into pairs. Then we replace each pair with one block variable whose value is proportional to the sum of the two original spins. But now we have a problem, because this sum may take three values, $+2$, 0 or -2, depending on the values of the original spins; this violates the condition above. It is for this reason that we will be adopting the majority rule scheme when we come to tackle the Ising model by a real-space renormalization technique.

5.3 The renormalization of the Hamiltonian

If $P(\{s_i\})$ is the probability that the spins in our model take on some particular set of values $\{s_i\}$, then we can define an **effective Hamiltonian** for our model by the equation

$$P(\{s_i\}) = \frac{1}{Z} \exp[-\mathcal{H}(\{s_i\})]. \qquad (5.5)$$

The effective Hamiltonian will in general depend on the temperature[1] and on a selection of parameters, such as spin–spin interaction. Presumably we can now work out what the corresponding probability distribution $P^{(1)}(\{\sigma_i^{(1)}\})$ is for the block variables. Indeed, since we know from (5.1) how the variables $\sigma_i^{(n+1)}$ are defined in terms of the $\sigma_i^{(n)}$, we can work out the probability distribution $P^{(n+1)}(\{\sigma_i^{(n+1)}\})$ of the block variables at the $(n+1)^{\text{th}}$ stage in the renormalization, provided only that we know $P^{(n)}(\{\sigma_i^{(n)}\})$—we simply sum over the probabilities of all the configurations $\{\sigma_i^{(n)}\}$ consistent with the configuration $\{\sigma_k^{(n+1)}\}$ of the block variables. Given such a probability distribution we can define an **effective Hamiltonian** for the system after

---

[1] Throughout this chapter we write our Boltzmann factors in the form $\exp(-\mathcal{H})$, and refer to $\mathcal{H}$ as the effective Hamiltonian. For the starting system, the temperature dependence of $\mathcal{H}$ is very simple—we can write $\mathcal{H} = \beta H$ where H is independent of temperature. However, when we renormalize the effective Hamiltonian its temperature dependence can become much more involved than this, so one should try to avoid carrying around any prejudices about what form the temperature dependence might take.

n iterations of the renormalization transformation by analogy with equation (5.5):

$$P^{(n)}(\{\sigma_i^{(n)}\}) = \frac{1}{Z^{(n)}} \exp[-\mathcal{H}^{(n)}(\{\sigma_i^{(n)}\})], \qquad (5.6)$$

where

$$Z^{(n)} \equiv \sum_{\{\sigma_i^{(n)}\}} \exp[-\mathcal{H}^{(n)}(\{\sigma_i^{(n)}\})]. \qquad (5.7)$$

This ties down $\mathcal{H}^{(n)}$ to within an additive constant. We fix this constant by specifying some normalization condition on $\mathcal{H}^{(n)}$, such as

$$\mathcal{H}^{(n)}(\{\sigma_i^{(n)}\})\Big|_{\sigma_i^{(n)}=0} = 0. \qquad (5.8)$$

Many other such conditions are possible. This one will prove particularly convenient for the calculation of the free energy in §5.6.3, but other choices may prove more convenient for certain systems. For the Ising model, for instance, it goes against the grain to set all the block variables $\{\sigma_i^{(n)}\}$ to zero, since they normally take only the values ± 1. Instead, it is common to adopt the condition

$$\sum_{\{\sigma_i^{(n)}\}} \mathcal{H}^{(n)}(\{\sigma_i^{(n)}\}) = 0. \qquad (5.9)$$

Any condition will do, as long as it ties down the constant term in the effective Hamiltonian unambiguously.

With $\mathcal{H}^{(n)}$ defined in this way, it is clear that the thermal expectation of any function X of the variables $\sigma_i^{(n+1)}$ will yield the same value, whether we evaluate it using $\mathcal{H}^{(n+1)}$, or $\mathcal{H}^{(n)}$. That is, if we evaluate the quantity

$$\langle X \rangle = \frac{1}{Z^{(n+1)}} \sum_{\{\sigma^{(n+1)}\}} X(\{\sigma_i^{(n+1)}\}) \exp[-\mathcal{H}^{(n+1)}(\{\sigma_i^{(n+1)}\})], \qquad (5.10)$$

we will get exactly the same answer as if we take the sum over the larger set of states of all possible values of the variables $\sigma_i^{(n)}$, and work out

$$\langle X \rangle = \frac{1}{Z^{(n)}} \sum_{\{\sigma_i^{(n)}\}} X(\{\sigma_i^{(n+1)}\}) \exp[-\mathcal{H}^{(n)}(\{\sigma_i^{(n)}\})]. \qquad (5.11)$$

Thus, when we write down a thermal expectation value there will be no need for us to indicate which effective Hamiltonian we are using—all thermal expectations are equivalent. This point is crucial to the arguments of §5.4.

When performing a true real-space renormalization calculation, we also impose one more condition on the effective Hamiltonian $\mathcal{H}^{(n+1)}$—we stipulate that it must take the same functional form as $\mathcal{H}^{(n)}$, so that the model

5.3 The renormalization of the Hamiltonian

is exactly the same at every stage, except for a change in the parameters appearing in the effective Hamiltonian. "But this is not possible", you will be saying. "How can we satisfy this condition when the whole system and its effective Hamiltonian are already completely specified?" You are completely right—this last condition is, in almost all cases, impossible to meet exactly. There do exist models for which it can be done (e.g., the hierarchical model—see Dyson 1969, 1971, Baker 1972, Collet and Eckmann 1978), but in most cases the best we can do is to approximate the true renormalized effective Hamiltonian $\mathcal{H}^{(n+1)}$ as we defined it above, with a function of the same form as $\mathcal{H}^{(n)}$. In doing this, we choose the parameters appearing in the function to make the approximation as accurate as possible, but a degree of approximation is inevitable.

Nonetheless, if we are to proceed, we must by one means or another force our renormalized effective Hamiltonian into the same functional form as the original so that, for example, the renormalized (zero-field) Ising Hamiltonian is just another Ising Hamiltonian but with a different temperature. The renormalization of the Hamiltonian then is some mapping which takes $\mathcal{H}^{(n)}$ and maps it onto $\mathcal{H}^{(n+1)}$:

$$\mathcal{H}^{(n+1)} = Q(\mathcal{H}^{(n)}). \tag{5.12}$$

The only effect of the transformation Q is to change the values of the parameters appearing in $\mathcal{H}$. The new values of these parameters depend only on the old values, and on nothing else, though often this dependence will involve complicated non-linear functions.

Equation (5.12) is perhaps a little misleading; it relates the *functions* $\mathcal{H}(\{\sigma_i\})$ before and after the renormalization, though all that actually changes from one stage to another is the value of the parameters. Let us instead adopt another notation. Suppose we call the parameters in the effective Hamiltonian $H_1, H_2 \ldots$. Then we can make a vector $\mathbf{H}$ out of them, and write the renormalization as some (probably non-linear) operation $\mathbf{R}$, acting on this vector:

$$\mathbf{H}^{(n+1)} = \mathbf{R}(\mathbf{H}^{(n)}). \tag{5.13}$$

Amongst the parameters H_n we must, in general, include the temperature—the effective Hamiltonian is temperature dependent.

It is often helpful to think of a space of Hamiltonians—a finite-dimensional space in which a point with position vector $\mathbf{H}$ represents the effective Hamiltonian with the corresponding values of the parameters. In this space the parameters H_n take on the rôle of coordinates. In the case of the Ising model in zero external field, for example, we would have a one-dimensional space, the only free parameter in the Hamiltonian being the quantity $\beta \mathcal{J}$. For the Ising model in non-zero field on the other hand, we would have a two-dimensional space, the two coordinates being $\beta \mathcal{J}$ and the field B (or any two independent functions of these). Our renormalization transformation $\mathbf{R}$

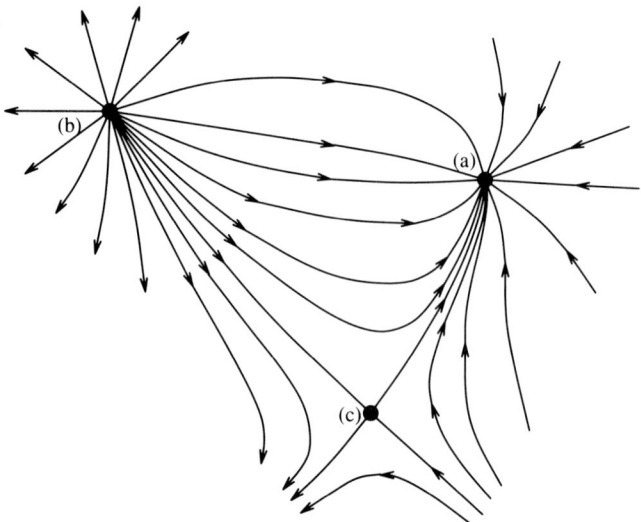

Figure 5.4 Illustration of the lines of flow about the three types of fixed point: (a) an attractive fixed point, (b) a repulsive one and (c) a mixed fixed point.

moves us around in this space by discrete jumps. If we start at some initial point $\mathbf{H}^{(0)}$ and iterate equation (5.13) we will be propelled in steps along some trajectory through our space. We illustrate this motion by diagrams such as Figure 5.4, in which, for clarity, successive dots have been joined to form smooth curves. At each step the values of the parameters appearing in the effective Hamiltonian are replaced by new ones, which we calculate from the old values according to equation (5.13).

In what follows we deal almost exclusively with the vectors $\mathbf{H}$ rather than the effective Hamiltonian functions $\mathcal{H}$, and frequently refer to the vectors themselves as effective Hamiltonians.

5.3.1 Fixed points

In general, there will be **fixed points** in our space. A fixed point is a point that is mapped onto itself by the transformation $\mathbf{R}$:

$$\mathbf{H} = \mathbf{R}(\mathbf{H}). \tag{5.14}$$

We may classify fixed points as being 'attractive', 'repulsive' or 'mixed'. An **attractive fixed point**, as the name suggests, is one towards which all effective Hamiltonians in some neighbourhood of the point tend upon iteration of equation (5.13). And a **repulsive fixed point** is one towards which none of the effective Hamiltonians in the neighbourhood of the point tend upon iteration. (Or, if you like, they are all repelled.) This does

5.3 The renormalization of the Hamiltonian

not take care of all our fixed points however. In fact, as it will turn out, all the really interesting fixed points are **mixed**, i.e., neither attractive nor repulsive. Such a point will attract an effective Hamiltonian which is close to it in one direction, but repel one close to it in another. Figure 5.4 depicts the trajectories of effective Hamiltonians in the vicinity of the three types of fixed point in a two-dimensional Hamiltonian space.

We can determine the nature of the fixed points of a given system as follows. First we solve equation (5.14) to find the stationary points. (This may in fact be no mean feat, but let us assume for the moment that it can be done.) Then suppose we are interested in exploring the nature of a fixed point at $\mathbf{H} = \mathbf{H}_0$. We write the effect of $\mathbf{R}$ on an effective Hamiltonian close to $\mathbf{H}_0$ as

$$\mathbf{H}_0 + \delta\mathbf{H}' = \mathbf{R}(\mathbf{H}_0 + \delta\mathbf{H}) = \mathbf{R}(\mathbf{H}_0) + \mathbf{M}\delta\mathbf{H} + O(\delta\mathbf{H}^2), \qquad (5.15)$$

where we have expanded $\mathbf{R}$ in a Taylor series about $\mathbf{H}_0$. $\mathbf{M}$ is an ordinary matrix whose elements are given by

$$M_{ij} = \frac{\partial R_i(\mathbf{H})}{\partial H_j}\bigg|_{\mathbf{H}=\mathbf{H}_0}. \qquad (5.16)$$

Since $\mathbf{H}_0$ is a stationary point of $\mathbf{R}$ we conclude that

$$\delta\mathbf{H}' = \mathbf{M}\delta\mathbf{H} + O(\delta\mathbf{H}^2). \qquad (5.17)$$

Thus, as we come infinitesimally close to the fixed point at $\mathbf{H}_0$, our trajectory is given completely by a single matrix, or, if you like, by its eigenvalues and eigenvectors. If $\mathbf{M}$ has an eigenvalue $\lambda_R > 1$, the fixed point will be repulsive in the direction of the corresponding eigenvector. And similarly, if it has an eigenvalue $\lambda_A < 1$, the fixed point will be attractive in the corresponding direction.[2] So to identify $\mathbf{H}_0$ as attractive, repulsive or mixed, we need only solve for the eigenvalues of $\mathbf{M}$.

For any system which undergoes a single phase transition, the identities of two of the fixed points are clear. One, the **high-temperature fixed point**, is the effective Hamiltonian for the system as $T \to \infty$. In this case all our variables assume random values and are completely uncorrelated. When we block such a system, it is clear that, for any sensible definition (5.1), the block variables will still be random and uncorrelated so their probability distribution does not change. Therefore the effective Hamiltonian cannot have changed either and we are at a fixed point. Another fixed point is the **low-temperature fixed point**, which is the effective Hamiltonian for the

---

[2] In the event that one of the eigenvalues is exactly one, the fixed point is said to be **marginal** in the direction of the corresponding eigenvector. In this case we need to consider the next term in the Taylor expansion of the renormalization transformation in order to decide the true nature of the fixed point.

system when $T \to 0$. In this case the system is in its ground state, in which complete order reigns—for example, in a ferromagnet all spins are aligned.[3] When we block this state the new block variables are also perfectly ordered, so the probability distribution, and hence the effective Hamiltonian, is the same before and after the renormalization and we are at a fixed point.

Both of these fixed points are attractive. To see this consider first any starting Hamiltonian for a temperature above T_c. When we iterate equation (5.13), the correlation length shrinks by a factor of b at each iteration, because in renormalizing the lattice we divide every length, including the correlation length, by b:

$$\xi' = \frac{\xi}{b}. \tag{5.18}$$

So as $n \to \infty$, ξ tends to zero, leaving the block variables completely uncorrelated. These block variables now have the same probability distribution as those at the high-temperature fixed point, and so we conclude that the effective Hamiltonian must also be that of the high-temperature fixed point. So every point corresponding to $T > T_c$ is attracted (eventually) to the high-temperature fixed point.

Now let us take a starting point below T_c. Below T_c we will have some non-zero average value for our spins, a 'spontaneous magnetization'. As we block these spins into larger and larger block variables, we are averaging over more and more of them until, eventually, as $n \to \infty$ all the block variables will become equal.[4] This gives them the same probability distribution as the variables at the low-temperature fixed point, and so, we conclude, the effective Hamiltonian must also be that of the low-temperature fixed point. So every point corresponding to $T < T_c$ is attracted (eventually) to the low-temperature fixed point.

And what happens at T_c? Well, the effective Hamiltonians that flow towards the high-temperature fixed point, and those that flow towards the low, must be separated by some plane (or line) across our Hamiltonian space which we call the **critical surface**.[5] Each point on this surface corresponds to an effective Hamiltonian which is at the critical temperature appropriate

---

[3] As in Chapter 3, we exclude from consideration systems without a well-ordered ground state, such as a frustrated anti-ferromagnet.

[4] This argument does not work if we define our block variables by the decimation procedure described in §5.1. This partly explains why decimation gives poor results for the Ising model.

[5] It could turn out of course that the critical surface is not simply connected. Indeed there are much worse pathologies it *could* display even than this. An example with which the reader may be familiar is that of a Hamiltonian depending on two parameters x and y, which transform under renormalization according to the equations

$$x' = x^2 + y^2 + a,$$
$$y' = 2xy + b.$$

Considered in the complex plane of the variable $z \equiv x + iy$, the critical surface for such a system is the Julia set characterized by the constant $c \equiv a + ib$, which is in general fractal.

5.3 The renormalization of the Hamiltonian

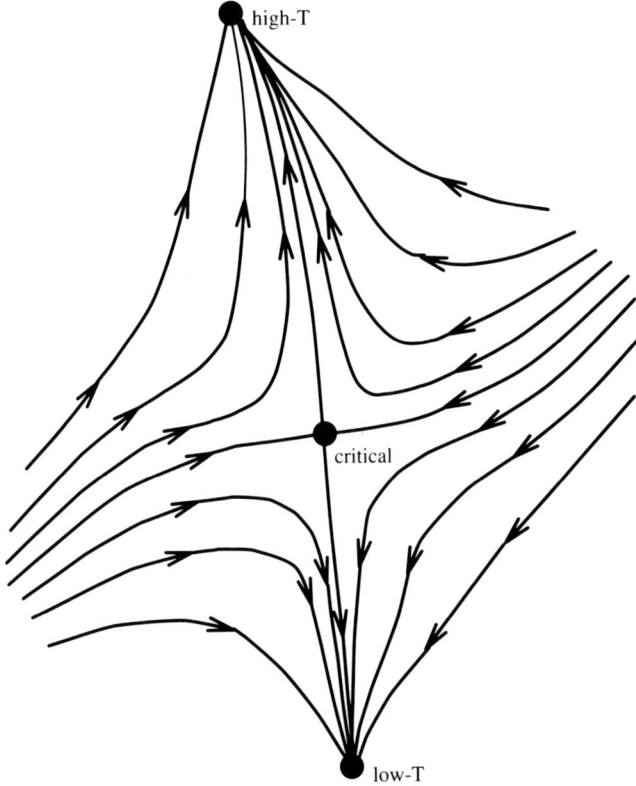

Figure 5.5 A portion of the Hamiltonian space of our system, showing the high- and low-temperature fixed points (both attractive) and the critical fixed point (mixed), and the lines of flow around them.

to the values of the other parameters at that point. One must be careful not to think of the critical surface as an isothermal surface, $T = T_c$. In general, the critical temperature will be different at different points on the critical surface.

Now imagine what happens if we choose a point on the critical surface as our starting Hamiltonian and iterate equation (5.13). Neither of the arguments given above applies in this case, and **H** tends neither to the high- nor to the low-temperature fixed point. Instead it stays within the critical surface. But within this restriction there are still a number of possibilities:

(i) as $n \to \infty$, **H** tends to a finite limit, $\mathbf{H}^*$;
(ii) **H** disappears off to infinity as $n \to \infty$;

In this book, however, we ignore such unpleasant possibilities and treat the case just of a single, simple critical surface across our Hamiltonian space.

(iii) **H** falls into a closed cycle of values, moving round and round without ever tending to a limit;

(iv) **H** moves about chaotically on the critical surface without ever reaching a limit.

The last two possibilities occur in the study of 'critical dynamics' using real-space renormalization, but this topic lies beyond the scope of this book. For the equilibrium systems we are concerned with here, the behaviour of the system always falls into one or other of categories (i) and (ii). Category (ii), however, is not really distinct from category (i). Both have a fixed point which is attractive within the critical surface, but in (ii) it is at infinity. By choosing a new set of coordinates for our Hamiltonian space, or changing parameters appearing in the definition of the block variables (such as the parameter $A^{(n)}$ appearing in equation (5.3)), we can always make this fixed point finite. (For example, we could simply invert the coordinate or coordinates that measure our progress across the critical surface.) Let us assume this has been done. We have then identified a new fixed point on the critical surface at $\mathbf{H}^*$. This is the **critical fixed point**. Within the critical surface it is attractive.[6] Outside it, on the other hand, we know that all effective Hamiltonians are attracted either to the high- or to the low-temperature fixed points. So, we reason, the critical fixed point must be a mixed point—repulsive along the direction out of the critical surface, and attractive along all the directions within the critical surface. The matrix **M** corresponding to this point has only one eigenvalue which is greater than one. The critical fixed point is by far the most interesting of the fixed points, since its properties tell us directly about the critical behaviour of our system.

5.3.2 The calculation of ν

Take a look at Figure 5.5, and imagine what would happen if we were to take our starting Hamiltonian and vary the temperature while keeping all other parameters fixed (spin–spin interaction for example). This would move us along a line—usually referred to as a **physical trajectory**—which, as we approached and passed T_c, would approach and then intersect with the critical surface. If we were to take one of these starting Hamiltonians with the temperature just fractionally above T_c and the external field zero, and iterate equation (5.13), our effective Hamiltonian would pass along a trajectory that skims close to the critical surface, lingers for a while near the critical fixed point (where, since **H** is nearly equal to $\mathbf{H}^*$, it is very little affected by **R**), and then, eventually, turns away from the critical surface and heads out towards the high-temperature fixed point. We can now ask, when, roughly,

---

[6] This attraction within the critical surface can still take a number of different forms. For example, the decimation treatment of the two-dimensional Ising model considered in Problem 5.2 gives rise to a matrix **M** for the critical fixed point which has a negative eigenvalue. This means that the trajectory of a Hamiltonian starting on the critical surface dots back and forth from one side of the critical fixed point to the other, rather than tending smoothly towards it.

5.3 The renormalization of the Hamiltonian

does this 'turning away' from the critical surface happen? There are two answers to this question. One is a simple physical answer: as we approach the critical temperature, the correlation length diverges (see §1.3) and so the regime close to the critical surface is one in which the correlation length is large by comparison with the lattice parameter. The regime out towards the high-temperature fixed point on the other hand is one in which the site variables are random and almost uncorrelated. In other words the correlation length is very small by comparison with the lattice parameter. The crossover between these regimes, the 'turning point' of the trajectory, is where the correlation length and the lattice parameter are roughly the same size. From equation (5.18) we know that the correlation length after n iterations of the renormalization transformation will be

$$\xi^{(n)} = \frac{\xi}{b^n}, \tag{5.19}$$

where by ξ we mean $\xi^{(0)}$. So we can write the condition for the turning point as

$$\frac{\xi}{ab^n} = u, \tag{5.20}$$

where a is the lattice parameter and u is a number of order unity. (Its exact value turns out to be unimportant in the following analyses.)

However, there is another way of looking at this problem; there is also a mathematical answer we can give to the question of where the turning point is. For simplicity we will consider again a system with only a two-dimensional Hamiltonian space. (The external field B we will keep at zero.) Suppose then that we have found the critical fixed point and expanded the transformation $\mathbf{R}$ about it as we did above to find the matrix $\mathbf{M}$. Since the critical fixed point is a mixed point, one of $\mathbf{M}$'s eigenvalues, which we will denote by λ_R, must be greater than unity ('repulsive'), and the other λ_A must be less than unity ('attractive'). We also find the corresponding normalized eigenvectors, $\mathbf{H}_R$ and $\mathbf{H}_A$. Now if we take an effective Hamiltonian very close to the critical fixed point, we can express its difference from $\mathbf{H}^*$ as a linear combination $\mathbf{H} - \mathbf{H}^* = x_A \mathbf{H}_A + x_R \mathbf{H}_R$ of the two eigenvectors. Then, as long as $\mathbf{H}$ remains in the vicinity of $\mathbf{H}^*$, we can write its value m iterations of equation (5.13) later as

$$\mathbf{H} = \mathbf{H}^* + x_A \lambda_A^m \mathbf{H}_A + x_R \lambda_R^m \mathbf{H}_R. \tag{5.21}$$

(Once $\mathbf{H} - \mathbf{H}^*$ becomes large this equation is no longer valid, but let us work for the moment in the region where it is small.) If we start with our Hamiltonian not in the immediate vicinity of $\mathbf{H}^*$, but at some other point very close to the critical surface, then we will have to iterate our renormalization transformation a certain number of times r, say, to bring $\mathbf{H}$ into the region where equation (5.21) is valid to the degree of accuracy we desire. In this case $m = n - r$ (n being, as usual, our total number of iterations) and substituting this into equation (5.21) we get

$$\mathbf{H}^{(n)} = \mathbf{H}^* + x_A \lambda_A^{n-r} \mathbf{H}_A + x_R \lambda_R^{n-r} \mathbf{H}_R. \tag{5.22}$$

As we let our starting temperature T approach the critical temperature,[7] holding all the other parameters in the problem constant, x_R becomes smaller and smaller until, at exactly T_c, it vanishes, with the result that $\mathbf{H} \to \mathbf{H}^*$ as $n \to \infty$. So we should be able to perform a Taylor expansion of x_R about T_c and, assuming we are not so unlucky that the coefficient of $T - T_c$ turns out to be exactly zero, it will always be a good approximation to write $x_R = y_R(T - T_c)$ for T sufficiently close to T_c. x_A on the other hand is independent of $T - T_c$ in this limit, so we get

$$\mathbf{H}^{(n)} = \mathbf{H}^* + x_A \lambda_A^{n-r} \mathbf{H}_A + y_R(T - T_c) \lambda_R^{n-r} \mathbf{H}_R. \tag{5.23}$$

The turning point, where the trajectory of $\mathbf{H}$ swings away from the critical surface and out towards the high-temperature fixed point, is thus given by an expression of the form[8]

$$y_R |T - T_c| \lambda_R^{n-r} = v, \tag{5.24}$$

where v is another number of order unity. (Again, it doesn't matter what this number is.) Eliminating n between this equation and equation (5.20) we find

$$\xi = au \left(\frac{v\lambda_R^r}{y_R}\right)^\nu |T - T_c|^{-\nu}, \tag{5.25}$$

with

$$\nu = \frac{\log b}{\log \lambda_R}. \tag{5.26}$$

Equation (5.25) shows that ν is the critical exponent governing the divergence of the correlation length with temperature as we approach T_c from above. As we can see, its value is independent of all the unknown factors in the calculation (u, v, r, y_R etc.) and is given very simply in terms of the scaling factor b of the renormalization transformation, and the largest eigenvalue λ_R of the matrix $\mathbf{M}$ associated with the critical fixed point.

We can now repeat the entire argument of this section for a temperature just fractionally below the critical temperature of our starting Hamiltonian.

---

[7] Bear in mind that T is one of the parameters that gets renormalized—it is different after each iteration. We use the letter T exclusively to denote the temperature of the initial unrenormalized system, i.e., $T \equiv T^{(0)}$. We will never in fact need to refer to $T^{(n)}$, the temperature after n iterations of equation (5.13).

[8] By introducing the modulus signs $|\ldots|$ here, we allow for the generalization of this argument to the case $T < T_c$ (see the end of this section). For the case $T > T_c$ they are, of course, unnecessary.

In this case the turning point will be where the trajectory of the effective Hamiltonian turns away from the critical surface and heads out towards the low-temperature fixed point, and the exponent we will calculate will be ν', which governs the divergence of the correlation length with temperature as we approach T_c from below. The argument, however, follows through exactly as above, and though some quantities (u, v, x_A and so on) may take different values, the final result is the same: $\nu' = \log b / \log \lambda_R$. And b and λ_R still do have the same values, so we establish another very useful result:

$$\nu' = \nu. \tag{5.27}$$

5.4 The renormalization of B, M, χ and G_c

Hitherto, all our results have been very general. They have not assumed any particular form for the definition (5.1) of the block variables, and so, in theory, our results for ν and ν' apply to any system for which we can define 'sensible' block variables (see §5.2). The calculation of the remaining five critical exponents, however, depends on the details of how the block variables are defined and so to progress any further we must choose one particular form.[9] The form we will use to illustrate the argument is the linear form (5.3). This form we choose for its simplicity of manipulation, and because historically it was one of the first definitions to be used in this type of calculation.[10] We will give examples of the use of other (non-linear) definitions such as the majority rule scheme later in the chapter when we discuss the application of real-space renormalization techniques to various well-known models.

Having chosen (5.3) as our definition of the block variables, it still remains for us to fix the constant $A^{(n)}$. This must be done carefully if our calculations are to yield any results. Specifically, we shall find it necessary to choose $A^{(n)}$ so that the mean-square value of any one variable $\sigma_i^{(n)}$ in our system is finite. This is not a trivial point, because in many of our subsequent arguments we will consider how our system behaves in the limit $n \to \infty$ in which we have iterated equation (5.13) infinitely many times. And our arguments would hold no water at all if it were possible that all the block variables became infinite, or zero in this limit. It is conventional to write the normalization constant in terms of a new variable ω as $A^{(n)} = b^{-d\omega}$ so that

$$\sigma_k^{(n+1)} = \frac{1}{b^{d\omega}} \sum_{i \in S_k} \sigma_i^{(n)}. \tag{5.28}$$

---

[9] But see §§5.4.2 and 5.7.
[10] It does have disadvantages over the non-linear schemes however. We will come to these soon.

The quantity w is allowed to vary with temperature and with the other parameters of the problem. In addition to keeping the mean-square value of each individual spin finite, we would also like w to be a smoothly varying function of these parameters, and both of these conditions can be satisfied if we choose w such that the mean-square value $\langle \sigma_i^{(n)2} \rangle$ of each spin is a constant of the renormalization transformation. Let this then be the defining condition on w. It is not the only possible definition, but one can show that all definitions satisfying the conditions above give the same answers ultimately. We have chosen this one for its simplicity.

5.4.1 The value of w

It is in general not easy to calculate the value of the number w appearing in equation (5.28). However, we can calculate its value at the high- and low-temperature fixed points with relative ease.

At the high-temperature point our block variables are completely uncorrelated. They are random. This means that as we add them together to form larger and larger block variables, the distribution of the latter will, by the central limit theorem, approach closer and closer to a Gaussian, whose mean-square value Δ is proportional to the number of variables in each block. As $n \to \infty$ the distribution becomes the product of factors of the form

$$p^{(n)}(\sigma_i^{(n)}) = \frac{1}{\sqrt{2\pi\Delta}} \exp\left(-\frac{\sigma_i^{(n)2}}{2\Delta}\right). \tag{5.29}$$

Now we block these variables in groups of b^d according to equation (5.28). The mean-square value of the sum $\sum_{i \in S_k} \sigma_i^{(n)}$ is $b^d \Delta$. So the mean-square value of $\sigma^{(n+1)}$ is $b^{d(1-2w_h)}\Delta$, where w_h is the value of w at the high-temperature fixed point. Hence in order to make the mean-square value of each block variable at the $(n+1)^{\text{th}}$ stage exactly the same as that at the nth, we need to put $w_h = \frac{1}{2}$. This value is independent of the form of our effective Hamiltonian, so it is the same for all models.

At the low-temperature fixed point of a ferromagnetic model, all our spin variables will be 'lined up'—they will all assume the same value. Clearly then, if we block them in groups of b^d, we must also divide by b^d to render the new mean-square value the same as the old. Thus the value of w at the low-temperature fixed point is $w_l = 1$.[11]

At all points between $T = 0$ and $T = \infty$ our block variables may be expected to be more aligned than at the high-temperature point, but less so than at the low-temperature one, so w will presumably be in the range $\frac{1}{2} \leq w \leq 1$.

---

[11] For frustrated anti-ferromagnets this is not the case, but these models are anyway beyond the scope of this book.

5.4 The renormalization of B, M, χ and G_c

The only other value of ω which will be important to us is the value ω_c at the critical fixed point, which will appear in many of our formulae for critical exponents. Unfortunately, ω_c does *not* take a universal value, and no method exists for calculating it within the renormalization scheme (see Bell and Wilson 1974). Where its value is known, e.g. in the Gaussian model for which it takes the value $\omega_c = \frac{1}{2} + 1/d$ (see Parisi 1988), the value has been found from calculations using techniques other than real-space renormalization. This difficulty with ω_c is a problem peculiar to linear renormalization schemes and we discuss it further in the next section.

5.4.2 Non-zero external field

So far we have only discussed a system in zero external field B. When B is non-zero, our space of effective Hamiltonians is one dimension larger—there is an extra axis marked with the possible values of the external field, and an extra term $B^{(n)} \sum_i \sigma_i^{(n)}$ appears in the effective Hamiltonian $\mathcal{H}$. The reader will recall that we required of our renormalized effective Hamiltonian that it assume the exact same functional form as the original, the only changes allowed being changes in the values of the parameters appearing in it (the components of the vector $\mathbf{H}$). Assuming we stick to this rule, there must then appear in $\mathcal{H}^{(n+1)}$ a term $B^{(n+1)} \sum_i \sigma_i^{(n+1)}$, which, using equation (5.28), we can write as

$$B^{(n+1)} \sum_i \sigma_i^{(n+1)} = \frac{B^{(n+1)}}{b^{d\omega}} \sum_i \sigma_i^{(n)}. \tag{5.30}$$

Comparing this with the corresponding term in $\mathcal{H}^{(n)}$ we now conclude that we will get the correct probability distribution of the block variables if we put

$$B^{(n+1)} = b^{d\omega} B^{(n)}. \tag{5.31}$$

Thus, if we allow the external field to be non-zero, we simply increase the dimension of our space by one, and add another **magnetic eigenvalue** λ_B to the matrix $\mathbf{M}$ which takes the value

$$\lambda_B = b^{d\omega}, \tag{5.32}$$

which is always greater than one. Notice however, that the eigenvector corresponding to this eigenvalue, which lies in the B-direction in our Hamiltonian space, does not 'mix' with the other eigenvectors. In other words, upon renormalization the new value of the magnetic field depends only on the old one and not the other parameters like temperature.

This result that the renormalized magnetic field is independent of all other parameters in $\mathcal{H}$ is a particular example of a more general result: 'odd' and 'even' parameters do not mix when we renormalize.

By an **odd parameter**, we mean one which multiplies some function of the block variables which changes sign if we change the sign of each variable, such as the sum $\sum_i \sigma_i^{(n)}$ multiplied by B above. An **even parameter** on the other hand is one multiplying a combination of block variables which does not change sign in this way. Examples are the spin–spin coupling in the Gaussian and Ising models. Now if, like the Gaussian and Ising models in zero field, the effective Hamiltonian is invariant under a change of sign of all block variables (i.e., all the parameters are even), then renormalization cannot produce any odd parameters if the renormalized effective Hamiltonian is to have the same form as the original one. Similarly, if we had an effective Hamiltonian composed only of odd functions of the block variables[12] (i.e., only odd parameters) we could not produce any even ones by renormalizing. In the general case therefore, where both types of parameter are present (e.g., the Ising model in non-zero external field) the even and odd parameters transform only amongst themselves and not from even to odd, or *vice versa*.

We will show in §5.5.1 that the exponent η is related to ω_c by

$$\eta = d + 2 - 2d\omega_c. \tag{5.33}$$

If we use (5.32) to eliminate $d\omega_c$ from this expression, we get

$$\eta = d + 2 - 2\frac{\log \lambda_B}{\log b}, \tag{5.34}$$

where λ_B is now specifically the eigenvalue of $\mathbf{M}$ at $\mathbf{H}^*$. It turns out that this formula is more general that the model we have used to deduce it. It applies for any model and renormalization scheme, if λ_B is taken to be the largest eigenvalue of the matrix describing the transformation of the odd parameters amongst themselves (see Niemeijer and van Leeuwen 1976). (In fact λ_B will be the only eigenvalue greater than one.) Thus, by calculating the largest eigenvectors of the matrices for the linearized transformations of the even and odd parameters near to the critical fixed point, we can calculate two critical exponents. As discussed in Chapter 1, and demonstrated for the present model in the following sections, the rest of the exponents are then related to these two by simple scaling laws.

For the linear renormalization scheme we have been discussing, we cannot find an answer for η in this way. The reason should be clear from equation (5.33). η is given in terms of ω_c, which as we have said cannot be calculated within the renormalization scheme. This is a serious problem with linear renormalization schemes, and it is the main reason that they have fallen into disuse in recent years. Non-linear schemes such as decimation or majority rule do not contain unknown parameters like ω_c, and so allow us to calculate the magnetic eigenvalue as well as the 'thermal eigenvalue' λ_R of §5.3.2.

---

[12] An example is the triple-spin interaction model of Baxter and Wu (1973). We give little consideration to models of this type however, since they are clearly non-isotropic.

5.4 *The renormalization of B, M, χ and G_c*   131

It is not time to abandon the linear renormalization scheme yet however. Even without knowing the value of ω_c, many interesting results about the critical exponents of our system can be deduced. In particular, we can prove the four scaling laws.

5.4.3 The renormalization of M, χ and G_c

We have already seen, in §5.3.1, how the correlation length ξ changes when we renormalize the lattice—it has the dimensions of a length and so it must scale, like all other lengths on the lattice, by a factor of b:

$$\xi^{(n+1)} = \frac{\xi^{(n)}}{b}. \tag{5.35}$$

Knowing how the block variables $\sigma_i^{(n+1)}$ are related to the $\sigma_i^{(n)}$, we can also deduce what happens to many of the other physically interesting variables. To begin with we can define the magnetization after n renormalizations of the lattice:

$$m^{(n)} = \langle \sigma_k^{(n)} \rangle, \tag{5.36}$$

where, assuming we have a translationally invariant system, we can evaluate the thermal average on any site k. Now using equation (5.28) we can write

$$m^{(n+1)} = \langle \sigma_k^{(n+1)} \rangle = \frac{1}{b^{d\omega}} \sum_{i \in \mathcal{S}_k} \langle \sigma_i^{(n)} \rangle. \tag{5.37}$$

But for a translationally invariant system each of the b^d terms in this sum is the same and equal to $m^{(n)}$. So the sum may be replaced with $b^d m^{(n)}$ and

$$m^{(n+1)} = b^{d(1-\omega)} m^{(n)}. \tag{5.38}$$

Next, if we define the connected correlation of our block variables by

$$G_c^{(n)}(i,j) = \langle \sigma_i^{(n)} \sigma_j^{(n)} \rangle_c, \tag{5.39}$$

then

$$G_c^{(n+1)}(i,j) = \frac{1}{b^{2d\omega}} \sum_{\substack{k \in \mathcal{S}_i \\ l \in \mathcal{S}_j}} \langle \sigma_k^{(n)} \sigma_l^{(n)} \rangle_c. \tag{5.40}$$

In general this expression is not easy to evaluate, but in the special case in which the blocks i and j are far apart (by comparison with the size of an individual block), all the correlations $\langle \sigma_k^{(n)} \sigma_l^{(n)} \rangle_c$ can be assumed to be equal. Since there are b^{2d} of them in the sum, we can then just replace the sum with $b^{2d} G_c^{(n)}(k,l)$, and get

$$G_c^{(n+1)}(i,j) = b^{2d(1-\omega)} G_c^{(n)}(k,l), \qquad k \in \mathcal{S}_i, l \in \mathcal{S}_j, \tag{5.41}$$

where k labels any site in $\mathcal{S}_i$ and l any site in $\mathcal{S}_j$. Note that the sites k and l, which are on the unrenormalized lattice, are b times as far apart as the sites i and j, since all lengths are reduced by that factor when we renormalize. To emphasize this it helps to write G_c as a function of the separation x of the two sites i and j, which is legitimate for a translationally invariant system. In that case

$$G_c^{(n+1)}(x) = b^{2d(1-\omega)} G_c^{(n)}(bx), \qquad x \to \infty. \qquad (5.42)$$

Finally, by analogy with equation (2.28), we can define the susceptibility of our renormalized model by

$$\chi^{(n)} = \sum_i \langle \sigma_0^{(n)} \sigma_i^{(n)} \rangle_c = \frac{1}{N^{(n)}} \sum_{ij} \langle \sigma_i^{(n)} \sigma_j^{(n)} \rangle_c, \qquad (5.43)$$

where $N^{(n)}$ is the number of sites on the lattice after n renormalizations. Then, using equations (5.28) and (5.40), and the fact that $N^{(n+1)} = N^{(n)}/b^d$ (see §5.1), it is straightforward to show that

$$\chi^{(n+1)} = b^{d(1-2\omega)} \chi^{(n)}. \qquad (5.44)$$

5.4.4 Critical exponents for the renormalized model

The renormalized quantities that we have defined above—$\xi^{(n)}$, $m^{(n)}$, $\chi^{(n)}$ and $G_c^{(n)}$—are the genuine physical values of the correlation length, magnetization, connected two-point correlation function and susceptibility for a system of variables $\sigma_i^{(n)}$ governed by a Hamiltonian $\mathcal{H}^{(n)}$, with an applied external field of magnitude $B^{(n)}$ coupling linearly to the variables. In general, this system bears little resemblance to the real system in which we are interested, the one governed by the Hamiltonian $\mathcal{H}^{(0)}$, because the renormalization process changes the values of all the parameters appearing in the Hamiltonian and thus changes the behaviour of the system. However, it is simple to demonstrate that the critical exponents for the two systems are the same. We give the argument only for one exponent, δ, but its generalization to the others is straightforward.[13]

δ describes how the magnetization varies with the external field B as $B \to 0$ at the critical temperature (see §1.1.1):

$$m \sim B^{1/\delta}; \qquad B \to 0, \quad T = T_c. \qquad (5.45)$$

---

[13] The only exponents for which this argument is not applicable are ν, for which the arguments of §5.3.2 demonstrate that the value is universal, and α, the exponent governing the specific heat. In §5.6.5 we calculate α from the true Hamiltonian $\mathcal{H}^{(0)}$, by a method different to that used for the other exponents, which does not rely an analogous argument.

5.5 The critical exponents for $T = T_c$

Now suppose we calculate the value of δ at the nth stage in the renormalization:

$$m^{(n)} \sim B^{(n)\,1/\delta}. \tag{5.46}$$

Using equations (5.31) and (5.38) we can write

$$B^{(n)} = B^{(0)} \prod_{i=1}^{n} b^{d\omega^{(i)}}, \qquad m^{(n)} = m^{(0)} \prod_{i=1}^{n} b^{d(1-\omega^{(i)})}, \tag{5.47}$$

where $\omega^{(i)}$ indicates the value of ω at the ith stage of the renormalization. Substituting these expressions into equation (5.46), it is clear that $m^{(0)} \sim B^{(0)\,1/\delta}$ where δ takes the same value as above. So δ is independent of n.

Similar arguments demonstrate that the same is true of the other critical exponents (see Problem 5.4). This means that we are at liberty to study our model at any stage in the renormalization to extract the critical exponents, and these will still be the correct exponents for the unrenormalized model.

5.5 The critical exponents for $T = T_c$

In §5.4 we calculated the critical exponent ν, which relates the divergence of the correlation length to the temperature. With the results derived in the last section, we can now calculate the remaining five critical exponents. Let us start with the two exponents η and δ describing the behaviour of the system at the critical temperature. As we have argued above, we can calculate the exponents for our system after any number n of iterations of the renormalization transformation, equation (5.13), and they will be the same as the exponents for the unrenormalized system. In order to calculate η and δ, we will work in the limit $n \to \infty$. (We will make a different choice when we come to work out the exponents for $T \neq T_c$.) In this limit, since our starting Hamiltonian is on the critical surface, $\mathbf{H}$ tends to the critical fixed point, and so becomes stationary—as $n \to \infty$, $\mathbf{H}^{(n+1)} = \mathbf{H}^{(n)} = \mathbf{H}^*$. The systems governed by the effective Hamiltonians at successive stages in the renormalization are then the same system.

5.5.1 The exponent η

The critical exponent η (see §1.3) describes the variation at the critical temperature of the connected two-point correlation function $G_c(x)$ with the separation x of the two points, as $x \to \infty$:

$$G_c(x) \sim x^{-(d-2+\eta)}; \qquad x \to \infty, \quad T = T_c. \tag{5.48}$$

This implies that, for any two values of x, we must have

$$\frac{G_c(x_1)}{G_c(x_2)} = \left(\frac{x_1}{x_2}\right)^{-(d-2+\eta)}. \tag{5.49}$$

In the previous section, however (equation (5.42)), we already deduced that

$$G_c^{(n+1)}(x) = b^{2d(1-\omega)} G_c^{(n)}(bx), \tag{5.50}$$

when x is very large compared with the lattice size. Now, as we have argued above, as we let $n \to \infty$, the systems governed by effective Hamiltonians at successive stages in the renormalization are the same system. Therefore the correlation functions $G_c^{(n)}$ and $G_c^{(n+1)}$ also apply to the same system—the upper index is irrelevant. So we can just substitute from this equation into equation (5.49) and get

$$b^{2d(1-\omega_c)} = b^{d-2+\eta}, \tag{5.51}$$

or

$$\eta = d + 2 - 2d\omega_c. \tag{5.52}$$

This is the result we used in §5.4.2 in the discussion of the calculation of η using the magnetic eigenvalue. Note that we have set $\omega = \omega_c$, because we are at the critical fixed point.

5.5.2 The exponent δ

By a similar trick, we can calculate δ (see §1.1.1), the exponent which describes how the magnetization varies with the external field B as $B \to 0$ at the critical temperature:

$$m \sim B^{1/\delta}; \quad B \to 0, \quad T = T_c. \tag{5.53}$$

For any two (infinitesimal) values B_1 and B_2 of the external field and the corresponding values of the magnetization, we may write

$$\frac{m_1}{m_2} = \left(\frac{B_1}{B_2}\right)^{1/\delta}. \tag{5.54}$$

When we are at the critical fixed point and $\mathbf{H}$ is stationary, equation (5.31) tells us about just such a pair of values of the external field, and equation (5.38) does the same for the magnetization. Substituting these results into equation (5.54) then gives

$$b^{d(1-\omega_c)} = b^{d\omega_c/\delta}, \tag{5.55}$$

or

$$\delta = \frac{\omega_c}{1-\omega_c} = \frac{d+2-\eta}{d-2+\eta}, \tag{5.56}$$

where we have used (5.52).

5.6 The critical exponents for $T \neq T_c$

To calculate the values of the remaining three ($T \neq T_c$) critical exponents, we study the model in the region of the 'turning point' (see §5.3.2), where $\mathbf{H}^{(n)}$ passes closest to the critical fixed point. As we let our starting temperature, the temperature appropriate to the Hamiltonian before we start renormalizing it (the 'real' temperature, if you like), tend towards T_c, the trajectory of $\mathbf{H}^{(n)}$ through the Hamiltonian space will pass closer and closer to the critical fixed point before it swings out towards either the high- or low-temperature point. At the turning point, where it is closest to $\mathbf{H}^*$, the effective Hamiltonian will move very slowly; its value is very close to $\mathbf{H}^*$ and so it is very nearly a stationary point of the renormalization transformation, equation (5.13). In this region successive values of the effective Hamiltonian can be made as close to one another as we like simply by letting the starting temperature T come sufficiently close to the critical temperature. So we may for all practical purposes consider the systems governed by $\mathcal{H}^{(n)}$ and by $\mathcal{H}^{(n+1)}$ to be the same system. In particular, they will have the same critical behaviour, by which we mean that the variation of critical quantities like ξ and m with the temperature as $T \to T_c$ will be the same for the two models, including the constants of proportionality. (We already know that the critical exponents are the same for all these systems—see §5.4.4.)

This does not mean, however, that the numerical values of measured quantities stay the same as we renormalize. Indeed we already know from equations such as (5.35) and (5.38) that they do not. We are very close to the critical fixed point, in a regime where many measurable quantities change very drastically in response to small changes in the parameters H_n appearing in the effective Hamiltonian. We are already familiar with this behaviour as far as changes in temperature are concerned (remember that the effective Hamiltonian is temperature-dependent)—it is the most fundamental idea of critical phenomena that measured quantities like the correlation length and the magnetization are very sensitive to small temperature variations near T_c. What may be less familiar is the idea that they can also be sensitive to changes in any of the other parameters in the effective Hamiltonian. Perhaps the simplest way to think of this is to recall that the critical temperature is, in general, a function of the other parameters, so that a change in one of them can change T_c and so, indirectly, bring you closer to the critical point, even though the temperature itself may stay the same.

So the idea here is that the systems governed by the nth and $(n+1)^{\text{th}}$ effective Hamiltonians display the same behaviour in the critical regime, right down to constants of proportionality. The magnetization, for instance, will satisfy a relation of the form $m = k(T - T_c)^\beta$, where the constants k and β take the same values for the two different effective Hamiltonians. In the process of renormalizing the model we also change the temperature, and usually T_c as well, and this causes changes in the values of observable quantities as described by equations (5.35), (5.38), (5.44) and (5.42). We do not know exactly how T and T_c vary as we renormalize, but we can still evaluate the remaining critical exponents. As the next subsection shows, the trick is to use ξ as a surrogate for T.

5.6.1 The exponent β

Of the exponents for $T \neq T_c$, the simplest to calculate is β, which describes how the spontaneous magnetization varies with the temperature as we approach T_c from below (see §1.1.1):

$$m \sim |T - T_c|^\beta, \qquad B = 0, \; T \to T_c \text{ from below.} \tag{5.57}$$

As we have said, we know very little about how T and T_c vary when we renormalize, but we can skip around the problem by eliminating $T - T_c$ in favour of the correlation length ξ, which varies with temperature according to

$$\xi \sim |T - T_c|^{-\nu}, \tag{5.58}$$

where ν is given by equation (5.26). So

$$m \sim \xi^{-\beta/\nu}, \qquad \xi \to \infty. \tag{5.59}$$

Now for any two values of the correlation length and the corresponding values of the magnetization, we can write

$$\frac{m_1}{m_2} = \left(\frac{\xi_1}{\xi_2}\right)^{-\beta/\nu}. \tag{5.60}$$

When we are close to the critical fixed point, however, equation (5.35) describes just such a pair of values for the correlation length, and equation (5.38) the corresponding ones for the magnetization. Substituting these into equation (5.60) then gives

$$b^{d(1-\omega_c)} = b^{\beta/\nu}, \tag{5.61}$$

or

$$\beta = \nu d(1 - \omega_c) = \tfrac{1}{2}\nu(d + \eta - 2), \tag{5.62}$$

where we have made use of equation (5.52). Also, we have put $\omega = \omega_c$ since (as we said in §5.4) we are assuming that $\omega \to \omega_c$ as $\mathbf{H} \to \mathbf{H}^*$.

5.6 The critical exponents for $T \neq T_c$

5.6.2 The exponent γ

The critical exponent γ (see §1.1.1) describes how the magnetic susceptibility χ varies with the temperature as we approach T_c from above:

$$\chi \sim |T - T_c|^{-\gamma}, \qquad T \to T_c \text{ from above.} \tag{5.63}$$

Again, we make use of equation (5.58) to write this as

$$\chi \sim \xi^{\gamma/\nu}, \tag{5.64}$$

and so

$$\frac{\chi_1}{\chi_2} = \left(\frac{\xi_1}{\xi_2}\right)^{\gamma/\nu}, \tag{5.65}$$

for any two values of the correlation length, and the corresponding values of the susceptibility. Then, using equations (5.35) and (5.44), we get

$$b^{d(1-2\omega_c)} = b^{-\gamma/\nu}, \tag{5.66}$$

or

$$\gamma = \nu d(2\omega_c - 1) = \nu(2 - \eta). \tag{5.67}$$

The exponent γ' describes how the magnetic susceptibility χ varies with the temperature as we approach T_c from below. However, the argument we have just given applies equally well below T_c as it does above. So we conclude that

$$\gamma' = \gamma. \tag{5.68}$$

5.6.3 The exponent α

Finally, we come to the critical exponent α (see §1.1.1) which describes how the specific heat diverges with temperature as we approach the critical point from above (if indeed it diverges at all, for there are plenty of models in which it does not). This exponent proves considerably harder to calculate than the others because the specific heat is not in any simple way related to the block variables $\sigma_i^{(n)}$. We proceed by first calculating the (dimensionless) free energy per site for the renormalized model after n iterations of equation (5.13):

$$\hat{f}^{(n)} \equiv \frac{\beta F^{(n)}}{N^{(n)}} = -\frac{1}{N^{(n)}} \log Z^{(n)}. \tag{5.69}$$

$N^{(n)}$ is, once again, the number of sites left in the lattice after n iterations, and the partition function $Z^{(n)}$ is as defined by equation (5.7). Luckily, we don't actually need to perform the sum over states appearing in (5.7) to find

$\hat{f}^{(n)}$. If we set all the variables $\sigma_i^{(n)}$ equal to zero in equation (5.6), then using equation (5.8) we find that

$$P^{(n)}(\{0\}) = \frac{1}{Z^{(n)}} \tag{5.70}$$

so that

$$\hat{f}^{(n)} = \frac{1}{N^{(n)}} \log[P^{(n)}(\{0\})]. \tag{5.71}$$

We can calculate the probability $P^{(n+1)}(\{0\})$ if we know $P^{(n)}(\{\sigma_i^{(n)}\})$ simply by summing the probabilities of all the configurations of the $\sigma_i^{(n)}$ which give $\sigma_i^{(n+1)} = 0$:

$$P^{(n+1)}(\{0\}) = \sum_{\sigma_i^{(n+1)}=0} P^{(n)}(\{\sigma^{(n)}\}) = \frac{1}{Z^{(n)}} \sum_{\sigma_i^{(n+1)}=0} \exp[-\mathcal{H}^{(n)}(\{\sigma_i^{(n)}\})]. \tag{5.72}$$

And so, taking logarithms and dividing by $N^{(n+1)}$, we get

$$\hat{f}^{(n+1)} = -\frac{1}{N^{(n+1)}} \log Z^{(n)} + \frac{1}{N^{(n+1)}} \log \sum_{\sigma_i^{(n+1)}=0} \exp[-\mathcal{H}^{(n)}(\{\sigma_i^{(n)}\})]$$

$$= b^d(\hat{f}^{(n)} - g^{(n)}), \tag{5.73}$$

where

$$g^{(n)} \equiv -\frac{1}{N^{(n)}} \sum_{\sigma_i^{(n+1)}=0} \exp[-\mathcal{H}^{(n)}(\{\sigma_i^{(n)}\})]. \tag{5.74}$$

Here we have made use of the fact that $N^{(n+1)} = N^{(n)}/b^d$ (see §5.1). Thus the free energy $\hat{f}^{(n)}$ does not renormalize in the simple multiplicative fashion of the quantities we have considered before (magnetization, external field, etc.). It follows a more complicated law which involves additive as well as multiplicative changes. This behaviour is characteristic of the free energy, and we will see it again in §12.2.3. Notice also that the new quantity $g^{(n)}$ which we have just defined is an intensive one—for a large system it is independent of the size of the system. We will soon see the importance of this observation.

By using equation (5.73) iteratively it is easily shown that the true free energy $f = k_B T \hat{f}^{(0)}$ of our system is given by[14]

$$f = k_B T \sum_{n=0}^{\infty} b^{-nd} g^{(n)}. \tag{5.75}$$

---

[14] We also require that $\lim_{n\to\infty} b^{-nd}\hat{f}^{(n)} = 0$. However, as we tend to the fixed point, the free energy per site tends to a constant, so this requirement is always satisfied.

5.6 The critical exponents for $T \neq T_c$

It is perhaps more physical if we write g as a function of the correlation length $\xi^{(n)} \equiv \xi/b^n$ rather than of n. In fact, since g is dimensionless, and there are only two parameters in the problem with the dimensions of length, it must be a function of $a/\xi^{(n)}$, a being the lattice parameter. Let us call this combination of variables x. Then equation (5.75) becomes

$$f = k_B T \left(\frac{\xi}{a}\right)^{-d} \sum_{n=0}^{\infty} x_n^{-d} g(x_n), \tag{5.76}$$

with $x_n \equiv a/\xi^{(n)} = ab^n/\xi$ for $n \geq 0$. The difference between x_n and x_{n+1} is $ab^n(b-1)/\xi$, which becomes smaller and smaller as we approach the critical temperature and $\xi \to \infty$. So as we take this limit, the sum in equation (5.76) can be better and better approximated by an integral. The appropriate measure for this integral is

$$\frac{dn}{dx} dx = \frac{dx}{x \log b},$$

and so our expression for f becomes

$$f = k_B T \left(\frac{\xi}{a}\right)^{-d} [\log b]^{-1} \int_{a/\xi}^{\infty} x^{-d} g(x) \frac{dx}{x}. \tag{5.77}$$

As $T \to T_c$ the bottom limit of this integral becomes zero, and the integral becomes independent of the correlation length. Thus f goes as ξ^{-d} and so $f \sim |T - T_c|^{\nu d}$ and hence in the limit as $T \to T_c$ the specific heat C obeys

$$C = T \frac{d^2 f}{dT^2} \sim |T - T_c|^{\nu d - 2}, \tag{5.78}$$

and

$$\alpha = 2 - \nu d. \tag{5.79}$$

There is nothing in this argument which requires that we be above the critical temperature. It is equally applicable to the exponent α' describing the divergence (if any) of the specific heat with temperature as we approach T_c from below. Thus, we can immediately conclude that

$$\alpha' = \alpha.$$

This is the last of our critical exponents. In Table 5.1 we have gathered together the expressions for each of the four critical exponents α, β, γ and δ in terms of η and ν. The results given in this table reduce the problem of solving for the critical exponents to the problem of finding just these two.

Table 5.1. Summary of the relations between the critical exponents

Exponent	Definition	Value
α	$c_B \sim |T - T_c|^{-\alpha}$, $T \to T_c$, $B = 0$	$2 - \nu d$
β	$m \sim |T - T_c|^{\beta}$, $T \to T_c$ from below, $B = 0$	$\frac{1}{2}\nu(d + \eta - 2)$
γ	$\chi \sim |T - T_c|^{-\gamma}$, $T \to T_c$, $B = 0$	$\nu(2 - \eta)$
δ	$m \sim B^{1/\delta}$, $B \to 0$, $T = T_c$	$\frac{d+2-\eta}{d-2+\eta}$

NOTES: See equations (5.56), (5.62), (5.67) and (5.79).

5.7 The scaling laws

We have seen that we can write all the critical exponents in terms of just two numbers, ν and either ω_c or η. Also we have seen that there is no distinction between exponents measured above the critical temperature and below it, so we can drop the primed notation used to distinguish them. Using the results summarized in Table 5.1, it is now simple to show that

$$\begin{aligned} 2\beta + \gamma &= 2 - \alpha, & \text{(Rushbrooke's law)} \\ 2\beta\delta - \gamma &= 2 - \alpha, & \text{(Griffiths' law)} \\ \gamma &= \nu(2 - \eta), & \text{(Fisher's law)} \\ \nu d &= 2 - \alpha, & \text{(Josephson's law)}. \end{aligned} \tag{5.80}$$

These four equations are exactly the scaling laws which we derived from the Widom and Kadanoff homogeneity hypotheses in §1.5.1. Although in this form they are not particularly useful equations—for the purposes of calculation the formulae in Table 5.1 are probably more convenient—they have considerable historical interest because they were all suggested before it was possible to prove their validity.

We have deduced the scaling laws by assuming a particular form (5.28) for the definition of the block variables. However, they also hold for other definitions of the block variables. Indeed, it can be shown that they hold for any sensible definition of the block variables. The proof is as follows.

Suppose that, as in §5.3, we have an effective Hamiltonian $\mathcal{H}$ which is a function of the block variables σ_i and a set of parameters $\{H_i\}$. And suppose we have constructed a renormalization procedure which maps this effective Hamiltonian onto another of exactly the same functional form, but with changed values of the parameters H_i. The block variables for this transformation can be defined by any of the methods we have seen in this chapter, or by one of the many other methods to be found in the literature. We now deduce, for this completely general problem, the equivalent of equation (5.73), which tells us how the free energy changes when we renormalize.

The Boltzmann factor for the renormalized system can be written as

$$e^{-\mathcal{H}'(\{\sigma'_i\})} = e^{-\beta G} \sum_{\substack{\text{configurations } \{\sigma_i\} \\ \text{consistent with } \{\sigma'_i\}}} e^{-\mathcal{H}(\{\sigma_i\})}. \tag{5.81}$$

5.7 The scaling laws

This equation defines the renormalized effective Hamiltonian. The constant $e^{-\beta G}$ is independent of $\{\sigma'_i\}$ (though it will in general depend on $\{H'_i\}$) and is chosen so as to satisfy the normalization condition (5.8), or whatever the corresponding condition is in this system. Now, summing both sides of (5.81) over all possible configurations $\{\sigma'_i\}$ of the renormalized system, taking logarithms and dividing by β, we find that

$$F' = F - G. \tag{5.82}$$

The free energy F is a extensive function of the parameters H_i appearing in $\mathcal{H}$, so we can write it in the form $\beta F = N\hat{f}(\{H_i\})$, where the function $\hat{f}$ is the dimensionless free energy per site introduced in the last section. Similarly, we can write $\beta F' = N'\hat{f}(\{H'_i\})$ where $\hat{f}$ is the same function again, though its argument is different this time, and $N' = N/b^d$ is the number of sites on the renormalized lattice. And clearly if F and F' are both proportional to N, then G must be too, so we can define $\beta G = Ng(\{H_i\})$, where g is an intensive quantity. With these definitions, equation (5.82) becomes

$$\hat{f}(\{H'_i\}) = b^d \left(\hat{f}(\{H_i\}) - g(\{H_i\}) \right). \tag{5.83}$$

Now let us consider what happens when we are close to the critical fixed point. In this region the parameters H'_i can be expressed as linear combinations of the H_i, as described in §5.3.1. The relation between the two sets of parameters is described by the matrix $\mathbf{M}$ according to equation (5.17). Let us transform to the basis in which $\mathbf{M}$ is diagonal. We will denote the parameters in this basis by x_i. (See for example equation (5.21).) In this basis equation (5.17) becomes simply

$$x'_i = \lambda_i x_i, \qquad \text{(no sum over } i\text{)}, \tag{5.84}$$

where λ_i is the ith eigenvalue of $\mathbf{M}$. This means that we can write (5.83) as

$$\hat{f}(\lambda_1 x_1, \lambda_2 x_2, \ldots) = b^d \left(\hat{f}(x_1, x_2, \ldots) - g(x_1, x_2, \ldots) \right). \tag{5.85}$$

To progress beyond this point we have to make a further assumption. We assume that all the singular behaviour in the free energy[15] of the renormalized system comes from the term $\hat{f}(x_1, x_2, \ldots)$ on the right-hand side of (5.85), and not from $g(x_1, x_2, \ldots)$. This assumption, of the **regularity** of the renormalization transformation at the critical fixed point, is vital to the success of the renormalization approach, but we have no *a priori* reason

---

[15] By 'singular behaviour' we do not mean to imply that the free energy actually has any singularities (i.e., poles) at the critical temperature. Rather, we are using the expression in the sense of §1.5 to refer to the terms in $\hat{f}$ which go like a non-integer power of $|T - T_c|$.

to believe in its validity. Some renormalization transformations are regular, others are not. It can, for instance, be shown that any transformation like (5.3), in which the renormalized block variables depend linearly on the original variables, is regular (see Niemeijer and van Leeuwen 1976). We must try to use only regular transformations, though for many of the rather complicated transformations used in real-space renormalization calculations it is extremely difficult to decide whether this condition is satisfied. For the moment, however, let us proceed under the assumption that we have a regular transformation. Then, writing the singular terms in $\hat{f}$ as $\hat{f}_{\text{sing}}$, we have

$$\hat{f}_{\text{sing}}(\lambda_1 x_1, \lambda_2 x_2, \ldots) = b^d \hat{f}_{\text{sing}}(x_1, x_2, \ldots). \tag{5.86}$$

Consider now the simple case we had before of a system in zero external field B (see §5.3). There is then one eigenvalue (say λ_1) which is greater than unity. Its eigenvector points out of the critical surface. The coordinate x_1 in this direction is proportional to $T - T_c$ near the critical fixed point. Suppose that, as we approach the critical fixed point by varying x_1, the singular part $\hat{f}_{\text{sing}}$ of the free energy varies as $x_1^{a_1}$. Putting every x_i except x_1 equal to zero in equation (5.86), we get

$$(\lambda_1 x_1)^{a_1} = b^d x_1^{a_1}, \tag{5.87}$$

and so

$$\lambda_1 = b^{y_1}, \tag{5.88}$$

where

$$y_1 = \frac{d}{a_1}. \tag{5.89}$$

We can run through a similar argument for the case in which $T = T_c$ but $B \neq 0$. Here again, there is one eigenvalue (the magnetic one), say λ_2, which is greater than zero. The coordinate x_2 in the direction of the corresponding eigenvector is proportional to the size of the external field B near the critical fixed point. If, as we approach the critical fixed point by varying x_2, $\hat{f}_{\text{sing}} \sim x_2^{a_2}$, then

$$\lambda_2 = b^{y_2}, \tag{5.90}$$

where

$$y_2 = \frac{d}{a_2}. \tag{5.91}$$

Putting these results together, we can write equation (5.86) as

$$\hat{f}_{\text{sing}}(b^{y_1} x_1, b^{y_2} x_2, 0, 0, \ldots) = b^d \hat{f}_{\text{sing}}(x_1, x_2, 0, 0, \ldots). \tag{5.92}$$

Setting $b = x_1^{-1/y_1}$, this now becomes

$$x_1^{d/y_1} \hat{f}_{\text{sing}}(1, x_2/x_1^{y_2/y_1}, 0, 0, \ldots) = \hat{f}_{\text{sing}}(x_1, x_2, 0, 0, \ldots). \tag{5.93}$$

This is now in exactly the form of the Widom homogeneity hypothesis, equation (1.17). But, as we demonstrated in Chapter 1, from this hypothesis the scaling laws (5.80) follow immediately. Consequently, they must apply to any renormalization transformation which is regular at the critical fixed point.

5.8 Percolation

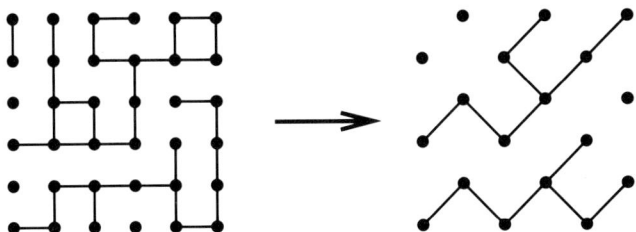

Figure 5.6 Illustration of the renormalization of the percolation problem on a square lattice. Bridges are erected on the new lattice along bonds that were connected by two bridges on the old.

5.8 Bond percolation in two dimensions

We now give a number of examples of calculations of the critical behaviour of models by real-space renormalization using the techniques discussed above. Much of our effort will be devoted to the study of the Ising model in one and two dimensions, for which the known exact solutions (see §§3.2.1 and 3.4.2) allow us to gauge the success of the renormalization method. Before we look at these examples however, let us first study a much simpler one, perhaps the simplest there is—that of bond percolation on a two-dimensional square lattice, introduced in §3.1.7.

We renormalize the lattice by knocking out every second site, as depicted in Figure 5.6. Then we join two sites on the renormalized lattice with a bridge if there were two bridges on the old lattice joining those two sites. In this way, we ensure that the size of each cluster on the new lattice is approximately the same as the corresponding cluster on the old. Then we complete the renormalization of the system by scaling all the lengths down by a factor of $b = \sqrt{2}$. Let us work out the probability that any particular bond on the new lattice has a bridge along it. In Figure 5.7 we show all the configurations of bonds around a square which give rise to a bond on the renormalized lattice. If the probability of each of the bonds on the square is p, then we can calculate the probability of each of these configurations, and sum them to get p'. The result is

$$p' = 2p^2 - p^4. \tag{5.94}$$

This is the equivalent of equation (5.13) for this problem. p is the only parameter appearing in the percolation problem, so we have only a one-dimensional space, and finding the fixed points is simply a matter of solving equation (5.94) for $p' = p$. Two of the solutions are $p = 1, 0$, which are, if you like, the 'high-' and 'low-temperature' fixed points for the problem. The other two solutions are $p = \frac{1}{2}(-1 \pm \sqrt{5})$. The negative one we can clearly discount, so we have

$$p^* = \frac{\sqrt{5}-1}{2} = 0.618\ldots \tag{5.95}$$

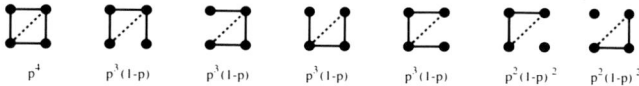

Figure 5.7 The configurations of bridges around a square on the unrenormalized lattice that can give rise to a bridge along the dotted line on the renormalized lattice. Each configuration is labelled with the probability of its occurrence.

The exact result is $\frac{1}{2}$. We can show this by the following argument. We build up another lattice by putting sites in the middle of all the squares on our present lattice (see Figure 5.8(a)) and erect a bridge between every pair of adjacent sites on the new lattice which are not already separated by one of the old bridges (see Figure 5.8(b)). In this way we produce a new percolation system based on a lattice of exactly the same size and shape as the old one, but on which bridges exist with probability

$$q = 1 - p, \tag{5.96}$$

where p is the corresponding probability on the original lattice.

Now, if the original percolation system is above its percolation threshold, $p > p^*$, so that there exists a cluster of joined sites of infinite size (on an infinite lattice), then the new system must be below its percolation threshold; there cannot be a path that stretches all the way from one side of the new lattice to the other, because such a path would somewhere have to cross the infinite cluster on the old lattice, and this is forbidden by the rules we used to place the bridges on the new lattice. Conversely, if we are ignorant of the state of the original system, but we know that the new one is below the percolation threshold, we can immediately conclude that the old system must be above its threshold—there must be a cluster of infinite size on the old lattice to stop the sites on the new one from connecting to form an infinite cluster. Thus

$$p > p^* \Leftrightarrow q < q^*. \tag{5.97}$$

Similarly, if the new system is above its percolation threshold, the old one must be below *its* threshold, and *vice versa*:

$$q > q^* \Leftrightarrow p < p^*. \tag{5.98}$$

From these two relations we conclude that when one system is *at* the percolation threshold, so must the other one be. Taking equation (5.96) at the threshold then, we get

$$q^* = 1 - p^*. \tag{5.99}$$

But since this new system is the same as the old one, it must have the same percolation threshold, $q^* = p^*$, so from equation (5.99)

$$p^* = \tfrac{1}{2}. \tag{5.100}$$

5.8 Percolation

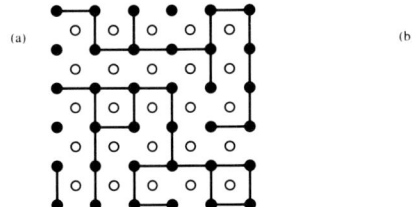 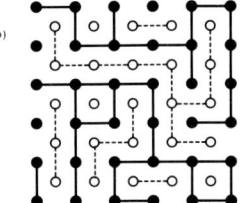

Figure 5.8 (a) The construction of the 'dual lattice' for the percolation problem on a square lattice. (b) We populate the dual lattice by building bridges between every pair of sites not separated by a bridge on the original lattice.

The result $p^* = 0.618\ldots$ of our renormalization calculation is thus not very accurate. Below we discuss the reasons for this, but first let us see if we can calculate a critical exponent for this problem. There is, in fact, only one critical exponent for the percolation problem, the exponent ν which describes the divergence of the mean cluster size ξ (see §3.1.7) with the probability p as we approach the critical point:[16]

$$\xi \sim (p - p^*)^{-\nu}, \qquad p \to p^*. \tag{5.101}$$

To get this exponent we need to linearize equation (5.94) for $p'(p)$ about the point $p = p^*$. That is, we write

$$\begin{aligned} p' - p^* &= \left.\frac{dp'}{dp}\right|_{p=p^*} (p - p^*) + \cdots \\ &= 4p^*(1 - p^{*2})(p - p^*) + \cdots. \end{aligned} \tag{5.102}$$

This is the equivalent of equation (5.17). In this case, because we have only a one-dimensional parameter space, the matrix **M** of equation (5.16) is just a scalar, and so has only the one eigenvalue $\lambda_R = 4p^*(1-p^{*2})$. This eigenvalue is greater than unity. (It is about 1.528.) Using this value of λ_R and the value $b = \sqrt{2}$ of the scaling factor, equation (5.26) yields $\nu = 0.818\ldots$. However, equation (5.26) was derived for statistical mechanical systems governed by a Hamiltonian, and not for a percolation system, so you might justifiably say that identifying p with the temperature and ξ with the correlation length is a bit dubious, and that equation (5.26) is not really applicable here. To allay these fears, let us go over the calculation of ν explicitly for the percolation problem.

Suppose the cluster size ξ diverges according to equation (5.101) as we approach the critical point. Then, in the region close to the critical point,

---

[16] Other exponents, analogous to those for a magnetic model can be defined for the percolation problem, though the definitions are rather devious. For an account, see Essam (1972).

Figure 5.9 Approximations arising in the renormalization scheme we used for the decimation problem. The sites marked A and B, which are connected on the original lattice, become unconnected on renormalization. This effect has a tendency to decrease the average cluster size ξ. Also, correlations arise between the probabilities of adjacent bonds having bridges on them. The presence or absence of a bridge on the bond marked 1, for example, affects the probabilities of there being bridges on both 1' and 2'.

any two values of the parameter p and the corresponding values of ξ are related by

$$\frac{\xi_1}{\xi_2} = \left(\frac{p_1 - p^*}{p_2 - p^*}\right)^{-\nu}. \tag{5.103}$$

Let us take the values p and p' of equation (5.94) to be our two values p_1 and p_2. If ξ is the cluster size for probability p, then the corresponding quantity for probability p' must be $\xi' = \xi/b$, because we know that all lengths scale by a factor of b when we renormalize. Using these results in equation (5.103) gives, in the limit as we approach p^*,

$$\frac{1}{b} = \left(\frac{p' - p^*}{p - p^*}\right)^{-\nu} = \left[4p^*(1 - p^{*2})\right]^{-\nu}, \tag{5.104}$$

where we have used (5.102). Rearranging for ν and sticking in our value for p^*, we again get

$$\nu = 0.818\ldots. \tag{5.105}$$

Series expansions (Dunn et al. 1975) give $\nu = 1.34 \pm 0.02$, and the exact result is believed to be $\nu = \frac{4}{3}$, so again our calculation is not impressively accurate. What's more, our value for ν is not at all heavily dependent on p^*. We do not dramatically improve our results if in equation (5.104) we use the exact value $p^* = \frac{1}{2}$—we get $\nu = 0.855\ldots$. Evidently there is a fundamental flaw in our approach. On the face of it however, it is not entirely clear what this flaw could be. To be sure, we have made approximations in our working. In particular we have taken a very simple criterion for connecting sites on the new lattice, which does not necessarily always preserve the typical cluster size ξ exactly. However, one might imagine that as $\xi \to \infty$ this approximation would become less and less significant. What then, is the source of the error? Well, there are actually two significant sources.

(i) Because of the way we assign bonds on the renormalized lattice, no two sites can be connected which were not also connected on the original lattice. However, under unfavourable conditions, we may have two sites unconnected on the new lattice which were connected on the old. An example of such unfavourable conditions is shown in Figure 5.9. Here, after renormalization of the lattice the sites marked A and B are no longer connected, although they were on the original lattice.

(ii) The renormalized system is not really a true percolation system, because the probabilities of there being bridges on adjacent bonds are no longer independent. To see this consider Figure 5.9 again. The presence of the bridge marked 1 affects the chances of there being bridges across both bonds 1' and 2' of the renormalized lattice. This gives rise to correlations in the presence or absence of such adjacent bridges when we renormalize—if we work out the probability of every configuration of the two squares which gives rise to bridges on 1' and 2', we find that if 1' is present on the renormalized lattice, there is an increased probability of 2' being there as well. Thus we are making an approximation by assuming that the renormalized system can be represented as a simple percolation problem with independent probability of the occupation of each bond.

Both these sources of error can be overcome by using a more sophisticated model with more parameters, which is able to keep track of these effects. Young and Stinchcombe (1975), for example, used a six-parameter renormalization scheme and arrived at the result $\nu = 1.313\ldots$ which is in fair agreement with the true answer of $1.333\ldots$.

Now let us move on to a harder problem, that of the Ising model.

5.9 The Ising model

Following the precedent of the last section on the percolation problem, we will give only a basic treatment of the Ising model, and indicate how this approximation fails, and how it may be improved. We are at the moment only interested in the Ising model with nearest-neighbour interactions. Nonetheless, we will find it profitable to develop the general theory for more complicated Ising-type problems first.

Consider then, a lattice of N sites, with an Ising block variable $\sigma_i = \pm 1$ defined on each site. We denote by $\{\mathcal{T}_m\}$ all the subsets of these spins, so that $\mathcal{T}_1$, for example, might denote the set consisting of the two spins σ_1, σ_2, and $\mathcal{T}_2$ might denote the four spins $\sigma_1, \sigma_2, \sigma_8, \sigma_9$. Then the most general Ising-type Hamiltonian may be written

$$\mathcal{H}(\{\sigma_i\}) = \beta \sum_m \mathcal{J}_m \prod_{i \in \mathcal{T}_m} \sigma_i$$
$$= \sum_m H_m t_m, \qquad (5.106)$$

where

$$H_m \equiv \beta \mathcal{J}_m, \qquad t_m \equiv \prod_{i \in \mathcal{T}_m} \sigma_i. \qquad (5.107)$$

There are many, many of these sets $\mathcal{T}_m$. One tends usually to think of those that consist of groups of nearest neighbours—pairs of adjacent spins, blocks of four on a square lattice, and so on. But the formalism is more general than this. It allows for sets consisting of odd spins, possibly widely separated, picked from anywhere on the lattice. And with every such set, there is associated a coupling constant H_m. To render the formalism tractable, we need to cut down on the number of such coupling constants. One way is to stick to homogeneous, isotropic models. If we do this, the number of independent interaction constants is much reduced, because all two-spin, nearest-neighbour interactions will be the same, all four-spin block interactions will be the same, and so forth. Note also that, for a system in zero external field B, all coupling constants corresponding to sets consisting of an odd number of spins (the 'odd parameters' of §5.4.2) must be zero. The reason is that the variables t_m corresponding to such sets change sign if we reverse every spin in the set. If the corresponding coupling constant were non-zero, this would mean that the effective Hamiltonian would change its value if we were to reverse the direction of every spin on the lattice, and that is forbidden for an isotropic model. Later, we will reduce the number of non-zero coupling constants still further, by discretely ignoring almost all of the longer-range ones. For the moment however, let us see how far we can get with the completely general model.

It is not hard to show (see Problem 5.5) that

$$\sum_{\{\sigma_i\}} t_m t_n = 2^N \delta_{mn}. \tag{5.108}$$

So, multiplying (5.106) by t_n and summing over all the configurations of the system, we recover the interaction constants:

$$H_n = 2^{-N} \sum_{\{\sigma_i\}} \mathcal{H}(\{\sigma_i\}) t_n. \tag{5.109}$$

It is this relation that allows us to calculate the renormalization laws for the parameters H_m from the renormalized effective Hamiltonian. The idea is as follows. First, we get the renormalized effective Hamiltonian from equation (5.81), which gives us

$$-\mathcal{H}'(\{\sigma_i'\}) = -G + \log \sum_{\substack{\text{configurations } \{\sigma_i\} \\ \text{consistent with } \{\sigma_i'\}}} e^{-\mathcal{H}(\{\sigma_i\})}. \tag{5.110}$$

Now we use equation (5.109):

$$H_n' = 2^{-N'} \sum_{\{\sigma_i'\}} \mathcal{H}'(\{\sigma_i'\}) t_n'$$

$$= 2^{-N'} \sum_{\{\sigma_i'\}} \left[G - \log \sum_{\substack{\text{configurations } \{\sigma_i\} \\ \text{consistent with } \{\sigma_i'\}}} e^{-\mathcal{H}(\{\sigma_i\})} \right] t_n'. \tag{5.111}$$

5.9 The Ising model

The first term on the right-hand side is zero, because there are exactly as many terms in the sum for which $t'_n = -1$ as there are for which $t'_n = +1$. So we are left with

$$H'_n = -2^{-N'} \sum_{\{\sigma'_i\}} t'_n \log \sum_{\substack{\text{configurations } \{\sigma_i\} \\ \text{consistent with } \{\sigma'_i\}}} e^{-\mathcal{H}(\{\sigma_i\})}. \quad (5.112)$$

In the standard Ising model, the only non-zero H_m in the Hamiltonian are those providing the nearest-neighbour interactions. However, this simplicity vanishes when we renormalize. On the square lattice, for example, there is no reason why, when we calculate the 'four-spins-in-a-square' interaction using (5.112), we should necessarily get a zero result. It is the neglect of these longer-range interactions which makes the simplest renormalization schemes inaccurate. As we have said, such approximations are an inevitable part of the real-space renormalization scheme, though we will shortly see how, by treating the Ising model as a special case of a more complicated model, we *can* systematically improve on our results. For the moment, however, let us make the most basic approximation of ignoring all interactions in the renormalized model, except the two-spin nearest-neighbour one, which is conventionally denoted K. In this case our renormalized model takes the same form as the original one, as we require that it should, and all we need to do is decide on a scheme for defining our block variables, and then evaluate (5.112) for the case in which t'_n is a set containing just two, nearest-neighbour block spins on the renormalized lattice. Let us call the nearest-neighbour interaction so derived K'.

The simplest and commonest scheme for defining the block variables is the majority rule of §5.2. That is, a variable σ'_i takes the value ± 1 of the majority of the spins σ_i which fall within that block on the original lattice. As an easy first example we might try to apply this definition to the one-dimensional case. However, we immediately run into a problem. The natural choice of renormalization scheme for the one-dimensional lattice is one in which we group the spins into pairs. But if we make this choice it becomes unclear what value we should assign to a block variable composed of one up spin and one down spin. We could get out of this easily by deciding to group the spins in threes instead of twos, but, because it will help us with our treatment of the two-dimensional Ising model below, let us resolve this difficulty in another way. Given a block containing one up spin and one down spin, we just decide at random that the block variable should be up or down, with equal probability. We account for this in equation (5.112) by introducing a factor of a half in front of the Boltzmann factors in the second sum for every such ambiguous block variable; the Boltzmann factor is proportional to the probability that the renormalized system will find itself in a particular state, and we are halving this probability by allowing the same configuration of original spins to correspond to two block-spin states with equal likelihood.

Now we need to perform the sums to calculate K' in terms of K. Unfortunately, this turns out to be impossible to do in closed form, even for the one-dimensional case, and we are forced to turn to numerical methods.[17] The best we can do in this case is to take a manageably small finite lattice, and do the calculation explicitly. Working on a finite lattice has the disadvantage that it tends to 'spread out' the critical point. As Figure 4.1 illustrates, on a finite lattice there is not really a critical point at all—the phase transition takes place gradually over a range of the parameter driving the problem (K in the Ising case). This effect is very easily seen in the percolation model. There the critical point is defined to be the point at which a 'cluster' stretching all the way across the lattice establishes itself as we increase the bond probability p. On an infinite lattice this point is perfectly well defined. But on a finite lattice the two sides of the lattice become joined by a single cluster as soon as the average cluster size ξ becomes larger than the linear size L of the system. So the phase transition becomes indistinct. It takes place gradually over the region between the value of p at which ξ first becomes greater than L, and the value at which it really diverges on an infinite lattice. In a similar way, the critical point of a thermodynamic model becomes smeared over a region roughly delimited by $\xi \geq L$, where ξ is now the correlation length. In order to minimize this effect, we normally want to make the lattice we are working on as large as possible. For the purposes of illustrative example however, a small lattice will do for the moment.

Figure 5.10 shows the results of a calculation of K' for an Ising chain of twelve sites with periodic boundary conditions. The critical point would be given by solving for $K' = K$ (which is marked in as a dashed line on the figure). Clearly, there is no phase transition in the range of K illustrated on this diagram. Further exploration to higher values of K makes it look extremely unlikely that there will be a transition at all. The only fixed points appear to be the high- and low-temperature ones at $K = 0$ and $K = \infty$, which is what we should expect for this model (see §3.2.1).

Figure 5.11 shows two plots of $K'(K)$ for the two-dimensional problem. The full curve was obtained from a 4×2 lattice with periodic boundary conditions. There is now a critical point, at $K^* = 0.1539\ldots$. If we set the spin–spin interaction $\mathcal{J}$ to one, this corresponds to a critical temperature of $\beta_c = 0.1539\ldots$. The correct result from §3.4.2 is $\beta_c = 0.4407\ldots$, so again our simple renormalization scheme is not at all accurate. Below we discuss how it can be improved, but first, let us see if we can calculate a critical exponent for this system. We can calculate the gradient of the curve $K'(K)$ at the critical point numerically. As in equation (5.102) this gradient is exactly the one eigenvalue of the 'matrix' $\mathbf{M}$ for this problem, since we again have only

---

[17] This is why we did the percolation problem first. At least you can actually do the sums there. In Problem 5.2 however, we consider a different approach to the two-dimensional Ising model using a decimation scheme which can be solved analytically.

5.9 The Ising model

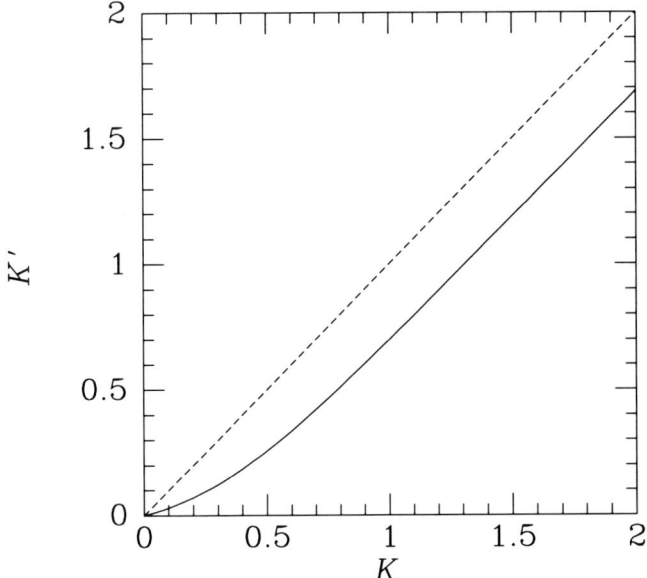

Figure 5.10 The renormalized two-spin interaction parameter K' for the one-dimensional Ising model as a function of the original interaction K.

a one-dimensional Hamiltonian space. Bearing in mind that $b = 2$, we find

$$\lambda = \frac{\mathrm{d}K'}{\mathrm{d}K} = 1.543\ldots \quad \Rightarrow \quad \nu = \frac{\log b}{\log \lambda} = 1.598\ldots$$

The exact answer is $\nu = 1$.

How are we to improve on these results? We have two principal courses of action. The first is to work on a larger lattice. As we saw in §4.1, it is not practical explicitly to count the states of lattices with more than a handful of sites on any computer, no matter how powerful, but we can certainly improve on a 4×2 lattice. The dotted curve in Figure 5.11 shows $K'(K)$ for a 4×4 lattice with periodic boundary conditions—each point on this curve required $2^8 = 256$ times as many machine operations as a point on the curve for the 4×2 lattice. The critical point is now at $K^* = 0.2268\ldots$, which is an improvement on the result from a 4×2 lattice, but the slope of $K'(K)$ at K^* is not much changed: $\mathrm{d}K'/\mathrm{d}K = 1.56\ldots$. So our improved value for the critical exponent, $\nu = 1.56\ldots$, is still far from the correct value $\nu = 1$ and represents a poor return on a 256-fold increase in the computing time employed.

The other way to improve our calculation is to include some of the longer-range interactions between spins of which we spoke earlier in this section. As we said, there is no reason to suppose that if we calculate the

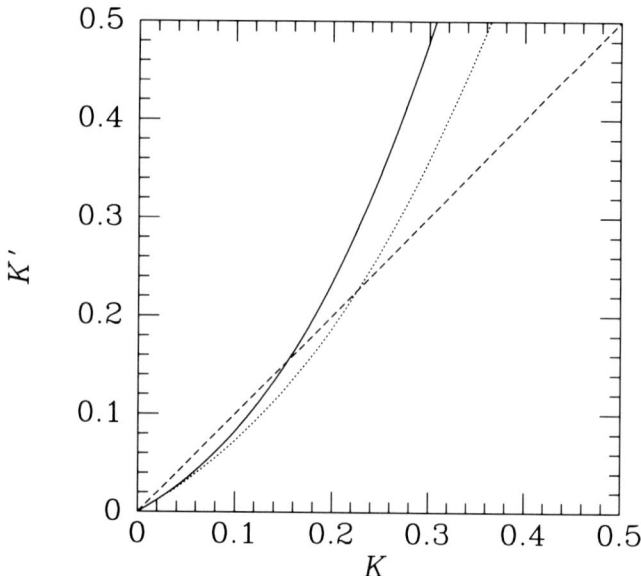

Figure 5.11 The renormalized two-spin interaction parameter K' for the two-dimensional Ising model as a function of the original interaction K. The full and dotted curves are for 4×2 and 4×4 lattices, respectively. The critical points, where $K' = K$, are at $K^* = 0.1539\ldots$, and $K^* = 0.2268\ldots$. The gradients of the curves $K'(K)$ at these points yield $\nu = 1.598\ldots$ and $\nu = 1.56\ldots$. The exact values are $K^* = 0.4407\ldots$ and $\nu = 1$.

interaction parameter between, say, diagonal-neighbour spins on the square lattice, we will get zero. In general we will not, and we ought somehow to allow for this. However, if we do put in these interactions, our renormalized model is no longer going to be a simple Ising model, as we stipulated it should be in our original discussion of the renormalization technique. This is not actually a problem. We simply study a more general model. For example, quite a bit of work has been done on the Ising-type problem on a square lattice with three different interaction parameters—the nearest-neighbour one, the diagonal-neighbour one, and the four-spins-in-a-square interaction (see Figure 5.12), which we will denote by K, L and N.[18] Higher-order interactions yet are just ignored. We can now study the motion of the effective Hamiltonian $\mathbf{H}(K, L, N)$ in this three-dimensional Hamiltonian space, and look for its stationary points.

For example, Nauenberg and Nienhuis (1974) calculated the renormalization transformation of the three-parameter Ising Hamiltonian $\mathbf{H}(K, L, N)$ on a 4×4 lattice and found a critical fixed point at $K = 0.307$, $L = 0.084$, $N = -0.004$. Linearizing their renormalization transformation about this point and then diagonalizing it, they found, as expected, two attractive

---

[18] The letter M is not used to avoid confusion with magnetization.

5.10 Monte Carlo renormalization

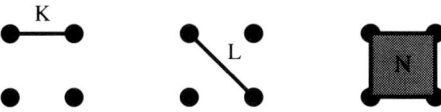

Figure 5.12 The K, L and N interactions for the generalized Ising model discussed in the text.

eigenvalues (that is, ones less than unity) and one repulsive one, $\lambda_R = 1.914$. Given that the scale factor b is 2 for this renormalization scheme, we can plug this eigenvalue into equation (5.26), and get $\nu = 1.07\ldots$, which is much closer to the exact value $\nu = 1$ than our previous approximation, in which we set L and N to zero.

There is a limit to how far we can go in improving this style of calculation however, because of the rapid increase in computing power necessary to sample larger lattices. Even if we continue to work with the smallest lattices possible, the inclusion of longer-range interactions will force us to bigger systems, and there is a limit to what could be done even with computers many times faster than those available today. Better results than those quoted above have been reached however, by a combination of real-space renormalization and Monte Carlo simulation. This elegant development, which in one stroke eliminates the problems of finite lattice size mentioned here, and the problems of critical fluctuations in Monte Carlo simulations (see §4.7), is the subject of the next section.

5.10 Monte Carlo renormalization

The idea behind Monte Carlo renormalization is to express the matrix $\mathbf{M}$ that represents the leading term in the expansion of the transformation $\mathbf{R}(\mathbf{H})$ about the critical fixed point, as a thermal average of some quantity, and then to calculate it using a standard Monte Carlo method, such as those described in §4.6. This allows us to study much larger lattices than those possible by the direct summation methods of the last section. The most general realization of this goal was given by Swendsen, and his idea is as follows (Swendsen 1979a, 1982).

We first decide on some renormalization transformation appropriate to the system we are studying, such as the additive blocking transformation used for the continuous-spin model of §§5.4–5.6, or the majority rule or decimation scheme for Ising-type models. At each stage of the renormalization we express the Hamiltonian as a sum

$$\mathcal{H} = \sum_n H_n T_n. \tag{5.113}$$

Here we have decomposed $\mathcal{H}$ into parameters H_n, which are things like the temperature and the external field which in general change at each stage of the renormalization process, and some invariant functions $T_n(\{\sigma_i\})$ of the values taken by the (renormalized) order parameter on the (renormalized) lattice.[19] For example, in a simple application of Monte Carlo renormalization to the two-dimensional Ising lattice, we might have (see Figure 5.12)

$$H_1 = K, \qquad H_2 = L, \qquad H_3 = N \qquad (5.114)$$

and

$$T_1 = \sum_{\substack{\text{nearest} \\ \text{neighbours}}} \sigma_i \sigma_j, \quad T_2 = \sum_{\substack{\text{secondnearest} \\ \text{neighbours}}} \sigma_i \sigma_j, \quad T_3 = \sum_{\text{in a block}} \sigma_i \sigma_j \sigma_k \sigma_l. \qquad (5.115)$$

One of the strengths of Monte Carlo renormalization is that, even though one is always obliged to make the approximation of including only finitely many of the parameters H_m, it is comparatively easy to assess the error to which this gives rise by increasing the number of H_m included.

When we renormalize the system with the Hamiltonian of (5.114), we obtain a Hamiltonian of the form

$$\mathcal{H}' = \sum_m H'_m T'_m, \qquad (5.116)$$

where the primes on the T_m indicate that they are evaluated using the values of the new block variables σ'_i. We now focus our attention on the thermal averages $\langle T_m \rangle$ and $\langle T'_m \rangle$, and their derivatives with respect to the parameters H_m, H'_m. We write

$$\frac{\partial \langle T'_m \rangle}{\partial H_l} = \sum_n \frac{\partial H'_n}{\partial H_l} \frac{\partial \langle T'_m \rangle}{\partial H'_n}. \qquad (5.117)$$

This equation contains no physics; it is simply a statement of the 'chain rule' of partial differential calculus. In matrix form it reads

$$\mathbf{A} = \mathbf{MB}, \qquad (5.118)$$

where $\mathbf{M}$ is precisely the matrix

$$M_{ij} = \left. \frac{\partial H'_i}{\partial H_j} \right|_{\mathbf{H}=\mathbf{H}_c} \qquad (5.119)$$

we are anxious to calculate (see equation (5.16)), and the matrices $\mathbf{A}$ and $\mathbf{B}$ are defined by

$$\begin{aligned} A_{lm} &\equiv \frac{\partial \langle T'_m \rangle}{\partial H_l} = \langle T_l T'_m \rangle - \langle T_l \rangle \langle T'_m \rangle, \\ B_{mn} &\equiv \frac{\partial \langle T'_m \rangle}{\partial H'_n} = \langle T'_m T'_n \rangle - \langle T'_m \rangle \langle T'_n \rangle. \end{aligned} \qquad (5.120)$$

---

[19] Equation (5.113) is very similar to equation (5.106), but the quantities T_n are defined in a slightly different fashion to the t_n, being sums over large numbers of products of nearest neighbours, rather than just single products.

5.10 Monte Carlo renormalization

The equalities on the right-hand side of these equations follow from the linear response theorem, equation (2.28), though with one subtlety: in order to prove both equalities above (the one for **A** and the one for **B**), we must write

$$\langle T'_m \rangle = \frac{\sum [T'_m e^{-\mathcal{H}'}]}{\sum e^{-\mathcal{H}'}} = \frac{\sum [T'_m e^{-\mathcal{H}}]}{\sum e^{-\mathcal{H}}}. \tag{5.121}$$

The sums in the middle term here are over all states of the blocked system, and the sums on the right-hand side are over all states of the original, unblocked one. The second equality in this equation comes from the condition (5.11) that the average of a quantity using the renormalized Hamiltonian should give the same answer as the average using the original one. On differentiating these expressions we get equation (5.120).

Monte Carlo renormalization consists in using one of the Monte Carlo schemes of §4.6 to calculate **A** and **B** from equations (5.120), and then calculating $\mathbf{M} = \mathbf{AB}^{-1}$. The critical exponent ν then follows from **M**'s largest eigenvalue.

In general it turns out that, for a given level of computing power, Monte Carlo renormalization not only allows one to keep more interactions H_m than is possible in direct renormalization calculations such as those at the end of §5.9, but it also allows us to work on a much larger lattice. Thus it gets around both of the difficulties mentioned at the end of the last section.

The method does have disadvantages, however. One is that its results are subject to statistical fluctuations. These however are not nearly such a problem as they would have been had we tried to evaluate the derivative defining **M** (equation (5.119)) using a direct Monte Carlo simulation of our system. For when trying to calculate a derivative of numerical data, any noise or fluctuations in those data give rise to worse noise in the derivative, and the very large fluctuations near the phase transition would thus make our value for **M** unusably noisy. Monte Carlo renormalization disposes of this problem by directly calculating the derivative itself. This derivative will still possess fluctuations, which will become larger as we approach the critical point. But these can be estimated, and so the accuracy of the results for the critical exponents can be gauged.

A more serious problem is that, since the matrix **M** is to be evaluated at $\mathbf{H} = \mathbf{H}^*$ (see equation (5.119)), we need to know where the critical point is before we can perform the calculation. The renormalization calculations of §5.9, in which $\mathbf{H}'$ was directly evaluated as a function of **H**, automatically gave us this information, but when we use Monte Carlo renormalization we must find the critical fixed point by some other means. However, it is not necessary to know the fixed point very accurately, since it usually turns out that **M** is a rather slowly varying function of **H** near $\mathbf{H}^*$. A method for determining its position approximately was given by Swendsen (1979b) in his treatment of the two-dimensional Ising model on a square lattice. He kept three interaction parameters only, the K, L and N parameters of §5.9, and used a scheme in which the spins were blocked in square groups of nine spins and then a poll taken to decide the direction of the block spin by majority rule. The position of the critical fixed point he found by experimentation. He calculated the largest eigenvalue of **M** for two consecutive iterations in the renormalization procedure. Exactly at the critical point, these should take the same value, but slightly away from it the eigenvalue will drift from one stage in the calculation to the next. By minimizing this drift, he was able to get close enough to the fixed point to obtain good results: he found $\nu = 0.998$, very close to the exact value $\nu = 1$ (see §3.4.3).

By a similar technique, Blöte and Swendsen (1979) calculated critical exponents for the Ising model in three dimensions on a cubic lattice. They included in this calculation first- to fourth-nearest-neighbour interactions and four-spin interactions. They found $\nu = 0.637$, which compares favourably with the best value $\nu = 0.638 \pm 0.002$ from high-temperature series—see §3.4.4. Another interesting calculation is that of Sahni and Banavar (1981), who applied Monte Carlo renormalization to the Ising model on another three-dimensional lattice—the face-centred cubic lattice. Their best result was $\nu = 0.613$. Although these particular calculations provided little in the way of error estimates on the results, this last figure lends further credence to the idea of universality introduced in §1.4; it appears plausible, at least, that in reality the exponent ν takes the same value on these two different three-dimensional lattices.

Problems

5.1 Consider the one-dimensional ferromagnetic Ising model with nearest-neighbour interactions only. Write down the probabilities that two nearest-neighbour spins will be parallel and anti-parallel in terms of the 'dimensionless coupling constant' $K \equiv -\beta \mathcal{J}$, where β is the inverse temperature and $\mathcal{J}$ the spin–spin interaction. ($\mathcal{J}$ is negative for ferromagnetic models, so $K > 0$.) Now perform a decimation of the lattice, removing every second spin whilst keeping the remaining ones just as they were, and calculate the new probability that nearest-neighbours are parallel or anti-parallel. Thus derive an expression for K', the dimensionless coupling constant of the decimated system. This is an exact renormalization equation. Convince yourself that it has no fixed points except at infinite temperature and at absolute zero, and therefore that the one-dimensional Ising model has no phase transition.

5.2 Consider the two-dimensional Ising model on a square lattice with nearest- and next-nearest-neighbour interactions only (the interactions denoted K and L in Figure 5.12). Perform a decimation of this system by knocking out every second site in a 'checker-board' fashion. Rescaling the lattice by a factor of $b = \sqrt{2}$ will now take us back to the original lattice parameter. Calculate the interactions on the decimated lattice, keeping only terms up to $O(K^2)$ and $O(L)$. Show that there are only two such interactions

Problems

to this order:
$$K' = 2K^2 + L,$$
$$L' = K^2. \tag{5.122}$$

Find the fixed points for these renormalization equations and identify the critical one. Linearizing about this point, find a value for the exponent ν.

5.3 Using a computer repeat the blocking calculation of $K'(K)$ for the nearest-neighbour Ising model described in §5.9 and attempt to reproduce Figure 5.11 and the quoted results for K^* and ν.

5.4 Construct proofs along the lines of that presented in §5.4.4, that the values of the exponents β, γ and η are independent of the number of renormalizations we perform on the system.

5.5 By considering separately the case $m = n$ and $m \neq n$, prove the relation (5.108).

6
Mean-field theory

In the previous chapter we saw that it is possible to make a great deal of progress in the study of particular models by real-space renormalization methods. These methods are very elegant and can explain why critical exponents arise and how they are related to each other. However, they are also very specific: it requires considerable ingenuity to invent an accurate (and tractable) real-space scaling transformation for any given model, and even when one has succeeded it is not normally very helpful in tackling the next problem.

So, there is clearly scope for a method which is more generally and simply applicable, even at some cost in quantitative accuracy. One such method is **mean-field theory**, which we describe in this chapter. Historically, this technique was invented by P. E. Weiss (1865–1940) as a theory of magnetism, and for a long time it was the *only* theory of phase transitions. It therefore has a very important place in the development of our subject, and even today it is usually the first tool that is applied to sort out the essential physics of a new type of phase transition. For a while, some powerful arguments due to L. D. Landau (1908–1968) suggested that mean-field theory was essentially exact; we now know that this is not the case, except for sufficiently large values of the spatial dimensionality or for systems with infinite-range interactions. However, it is well worth studying, not just because it is a useful tool in its own right but also because it is the simplest possible approximation to a kind of model of models, or 'metamodel', called the Landau–Ginzburg model, which we shall find contains all the physics

6.1 The Ising model

necessary to describe what goes on near a continuous phase transition and which will be our subject from Chapter 7 onwards.

We start by introducing mean-field theory heuristically through the study of two of the models discussed in previous chapters, namely the Ising model of a ferromagnet and the percolation model. Next, we see how the theory can be used to model the non-ideal gas. Then we show that it provides an upper bound for the free energy of a system, and discuss the extent to which observable properties such as correlation functions and critical exponents can be extracted from the theory. Finally, in the light of the knowledge we already have about the physics of the second-order transition, we ask what is left out in mean-field theory and how we can include the missing features while retaining the wide applicability of the theory.

6.1 Mean-field theory of the Ising model

Consider the d-dimensional Ising model (see §3.1.1) in a uniform external magnetic field B. We begin by writing down the thermal average of a certain spin s_i given that the spins of its neighbours are constrained to take particular values:

$$\langle s_i \rangle = \frac{\exp\left[-\beta(\sum_j \mathcal{J}_{ij} s_j - B)\right] - \exp\left[\beta(\sum_j \mathcal{J}_{ij} s_j - B)\right]}{\exp\left[-\beta(\sum_j \mathcal{J}_{ij} s_j - B)\right] + \exp\left[\beta(\sum_j \mathcal{J}_{ij} s_j - B)\right]} \quad (6.1)$$

$$= -\tanh\left[\beta(\sum_j \mathcal{J}_{ij} s_j - B)\right].$$

Here, $\mathcal{J}_{ij}$ arises rather than $\frac{1}{2}\mathcal{J}_{ij}$ because each pair of sites is counted twice in the sum over i and j of equation (3.1). To get the true value of $\langle s_i \rangle$ we must now average over all the values that the neighbouring spins $\{s_j\}$ can take. The correct weight for these spin configurations involves the interaction of the neighbouring spins with *their* neighbours, and so on. In order to break into this ladder of successively more complicated equations for the thermal averages of more and more distant spins, we make a simple approximation. We see that the quantity $B - \sum_j \mathcal{J}_{ij} s_j$ acts in (6.1) just like an effective magnetic field at site i; our approximation will be to replace this field by its mean value $B - \sum_j \mathcal{J}_{ij} \langle s_j \rangle$. This is the origin of the name mean-field theory. Equation (6.1) therefore becomes

$$\langle s_i \rangle = -\tanh\left[\beta(\sum_j \mathcal{J}_{ij} \langle s_j \rangle - B)\right]. \quad (6.2)$$

We now adopt our usual form for the matrix $\mathcal{J}_{ij}$:

$$\mathcal{J}_{ij} = \begin{cases} \mathcal{J} & \text{if } i \text{ and } j \text{ are nearest neighbours,} \\ 0 & \text{otherwise.} \end{cases} \quad (6.3)$$

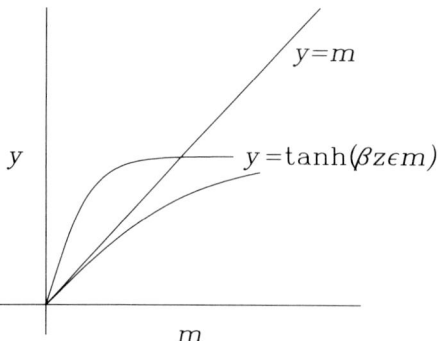

Figure 6.1 Graphical solution of the mean-field equation (6.4) for the Ising model.

We assume $\mathcal{J} < 0$, so that the model has a tendency for spins to align, and put $\epsilon = -\mathcal{J}$. Since the system is translationally invariant, we do not expect the average $\langle s_i \rangle$ of a spin to vary from site to site—it will be everywhere equal to the specific magnetization m. Equation (6.2) therefore becomes

$$m = \tanh[\beta(z\epsilon m + B)], \qquad (6.4)$$

where z is the number of nearest neighbours (the coordination number).

A graphical solution of (6.4) with $B = 0$ is shown in Figure 6.1. For small $z\beta\epsilon$, (6.4) has no non-zero solution for m. However, as $z\beta\epsilon$ increases, a solution with non-zero m appears. The critical value β_c of β at which a non-zero average value of the total spin becomes possible is given by

$$z\beta_c\epsilon = 1. \qquad (6.5)$$

Let us compare this result with what we already know about the Ising model from Chapters 3 and 5. Equation (6.5) predicts that β_c should depend on the geometry of the model only through the coordination number z and that it is non-zero for all $z \neq 0$. We know, however, that the spatial dimensionality is also crucial in determining the exact behaviour of the model. In §3.2 we showed that, in contrast to the predictions of our mean-field theory, there is no transition in the one-dimensional Ising model. In two dimensions we can claim slightly more success; equation (6.5) correctly predicts that there will be a transition to non-zero magnetization and the mean-field value for the transition temperature on a square lattice $\beta_c = 0.25/\epsilon$ is at least of the same order of magnitude as the exact result $\beta_c = 0.4407/\epsilon$. However, we shall see in §6.7 that the predictions of mean-field theory for the critical exponents of the two-dimensional Ising transition are substantially in error. The state of affairs for $d = 3$ is similar; although in this case we have no exact solution with which to compare our approximation, our prediction of $\beta_c = 0.133/\epsilon$ is only very roughly equal to the numerical result $\beta_c = 0.222$

6.2 Percolation

(see Ferrenberg and Landau 1991). In every case mean-field theory has underestimated β_c and therefore overestimated T_c. This is a general property of mean-field theory. Furthermore, the numerical values for the critical exponents in $d = 3$ differ from the exact $d = 2$ results, whereas we shall see in §6.7 that the corresponding mean-field predictions are independent of dimensionality.

6.2 Mean-field theory of percolation

We can apply mean-field techniques to the percolation model of §3.1.8 and §5.8. Recall that p is the probability that a 'bridge' joins any two sites on a lattice and that we are concerned to discover at what value of p there is a high enough density of bridges that it becomes possible to walk across them from one side of the lattice to the other.

We shall concentrate on the probability P that a bridge on a randomly chosen bond forms part of an infinite network stretching across the system. It is clear that a given bridge i will only form part of such a network if it has at least one neighbouring bridge. We say that one bridge is a 'neighbour' of another if it leads into one or other of the two sites that the bridge connects. Put another way, it will *not* form part of a network if its neighbouring bonds either have not been bridged or, if they have, are themselves not connected to the network. This gives us a relation connecting the probabilities that neighbouring bonds are part of the infinite cluster; if we make the (mean-field) assumption that these probabilities are independent for all bonds, this relation takes the form

$$1 - P_i = \prod_j (1 - pP_j), \quad j \text{ neighbouring } i, \quad (6.6)$$

where the product runs over neighbours j of the chosen bridge i and P_i is the probability that the ith bridge belongs to the infinite cluster. However, our assumption that all the P_i are independent of one another means that they must all be equal, $P_i = P$, so that equation (6.6) becomes

$$1 - P = (1 - pP)^z, \quad (6.7)$$

where z is the number of neighbouring bonds. A graphical solution of this equation is shown in Figure 6.2.

For all values of p, (6.7) has the trivial solution $P = 0$. However, for $p > p_c = 1/z$ there is a second, non-zero solution for P. Therefore, p_c represents the critical bridge concentration at which it is possible for an infinite cluster to exist.

Comparing this result with our previous calculations (see §5.8), we find that mean-field theory shows none of the sensitive dependence on the lattice geometry which we might expect. For example, we now predict the same

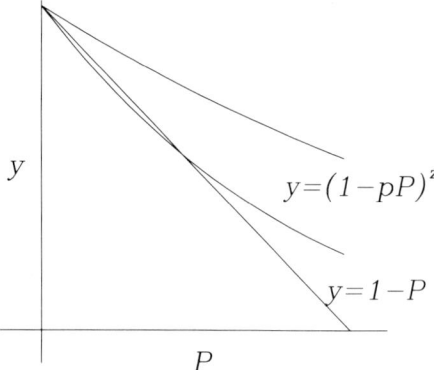

Figure 6.2 Graphical solution of the mean-field equation (6.7) for the percolation problem.

p_c for the cubic lattice in three dimensions and the triangular lattice in two dimensions, since both have $z = 6$. Just as with the Ising model, we also find a quite unphysical result in $d = 1$—we obtain a non-zero value for p_c even though we can see that in one dimension the loss of even one bond prevents percolation completely since there is only one possible path.

6.3 Mean-field theory of the non-ideal gas

In a real gas, all the atoms or molecules interact with each other through some potential $\Phi(\{\mathbf{x}_i\})$. For a system of N molecules or atoms whose internal degrees of freedom we neglect, we can write the partition function as

$$Z = \int d^d\mathbf{p}_1 d^d\mathbf{x}_1 \ldots d^d\mathbf{p}_N d^d\mathbf{x}_N \, \exp\left\{-\beta\left[\sum_i \frac{p_i^2}{2m} + \Phi(\{\mathbf{x}_i\})\right]\right\}, \quad (6.8)$$

where i labels the molecules. The mean-field theory of a non-ideal gas consists in supposing that the effect on each particle of all the others can be represented by an effective single-particle potential $\phi(\mathbf{x})$ in which all the molecules may be assumed to move. The total potential energy then decomposes into a sum of terms, one for each particle:

$$\Phi(\{\mathbf{x}_i\}) \approx \sum_i \phi(\mathbf{x}_i). \quad (6.9)$$

The integral in (6.8) now factorizes and we obtain

$$Z = \left[\int d^d\mathbf{p}\, d^d\mathbf{x} \, \exp\left\{-\beta[p^2/2m + \phi(\mathbf{x})]\right\}\right]^N. \quad (6.10)$$

6.3 The non-ideal gas

The momentum and position integrals now separate. The momentum integral will contribute a volume-independent term to the free energy and we can therefore neglect it for the purpose of calculating the pressure exerted by the gas.

The optimum form of the function $\phi(\mathbf{x})$ will depend on the original potential Φ. For the purposes of the present calculation, however, let us assume a particular simple form for it; let us suppose that it excludes each molecule from some volume V_{ex} of space by means of an infinite potential barrier (in order to model the 'hard-core repulsion' of the other molecules) and takes some finite value u elsewhere (which will be negative because of the long-range attractive forces acting between molecules). Therefore, to within a volume-independent factor coming from the momentum integration in (6.10),

$$Z = \left[(V - V_{\text{ex}})e^{-\beta u}\right]^N. \tag{6.11}$$

The free energy is then

$$F = -Nk_BT[\log(V - V_{\text{ex}}) - \beta u]. \tag{6.12}$$

The pressure may be found by differentiating with respect to volume at constant temperature:

$$p = -\left(\frac{\partial F}{\partial V}\right)_T = \frac{Nk_BT}{V - V_{\text{ex}}} - N\left(\frac{\partial u}{\partial V}\right)_T. \tag{6.13}$$

Recalling that u was supposed to originate from the attraction of a given molecule for all the others outside the excluded-volume region, it is clear that u should be proportional to the density of molecules in the gas, $u \propto N/V$. Presumably the excluded volume itself will be proportional to N. If we choose the constants of proportionality so that

$$u = -\frac{a}{N_A^2}\frac{N}{V}, \qquad V_{\text{ex}} = \frac{b}{N_A}N, \tag{6.14}$$

where N_A is Avogadro's constant, then (6.13) becomes

$$p = \frac{Nk_BT}{V - bn} - \frac{an^2}{V^2}, \tag{6.15}$$

where $n = N/N_A$ is the number of moles of gas in our sample. The reason for our apparently eccentric choice of proportionality constants is now clear; (6.15) is nothing but the van der Waals equation of state, recovered as the macroscopic form of mean-field theory for the non-ideal gas with a particularly simple *ansatz* for the mean-field potential.

6.4 A variational derivation of mean-field theory

So far, we have motivated mean-field theory by a common-sense appeal to the idea of interaction with some average, macroscopic 'field' (the magnetization for the Ising model, the percolation probability P in the percolation model, the effective potential in the non-ideal gas). We now show how we can put the method on a more rigorous footing mathematically and simultaneously gain more insight into its nature, at least for problems involving statistical mechanics.

First, we need to derive a general inequality for the free energy of any thermodynamic system which was first suggested by Peierls (1934). We shall follow in essence Feynman's (1955) proof.

The partition function is (equation (2.2))

$$Z = \sum_{\{s\}} \exp\bigl[-\beta H(\{s\})\bigr], \qquad (6.16)$$

where the set of variables $\{s\}$ describe the configuration of the system (they need not necessarily be spins). Now suppose that we break up the Hamiltonian into two parts:

$$H = H_0 + H_1. \qquad (6.17)$$

The choice of H_0 is governed only by the requirement that it should be possible to evaluate the corresponding partition function. Now consider the ratio of the true partition function to that of the system described by the Hamiltonian H_0:

$$\begin{aligned}\frac{Z}{Z_0} &= \frac{\sum_{\{s\}} \exp\bigl[-\beta(H_0 + H_1)\bigr]}{\sum_{\{s\}} \exp(-\beta H_0)} \\ &= \sum_{\{s\}} P_0(\{s\}) \exp(-\beta H_1) \\ &= \bigl\langle \exp(-\beta H_1) \bigr\rangle_0, \end{aligned} \qquad (6.18)$$

where P_0 is the correctly normalized Boltzmann probability factor for the system described by H_0:

$$P_0(\{s\}) = \frac{\exp\bigl[-\beta H_0(\{s\})\bigr]}{Z_0} \qquad (6.19)$$

and $\langle \ldots \rangle_0$ represents the average with respect to P_0.

For real arguments, the first and second derivatives of the exponential function are always positive. This implies (see Box 6.1) that for any function f

$$\langle \exp f \rangle \geq \exp \langle f \rangle. \qquad (6.20)$$

6.4 A variational derivation

Box 6.1: The convexity inequality

Let $g(x)$ be a function such as the exponential function for real arguments, whose first and second derivatives are everywhere positive, so that its slope and curvature are always upwards:

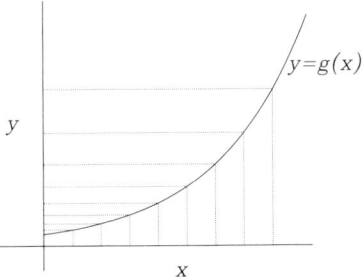

Consider any set of points $\{x_i\}$ along the x axis and the set $\{g(x_i)\}$ of their images under the function. Then, because of the curvature, the separation of the images of the points at the top end of the set is increased relative to the separation of the images of points at the bottom end, so that the 'centre of gravity' of the set of images always lies above the image of the original centre of gravity. The only exception is if all the original points happen to be coincident, in which case the centre of gravity of the image and the image of the centre of gravity are the same. We can write this as

$$\langle g(x) \rangle \geq g(\langle x \rangle), \tag{1}$$

where $\langle \ldots \rangle$ denotes an average over all the points.

Applying this result to (6.18), we find

$$\frac{Z}{Z_0} \geq \exp\left(-\beta \langle H_1 \rangle_0\right). \tag{6.21}$$

Taking logarithms of both sides, we obtain

$$\log Z - \log Z_0 \geq -\beta \langle H_1 \rangle_0, \tag{6.22}$$

or, remembering that $F = -k_B T \log Z$,

$$F \leq F_0 + \langle H_1 \rangle_0, \tag{6.23}$$

where F_0 is the free energy of the system governed by the Hamiltonian H_0.

This is known as the **Bogoliubov inequality**. To understand what it is telling us, it helps to rewrite the right-hand side using

$$F_0 = \langle H_0 \rangle_0 - T S_0, \tag{6.24}$$

where S_0 is the entropy corresponding to the probability distribution P_0:

$$S_0 = -k_B \sum_{\{s\}} P_0 \log P_0. \tag{6.25}$$

Substituting (6.24) into (6.23), we find

$$F \leq \langle H \rangle_0 - T S_0. \tag{6.26}$$

The right-hand side is therefore the free energy, evaluated using the *full* Hamiltonian $H = H_0 + H_1$ but with the simpler probability distribution P_0, associated with some other Hamiltonian H_0.

The Bogoliubov inequality is useful because, if we choose H_0 to be some simple approximation to the real Hamiltonian of a system, for which the right-hand side of the inequality can be exactly evaluated, then it gives us a rigorous upper bound on the free energy for the true Hamiltonian. In order to obtain good results, we would like H_0 to incorporate as much as possible of the physics, while still leaving the calculation simple. One choice for H_0 which is almost always tractable is one that decomposes into a sum of many terms, one for each of the problem's degrees of freedom. In other words, one for which the energy associated with each variable depends only on that variable's value and not on any of the other variables. But this is precisely the idea of mean-field theory—to replace the actual interaction between the parts of the system by a fictitious interaction with some external field or potential.

Our strategy for applying (6.23) is therefore this. We choose a trial Hamiltonian H_0 in which each variable in the system interacts not with the other variables, but with an effective external potential. We define this interaction in terms of one or more parameters $\lambda^{(i)}$. Then (6.23) tells us that, if we change the value of the $\lambda^{(i)}$, thereby altering *both* H_0 and H_1, in such a way as to minimize the quantity

$$F_{\text{var}}(\{\lambda^{(i)}\}) \equiv F_0(\{\lambda^{(i)}\}) + \langle H_1(\{\lambda^{(i)}\}) \rangle_0, \tag{6.27}$$

we shall obtain the best possible approximation to the full free energy with H_0 of this form.

Let us see how this works with the Ising model. Since the individual spins for this model can take only the values ± 1, the most general possible form of Hamiltonian which decomposes into a sum of functions of the individual spins and which satisfies the requirement of translational invariance can be expressed in terms of a single parameter λ in the form

$$H_0 = -\lambda \sum_i s_i. \tag{6.28}$$

6.4 A variational derivation

The partition function corresponding to H_0 is easily found:

$$Z_0(\lambda) = \left(2\cosh(\beta\lambda)\right)^N, \tag{6.29}$$

where N is the total number of spins.

The remaining part of the Hamiltonian, H_1, is therefore

$$H_1 = -\epsilon \sum_{\text{pairs } ij} s_i s_j + (\lambda - B)\sum_i s_i, \tag{6.30}$$

where $\epsilon = -\mathcal{J}$.

To evaluate the thermal average of H_1 in the canonical ensemble corresponding to H_0, we consider the two terms separately. In the first term, we can replace the values of s_i and s_j separately by their thermal averages since they are independent variables in the distribution P_0. The thermal average of each one is

$$\langle s\rangle_0 = \tanh(\beta\lambda) \tag{6.31}$$

and therefore the thermal average of the first term is

$$\tfrac{1}{2}Nz\epsilon\langle s\rangle_0^2 = \tfrac{1}{2}Nz\epsilon\tanh^2(\beta\lambda), \tag{6.32}$$

where $Nz/2$ is the number of nearest-neighbour pairs. The thermal average of the second term is even easier to calculate; it is $N(\lambda - B)\tanh(\beta\lambda)$. Equation (6.27) therefore becomes

$$F_{\text{var}}(\lambda) = N\left[-\frac{1}{\beta}\log\left[2\cosh(\beta\lambda)\right] - \tfrac{1}{2}z\epsilon\tanh^2(\beta\lambda) + (\lambda - B)\tanh(\beta\lambda)\right]. \tag{6.33}$$

To find the minimum, we differentiate with respect to λ and set the resulting expression equal to zero to obtain:

$$\lambda_{\min} - B = z\epsilon\tanh(\beta\lambda_{\min}). \tag{6.34}$$

The variational free energy when λ is given by (6.34) is

$$F_{\text{var}}(\lambda_{\min}) = -\frac{N}{\beta}\log\left[2\cosh(\beta\lambda_{\min})\right] + \frac{N(\lambda_{\min} - B)^2}{2z\epsilon}. \tag{6.35}$$

Now, by differentiating the free energy with respect to the external field B, we obtain the magnetization:

$$m = -\frac{1}{N}\frac{dF}{dB} = -\frac{1}{N}\left(\frac{\partial F}{\partial B} + \frac{\partial F}{\partial \lambda}\frac{\partial \lambda}{\partial B}\right)_{\lambda_{\min}}. \tag{6.36}$$

However, the second term in the brackets is zero because we are working at the minimum of F with respect to λ, and so

$$m = -\frac{1}{N}\frac{\partial F}{\partial B}$$
$$= \frac{\lambda_{\min} - B}{z\epsilon}. \qquad (6.37)$$

Using this to eliminate $\lambda_{\min}$ from equation (6.34), we find

$$m = \tanh[\beta(z\epsilon m + B)], \qquad (6.38)$$

which is exactly the same as our original mean-field equation (6.4). We have therefore shown that the best possible free energy of the form (6.27) with an F_0 corresponding to independent spins is exactly that of mean-field theory.

6.5 Correlation functions in mean-field theory

In this section we shall see how we can use mean-field theory to work out connected two-point correlation functions. However, we must first think carefully about just what it is that we want to calculate. In particular, we do not want $\langle s_i s_j \rangle_{c0}$, the connected correlation function for two spins in the trial probability distribution P_0 which we introduced in the previous section. This quantity will inevitably be zero, since the corresponding Hamiltonian H_0 includes no interactions between spins, so they cannot possibly be correlated. Instead let us take our variational expression (6.27), in which the free energy of the independent spins is corrected to give an approximation to the free energy of the interacting spins. Generalizing to allow for an external field B varying from site to site, we insert it into the linear response theorem (see §2.2):

$$\langle s_i s_j \rangle_c = -\frac{1}{\beta}\frac{\partial^2 F}{\partial B_i \partial B_j} = \frac{1}{\beta}\frac{\partial \langle s_i \rangle}{\partial B_j}. \qquad (6.39)$$

We should perhaps explain why we expect this procedure to give a better answer than just evaluating $\langle s_i s_j \rangle_{c0}$. We have optimized our variational parameters in such a way as to produce the best possible approximation to the true Hamiltonian (as measured by the value of the free energy). However, there is still a finite discrepancy between our trial probability distribution P_0 and the true equilibrium probability distribution P_{eq} for our original model. If we say loosely that this difference is of the order of some small number δP, then (above T_c at least) the difference between $\langle s_i s_j \rangle_c$ and $\langle s_i s_j \rangle_{c0}$ is

$$\langle s_i s_j \rangle_c - \langle s_i s_j \rangle_{c0} = \sum_{\{s\}}(P_{\text{eq}} - P_0)s_i s_j = O(\delta P). \qquad (6.40)$$

6.5 Correlation functions

On the other hand, we know from equation (6.23) that the exact free energy is the minimum with respect to all possible variations of H_0, and therefore of P_0. So, if we expand the free energy around the equilibrium probability distribution, there will be no term linear in $P_0 - P_{eq}$:

$$F_{\text{var}} = F_{\text{eq}} + O(\delta P^2). \tag{6.41}$$

Differentiating twice with respect to the external field B, we see that the connected correlation function calculated from F_{var} using equation (6.39) differs from its true value only by terms of order δP^2. Even if we could not see immediately that $\langle s_i s_j \rangle_{c0}$ would give zero, we should still expect that using the variational free energy in (6.39) would give a better answer.

Consider, for example, the Ising model in an inhomogeneous magnetic field. There is now one parameter $\lambda^{(i)}$ in H_0 for every site on the lattice and one may readily show (see Problem 6.1) that the values of $\lambda^{(i)}$ which minimize F_{var} satisfy

$$\lambda_{\min}^{(i)} - B_i = \epsilon \sum_{\text{neighbours } j} \tanh(\beta \lambda_{\min}^{(j)}), \tag{6.42}$$

where the sum on j runs over the nearest neighbours of site i. Equation (6.35) generalizes to (see Problem 6.1):

$$F_{\text{var}}(\{\lambda_{\min}^{(k)}\}) = \sum_i \left[-\frac{1}{\beta} \log\left[2\cosh(\beta \lambda_{\min}^{(i)})\right] + \tfrac{1}{2}(\lambda_{\min}^{(i)} - B_i)\tanh(\beta \lambda_{\min}^{(i)}) \right]. \tag{6.43}$$

Differentiating this with respect to B_i and using (6.42) to eliminate $\lambda_{\min}^{(i)}$, we find that the thermal average of spin i is

$$\langle s_i \rangle = \tanh\left[\beta\left(B_i + \epsilon \sum_{\text{neighbours } j} \langle s_j \rangle\right)\right]. \tag{6.44}$$

Equation (6.44) for $i = 1, \ldots, N$ constitutes a system of N non-linear simultaneous equations for the $\langle s_i \rangle$, which because of the inhomogeneous magnetic field are no longer all equal. Let us assume that $T > T_c$ and that the fields B_i are weak. Then the thermal averages $\langle s_j \rangle$ will be small and we can make a first-order Taylor expansion in the argument of the tanh to obtain:

$$\langle s_i \rangle = \beta\left(B_i + \epsilon \sum_{\text{neighbours } j} \langle s_j \rangle\right). \tag{6.45}$$

This is now a system of *linear* simultaneous equations for the $\langle s_i \rangle$. The interaction is invariant under translations, that is, $\mathcal{J}_{ij}$ is a function of $\mathbf{i} - \mathbf{j}$ only, where (as in §§3.3 and 3.4) $\mathbf{i}$ is the position vector of site i in units of the lattice constant. Hence we can produce N independent equations by taking discrete Fourier transforms—see §3.3 and Appendix F. For a cubic lattice in d dimensions, we find (cf. below equation (3.46))

$$\left(1 - 2\beta\epsilon \sum_{l=1}^{d} \cos(2\pi q_l/L)\right) \langle \tilde{s}_\mathbf{q} \rangle = \beta \tilde{B}_\mathbf{q}, \qquad (6.46)$$

where $\mathbf{q}$ is a vector with integer components and $L \equiv N^{1/d}$ is the number of sites in each direction across the system. Therefore,

$$\frac{\partial \langle \tilde{s}_\mathbf{q} \rangle}{\partial \tilde{B}_\mathbf{q}} = \frac{\beta}{1 - 2\beta\epsilon \sum_{l=1}^{d} \cos(2\pi q_l/L)}. \qquad (6.47)$$

However, Fourier transforming equation (6.39), we see that $\partial \langle \tilde{s}_\mathbf{q} \rangle / \partial \tilde{B}_\mathbf{q}$ is nothing but β times the connected spin–spin correlation in Fourier space. Hence we have

$$\widetilde{G}_c^{(2)}(\mathbf{q}) = \frac{1}{1 - 2\beta\epsilon \sum_{l=1}^{d} \cos(2\pi q_l/L)}. \qquad (6.48)$$

Below T_c, there is spontaneous magnetization m per site even in the absence of any applied magnetic field. If the applied field, and hence the changes $\delta \langle s_i \rangle \equiv \langle s_i \rangle - m$ in the thermal averages of the spins from their zero-field values, are small, then we can expand (6.44) about $\langle s_i \rangle = m$ to get

$$\delta \langle s_i \rangle = \beta \left(B_i + \epsilon \sum_{\text{neighbours } j} \delta \langle s_j \rangle \right) \text{sech}^2(\beta z \epsilon m), \qquad (6.49)$$

which has the same form as that above T_c, except for the presence of the additional sech^2 factor. Thus we arrive at an expression for $\widetilde{G}_c^{(2)}(\mathbf{q})$ below the critical temperature:

$$\widetilde{G}_c^{(2)}(\mathbf{q}) = \frac{\text{sech}^2(\beta z \epsilon m)}{1 - 2\beta\epsilon \, \text{sech}^2(\beta z \epsilon m) \sum_{l=1}^{d} \cos 2\pi q_l / L}. \qquad (6.50)$$

In the theory of critical phenomena we are most interested in the form of $\widetilde{G}_c^{(2)}(\mathbf{q})$ in the limit of small $\mathbf{q}$, since this encodes the long-range limit of the correlation function $G_c^{(2)}(j) = \langle s_0 s_j \rangle_c$. For small

$$k_l \equiv 2\pi q_l/L \qquad (6.51)$$

we can expand the cosines in (6.48) to order k_l^2. Recalling that for a cubic lattice $z = 2d$, we find

$$\widetilde{G}_c^{(2)}(k) = \left(\frac{\xi^2}{\beta\epsilon}\right)\left(\frac{1}{1 + k^2\xi^2}\right) \quad (T > T_c), \qquad (6.52)$$

where

$$\xi^2 \equiv \frac{\beta\epsilon}{1 - z\beta\epsilon} \quad (T > T_c). \tag{6.53}$$

Since $\widetilde{G}_c^{(2)}$ is a function of $k\xi$, its Fourier transform, the real-space connected correlation function $G_c^{(2)}(i)$, must be a function of i/ξ, and we infer that ξ is proportional to the system's correlation length.

Expanding the cosine on the right-hand side of equation (6.50) for $\widetilde{G}_c^{(2)}$ at $T < T_c$ in powers of k, and using (6.38) to replace sech^2 with $1 - m^2$, we find that (6.52) still applies but with ξ given by

$$\xi^2 = \frac{\beta\epsilon(1 - m^2)}{1 - z\beta\epsilon(1 - m^2)} \quad (T < T_c). \tag{6.54}$$

6.6 Infinite-range interactions

We might expect that mean-field theory would become exact if every spin interacted equally strongly with an infinite number of others, so that averaging over configurations of the spins with which it interacted would be equivalent to averaging over the whole sample. We shall now show that this is indeed the case. Suppose that we have a sample of N Ising spins, with equal interactions of strength $-\epsilon/N$ between each pair:[1]

$$H = -\frac{\epsilon}{2N} \sum_{i \neq j} s_i s_j = -\frac{\epsilon}{2N}\left[\left(\sum_i s_i\right)^2 - N\right], \tag{6.55}$$

where we have used the identity $s_i^2 \equiv 1$. The partition function is then

$$Z = \sum_{\{s\}} \exp\left(\tfrac{1}{2}\beta\epsilon\left[\frac{1}{N}\left(\sum_i s_i\right)^2 - 1\right]\right). \tag{6.56}$$

We now insert the identity

$$e^{a^2} = \frac{1}{\sqrt{2\pi}} \int_{-\infty}^{\infty} dy \, \exp\left(-\tfrac{1}{2}y^2 + \sqrt{2}ay\right) \tag{6.57}$$

with $a^2 = (\beta\epsilon/2N)(\sum_i s_i)^2$ into (6.56) and interchange the order of the sum and the integral to obtain

$$Z = \frac{e^{-\beta\epsilon/2}}{\sqrt{2\pi}} \int_{-\infty}^{\infty} dy \, e^{-y^2/2} \sum_{\{s\}} \exp\left[y\left(\frac{\beta\epsilon}{N}\right)^{1/2} \sum_i s_i\right]. \tag{6.58}$$

---

[1] The factor of $1/N$ has been introduced in order to ensure a well-behaved thermodynamic limit. Without it, the energy associated with the reversal of a single spin would grow linearly with the volume of the sample and a well-defined limit would be impossible.

Noticing that the second exponential in the integral can be factorized into a product of N terms, one for each spin, we can carry out the summation. Changing to a new variable $x \equiv y/\sqrt{N}$ we find

$$Z = e^{-\beta\epsilon/2}\sqrt{\frac{N}{2\pi}} \int_{-\infty}^{\infty} dx \left[2e^{-x^2/2} \cosh\left(\sqrt{\beta\epsilon}\,x\right) \right]^N. \qquad (6.59)$$

As N becomes large, the integrand becomes more and more dominated by the values of x near the maximum of the term in square brackets. This maximum occurs when

$$\tanh\left(x\sqrt{\beta\epsilon}\right) = \frac{x}{\sqrt{\beta\epsilon}}. \qquad (6.60)$$

The integral can now be evaluated by the method of steepest descent described in Appendix G. With x given by equation (6.60), the result is

$$\begin{aligned}\log Z &= N \log[2e^{-x^2/2}\cosh(\sqrt{\beta\epsilon}\,x)] + O(\log N), \\ &= N\log[2\cosh(\beta\lambda)] - \frac{N\beta}{2\epsilon}\lambda^2 + O(\log N),\end{aligned} \qquad (6.61)$$

where

$$\lambda \equiv \sqrt{\epsilon/\beta}\,x. \qquad (6.62)$$

If we remember that ϵ in our present problem is equivalent to $z\epsilon$ as we defined it in §6.4, we see that equations (6.60) and (6.61) are just the same as the corresponding equations, (6.34) and (6.35), for mean-field theory in the limit $z \to \infty$ with $B = 0$.

Before we move on, it is worth noticing that the limit we have been working in, in which each spin is coupled to infinitely many others, could also be obtained by increasing the spatial dimensionality. In a cubic lattice each spin has $2d$ nearest neighbours, so even with a nearest-neighbour interaction we can ensure that each spin interacts with infinitely many others simply by letting $d \to \infty$. We should expect, therefore, that mean-field theory is also exact in the limit of large d. In fact, as we shall see in Chapter 10, there is for each system a particular dimensionality, known as the (upper) critical dimensionality, above which mean-field theory gives exactly the right description of critical properties. For the Ising model and many of the other models of interest in the study of phase transitions, we shall see that this dimensionality is four.

6.7 Critical exponents in mean-field theory

In §1.3 we defined six critical exponents. Let us evaluate these exponents for the Ising model of a ferromagnet in the mean-field approximation. First, let us find the four exponents that describe the behaviour of the thermodynamic variables of the system, i.e., those which do not involve the correlation functions. All these may be found from equation (6.4). It is convenient to rewrite this equation in terms of the 'reduced variables'

$$t \equiv \frac{T - T_c}{T_c}, \qquad b \equiv \beta B, \tag{6.63}$$

so that (6.4) becomes

$$m = \tanh\left(\frac{m}{1+t} + b\right), \tag{6.64}$$

where we have made use of the result (6.5) that $z\beta_c \epsilon = 1$. To start with, let us examine the spontaneous magnetization in zero magnetic field just below T_c, that is for t small and negative. Expanding the tanh function to third order in its argument:

$$\tanh(x) = x - \tfrac{1}{3}x^3 + O(x^5), \tag{6.65}$$

we find

$$m = \frac{m}{(1+t)} - \tfrac{1}{3}\frac{m^3}{(1+t)^3}. \tag{6.66}$$

The non-zero solution for m is given by

$$m^2 = -3t(1+t)^2. \tag{6.67}$$

So, when $|t|$ is small

$$m \sim |t|^{1/2}. \tag{6.68}$$

This gives us the critical exponent $\beta = 1/2$.

Second, let us calculate the zero-field isothermal susceptibility as a function of temperature. Above T_c it is sufficient to keep only the first term in the Taylor expansion (6.65):

$$m = \frac{m}{(1+t)} + b. \tag{6.69}$$

Therefore

$$\chi \equiv \beta \frac{\partial m}{\partial b}\bigg|_{b=0} \sim 1/t. \tag{6.70}$$

For $T < T_c$, we must include the term in m^3 on the right-hand side of equation (6.69) and then evaluate the derivative at the non-zero value of m given by equation (6.67). When we are close to the critical temperature, the solution for m remains small and the first-order change in it is

$$\delta m = \frac{(1+t)b}{t} \sim 1/t. \tag{6.71}$$

We therefore have $\gamma = \gamma' = 1$.

Third, let us calculate the specific heat at zero magnetic field as a function of temperature. We can express this in terms of the free energy in the form:

$$C_B = T\left(\frac{\partial S}{\partial T}\right)_{B=0} = T\left(\frac{\partial^2 F}{\partial T^2}\right)_{B=0}. \tag{6.72}$$

In taking the derivative, we have in principle to include the dependence of F on T both explicitly and implicitly through the value of the variational parameter λ. However, our variational procedure guarantees that F is stationary with respect to variations of λ, so we need only consider the temperature dependence of λ when performing the second of our two differentiations. With this simplification, (6.72) becomes:

$$C_B = Nk_B\left[\frac{\lambda_{\min}}{k_B T^2} - \beta\frac{\partial \lambda_{\min}}{\partial T}\right]\tanh(\beta\lambda_{\min}). \tag{6.73}$$

Above the transition temperature, $\lambda_{\min} = 0$ and we see at once that $C_B = 0$. Below the transition, we already know that $m \sim |t|^{1/2}$ and that in zero field $m \propto \lambda$, so $\partial \lambda/\partial T \sim |t|^{-1/2}$. Therefore, as the transition is approached from below, C_B tends to a non-zero constant and there is a simple jump discontinuity in C_B at T_c. We therefore have $\alpha = \alpha' = 0$.

Now we calculate the magnetization that results when a magnetic field is applied exactly at the critical temperature. Expanding the tanh function to third order in its argument, we find

$$m = m + b - \tfrac{1}{3}m^3 - \tfrac{1}{3}b^3 + O(m^5, b^5). \tag{6.74}$$

Cancelling m from both sides, we find

$$b \sim m^3 \quad \text{for small } b, m, \tag{6.75}$$

which gives the critical exponent $\delta = 3$.

Finally we find the value of the exponent ν that governs the divergence of the correlation length at $T > T_c$. We argued above that for $T > T_c$ the correlation length is proportional to the quantity ξ defined by equation (6.53). Using (6.5) we find that ξ depends on temperature as

$$\xi \sim t^{-1/2} \quad \text{for } t \text{ small and positive.} \tag{6.76}$$

Hence the critical exponent $\nu = \tfrac{1}{2}$. A similar analysis of (6.54) shows that ν', which governs the divergence of the correlation length as T tends to T_c from below, is also equal to $\tfrac{1}{2}$.

6.7 Critical exponents

6.7.1 Calculating η from $\widetilde{G}_c^{(2)}(\mathbf{k})$

The exponent η is defined by equation (1.11) which says that

$$G_c^{(2)}(r) \sim \frac{1}{r^{d-2+\eta}}, \qquad r \to \infty, \ T = T_c. \tag{6.77}$$

This is not really a very convenient definition, since we only know the correlation function in Fourier space, and not in real space. Luckily, it proves possible to Fourier transform the definition (6.77) itself to give an alternative definition of η in terms of the behaviour of $\widetilde{G}_c^{(2)}(\mathbf{k})$ at small $\mathbf{k}$. This alternative definition will be useful in the following chapters on the Landau–Ginzburg model, as well as in the present mean-field calculation, so we go through its derivation here in detail. The argument runs as follows.

If $G_c^{(2)}(r)$ has the asymptotic behaviour described by equation (6.77), it must be possible to write it in the form

$$G_c^{(2)}(r) = f(r/a) \frac{a^p}{r^{d-2+\eta}}. \tag{6.78}$$

Here a is any constant quantity with the dimensions of a length, and p is some number chosen to give the overall expression the correct dimensions. It does not matter for the purposes of the argument what values these quantities take. f is a dimensionless function of the dimensionless variable r/a, which must tend to zero as its argument goes to zero in order that $G_c^{(2)}(0)$ be finite, and to a constant as its argument diverges so that $G_c^{(2)}$ assumes the given asymptotic form.

Now we Fourier transform to get $\widetilde{G}_c^{(2)}(k)$:

$$\widetilde{G}_c^{(2)}(k) = a^p \int d^d\mathbf{r}\, e^{-i\mathbf{k}\cdot\mathbf{r}} f(r/a) \frac{1}{r^{d-2+\eta}}. \tag{6.79}$$

It is simplest to evaluate the integral in d-dimensional spherical coordinates. If we choose the polar axis to lie in the direction of $\mathbf{k}$, all but one of the angular integrals can immediately be performed to give

$$\begin{aligned}
\widetilde{G}_c^{(2)}(k) &= a^p \Omega_{d-2} \int_{r=0}^{\infty} dr\, \frac{f(r/a)}{r^{-1+\eta}} \int_{\theta=0}^{\pi} d\theta\, \sin^{d-2}\theta\, e^{-ikr\cos\theta} \\
&= a^p \Omega_{d-2} [\tfrac{1}{2}(d-3)]!\, (-\tfrac{1}{2})! \int_0^{\infty} dr\, \frac{f(r/a)}{r^{-1+\eta}} \frac{J_{(d-2)/2}(kr)}{(kr/2)^{(d-2)/2}} \\
&= k^{-2+\eta} a^p (2\pi)^{d/2} \int_0^{\infty} dx\, x^{2-(d/2)-\eta} f\left(\frac{x}{ak}\right) \frac{J_{(d-2)/2}(x)}{x^{(d-2)/2}},
\end{aligned} \tag{6.80}$$

where Ω_{d-2} is the area of the unit $(d-2)$-sphere (see Problem 3.4), J_ν is a Bessel function, and in the third line we have made a change of variables

Table 6.1. Critical exponents in mean-field theory

Exponent	Definition	Conditions	Value
α	$C_B \sim \|T - T_c\|^{-\alpha}$	$B = 0$	0 (discontinuity)
β	$M \sim (T_c - T)^\beta$	$T < T_c$, $B = 0$	1/2
γ	$\chi_T \sim \|T - T_c\|^{-\gamma}$	$B = 0$	1
δ	$B \sim \|M\|^\delta$	$T = T_c$	3
ν	$\xi \sim \|T - T_c\|^{-\nu}$		1/2
η	$\widetilde{G}_c^{(2)}(k) \sim k^{-2+\eta}$	$T = T_c$	0

to the dimensionless variable of integration $x \equiv kr$. Now let us examine the limit $k \to 0$. In this limit the remaining integral will become independent of k, because it will be dominated by contributions from the regime in which the argument of f is large, and in this regime f is, as we have said, a constant. Thus the limiting behaviour of $\widetilde{G}_c^{(2)}(k)$ with k small is

$$\widetilde{G}_c^{(2)}(k) \sim k^{-2+\eta}. \tag{6.81}$$

At T_c, when $\beta = \beta_c = 1/z\epsilon$, equation (6.48) predicts that

$$\widetilde{G}_c^{(2)}(k) \sim k^{-2}, \tag{6.82}$$

where **k** is now defined by equation (6.51). Comparing this with (6.80), it follows that mean-field theory predicts $\eta = 0$.

The exponents predicted by mean-field theory are summarized in Table 6.1. It is interesting to see whether these results satisfy the scaling laws that we saw in §1.5 are obeyed experimentally. The answer is rather curious: substituting the values from Table 6.1 into the Rushbrooke, Fisher and Griffiths scaling equations (see (5.80)), one sees at once that they are satisfied. However, the Josephson scaling relation, in which the dimensionality d appears in addition to the critical exponents, is obeyed *only* if $d = 4$. This is related to the fact, to which we alluded at the end of §6.6, that mean-field theory gives a correct picture of critical behaviour only in $d \geq 4$.

6.8 What is missing from mean-field theory?

We have seen how mean-field theory can be used to describe a wide variety of models. However, comparison of the results for the critical exponents for the Ising model given in Table 6.1 with the exact results listed in Table 3.1 shows that mean-field theory is not accurate near the transition. Worse still, the mean-field theory of the Ising model predicts a phase transition at non-zero temperature even for $d = 1$, whereas we saw in §3.2 that the Ising chain has zero magnetization for all $T \neq 0$. And when a transition *does* occur, as in the $d = 2$ and $d = 3$ Ising models, mean-field theory overestimates T_c.

The feature of mean-field theory which makes it universally applicable is that it identifies the order parameter of the system and tries to describe it as simply as possible. However, in one important respect it goes too far: it assumes that one need only take account of configurations in which the order parameter is uniform, and therefore that every spin, bond or whatever behaves in an average manner, regardless of what its neighbours are doing. In other words, it neglects fluctuations in the order parameter in which nearby parts of the system, while remaining correlated with each other, do something different from the average. It is this neglect that is responsible for the consistent overestimation of T_c (or underestimation of β_c) in mean-field theory; long-wavelength fluctuations enable the system to enjoy almost all the energetic benefits of the ordered state without the entropy cost associated with macroscopic ordering. This is also the reason why mean-field theory fails to describe the correlation functions (and therefore the critical exponents) properly in systems with local interactions in low dimensionalities.

What is needed to improve on mean-field theory is a treatment which includes these fluctuations and in particular recognizes the significance of the divergence of the correlation length at the transition as a symptom of the importance of fluctuations of larger and larger wavelengths. The next chapter provides one, in the shape of the Landau–Ginzburg model.

Problems

6.1 Derive (6.43), which gives the variational free energy for the Ising model in the presence of an inhomogeneous magnetic field.

6.2 Consider a mean-field treatment of the Gaussian model of §3.1.6. Show that it resembles the exact answer in that the average value of each spin is zero above a critical temperature, while below the critical temperature mean-field theory becomes undefined.

6.3 Use the linear response theorem to show that the expression for the two-spin correlation function within the mean-field theory of the Gaussian model is identical to the mean-field result for the Ising model.

6.4 Show that the mean-field critical temperature and correlation functions computed for the Gaussian model in Problems 6.2 and 6.3 are in fact exact.

7
The Landau–Ginzburg model

Most of the rest of this book is devoted to study of the Landau–Ginzburg model. We devote so much space to this model because it is a kind of 'metamodel'—it captures the essence of many of the models we have discussed, and elucidates how their behaviour near a critical point depends (i) on the dimensionality d of the underlying lattice, and (ii) on the dimensionality D of the order parameter. A toy such as the Landau–Ginzburg model that allows one to vary d and D with a minimum of fuss (and even make them non-integer!) reveals just how much phase transitions depend on D and d, and how little on the underlying lattice.

Indeed, the Landau–Ginzburg model tries to eliminate the lattice altogether by making the order parameter into a continuous D-component field $\boldsymbol{\phi}(\mathbf{x})$. We say 'tries' because we shall find in Chapter 8 that the lattice does not completely disappear from view. It simply recedes into a hazy fuzz characterized by a single number Λ.

7.1 Formulation of the Landau–Ginzburg model

You may get a clearer picture of what the Landau–Ginzburg model is trying to do by imagining that you are using an optical laser to measure opalescence at a gas–liquid critical point. Light is scattered by fluctuations in the density $\rho(\mathbf{x})$, so you seek the probability functional $P[\delta\rho]$ that will return the probability associated with any given form $\delta\rho(\mathbf{x})$ of the difference between the

7.1 Formulation of the Landau–Ginzburg model

density and its mean value. Since one cannot see atoms with visible light, you argue that P should not involve details of the structure and disposition of individual gas molecules, but should be expressible in terms of macroscopic, phenomenological parameters such as the temperature $T(\mathbf{x})$, compressibility $\kappa_T(\mathbf{x})$, or whatever, averaged within volumes large compared with the wavelength of visible light—volumes which contain hundreds of millions of atoms. Correspondingly $\delta\rho(\mathbf{x})$ is to be interpreted as the mean density within a volume of this type centred on $\mathbf{x}$.

Einstein offered an early solution to this problem of finding the probability functional $P[\delta\rho]$—see Appendix I. His probability functional is

$$P[\delta\rho] \propto \exp\left[-\beta \int d^3\mathbf{x}\, h_{\mathrm{Ei}}(\delta\rho(\mathbf{x}))\right], \qquad (7.1)$$

where h_{Ei} is

$$h_{\mathrm{Ei}}(\delta\rho) \equiv \frac{1}{2\kappa_T}(\delta\rho/\rho)^2. \qquad (7.2)$$

We refer to a function $h(\delta\rho)$ that generates a probability functional in the way that h_{Ei} generates $P[\delta\rho]$ as a **Hamiltonian density**; the similarity of (7.1) to the usual Boltzmann factor gives rise to the Hamiltonian sobriquet, and it is called a density because it must be integrated over all space before its exponential is taken.

According to Einstein's formula (7.1), density fluctuations in an opalescent fluid have a Gaussian distribution with variance proportional to the isothermal compressibility κ_T. This diverges at the critical point $T = T_c$, so the fluctuations also diverge there. This prediction, interesting though it is, is unsatisfactory in three respects:
(i) It does not tell us how κ_T varies with $(T - T_c)$.
(ii) It is unlikely to be valid for large $\delta\rho$ since its derivation involves the assumption that the fluctuations in temperature and pressure are proportional to those in density (see Appendix I).
(iii) It is clearly invalid at and below T_c since there $\delta\rho$ is not infinite as Einstein's formula predicts, but $|\delta\rho| \lesssim \rho_{\mathrm{liquid}} - \rho_{\mathrm{gas}}$.

To see how we could overcome these difficulties and to anticipate a problem that does not arise at a gas–liquid critical point, we now examine the corresponding problem for the Heisenberg model of arbitrary dimensionality d. The order parameter $\boldsymbol{\phi}$ is now the local mean magnetization. That is, $\boldsymbol{\phi}$ is the mean magnetization in volumes smaller than the resolution of one's measuring instruments, but large enough to contain many atoms. In general $\boldsymbol{\phi}$ is a vector that can point in any direction. Consider first configurations in which $\boldsymbol{\phi}$ does not vary from point to point. Then for any given microstate the energy is a functional of $|\boldsymbol{\phi}|^2$, so the Hamiltonian density may be expanded

$$h = h_0 + \tfrac{1}{2}\mu'|\boldsymbol{\phi}|^2 + \tfrac{1}{4!}\lambda'(|\boldsymbol{\phi}|^2)^2 + \cdots, \qquad (7.3)$$

where μ' and λ' are parameters to be determined.

Now consider configurations in which the direction of $\boldsymbol{\phi}$ varies with position. If the Hamiltonian density (7.3) were valid in such a case, the energy of a configuration would be the same whether $\boldsymbol{\phi}$ changed direction rapidly or constantly pointed the same way. This is clearly unrealistic—h should increase when the magnetization $\boldsymbol{\phi}$ varies rapidly from point to point since then many nearest-neighbour dipoles must be misaligned. We represent this state of affairs by making h_0 a function of $\boldsymbol{\nabla}\boldsymbol{\phi}$. Since it should be an increasing function of $|\boldsymbol{\nabla}\boldsymbol{\phi}|^2$ we write[1]

$$h_0 = \tfrac{1}{2}\alpha'^2|\boldsymbol{\nabla}\boldsymbol{\phi}|^2 + \cdots, \tag{7.4}$$

where α' is a real constant. We shall find it sufficient to take only the first term in this series and the first three terms in the series of (7.3). The energy of a microstate compatible with $\boldsymbol{\phi}(\mathbf{x})$ is then

$$H = \int d^d\mathbf{x}\, h(\mathbf{x}) = \int d^d\mathbf{x}\left[\tfrac{1}{2}\alpha'^2|\boldsymbol{\nabla}\boldsymbol{\phi}|^2 + \tfrac{1}{2}\mu'|\boldsymbol{\phi}|^2 + \tfrac{1}{4!}\lambda'(|\boldsymbol{\phi}|^2)^2\right]. \tag{7.5}$$

We cannot yet compare $e^{-\beta H}$, with H given by (7.5), with Einstein's equation (7.1) because $e^{-\beta H}$ is the probability of just one of the microstates that are compatible with a given mean-magnetization field $\boldsymbol{\phi}(\mathbf{x})$, while Einstein's relation directly gives the probability of a given $\boldsymbol{\phi}(\mathbf{x})$. To obtain a relation equivalent to Einstein's we must sum over all microstates compatible with $\boldsymbol{\phi}(\mathbf{x})$. The number N of these microstates is the sum, over all the tiny volumes for which the mean magnetization $\boldsymbol{\phi}$ is defined, of the number of arrangements of that volume's spins which generate the given $\boldsymbol{\phi}$. The number of such arrangements clearly depends only on $|\boldsymbol{\phi}|$, so we may express the desired total number of microstates as

$$N = \int d^d\mathbf{x}\, e^{\omega(|\boldsymbol{\phi}|^2)}. \tag{7.6}$$

We now expand ω as

$$\omega(|\boldsymbol{\phi}|^2) = \text{constant} + \tfrac{1}{2}\omega_2|\boldsymbol{\phi}|^2 + \tfrac{1}{4!}\omega_4(|\boldsymbol{\phi}|^2)^2 + \cdots. \tag{7.7}$$

If we retain only the first three terms in this expansion, the probability of a given mean-magnetization field $\boldsymbol{\phi}(\mathbf{x})$ will be

$$\begin{aligned} P[\boldsymbol{\phi}] &\propto e^{-\int d^d\mathbf{x}\,(\beta h - \omega)} \\ &\propto e^{-\int d^d\mathbf{x}\,\mathcal{H}_{LG}}, \end{aligned} \tag{7.8}$$

---

[1] If $\boldsymbol{\phi}$ is a multicomponent object, $|\boldsymbol{\nabla}\boldsymbol{\phi}|^2 \equiv \sum_{\alpha=1}^{D}\sum_{i=1}^{d}(\partial\phi_\alpha/\partial x_i)^2$.

7.1 Formulation of the Landau–Ginzburg model

Table 7.1. Dimensions of quantities involved in the Landau–Ginzburg model

Variable	$\mathcal{H}$	$\boldsymbol{\phi}$	α	μ^2	λ	Λ
Dimension	L^{-d}	$L^{-d/2}$	L	L^0	L^d	L^{-1}

where

$$\mathcal{H}_{LG} \equiv \tfrac{1}{2}\alpha^2|\boldsymbol{\nabla}\boldsymbol{\phi}|^2 + \tfrac{1}{2}\mu^2|\boldsymbol{\phi}|^2 + \tfrac{1}{4!}\lambda(|\boldsymbol{\phi}|^2)^2;$$
$$\alpha \equiv \sqrt{\beta}\,\alpha'; \quad \mu^2 \equiv \beta\mu' - \omega_2; \quad \lambda \equiv \beta\lambda' - \omega_4. \tag{7.9}$$

Finally, we generalize $\mathcal{H}_{LG}$ to include a non-zero external magnetic field **B**. Such a field gives rise to an additional energy density $-\mathbf{B}\cdot\boldsymbol{\phi}$ (see Appendix C), so in an external field $\mathcal{H}_{LG}$ is

$$\begin{aligned}\mathcal{H}_{LG} &= \tfrac{1}{2}\alpha^2|\boldsymbol{\nabla}\boldsymbol{\phi}|^2 + \tfrac{1}{2}\mu^2|\boldsymbol{\phi}|^2 + \tfrac{1}{4!}\lambda(|\boldsymbol{\phi}|^2)^2 - \beta\mathbf{B}\cdot\boldsymbol{\phi} \\ &= \tfrac{1}{2}\alpha^2|\boldsymbol{\nabla}\boldsymbol{\phi}|^2 + \tfrac{1}{2}\mu^2|\boldsymbol{\phi}|^2 + \tfrac{1}{4!}\lambda(|\boldsymbol{\phi}|^2)^2 - \mathbf{J}\cdot\boldsymbol{\phi},\end{aligned} \tag{7.10}$$

where in the second line we have introduced the **source field**

$$\mathbf{J} \equiv \beta\mathbf{B}. \tag{7.11}$$

The probability functional of equation (7.8) and effective Hamiltonian density (7.10) define the **Landau–Ginzburg model**. The latter depends upon three phenomenological parameters: a characteristic length α that determines the importance of the gradient term, the quantity μ^2 whose temperature variation we shall find drives the phase transition, and a parameter λ, which must be positive since otherwise P would be unbounded for large $|\boldsymbol{\phi}|$ and thus unnormalizable. For historical reasons λ is referred to as the **coupling**. Notice that the temperature-dependent coefficient of the Landau–Ginzburg model is μ^2, not μ. This notation is not intended to imply that $\mu^2 > 0$. Indeed for a phase change to be possible we shall see that we require $\mu' < 0$. The square is a tiresome convention that goes back to the origin of expressions like (7.9) in particle physics.

The argument of an exponential must be dimensionless, so $\mathcal{H}_{LG}$ must have dimension L^{-d}. From this it follows that $\boldsymbol{\phi}$ has dimension $L^{-d/2}$, that μ^2 is dimensionless, and that λ has dimension L^d (see Table 7.1).

In general we expect the number e^ω of microstates to decrease with increasing $|\boldsymbol{\phi}|$ since when $|\boldsymbol{\phi}|$ is large the spins have to be highly coordinated. Hence we expect ω_2, like μ', to be negative, and μ^2 will be of the form

$$\mu^2 = A\frac{T - T_0}{T} \tag{7.12}$$

where A and T_0 are positive constants. Hence μ^2 is liable to change sign from negative at low T (large β), to positive at high T.

The Landau–Ginzburg effective Hamiltonian density (7.10) augments Einstein's density (7.2) by two terms: (i) a term $\propto |\nabla\boldsymbol{\phi}|^2$ that discourages rapid fluctuations in the order parameter, and (ii) a term $\propto |\boldsymbol{\phi}|^4$ that discourages fluctuations towards large $\boldsymbol{\phi}$. This second term enables the Landau–Ginzburg model to make sense below the critical temperature, and thus to represent the partially condensed phase. We shall see that the first term enables the Landau–Ginzburg model to make predictions concerning the structure of correlation functions.

It is natural to ask whether we would obtain significantly different predictions if we retained more terms than we did in the expansions (7.3), (7.4) and (7.7). For example, what about including in $\mathcal{H}$ a term proportional to $|\boldsymbol{\phi}|^6$? Well, the coefficient of such a term would have to be positive—a negative coefficient would allow $\mathcal{H}$ to become unboundedly large and negative for large $\boldsymbol{\phi}$, with the consequence that P could not be normalized. And in this case the new term would make no difference to the critical exponents we would predict since in Chapter 14 we shall find that critical exponents do not depend on exactly how $\mathcal{H}$ depends on $|\boldsymbol{\phi}|$ for large $|\boldsymbol{\phi}|$ so long as $\mathcal{H}$ includes a certain minimum set of terms, which are precisely the terms that make up (7.10). This result goes some way towards explaining the phenomenon of universality encountered in §1.4. The universality class to which a theory belongs, that is to say the set of critical exponents which it predicts, depends on d and D but not on the detailed form of $\mathcal{H}$.

The critical exponents and therefore the universality class are, however, liable to change if we make $\mathcal{H}$ depend on the components of $\boldsymbol{\phi}$ other than through the combination $|\boldsymbol{\phi}|$. For example, if we introduced into $\mathcal{H}$ a term $\sum_{\alpha=1}^{D} \phi_\alpha^4$, which is not invariant when $\boldsymbol{\phi}$ is rotated in its D-dimensional space, some of the critical exponents would change. Quite generally, the critical exponents of a theory prove sensitive to the group of transformations of the order parameter under which the effective Hamiltonian is invariant.

Before we explore the consequences of adopting the Landau–Ginzburg probability functional (7.8), we should remark on the close connection between $\boldsymbol{\phi}$ and the block variables introduced in §5.2. We have here introduced $\boldsymbol{\phi}$ as a phenomenological variable to be interpreted as the 'local density in excess of the mean' or the 'local magnetization', etc. But this is just what a block variable becomes when the block is large enough to be considered of macroscopic size. The Landau–Ginzburg model is widely applicable because to a sufficient approximation it is usually possible to express the effective Hamiltonian of block variables in the form (7.9).

Finally, we should point out that we can rewrite the Hamiltonian of the Ising model in a form which closely resembles (7.10). This is surprising given that the order parameter of the Ising model can take only the values ± 1 while $\boldsymbol{\phi}$ is intrinsically a continuous variable. But Appendix K explains how it may be done.

7.2 Landau theory

The partition function of the Landau–Ginzburg model is

$$Z_{LG} = \int \mathcal{D}\boldsymbol{\phi}\, e^{-\int d^d\mathbf{x}\, \mathcal{H}_{LG}(\boldsymbol{\phi})}. \tag{7.13}$$

Here the notation $\int \mathcal{D}\boldsymbol{\phi}$ means that the integrand is to be summed over all admissible values of the order-parameter field $\boldsymbol{\phi}$. We postpone to Chapter 8 discussion of which fields are 'admissible'. There we shall find that summing over all admissible fields is anything but trivial. In fact, nobody knows how to do this exactly for non-zero λ and we shall have to fall back on approximation schemes. Here we investigate the crudest possible approximation to the integral over $\boldsymbol{\phi}$ of (7.13), namely that the sum is dominated by its largest term to the extent that we may replace the entire sum by this largest term. The theory of phase transitions to which this gives rise is called 'Landau theory' because Landau (1937) derived the same results from a rather different point of view.

The approximation we wish to investigate is

$$Z_{LG} \simeq A Z_L, \tag{7.14}$$

where A is an uninteresting constant, Z_L is defined by

$$Z_L \equiv e^{-\int d^d\mathbf{x}\, \mathcal{H}_{LG}(\boldsymbol{\phi}_0)}, \tag{7.15}$$

and the special field $\boldsymbol{\phi}_0$ is that for which the integral in the exponent of (7.15) takes its minimum value.

From equation (7.10) it is clear that the integral of (7.15) is minimized when $\boldsymbol{\nabla}\boldsymbol{\phi} = 0$, i.e. by a constant field. So in this approximation we drop from $\mathcal{H}_{LG}$ the term proportional to $|\boldsymbol{\nabla}\boldsymbol{\phi}|^2$.

Consider now the case $\mathbf{B} \neq 0$; it is easy to see from equation (7.10) that for a given value of $|\boldsymbol{\phi}|$, $\mathcal{H}_{LG}$ is in this case least when $\boldsymbol{\phi}$ is parallel to $\mathbf{B}$. Consequently, when $\mathbf{B} \neq 0$ we can take the direction of $\boldsymbol{\phi}$ for granted and concentrate on its magnitude $\phi \equiv |\boldsymbol{\phi}|$.

In view of these simplifications the integral of (7.15) becomes

$$\int d^d\mathbf{x}\, \mathcal{H}_{LG}(\boldsymbol{\phi}_0) = V\left(\tfrac{1}{2}\mu^2 \phi_0^2 + \tfrac{1}{4!}\lambda \phi_0^4 - \beta B \phi_0\right), \tag{7.16}$$

where V is the system's volume. So in this approximation, the Helmholtz free energy per unit volume is

$$f \simeq \frac{\text{constant}}{\beta} + \frac{\mu^2}{2\beta}\phi_0^2 + \frac{\lambda}{4!\,\beta}\phi_0^4 - B\phi_0. \tag{7.17}$$

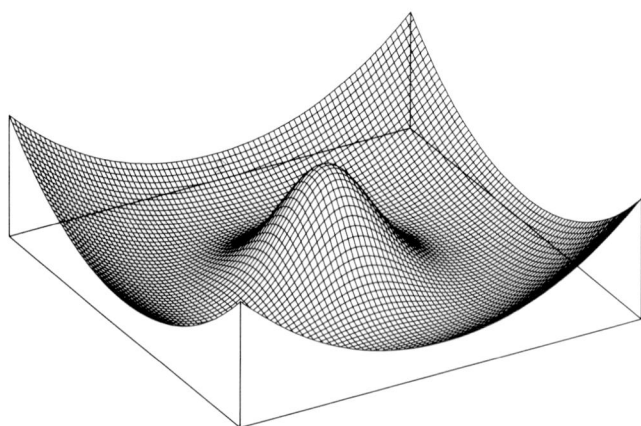

Figure 7.1 When the order parameter $\boldsymbol{\phi}$ has dimensionality $D \geq 2$ infinitely many fields $\boldsymbol{\phi}_0$ minimize $\mathcal{H}_{LG}$ when there is no external field ($B = 0$). Here $\mathcal{H}_{LG}(\boldsymbol{\phi}_0)$ is plotted for $B = 0$ and $T < T_c$ in the case $D = 2$.

Since ϕ_0 is determined by the condition that it minimize (7.16), it is a function of B and therefore so is f.

When $B = 0$, the situation can be more complex because we cannot take the direction of $\boldsymbol{\phi}$ for granted. However, if $\mu^2 > 0$, $\mathcal{H}_{LG}$ is minimized by $\boldsymbol{\phi}_0 = 0$, which points in no direction. So whenever $\mu^2 > 0$ the vectorial nature of the order parameter causes no difficulty, independently of the magnitude of B.

The really tricky case is that in which $\mu^2 < 0$ and $B = 0$, for then the function in (7.16) has more than one minimum (two if $\boldsymbol{\phi}$ has one component, infinitely many otherwise—see Figure 7.1), which are symmetrically placed about $\boldsymbol{\phi} = 0$. Do we include the contributions from all the minima when calculating Z_L, or does $\boldsymbol{\phi}$ remain pointing in one direction? In the language of §2.3 we are asking whether the symmetry of the system is spontaneously broken. Landau theory cannot answer this question, because to do so would require knowledge of the fluctuations in $\boldsymbol{\phi}$ as $B \to 0$ and Landau theory starts out by assuming that fluctuations are negligible. Here we simply assume that the symmetry *is* spontaneously broken, and so only one direction of $\boldsymbol{\phi}$ contributes to Z_L. (We shall return to this matter in Chapter 13.) Given this assumption, the Helmholtz free energy in zero field is the limit of (7.17) as $B \to 0$.

Now let's consider the system in the absence of an external field. Differentiating (7.16) shows that ϕ_0 satisfies

$$\mu^2 \phi_0 + \tfrac{1}{3!}\lambda \phi_0^3 = 0 \quad \Rightarrow \quad \begin{cases} \phi_0 = 0 & \text{if } \mu^2 \geq 0 \\ \phi_0^2 = -6\mu^2/\lambda & \text{if } \mu^2 < 0, \end{cases} \tag{7.18}$$

where we have used our knowledge that $\lambda > 0$. Thus in this approximation

7.2 Landau theory

a phase change takes place as μ^2 passes through zero: at high temperatures $\mu^2 > 0$ and the order parameter vanishes, while the order parameter is non-zero below the critical temperature T_c at which μ^2 changes sign; $\mu^2(T_c) = 0$. Just below the critical point $\phi_0 \propto |\mu^2|^{1/2}$, and from equation (7.12) we know that $\mu^2 \propto (T - T_c)$ near T_c, so we have[2]

$$\phi_0 \propto (T_c - T)^{1/2} \quad \text{just below } T_c. \tag{7.19}$$

Thus the critical exponent β, which is defined by the relation $\phi_0 \propto (T_c - T)^\beta$, has value $\beta = \frac{1}{2}$.

Inserting our solution (7.18) for ϕ_0 into equation (7.17), we have

$$\beta f = \text{constant} + \begin{cases} 0 & \text{if } T > T_c \\ -3\mu^4/2\lambda & \text{if } T < T_c. \end{cases} \tag{7.20}$$

The internal energy $u = \mathrm{d}(\beta f)/\mathrm{d}\beta$ is therefore

$$u = \begin{cases} 0 & \text{if } T > T_c \\ -\dfrac{3\mu^2}{\lambda}\dfrac{\mathrm{d}\mu^2}{\mathrm{d}\beta} + \dfrac{3\mu^4}{2\lambda^2}\dfrac{\mathrm{d}\lambda}{\mathrm{d}\beta} & \text{if } T < T_c, \end{cases} \tag{7.21}$$

which is continuous at $T = T_c$ ($\mu^2 = 0$). Thus no latent heat is associated with this phase transition. There is a discontinuity in the specific heat, however:

$$k_B c_V\big|_{\mu^2=0} = -\beta^2 \dfrac{\mathrm{d}u}{\mathrm{d}\beta}\bigg|_{\mu^2=0} = \begin{cases} 0 & \text{if } T = T_c+ \\ \dfrac{3\beta^2}{\lambda}\left(\dfrac{\mathrm{d}\mu^2}{\mathrm{d}\beta}\right)^2 & \text{if } T = T_c-. \end{cases} \tag{7.22}$$

In light of this result one says that the critical exponent α, which is defined by the relation $c_V \propto |T - T_c|^{-\alpha}$, takes the value $\alpha = 0$ on both sides of the critical point.[3]

To obtain the critical exponents γ and δ we need to consider the system in the presence of an external field. For non-zero $\mathbf{B}$ the constant field $\boldsymbol{\phi}_0$ that minimizes f is parallel to $\mathbf{B}$, and by equation (7.16) its modulus satisfies

$$\mu^2 \phi_0 + \tfrac{1}{3!}\lambda \phi_0^3 - \beta B = 0. \tag{7.23}$$

At the critical temperature $\mu^2 = 0$, ϕ_0 and B are related by

$$\phi_0^3 = \dfrac{6\beta}{\lambda} B, \tag{7.24}$$

---

[2] Equations (7.9) show that λ, like μ, is temperature-dependent. However, λ is not small near the transition, so its dependence on T is unimportant.

[3] But notice the difference between this case of a discontinuity in c_V and the logarithmic divergence of c_V for the $d = 2$ Ising model—the latter is also characterized by '$\alpha = 0$'.

so the critical exponent δ, which is defined by the relation $\phi_0 \propto B^{1/\delta}$ at $T = T_c$, has value $\delta = 3$.

Differentiating equation (7.23) with respect to B, we obtain the susceptibility χ:

$$\left(\mu^2 + \tfrac{1}{2}\lambda\phi_0^2\right)\chi = \beta. \tag{7.25}$$

Setting $\mathbf{B} = 0$ and inserting $\phi_0 = 0$ or $|\phi_0|^2 = -6\mu^2/\lambda$ as appropriate, we have

$$\chi\big|_{\mathbf{B}=0} = \begin{cases} \dfrac{\beta}{\mu^2} & \text{if } T > T_c \\ \dfrac{\beta}{2|\mu^2|} & \text{if } T < T_c. \end{cases} \tag{7.26}$$

Hence the zero-field susceptibility diverges as $|T - T_c|^{-1}$ on both sides of the transition and the critical exponent γ, which is defined by the relation $\chi \propto |T - T_c|^{-\gamma}$, has value $\gamma = 1$.

The remaining critical exponents, η and ν, lie beyond the scope of Landau theory: they involve the correlations $\langle\phi(\mathbf{x})\phi(\mathbf{x}')\rangle$ and Landau theory can make no predictions about correlations since it is based on the assumption that the order parameter ϕ takes the same value ϕ_0 at all points.

The alert reader will have noticed that the exponents we have recovered from Landau's approximation ($\alpha = 0$, $\beta = \tfrac{1}{2}$, $\gamma = 1$, $\delta = 3$) are exactly those predicted by the mean-field theory of Chapter 6. This should not come as a great surprise since Landau theory, like mean-field theory, assumes that the order parameter can be characterized by a single value at any temperature. Actually the order parameter fluctuates wildly near T_c and one needs to develop a theory that characterizes these fluctuations more precisely than through their mean: the average of a function g of a random variable x does not normally coincide with $g(\langle x \rangle)$. Hence to evaluate thermal averages of quantities related to ϕ, we may need to know more than just the mean value ϕ_0.

The Landau–Ginzburg partition function (7.13) fully specifies ϕ's probability distribution, and thus, in principle, enables us to predict the effects on critical exponents of the large fluctuations expected near T_c. In this section we have explored a spectacularly crude approximation to the mysterious integral over order-parameter fields that is implied by the symbol $\mathcal{D}\phi$ in (7.13), namely equating the integral to the largest value of its integrand. In the next chapter we develop a more sophisticated approach to the evaluation of functional integrals like that of (7.13). In subsequent chapters this approach will enable us to obtain from (7.13) results in reasonable agreement with experiment.

Problems

7.1 Show that the effective Hamiltonian density $\mathcal{H} = \frac{1}{2}\phi^2 + \frac{1}{3}a\phi^3 + \frac{1}{4}b\phi^4$, where $a(T), b(T) > 0$, yields a non-zero mean value of ϕ only when $a^2 > 4b$. Show further that the transition to non-zero ϕ is then first-order.

7.2 Show that in Landau theory the ferromagnetic transition is continuous even at non-zero B.

7.3 Show that the effective Hamiltonian density $\mathcal{H} = \frac{1}{2}\mu^2\phi^2 + \frac{1}{6}\lambda\phi^6$ yields critical exponents $\alpha = \frac{1}{2}$, $\beta = \frac{1}{4}$, $\gamma = 1$ and $\delta = 5$. Show that mean-field values of the critical exponents are restored if a ϕ^4 term is present in $\mathcal{H}$.

8
Diagrammatic perturbation theory

In the last chapter we introduced the Landau–Ginzburg model and calculated its partition function in a crude approximation known as Landau theory. In this chapter we develop a systematic method for calculating the Landau–Ginzburg partition function, the associated correlation functions, and the Helmholtz and Gibbs free energies. In subsequent chapters we shall use the techniques developed here to find the critical exponents for the Landau–Ginzburg model.

We begin §8.1 by solving the Landau–Ginzburg model with the coefficient of ϕ^4 in its Hamiltonian density (7.10) set equal to zero. This turns out to be a straightforward exercise and can be done exactly. Then in §§8.2 and 8.3 the partition function and correlation functions for the Landau–Ginzburg model with the ϕ^4 term are expressed in terms of the corresponding quantities for the model without this term. This cannot be done in closed form, but we will obtain series expansions of these quantities in positive integral powers of the coefficient of ϕ^4. In §8.4 we investigate the Gibbs free energy for this model.

The mathematical techniques needed to do all this are very similar to those used in quantum field theory; namely, functional differentiation and integration (see Appendix L). However, the physics is completely different. In quantum field theory one is usually working in a continuum consisting of three space dimensions plus one time dimension, and one calculates things such as scattering cross-sections or particle lifetimes. In statistical physics we are calculating the thermal average of measurable quantities in a model

8.1 The Gaussian partition function

The Landau–Ginzburg model with no ϕ^4 term is called the Gaussian model, because it is a continuum version of the model of the same name introduced in §3.1.6. Setting $\lambda = 0$ in equation (7.10), the partition function Z_G for this model is

$$Z_G[J] \equiv \int \mathcal{D}\phi\, e^{-H_G[J]}$$
$$= \int \mathcal{D}\phi\, \exp\left[-\int d^d x \left(\tfrac{1}{2}\alpha^2 |\boldsymbol{\nabla}\phi|^2 + \tfrac{1}{2}\mu^2 \phi^2 - J\phi\right)\right]. \qquad (8.1)$$

In this chapter we only consider the case in which the order parameter ϕ has a single component, for simplicity; $\phi(\mathbf{x})$ is a single real function of position $\mathbf{x}$. The symbol '$\mathcal{D}\phi$' denotes a 'functional integral' over the order parameter $\phi(\mathbf{x})$. It means a sum over all the innumerable forms that the function $\phi(\mathbf{x})$ can take.

The source field $J(\mathbf{x})$ plays a dual rôle. As well as allowing us to study the model in an external magnetic field, it also enables us to calculate correlation functions of the order parameter ϕ by differentiation. It follows from the form of the partition function (8.1), and from equations (2.32) and (2.38), that

$$G^{(n)}(\mathbf{x}_1, \ldots, \mathbf{x}_n; J) = \frac{1}{Z_G[J]} \frac{\delta}{\delta J(\mathbf{x}_1)} \cdots \frac{\delta}{\delta J(\mathbf{x}_n)} Z_G[J]$$
$$G_c^{(n)}(\mathbf{x}_1, \ldots, \mathbf{x}_n; J) = \frac{\delta}{\delta J(\mathbf{x}_1)} \cdots \frac{\delta}{\delta J(\mathbf{x}_n)} \log Z_G[J]. \qquad (8.2)$$

The symbol $\delta/\delta J(\mathbf{x})$ denotes a functional derivative—see Appendix L (and equation (L.6) in particular) for an explanation of functional differentiation and integration.

Before working out the functional integral (8.1) we must decide exactly which fields $\phi(\mathbf{x})$ we are to sum over—which fields are 'admissible'. To sum over every possible function $\phi(\mathbf{x})$ would surely be wrong, because the physical systems with which we are concerned are made of atoms, and the order parameter $\phi(\mathbf{x})$ is certainly not defined on scales of less than atomic dimensions. Indeed, we saw in Chapter 7 that the Landau–Ginzburg model is to be regarded as an effective model, obtained by averaging over volumes which, though small, contain many atoms. Now 'most' of the members of the infinite set of functions $\phi(\mathbf{x})$ vary ridiculously quickly from point to point; there are many more ways of drawing a wiggly line than of drawing a smooth one. It makes no sense to consider a field which varies rapidly on scales shorter than atomic dimensions, and such fields should be excluded from the partition function (8.1).

A simple way of doing this is to introduce the quantity $\tilde{\phi}(\mathbf{k})$, of which the order parameter $\phi(\mathbf{x})$ is the Fourier transform:

$$\phi(\mathbf{x}) = \int \frac{d^d\mathbf{k}}{(2\pi)^d}\, e^{i\mathbf{k}\cdot\mathbf{x}}\tilde{\phi}(\mathbf{k}) \equiv \int d^d\check{\mathbf{k}}\, e^{i\mathbf{k}\cdot\mathbf{x}}\tilde{\phi}(\mathbf{k}). \tag{8.3}$$

As the combination $d^d\mathbf{k}/(2\pi)^d$ will occur often, we write it as $d^d\check{\mathbf{k}}$. Since $\phi(\mathbf{x})$ is real, $\tilde{\phi}(\mathbf{k})$ must satisfy $\tilde{\phi}(-\mathbf{k}) = \tilde{\phi}^*(\mathbf{k})$. The wavevector $\mathbf{k}$ has dimensions of inverse length. From Table 7.1, $\phi(\mathbf{x})$ has dimension $L^{-d/2}$, so the dimension of $\tilde{\phi}(\mathbf{k})$ is $L^{+d/2}$. Then in the functional integral (8.1) we sum over only those functions $\phi(\mathbf{x})$ which can be written

$$\phi(\mathbf{x}) = \int_{k=0}^{\Lambda} d^d\check{\mathbf{k}}\, e^{i\mathbf{k}\cdot\mathbf{x}}\tilde{\phi}(\mathbf{k}), \tag{8.4}$$

where Λ is a suitable number called the **cutoff parameter**, or **cutoff**. A function $\phi(\mathbf{x})$ that can be expressed in the form (8.4) is smooth on distance scales $\lesssim \Lambda^{-1}$, and so restricting the sum to functions of this form eliminates the unwanted rapidly varying fields. The field $\tilde{J}(\mathbf{k})$, of which $J(\mathbf{x})$ is the Fourier transform,[1] is defined in a similar way.

The transformation to $\tilde{\phi}(\mathbf{k})$ not only makes the functional integral (8.1) well defined, but also makes it analytically tractable, because it breaks it up into a product of ordinary integrals over the different components of $\tilde{\phi}(\mathbf{k})$. The result (derived in Appendix L) is

$$Z_G[\tilde{J}] = Z_G[0] \times \exp\left[\tfrac{1}{2}\int^{\Lambda} d^d\check{\mathbf{k}}\, \frac{\tilde{J}(\mathbf{k})\tilde{J}(-\mathbf{k})}{\alpha^2 k^2 + \mu^2}\right],$$
$$\text{where}\quad Z_G[0] = \exp\left[-\tfrac{1}{2}V\int^{\Lambda} d^d\check{\mathbf{k}}\, \log(\alpha^2 k^2 + \mu^2)\right]. \tag{8.5}$$

$Z_G[\tilde{J}]$ factorizes into two pieces. One of these contains the $\tilde{J}$-dependence and therefore gives the correlation functions when differentiated (see equation

---

[1] You may wonder why we have used this turn of phrase when introducing $\tilde{\phi}$ and $\tilde{J}$, rather than saying that they are the Fourier transforms of ϕ and J respectively. The point is that there are functions $\phi(\mathbf{x})$ and $J(\mathbf{x})$ which do not have Fourier transforms (this will be true for 'most' functions in an infinite volume, and it can also happen in a finite volume if $\phi(\mathbf{x})$ or $J(\mathbf{x})$ are sufficiently badly behaved). By only considering functions which can be written in the form (8.4) we are ignoring such functions.

8.1 Gaussian partition function

(8.2)). The other piece is independent of $\tilde{J}$, and does not contribute to the correlation functions. The integral in the $\tilde{J}$-dependent piece can be rewritten as

$$\int^\Lambda d^d\tilde{k}\, \frac{\tilde{J}(k)\tilde{J}(-k)}{\alpha^2 k^2 + \mu^2} = \int d^d\mathbf{x}\, d^d\mathbf{y}\, J(\mathbf{x})\Delta(\mathbf{x}-\mathbf{y})J(\mathbf{y}), \tag{8.6}$$

where

$$\Delta(\mathbf{x}-\mathbf{y}) \equiv \int^\Lambda d^d\tilde{k}\, \frac{e^{i\mathbf{k}\cdot(\mathbf{x}-\mathbf{y})}}{\alpha^2 k^2 + \mu^2}, \quad \mu^2 > 0. \tag{8.7}$$

The properties of the function Δ are discussed below, where we shall see that it is the connected two-point correlation function for this model. This expression for Δ is defined only if μ^2 is positive. If μ^2 is negative the fluctuations in $\phi(\mathbf{x})$ are unbounded, since the exponent in (8.1) is not bounded above, and the Gaussian model is nonsense.

From (8.5), the Helmholtz free energy $F_G[J]$ of this model is

$$F_G[J] = -\beta^{-1} \log Z_G[J] \tag{8.8}$$

$$= \beta^{-1} \left(\tfrac{1}{2} V \int^\Lambda d^d\tilde{k}\, \log(\alpha^2 k^2 + \mu^2) - \tfrac{1}{2} \int d^d\mathbf{x}\, d^d\mathbf{y}\, J(\mathbf{x})\Delta(\mathbf{x}-\mathbf{y})J(\mathbf{y}) \right).$$

The first term depends strongly on the value chosen for Λ, because increasing Λ increases the number of degrees of freedom of the model, and in classical statistical mechanics all these degrees of freedom contribute to the free energy. The second term does not depend strongly on Λ if $J(\mathbf{x})$ is smooth on the scale of Λ^{-1}.

The introduction of the cutoff parameter Λ is conceptually correct. The rapidly varying functions are unphysical and should not be included in the sum-over-states. But we have just seen that the Helmholtz free energy is dominated by rapidly varying functions. This is embarrassing, because our cutoff procedure is almost the crudest possible, taking no account of the *actual* short-distance structure of the system. If rapidly varying functions are so important, surely more care should have been taken when dealing with them?

One way out of this difficulty would be to note that it is only the $\tilde{J}$-independent part of the Helmholtz free energy that depends strongly on Λ. The $\tilde{J}$-dependent part, which contains the information about correlation functions, does not depend strongly on Λ. So perhaps the Λ-dependence is not so important after all. But this is only true in the Gaussian model. When in the next section we include the term $\int d^d\mathbf{x}\, \frac{\lambda}{4!}\phi^4$ in the Hamiltonian, all the correlation functions will depend strongly on Λ. The dependence of physical quantities on the detailed short-distance properties of the model is real and natural. However, we are not interested in the consequences of a particular microscopic model. We are interested in the relationships between different *macroscopic* quantities—the dependence of the correlation length on temperature, for example. It is logically possible that if we eliminate the microscopic parameters α, μ, λ, and Λ of our model in favour of macroscopic parameters, such as the susceptibility, correlation length, etc., and then express all other quantities in terms of these new variables, then the relationships between macroscopic quantities that we obtain might prove to be independent of the details of the model at short distances. In the next chapter we shall see that this is indeed the case for the Landau–Ginzburg model. So we will be justified a *posteriori* in using our simple model for the short-distance behaviour of the system, since the results we want (in particular, the critical exponents) will be shown to be independent of it.

8.1.1 Correlation functions in the Gaussian model

Now let us calculate the connected correlation functions for the Gaussian model. From equations (8.2) and (8.8) it follows that

$$
\begin{aligned}
G_c^{(1)}(\mathbf{x}) &\equiv \langle \phi(\mathbf{x}) \rangle \\
&= \int d^d \mathbf{z}\, \Delta(\mathbf{x} - \mathbf{z}) J(\mathbf{z}); \\
G_c^{(2)}(\mathbf{x}, \mathbf{y}) &= \langle \left(\phi(\mathbf{x}) - \langle \phi(\mathbf{x}) \rangle\right) \left(\phi(\mathbf{y}) - \langle \phi(\mathbf{y}) \rangle\right) \rangle \\
&= \Delta(\mathbf{x} - \mathbf{y}).
\end{aligned}
\qquad (8.9)
$$

All higher-order connected correlation functions vanish. The function $G_c^{(1)}$, which is simply the thermal average of ϕ, vanishes if $J(\mathbf{x})$ does. This would be expected from the symmetry of H_G under $\phi \to -\phi$ when $J = 0$. The 'source' of a non-zero value for $\langle \phi(\mathbf{x}) \rangle$ is the function $J(\mathbf{x})$, which explains its name. $G_c^{(2)}$, which measures the correlation between the fluctuations in the order parameter at the points $\mathbf{x}$ and $\mathbf{y}$, is the function $\Delta(\mathbf{x} - \mathbf{y})$ that was defined in (8.7). This function depends only on the modulus of its argument. It is given in Table 8.1 for spaces of one to four dimensions, in the limit of large $\alpha\Lambda/\mu$. Physically, this limit corresponds to the 'atomic structure' (on length scale Λ^{-1}) being much smaller than the correlation length $\alpha\mu^{-1}$. $\Delta(\mathbf{x})$ is plotted in Figure 8.1.

The connected two-point correlation function $G_c^{(2)}$ of a general model is called the **propagator**, and $G_c^{(2)}$ for the Gaussian model is called the **bare propagator**. These terms come from quantum field theory, but they are in common use in statistical physics.

We would expect the correlation between the field fluctuations at two points to decrease as the points move further apart. The graphs in Figure 8.1 show that this is what happens. In all four cases (and in higher dimensions, too) the correlation falls off exponentially with distance if $\mu \neq 0$. The correlation length is $\alpha\mu^{-1}$. At separations of more than a few times $\alpha\mu^{-1}$, the fluctuations in the order parameter are essentially uncorrelated. As μ

8.1 Gaussian partition function

Table 8.1. The function Δ

No. of dimensions	$\Delta(\mathbf{x})$ for $\Lambda = \infty$	$\Delta(0)$ for $\Lambda \gg \mu$
1	$\dfrac{1}{2\alpha\mu} e^{-\mu x/\alpha}$	$\dfrac{1}{2\alpha\mu}$
2	$\dfrac{K_0(\mu x/\alpha)}{2\pi\alpha^2}$	$\dfrac{1}{2\pi\alpha^2} \log\left(\dfrac{\Lambda\alpha}{\mu}\right)$
3	$\dfrac{e^{-\mu x/\alpha}}{4\pi\alpha^2 x}$	$\dfrac{\Lambda}{2\pi^2\alpha^2}$
4	$\dfrac{\mu K_1(\mu x/\alpha)}{4\pi^2\alpha^3 x}$	$\dfrac{\Lambda^2}{16\pi^2\alpha^2}$

NOTES: The first column shows the function $\Delta(\mathbf{x})$ as a function of $x \equiv |\mathbf{x}|$ in the limit $\Lambda \gg \mu/\alpha$ and $x \gg \Lambda^{-1}$. The second column shows the value of $\Delta(0)$, with $\Lambda \gg \mu/\alpha$ still. K_0 and K_1 are modified Bessel functions.

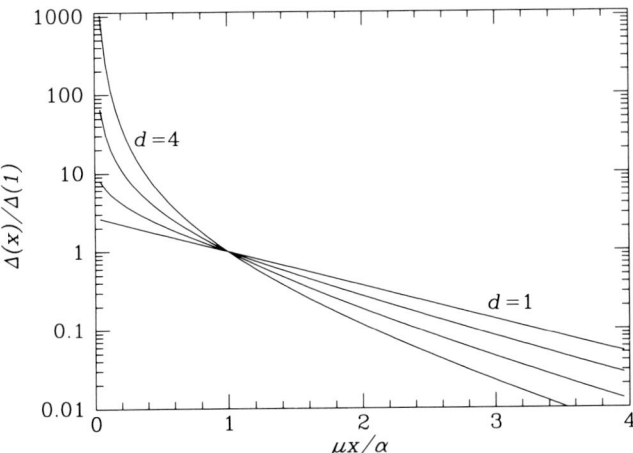

Figure 8.1 The two-point connected correlation function $\Delta(\mathbf{x})$ of the one-, two-, three-, and four-dimensional Gaussian models. On the extreme left the lower curve is for the one-dimensional model and the upper curve is for the four-dimensional model.

decreases the correlation length increases, diverging as $\mu \to 0$. When this happens the correlation length is limited only by the physical system size. The fluctuations are no longer microscopic, and the model is at a critical point.

The behaviour of $\Delta(\mathbf{x})$ as $x \to 0$ depends on the dimensionality of the system. In $d = 1$, $\Delta(\mathbf{x})$ does not depend strongly on Λ as $x \to 0$. But for $d \geq 2$ it does, and the Λ-dependence grows stronger as the dimension increases. The value of $\Delta(0)$ in the limit of large Λ is shown in the second column of Table 8.1. Can we understand what causes the strong dependence of $\Delta(0)$ on Λ, and why the dependence grows with dimension?

At first sight such dependence seems very peculiar, since it means that in the absence of a cutoff the value of $\langle\phi(\mathbf{x})\phi(\mathbf{x})\rangle$ would be infinite. This in turn would mean that there are infinite fluctuations in the value of the order parameter at each point of the system, whereas the exponential in the partition function (8.1) might be expected to suppress such fluctuations. To see why it does not, consider a small 'cube' of space of side l and thus of volume l^d. If the mean square value of the order parameter in this volume is $\bar\phi^2$, then the contribution of this volume to the $\int d^d\mathbf{x}\frac{1}{2}\mu^2\phi^2$ term in the Hamiltonian for the Gaussian model (see equation (8.1)) is only $\frac{1}{2}\mu^2\bar\phi^2 l^d$. This vanishes as $l\to 0$, and does so more rapidly the bigger d is. So fluctuations involving very small volumes are not suppressed by this term in the Hamiltonian. If this were the only term they would become very large, diverging as $l\to 0$.

But there is another term in the Hamiltonian of the Gaussian model—the gradient term $\int d^d\mathbf{x}\frac{1}{2}\alpha^2|\nabla\phi|^2$. This term tries to stop the order parameter varying rapidly with position. In doing this it acts to prevent the divergence in $G_c^{(2)}$ by 'connecting together' the values of the order parameter over a region of the system. So it tries to put a lower limit on the volume that can take part in a fluctuation, and thus to prevent the large short-distance fluctuations which we have just seen.[2] But is it strong enough?

Consider a fluctuation in the order parameter of magnitude $(\bar\phi^2)^{1/2}$ taking place over a volume l^d, when the order parameter is smooth on distances shorter than l. Then the gradient term in the Hamiltonian gets a contribution of $(\alpha^2\bar\phi^2/l^2)l^d$ from the volume l^d; the other term in the Hamiltonian is of order $(\bar\phi^2\mu^2)l^d$. When $\mu l/\alpha \ll 1$ the gradient term dominates, and vice versa. This means that fluctuations on scales longer than the correlation length are governed by the term $\int d^d\mathbf{x}\frac{1}{2}\mu^2\phi^2$; fluctuations on scales shorter than the correlation length are governed by the gradient term. Consequently, for $l \ll \alpha\mu^{-1}$ we see that $\bar\phi^2 \sim (1/\alpha^2 l^{d-2})$, which diverges if $d>2$. So fluctuations in the order parameter do grow indefinitely on short scales (as long as we do not approach the cutoff), the more so the higher the dimension. In two dimensions $\bar\phi^2$ is constant as $l\to 0$, but this is still enough to cause a divergence in $G_c^{(2)}$ because it contains contributions from all values of l^{-1} (i.e., from all possible fluctuation wavevectors). In one dimension $\bar\phi^2 \to 0$ as $l\to 0$, and $G_c^{(2)}$ remains finite.

The reason for looking so closely at the Gaussian model is that the correlation functions of the full Landau–Ginzburg model can be expressed in terms of those of the Gaussian model. In the next section we shall see how this may be done.

---

[2] The cutoff parameter Λ also puts a lower limit on the volume that can take part in a fluctuation, by only allowing fields which are smooth on distances shorter than Λ^{-1}. Here we are supposing that Λ^{-1} is much shorter than the length l, so that the cutoff does not enter the discussion.

8.2 The partition function for the full Landau–Ginzburg model

The partition function Z for the full Landau–Ginzburg model is

$$Z[J] = \int \mathcal{D}\phi \, \exp\left[-\int d^d\mathbf{x} \, \left(\tfrac{1}{2}\alpha^2|\boldsymbol{\nabla}\phi|^2 + \tfrac{1}{2}\mu^2\phi^2 + \tfrac{\lambda}{4!}\phi^4 - J\phi\right)\right], \quad (8.10)$$

where the functional integration is over only those functions $\phi(\mathbf{x})$ which are 'sufficiently smooth', in the sense of the previous section. This functional integral cannot be done in the same way as (8.1). The trick that worked so well for the Gaussian model relied on the Hamiltonian splitting into a sum of separate pieces, each depending on only one Fourier component of $\widetilde{\phi}(\mathbf{k})$. But the quartic term prevents this. When written in terms of $\widetilde{\phi}(\mathbf{k})$ the quartic term becomes

$$\int^\Lambda d^d\check{\mathbf{k}}_1 d^d\check{\mathbf{k}}_2 d^d\check{\mathbf{k}}_3 d^d\check{\mathbf{k}}_4 \, \widetilde{\phi}(\mathbf{k}_1)\widetilde{\phi}(\mathbf{k}_2)\widetilde{\phi}(\mathbf{k}_3)\widetilde{\phi}(\mathbf{k}_4) \, (2\pi)^d \delta(\mathbf{k}_1 + \mathbf{k}_2 + \mathbf{k}_3 + \mathbf{k}_4).$$

This contains cross-terms between different Fourier components of ϕ, and the functional integral does not factorize into a product of integrals. There is no obvious way of doing it, and we are stuck.

8.2.1 The Feynman rules

Fortunately a systematic approximation scheme for calculating $Z[J]$ exists. This was developed by R. P. Feynman (1918–1988) (amongst others) for calculations in quantum field theory. It leads to the representation of functional integrals such as (8.10) by an (infinite) set of diagrams, called **Feynman diagrams**. The **Feynman rules** tell you which diagrams to draw, and how to assign an algebraic expression to each one. The functional integral is then given by the sum of the expressions corresponding to the diagrams. The rules for drawing the diagrams are very intuitive, and the rules for the expressions are more-or-less mechanical, so the whole scheme is very useful.

Our objective is not to find $Z[J]$ itself, so much as $\log Z[J]$, and its derivatives the connected correlation functions. But $Z[J]$ can be simply written in terms of $Z_G[J]$, whereas this is not directly true for their logarithms. So the next few pages will be spent calculating $Z[J]$. After this has been done a beautiful set of rules will emerge for finding $\log Z[J]$ and its derivatives.

The trick is to expand the exponential of the quartic term in (8.10) as a power series in λ, and then to exchange the order of the sum and the functional integral. This gives

$$Z[J] = \int \mathcal{D}\phi\, e^{-H_G} \left[\sum_{n=0}^{\infty} \frac{1}{n!} \left(-\frac{\lambda}{4!} \int d^d\mathbf{z}\, \phi^4 \right)^n \right]$$

$$= \sum_{n=0}^{\infty} \frac{1}{n!} \int \mathcal{D}\phi\, e^{-H_G} \left[\left(-\frac{\lambda}{4!} \int d^d\mathbf{z}\, \phi^4 \right)^n \right] \quad (8.11)$$

$$= Z_G[J] \times \sum_{n=0}^{\infty} \frac{1}{n!} \left\langle \left(-\frac{\lambda}{4!} \int d^d\mathbf{z}\, \phi^4 \right)^n \right\rangle_{\text{Gaussian}}.$$

(H_G is the Hamiltonian for the Gaussian model, defined in (8.1).) The exciting thing about the above expression is that it expresses Z for the full Landau–Ginzburg model in terms of an infinite series of thermal averages of powers of the order parameter evaluated in the *Gaussian* model, which, as we have seen in the previous section, are very easy to work out. So it looks as though it should be possible to find $Z[J]$ as accurately as desired without having to do too much hard work. Sadly, this isn't quite the case...

First it should be noted that the exchange of the sum and integral in the second line of (8.11) is not valid. This is not a pedantic mathematical objection; the series on the right-hand side of (8.11) does not converge. (If you are sceptical about this, an example of a one-dimensional integral which displays the same problem is given in Problem 8.2.) In fact no expansion of $Z[J]$ in powers of λ about $\lambda = 0$ exists. To see this, let us allow λ to become complex, and regard $Z[J]$ as a function of this complex variable. We know from the theory of functions of a complex variable that an analytic function of λ can be expanded in powers of $\lambda - \lambda_0$ about a point λ_0 out to $|\lambda - \lambda_0| = |\lambda_s - \lambda_0|$, where λ_s is the nearest point to λ_0 in the complex plane at which the function is singular (non-analytic)—$|\lambda_s - \lambda_0|$ is the radius of convergence of the power series. In (8.11) we are seeking an expansion of $Z[J]$ about the Gaussian theory, so $\lambda_0 = 0$. But $Z[J]$ is singular for negative real λ, no matter how close λ is to zero, since then the exponent of (8.10) can increase without limit. So there are singularities in $Z[J]$ arbitrarily close to $\lambda = 0$ and the radius of convergence of a Taylor expansion of $Z[J]$ in λ about $\lambda = 0$ must be zero. However, this does not (necessarily) make the series expansion (8.11) worthless. We can still hope that the series is 'asymptotic'; that is, the fractional difference between the sum of a finite number of terms and the true answer can be made as small as desired by taking λ small enough. So useful information can still be extracted from the series for small λ. The perturbation expansions that we shall derive are believed to be asymptotic.

Now let us return to $Z[J]$. We must work out the Gaussian correlation functions appearing on the right-hand side of (8.11). Using the first of

8.2 Full Landau–Ginzburg partition function

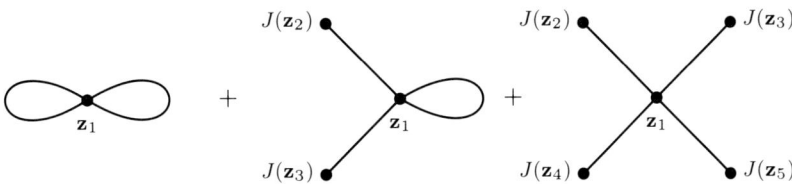

Figure 8.2 The O(λ) contribution to the partition function Z.

equations (8.2), we can write the last line of (8.11) in the form

$$Z[J] = \sum_{n=0}^{\infty} \frac{1}{n!} \left(-\frac{\lambda}{4!} \int d^d z \, \frac{\delta^4}{\delta J^4(z)} \right)^n Z_G[J]$$

$$\equiv \exp\left[-\frac{\lambda}{4!} \int d^d z \, \frac{\delta^4}{\delta J^4(z)} \right] Z_G[J]. \quad (8.12)$$

Writing $Z[J]$ in this form allows us to generate the terms in (8.11) systematically. The algebraic evaluation of (8.12) is straightforward, if rather tedious. From equations (8.5) and (8.6) we have

$$Z_G[J] = Z_G[0] \times \exp\left[\tfrac{1}{2} \int d^d x d^d y \, J(\mathbf{x}) \Delta(\mathbf{x}-\mathbf{y}) J(\mathbf{y}) \right]. \quad (8.13)$$

On expanding the exponential in (8.12), the O(λ) contribution to $Z[J]$ is (try it and see)

$$-\lambda \Big(\tfrac{1}{8} \int d^d z_1 \, \Delta^2(0) + \tfrac{1}{4} \int d^d z_1 d^d z_2 d^d z_3 \, \Delta(0) \Delta(\mathbf{z}_1-\mathbf{z}_2) J(\mathbf{z}_2) \Delta(\mathbf{z}_1-\mathbf{z}_3) J(\mathbf{z}_3)$$

$$+ \tfrac{1}{4!} \int d^d z_1 d^d z_2 d^d z_3 d^d z_4 d^d z_5 \, \Delta(\mathbf{z}_1-\mathbf{z}_2) J(\mathbf{z}_2) \Delta(\mathbf{z}_1-\mathbf{z}_3) J(\mathbf{z}_3)$$

$$\times \Delta(\mathbf{z}_1-\mathbf{z}_4) J(\mathbf{z}_4) \Delta(\mathbf{z}_1-\mathbf{z}_5) J(\mathbf{z}_5) \Big) Z_G[J], \quad (8.14)$$

all multiplied by $(-\lambda)$. This is a bit of a mess, but there is a much simpler diagrammatic notation for integrals such as these which occur in the expansion of $Z[J]$. Let us represent each coordinate in integrals like the above by a point, each $\Delta(\mathbf{x}-\mathbf{y})$ by a line—a **link**—between the points $\mathbf{x}$ and $\mathbf{y}$, and each occurrence of $J(\mathbf{z})$ by a blob (or **source-blob**, or **J-blob**) at $\mathbf{z}$. Then the integrals in (8.14) can be represented much more simply by the diagrams in Figure 8.2. The first diagram has one point—call it $\mathbf{z}_1$—and two links, each of which go from $\mathbf{z}_1$ back to $\mathbf{z}_1$. The position $\mathbf{z}_1$ is integrated over. So this gives $\int d^d z_1 \, \Delta^2(\mathbf{z}_1-\mathbf{z}_1)$, or $\int d^d z_1 \, \Delta^2(0)$, just as in (8.14). The other two diagrams work in the same way—try them.

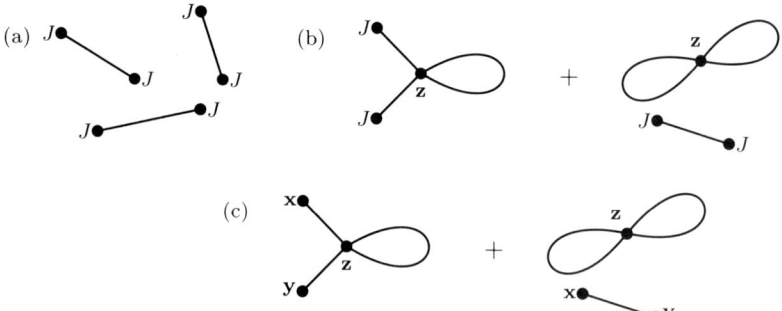

Figure 8.3 (a) The third term in the expansion of the Gaussian partition function. (b) The action of $\delta^4/\delta J^4(\mathbf{z})$ on this term. (c) The action of $\delta^2/\delta J(\mathbf{x})\delta J(\mathbf{y})$ on (b).

Not only are these diagrams more pleasant to look at than the integrals, but they also suggest that the process could be reversed, enabling us to write down the terms in the expansion of $Z[J]$ by drawing diagrams. To illustrate the idea, consider the algebraic expression for the $O(J^6)$ term in the expansion of $Z_G[J]$:

$$\frac{1}{3!}\left(\int d^d\mathbf{z}_1 d^d\mathbf{z}_2\, J(\mathbf{z}_1)\Delta(\mathbf{z}_1-\mathbf{z}_2)J(\mathbf{z}_2)\right)^3. \tag{8.15}$$

This term can also be represented by a diagram according to our rules, shown in Figure 8.3(a). There are three pairs of blobs joined by a link, one pair for each integral. This is an example of a **disconnected** diagram; such diagrams consist of two or more separate parts (in this case three parts). The algebraic expression corresponding to a disconnected diagram is the product of the expressions corresponding to each separate part. Diagrams consisting of only one part (such as the three diagrams in Figure 8.2) are called **connected** diagrams. Now suppose that the $\delta^4/\delta J^4(\mathbf{z})$ from the exponent in (8.12) acts on the term (8.15). The differentiation gives zero unless at least four of the Js have argument $\mathbf{z}$; that is, unless at least four of the blobs in the diagram are at the position $\mathbf{z}$. If they are, the differentiation removes four of the $J(\mathbf{z})$s leaving four lines coming from the point $\mathbf{z}$. This gives the two possible diagrams shown in Figure 8.3(b). So the effect of the fourth derivative on a diagram is to remove four blobs and to create a four-line **vertex**—a point where four links meet. From (8.12) we see that a factor of $-\lambda$ is associated with each such vertex, and that the position $\mathbf{z}$ of each such vertex should be integrated over. So to find the $O(\lambda^n)$ contribution to $Z[J]$, one simply draws all possible distinct diagrams containing n four-line vertices, joining all unused ends of links to the field $J(\mathbf{z})$. Then one translates each such diagram into an integral by putting in a $J(\mathbf{z})$ for each blob at position $\mathbf{z}$, a $\Delta(\mathbf{x}-\mathbf{y})$ for each link between the points $\mathbf{x}$ and $\mathbf{y}$, and a factor of $-\lambda$ for

8.2 Full Landau–Ginzburg partition function

each vertex. All position variables are integrated over. Every such diagram gives rise to a term that occurs in the expansion (8.12) of $Z[J]$ and all such terms in the expansion of $Z[J]$ correspond to such a diagram (since we have a definite prescription for changing any expression into a diagram). Although exactly the same in principle as evaluating (8.12) directly, this diagrammatic method of generating the terms in the expansion of $Z[J]$ is much easier in practice, because people prefer pictures to integrals.

The functional derivatives of $Z[J]$ with respect to J, which are related to the correlation functions $G^{(n)}$ (equation (8.2)), can be represented diagrammatically too. If a derivative $\delta/\delta J(\mathbf{x})$ acts on one of the diagrams in the expansion of $Z[J]$, it gives zero unless one of the source-blobs is at the point $\mathbf{x}$. If the source-blob is there (remember that the positions of the blobs are integrated over) the factor of $J(\mathbf{x})$ is removed. The effect of this single functional derivative on a diagram is thus to fix each of the links which are joined to blobs to the point $\mathbf{x}$, one after the other. So differentiating the diagrams in Figure 8.3(b) with respect to the source at the points $\mathbf{x}$ and $\mathbf{y}$ gives the diagrams shown in Figure 8.3(c). These diagrams are part of the $O(\lambda)$ contribution to $\delta^2 Z[J]/\delta J(\mathbf{x})\delta J(\mathbf{y})$. The points $\mathbf{x}$ and $\mathbf{y}$ are not integrated over when writing down the expressions for the diagrams in Figure 8.3(c). Higher derivatives of $Z[J]$ can be represented in the same way. Position variables which are not integrated over we call **external points**; those which are integrated over we call **internal points**.

8.2.2 The symmetry factor

Our plan, then, is to calculate $Z[J]$ by drawing diagrams that correspond to the terms generated by expanding the two exponentials in the expression

$$Z[J] = \exp\left[-\frac{\lambda}{4!}\int d^d\mathbf{z}\,\frac{\delta^4}{\delta J^4(\mathbf{z})}\right] \exp\left[\tfrac{1}{2}\int d^d\mathbf{x}d^d\mathbf{y}\,J(\mathbf{x})\Delta(\mathbf{x}-\mathbf{y})J(\mathbf{y})\right]Z_G[0]. \tag{8.16}$$

Translating a diagram into the corresponding algebraic expression is quite straightforward. The only tricky part comes when deciding what number should multiply the resulting expression. The rules for determining this number, called the **symmetry factor**, will be given below, together with some explanation, but really the only way to understand them is to play around with various diagrams yourself. As a last resort, you can always find out what the symmetry factor is for a given diagram by going back to equation (8.16) and working out the derivatives by hand.

What are the possible sources of numbers? First, there is the $\frac{1}{4!}$ associated with each occurrence of λ in (8.16). However, this always appears together with a fourth derivative with respect to $J(\mathbf{z})$. There must always be four $J(\mathbf{z})$s for it to act on, if it is not to give zero, and the $\frac{1}{4!}$ gets cancelled,

since
$$\frac{\delta^4}{\delta J^4(\mathbf{z})}\left[J(\mathbf{z}_1)J(\mathbf{z}_2)J(\mathbf{z}_3)J(\mathbf{z}_4)\right] = 4!\,\delta(\mathbf{z}_1 - \mathbf{z})\delta(\mathbf{z}_2 - \mathbf{z})\delta(\mathbf{z}_3 - \mathbf{z})\delta(\mathbf{z}_4 - \mathbf{z}). \tag{8.17}$$

This is why the coupling was written as $\frac{\lambda}{4!}$ in the first place. We can look on this 4! as coming from the 4! different ways of matching up the four derivatives with the four Js. The first derivative can act on any one of the four Js; that leaves three possible Js for the next derivative to act on, two for the one after that, and one for the last derivative. This gives a total of $4 \times 3 \times 2 \times 1 = 4!$ copies of the same term. In more complicated cases this way of looking at how derivatives generate numerical factors will be helpful.

A second source of numbers is the factors of one–half from the second exponent in (8.16). There is one of these for each link in a diagram. Normally these factors do not appear, since

$$\frac{\delta}{\delta J(\mathbf{x})}\frac{\delta}{\delta J(\mathbf{y})}\left[\tfrac{1}{2}\int d^d\mathbf{z}_1 d^d\mathbf{z}_2\, J(\mathbf{z}_1)\Delta(\mathbf{z}_1 - \mathbf{z}_2)J(\mathbf{z}_2)\right] = \Delta(\mathbf{x} - \mathbf{y}). \tag{8.18}$$

The factor of one half disappears, because the two derivatives can be matched up with the two Js in two ways. However, if the link goes from one internal point back to the same point, then the two different ways of assigning the derivatives to the Js have already been accounted for as part of the 4! generated by the fourth derivative associated with that point, since the two derivatives are both part of that one fourth derivative. So the factor of one–half survives, and we have our first rule:

> For every link with its two ends tied to the same internal point, there is a factor of one–half.

The third source of numbers is the factors from the expansion of the two exponents in (8.16). Expanding the first exponential gives factors of $1/n_v!$, where n_v is the number of vertices in the diagram; expanding the second exponential gives factors of $1/n_l!$, where n_l is the number of links in the diagram. Yet again these do not normally appear, since they are cancelled by the $n_l!n_v!$ ways of assigning the derivatives corresponding to the links and vertices. As an example, consider the diagram shown in Figure 8.4(a), which has $n_l = 4$ and $n_v = 1$. It is generated by the part of the term

$$\frac{\delta}{\delta J(\mathbf{x}_1)}\ldots\frac{\delta}{\delta J(\mathbf{x}_4)}\left[-\frac{\lambda}{4!}\int d^d\mathbf{z}\,\frac{\delta^4}{\delta J^4(\mathbf{z})}\right] \times \frac{1}{2^4 4!} \tag{8.19}$$

$$\times \int d^d\mathbf{z}_1\ldots d^d\mathbf{z}_8\, J(\mathbf{z}_1)\ldots J(\mathbf{z}_8)\,\Delta(\mathbf{z}_1 - \mathbf{z}_2)\Delta(\mathbf{z}_3 - \mathbf{z}_4)\Delta(\mathbf{z}_5 - \mathbf{z}_6)\Delta(\mathbf{z}_7 - \mathbf{z}_8)$$

in which each of the external derivatives (the first four in (8.19)) acts on Js attached to different Δs. The derivative with respect to $J(\mathbf{x}_1)$ can act on

8.2 Full Landau–Ginzburg partition function

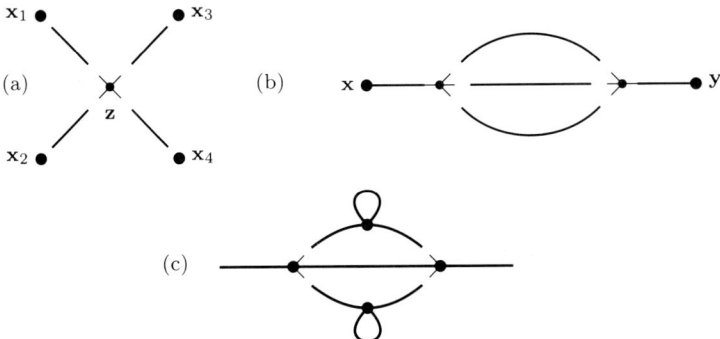

Figure 8.4 (a) A diagram contributing to $G^{(4)}(\mathbf{x}_1, \ldots, \mathbf{x}_4)$, discussed in the text. (b) A diagram where three links join the same two points. (c) A diagram where not all arrangements of links and vertices give different diagrams.

any one of the eight $J(\mathbf{z}_i)$. The derivative with respect to $J(\mathbf{x}_2)$ can then only act on one of six if the resulting term is to correspond to the diagram in Figure 8.4(a). For example, if the derivative with respect to $J(\mathbf{x}_1)$ acts on $J(\mathbf{z}_1)$, then the derivative with respect to $J(\mathbf{x}_2)$ cannot act on $J(\mathbf{z}_2)$ since this would correspond to a diagram containing a link joining $\mathbf{x}_1$ and $\mathbf{x}_2$. Then there are four possibilities for the derivative with respect to $J(\mathbf{x}_3)$, and just two for the last external derivative. The four derivatives with respect to $J(\mathbf{z})$ can act on the four remaining sources in 4! ways. So the number resulting from the differentiations for this diagram is

$$8 \times 6 \times 4 \times 2 \times 4! = 2^4 \times 4! \times 4!.$$

The 2^4 arises because each of the four links in the diagram can be either way around; it cancels the $\frac{1}{2^4}$ as we argued above that it should. The 4! from the derivatives with respect to $J(\mathbf{z})$ cancels the $\frac{1}{4!}$ in the first square bracket in (8.19), again as already argued. We see that there is another 4!, and that it cancels the second $\frac{1}{4!}$ in (8.19), which came from expanding the second exponential in (8.16). This 4! arose because the four external derivatives could be assigned to the four links in Figure 8.4(a) in 4! different ways. This is the cancellation of the $1/n_l!$ by the $n_l!$ ways of assigning the derivatives to the links, referred to at the start of this paragraph. A similar thing happens in diagrams with more than one vertex; the $n_v!$ gets cancelled in the same way.

But in some diagrams not all of the different ways of assigning the derivatives count. We have already seen that this happens when both ends of one link are joined to the same internal point. It also happens when more than one link joins the same two internal points, as in Figure 8.4(b). There are 3! different ways of 'connecting up' the three links between the two vertices, but these different connections are equivalent to rearrangements of the connections at each of the vertices. Therefore they have already been taken into account in the factor of 4! at each vertex. So there are only 5!/3! distinct ways of arranging the five links in the diagram, and the $1/n_l! = 1/5!$ is not completely cancelled; a factor of $1/3!$ remains. This gives us our second rule:

> For every pair of points which are directly joined by n links there is a factor of $1/n!$.

The two previous rules are special cases of a more general rule. In a given diagram with n_l links and n_v vertices there are $2^{n_l} n_l!$ ways of putting in the links, $n_v!$ ways of putting in the vertices, and $(4!)^{n_v}$ ways of connecting up the links at each vertex. If all these ways are different then these factors exactly cancel those coming from the expansion of the exponentials in (8.16). But if only a fraction f of these ways are different the cancellation is not complete and the diagram must be multiplied by f (f is less than one). By two arrangements of links and vertices being 'different' we mean that one cannot be changed into the other merely by swapping around the connections at the vertices. The first rule above covers the case in which a link goes to and from the same vertex, so that the two ways of connecting it have already been taken into account in the permutations of the connections at that vertex; the second rule above deals with more than one link joining a pair of vertices, so that the different ways of connecting up the links between the vertices are duplicated by the permutations of the connections at each vertex. Cases arise which are covered by neither of these two rules. Consider the diagram in Figure 8.4(c). If the links and vertices above the central line are swapped with those below the line, the new arrangement of links and vertices is the same as the one we started with. The change can be undone by altering the connections at the two vertices on the central line. In this diagram a factor of one half is needed to account for these two equivalent arrangements, in addition to the factor of $\frac{1}{4}$ given by the previous two symmetry factor rules.

The general rule to cover such cases, which is our third and final rule governing symmetry factors, is as follows:

> If there are n different ways of arranging the internal points (vertices and source-blobs) and links of a diagram, keeping external points fixed in position and without cutting any links, whilst leaving the appearance of the diagram unchanged, then the diagram should be multiplied by a factor of $1/n$. (Links may be slid over each other in the rearrangement process.)

In Figure 8.4(c) the two 'arms' above and below the axis can be swapped without changing the appearance of the diagram, and n is two. It is important that this swap can be made without cutting any links; if the leftmost link is exchanged with the rightmost link then the diagram also remains unchanged (obviously), but this swap does not count towards n because it cannot be carried out continuously. Notice that if the external points **x** and

y in the diagram are replaced by J-blobs then the diagram can be rotated through 180° without changing it. Because their positions are integrated over, the J-blobs can be moved about freely when rearranging the diagram. There are then four ways of arranging the internal points of the diagram whilst leaving it looking the same, and n is four.

The diagrams in Figure 8.2 provide a further illustration of the rules; moreover, the correct answers appear in equation (8.14). For example, the first diagram has a symmetry factor of $\frac{1}{8}$: two factors of $\frac{1}{2}$ from the links with their ends joined to the same vertex, and one factor of $\frac{1}{2}$ because the two links may be swapped over without changing the diagrams. The symmetry factors for the other two diagrams may be derived in a similar way.

This concludes the rules for the symmetry factor. While their derivation may have been obscure, the rules themselves are simple to use. A number of examples will be found in Problem 8.3.

8.3 The Helmholtz free energy of the Landau–Ginzburg model

We have just seen how to express the partition function for the Landau–Ginzburg model as the sum of an infinite number of correlation functions evaluated in the Gaussian model. We saw how to generate the terms in this sum by writing down diagrams, and how to assign the correct algebraic expression to each such diagram. However, what we would really like is a series for the Helmholtz free energy of the Landau–Ginzburg model, rather than its partition function. The free energy is a physically interesting quantity, and its derivatives are the connected correlation functions of the order parameter. It is proportional to the volume of the system, whereas the partition function grows exponentially with system size. Fortunately we can now obtain a series for the free energy, using the symmetry factor rules developed at the end of the last section.

Recall that diagrams can be classed as either connected diagrams, which are all in one piece, or as disconnected diagrams, which are the product of several connected diagrams (see §8.2.1). Clearly there are many fewer connected diagrams than disconnected diagrams. Now it turns out that the expansion of $\log Z[J]$ is given by the connected diagrams *only*. This can be seen from what we already know: $\log Z[J]$ should be proportional to the volume V. But a disconnected diagram with n separate pieces is proportional to V^n when all of its position variables are integrated over, because the value of the integrand does not change when any of the n connected subdiagrams are moved rigidly through space. Therefore only diagrams with one component—the connected diagrams—are proportional to V, and only these can contribute to $\log Z$. The proof that $\log Z$ is in fact equal to the

> **Box 8.1: $\log Z[J]$ consists of connected diagrams only**
>
> Let the index i label the members of the set of connected diagrams, and let C_i be the expression associated with the ith such diagram. Now any disconnected diagram can be completely described by the number of times n_i that each connected diagram i appears in it. Suppose that the first connected diagram appears three times, for example. Then the last rule for the symmetry factor above says that there will be a factor $1/3!$ multiplying the disconnected diagram due to the $3!$ ways of arranging the three copies of the first diagram while leaving the complete diagram unchanged. Similarly, if the second diagram appears four times, there will be a factor of $1/4!$. In general, the expression D associated with the general (possibly disconnected) diagram labelled by the set of numbers $\{n_i\}$ is
>
> $$D(\{n_i\}) = \prod_i \frac{(C_i)^{n_i}}{n_i!}. \qquad (1)$$
>
> The partition function Z is the sum over all possible expressions D_i:
>
> $$Z = \sum_{n_1=0}^{\infty} \sum_{n_2=0}^{\infty} \cdots D(\{n_i\})$$
>
> $$= \prod_{i=1}^{\infty} \left(\sum_{n_i=0}^{\infty} \frac{(C_i)^{n_i}}{n_i!} \right) \qquad (2)$$
>
> $$= \exp \sum_i C_i.$$
>
> In other words, $\log Z$ is the sum of the connected diagrams only.

sum of all such diagrams follows from the symmetry factor rules, and is given in Box 8.1.

This is a great simplification. There are many fewer connected than disconnected diagrams, and only a finite number of connected diagrams at each order in λ. Most importantly it also means that only the connected diagrams contribute to the connected correlation functions, since these are the derivatives of $\log Z$.

8.3.1 Feynman rules in wavevector space

We saw in §8.1 that the Fourier transform of the bare propagator Δ was much simpler than the bare propagator itself. In any number of dimensions this Fourier transform is just $(\alpha^2 k^2 + \mu^2)^{-1}$, whereas its position-space representations (see Table 8.1) are complicated and dimension-dependent. For this reason correlation functions of the order parameter in position space

8.3 The Helmholtz free energy

are rarely calculated. It is much more usual to calculate instead correlation functions of the function $\widetilde{\phi}(\mathbf{k})$ introduced in §8.1, of which $\phi(\mathbf{x})$ is the Fourier transform. Such correlation functions are known as **wavevector-space** or **k-space** correlation functions. To calculate them we first rewrite the source term in the Landau–Ginzburg Hamiltonian (see equation (8.1)) as

$$-\int d^d\mathbf{x}\, J(\mathbf{x})\phi(\mathbf{x}) = -\int \frac{d^d\mathbf{k}}{(2\pi)^d}\, \widetilde{J}(-\mathbf{k})\widetilde{\phi}(\mathbf{k}). \qquad (8.20)$$

Just as differentiating the partition function (8.1) with respect to $J(\mathbf{x})$ brings down a factor of $\phi(\mathbf{x})$ from the source term, we see from (8.20) that differentiating (8.1) with respect to $\widetilde{J}(-\mathbf{k})$ brings down a factor of $(2\pi)^{-d}\widetilde{\phi}(\mathbf{k})$.[3] So the connected correlation functions of $\widetilde{\phi}(\mathbf{k})$ are given by

$$\widetilde{G}_c^{(n)}(\mathbf{k}_1,\ldots,\mathbf{k}_n;\widetilde{J}) \equiv \langle \widetilde{\phi}(\mathbf{k}_1),\ldots,\widetilde{\phi}(\mathbf{k}_n)\rangle_{\text{connected}}$$
$$= (2\pi)^{dn} \frac{\delta}{\delta \widetilde{J}(-\mathbf{k}_1)} \cdots \frac{\delta}{\delta \widetilde{J}(-\mathbf{k}_n)} \log Z[\widetilde{J}]. \qquad (8.21)$$

From the definition of $\widetilde{\phi}(\mathbf{k})$, equation (8.4), it follows that these correlation functions are related to those of $\phi(\mathbf{x})$ by

$$G_c^{(n)}(\mathbf{x}_1,\ldots,\mathbf{x}_n) = \int d^d\check{\mathbf{k}}_1 e^{i\mathbf{k}_1\cdot\mathbf{x}_1}\ldots d^d\check{\mathbf{k}}_n e^{i\mathbf{k}_n\cdot\mathbf{x}_n}\, \widetilde{G}_c^{(n)}(\mathbf{k}_1,\ldots,\mathbf{k}_n). \qquad (8.22)$$

Hence the $\widetilde{G}_c^{(n)}$ are the Fourier transforms of the position-space correlation functions. However, it is not necessary to calculate the $\widetilde{G}_c^{(n)}$ by first finding the position-space correlation functions and then Fourier-transforming them. Instead, rules can be found which allow the $\widetilde{G}_c^{(n)}$ to be written down directly.

To this end consider the particular diagram contributing to $G_c^{(1)}$ shown in Figure 8.5(a). According to the diagram rules, the expression C corresponding to it is

$$C = \tfrac{1}{6}\lambda^2 \int d^d\mathbf{z}_1 d^d\mathbf{z}_2 d^d\mathbf{z}_3\, \Delta(\mathbf{x}-\mathbf{z}_1)\Delta^3(\mathbf{z}_1-\mathbf{z}_2)\Delta(\mathbf{z}_2-\mathbf{z}_3)J(\mathbf{z}_3). \qquad (8.23)$$

Now let us replace every occurrence of $\Delta(\mathbf{v}-\mathbf{w})$ in C with the equivalent expression

$$\Delta(\mathbf{v}-\mathbf{w}) = \int^\Lambda d^d\check{\mathbf{k}}\, \frac{e^{i\mathbf{k}\cdot(\mathbf{v}-\mathbf{w})}}{\alpha^2 k^2 + \mu^2} \qquad (8.24)$$

---

[3] Since $J(\mathbf{x})$ is real, $\widetilde{J}(-\mathbf{k})$ must satisfy $\widetilde{J}(-\mathbf{k}) = \widetilde{J}^*(\mathbf{k})$. It is important to note that when differentiating with respect to $\widetilde{J}(-\mathbf{k})$, this condition is *not* maintained. That is, only $\widetilde{J}(-\mathbf{k})$ is varied; $\widetilde{J}(\mathbf{k})$ is left unaltered. Varying only $\widetilde{J}(-\mathbf{k})$ brings down only $\widetilde{\phi}(\mathbf{k})$, which is what we want.

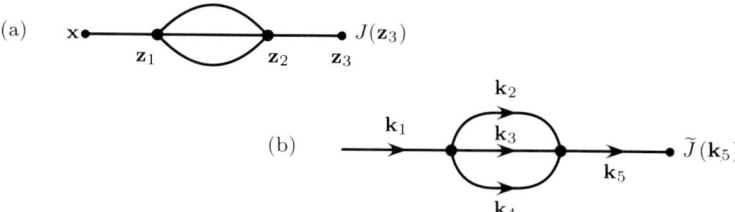

Figure 8.5 (a) A diagram contributing to $G_c^{(1)}$, discussed in the text. (b) The corresponding diagram contributing to the k-space correlation function $\tilde{G}_c^{(1)}$.

(see equation (8.7)). Then C becomes

$$C = \tfrac{1}{6}\lambda^2 \int^\Lambda \frac{d^d\tilde{\mathbf{k}}_1}{\alpha^2 k_1^2 + \mu^2} \cdots \frac{d^d\tilde{\mathbf{k}}_5}{\alpha^2 k_5^2 + \mu^2} e^{i\mathbf{k}_1\cdot\mathbf{x}} \qquad (8.25)$$

$$\times \int d^d\mathbf{z}_1 d^d\mathbf{z}_2\, e^{i(\mathbf{k}_2+\mathbf{k}_3+\mathbf{k}_4-\mathbf{k}_1)\cdot\mathbf{z}_1} e^{i(\mathbf{k}_5-\mathbf{k}_2-\mathbf{k}_3-\mathbf{k}_4)\cdot\mathbf{z}_2} \int d^d\mathbf{z}_3\, e^{-i\mathbf{k}_5\cdot\mathbf{z}_3} J(\mathbf{z}_3).$$

For every link i in the diagram there is now a wavevector $\mathbf{k}_i$, which is integrated over, and also a factor of $(\alpha^2 k_i^2 + \mu^2)^{-1}$. Now consider the internal four-point vertices, at $\mathbf{z}_1$ and $\mathbf{z}_2$. Integration over these positions leads to a δ-function involving the wavevectors arriving at each vertex:

$$\begin{aligned} \int d^d\mathbf{z}\, e^{i(\mathbf{k}_2+\mathbf{k}_3+\mathbf{k}_4-\mathbf{k}_1)\cdot\mathbf{z}} &= (2\pi)^d \delta(\mathbf{k}_2+\mathbf{k}_3+\mathbf{k}_4-\mathbf{k}_1) \\ \int d^d\mathbf{z}\, e^{i(\mathbf{k}_5-\mathbf{k}_2-\mathbf{k}_3-\mathbf{k}_4)\cdot\mathbf{z}} &= (2\pi)^d \delta(\mathbf{k}_5-\mathbf{k}_2-\mathbf{k}_3-\mathbf{k}_4). \end{aligned} \qquad (8.26)$$

So the sum of the wavevectors associated with the links attached to each vertex is zero. We shall refer to this fact as **wavevector conservation**. Next, consider the blob at $\mathbf{z}_3$. The integral over $\mathbf{z}_3$ in (8.25) replaces J with its Fourier transform $\tilde{J}$, the particular component being chosen by the wavevector of the link joined to the source. Integrating out the two δ-functions resulting from the $\mathbf{z}_1$ and $\mathbf{z}_2$ integrations in (8.25) gives

$$C = \tfrac{1}{6}\lambda^2 \int e^{i\mathbf{k}_1\cdot\mathbf{x}} \frac{d^d\tilde{\mathbf{k}}_1}{\alpha^2 k_1^2 + \mu^2} \frac{d^d\tilde{\mathbf{k}}_2}{\alpha^2 k_2^2 + \mu^2} \frac{d^d\tilde{\mathbf{k}}_3}{\alpha^2 k_3^2 + \mu^2} \qquad (8.27)$$

$$\times \frac{\Theta(\Lambda^2 - (\mathbf{k}_1-\mathbf{k}_2-\mathbf{k}_3)^2)}{\alpha^2(\mathbf{k}_1-\mathbf{k}_2-\mathbf{k}_3)^2 + \mu^2} \frac{\tilde{J}(\mathbf{k}_1)}{\alpha^2 k_1^2 + \mu^2}.$$

The function $\Theta(x)$ is the step function, being 1 when its argument is positive and zero otherwise. It appears because the magnitudes of the wavevectors

8.3 The Helmholtz free energy

in (8.25) all lie between 0 and Λ, and so if the δ-functions in (8.26) constrain k_4 to lie outside this range then the $\mathbf{k}_4$-integral in (8.25) must give zero. However, it will turn out that the quantities we are interested in calculating do not depend on the structure of Feynman integrals at large wavevectors and so the results obtained in subsequent chapters will not be affected by the omission of step functions in Feynman integrals. For this reason we will not include them from here on.

Equation (8.27) is now in the form of equation (8.22). Comparing the two equations we see that the contribution $\widetilde{C}$ of the diagram to the one-point correlation function $\widetilde{G}_c^{(1)}(\mathbf{k}_1)$ is

$$\widetilde{C} = \frac{1}{\alpha^2 k_1^2 + \mu^2} \int \frac{d^d \check{\mathbf{k}}_2}{\alpha^2 k_2^2 + \mu^2} \frac{d^d \check{\mathbf{k}}_3}{\alpha^2 k_3^2 + \mu^2} \frac{1}{\alpha^2 (\mathbf{k}_1 - \mathbf{k}_2 - \mathbf{k}_3)^2 + \mu^2} \frac{\widetilde{J}(\mathbf{k}_1)}{\alpha^2 k_1^2 + \mu^2}. \tag{8.28}$$

This example points the way to the general Feynman rules for finding the $\widetilde{G}_c^{(n)}$. The same diagrams are drawn as for the position-space correlation functions, but the expression assigned to each diagram is different. Each link has a wavevector $\mathbf{k}$ flowing along it, and an associated factor $(\alpha^2 k^2 + \mu^2)^{-1}$. The wavevector flowing into an external link is fixed by the particular correlation function being calculated; all other wavevectors are integrated over, subject to the constraint that wavevector is conserved at each vertex. We shall use '$\mathbf{k}_i$' for external, fixed wavevectors, and '$\mathbf{q}_i$' for wavevectors that are integrated over. A source-blob at the end of a link with wavevector $\mathbf{q}$ flowing into it contributes a factor $\widetilde{J}(\mathbf{q})$ to the diagram. The symmetry factor for a diagram is identical to that for the corresponding position-space diagram.

The rules for both position-space and k-space diagrams are summarized in Boxes 8.2 and 8.3. A number of examples to illustrate them will be found in the next section, and in the Problems at the end of the chapter.

A consequence of the conservation of wavevector at each vertex is that, if the source J vanishes, then not all of the δ-functions can be integrated away, and the correlation function vanishes unless the total of the wavevectors flowing in along its external legs is zero. For example,

$$\widetilde{G}_c^{(2)}(\mathbf{k}_1, \mathbf{k}_2) = \frac{(2\pi)^d \delta(\mathbf{k}_1 + \mathbf{k}_2)}{\alpha^2 k_1^2 + \mu^2} \tag{8.29}$$

in the Gaussian model. The presence of this δ-function reflects the translational invariance of the theory. It arises because (if $J = 0$) the position-space correlation functions do not depend on absolute position. Since correlation functions are most often calculated with $J = 0$, it is convenient to write $\widetilde{G}_c^{(n)}$ as

$$\widetilde{G}_c^{(n)}(\mathbf{k}_1, \ldots, \mathbf{k}_n) = (2\pi)^d \delta(\mathbf{k}_1 + \cdots + \mathbf{k}_n) \check{G}_c^{(n)}(\mathbf{k}_1, \ldots, \mathbf{k}_{n-1}). \tag{8.30}$$

The full n-point correlation function, a function of n wavevectors, is the product of the **reduced correlation function** $\check{G}_c^{(n)}(\mathbf{k}_1, \ldots, \mathbf{k}_{n-1})$, which

---

Box 8.2: Feynman rules in position space

To find the O(λ^n) contribution to $G_c^{(m)}(\mathbf{x}_1, \ldots, \mathbf{x}_m; J)$:

- Draw all distinct connected diagrams containing n four-link vertices, zero or more source-blobs, and m external points labelled $\mathbf{x}_1, \ldots, \mathbf{x}_m$, joining all these up with links (exactly four links per vertex, and one link per blob or point). If $m = 0$, the diagram contributes to $\log Z[J]$. Two diagrams are distinct if they cannot be deformed into each other by moving the vertices, blobs, and points, without cutting any links. (Sliding links over other links is allowed.) For each of these diagrams:
    - Assign to the diagram a factor $(-\lambda)^n$.
    - Label each of the internal points—vertices and blobs—with a position variable $\mathbf{z}_i$.
    - To each link assign a factor $\Delta(\mathbf{z}_2 - \mathbf{z}_1)$, where $\mathbf{z}_1$ and $\mathbf{z}_2$ are the points joined by the link. (The order of $\mathbf{z}_1$ and $\mathbf{z}_2$ is unimportant.)
    - Assign a factor $J(\mathbf{z})$ to each blob, where $\mathbf{z}$ is the position of the blob.
    - Integrate over the positions of all internal points.
    - Multiply by $\frac{1}{2}$ for each link whose two ends are joined to the same vertex.
    - Multiply by $(l!)^{-1}$ for each set of l links joining the same two vertices.
    - If the internal points of a diagram can be rearranged r ways and yet leave the diagram looking exactly the same, divide by r.
- Add up the resulting expressions for all these diagrams.

---

is a function of $n - 1$ wavevectors, and $(2\pi)^d \delta(\mathbf{k}_1 + \cdots + \mathbf{k}_n)$. Continuing the example of (8.29), the two-point connected correlation function in the Gaussian model can be written

$$\widetilde{G}_c^{(2)}(\mathbf{k}_1, \mathbf{k}_2) = (2\pi)^d \delta(\mathbf{k}_1 + \mathbf{k}_2) \widetilde{G}_c^{(2)}(\mathbf{k}_1), \tag{8.31}$$

where the reduced two-point correlation function $\widetilde{G}_c^{(2)}(\mathbf{k})$ is

$$\widetilde{G}_c^{(2)}(\mathbf{k}_1) = \frac{1}{\alpha^2 k_1^2 + \mu^2}. \tag{8.32}$$

Reduced correlation functions can only be defined when the source J is spatially uniform (in the usual case, $J = 0$ everywhere).

From here on we will distinguish full from reduced correlation functions by the number of their arguments. If no arguments are shown for some $\widetilde{G}_c^{(n)}$ then its type is unimportant.

8.3 The Helmholtz free energy

> **Box 8.3: Feynman rules in k-space**
>
> To find the $O(\lambda^n)$ contribution to $\widetilde{G}_c^{(m)}(\mathbf{k}_1, \ldots, \mathbf{k}_m; J)$:
> - Draw all distinct connected diagrams containing n four-link vertices, zero or more source-blobs, and m external points labelled $\mathbf{k}_1, \ldots, \mathbf{k}_m$, joining all these up with links (exactly four links per vertex, and one link per blob or point). If $m = 0$, the diagram contributes to $\log Z[J]$. Two diagrams are distinct if they cannot be deformed into each other by moving the vertices, blobs, and points, without cutting any links. (Sliding links over other links is allowed.) For each of these diagrams:
>   - Assign to the diagram a factor $(-\lambda)^n$.
>   - Assign a wavevector $\mathbf{q}_i$ to flow along each link and an associated factor $(\alpha^2 q_i^2 + \mu^2)^{-1}$, subject to conservation of wavevector at each vertex. The wavevector flowing along a link connected to an external point labelled $\mathbf{k}_i$ is equal to $\mathbf{k}_i$.
>   - Assign a factor $\tilde{J}(\mathbf{q})$ to a blob on the end of a link with wavevector $\mathbf{q}$ flowing into it.
>   - Integrate over all free wavevectors, including a factor of $(2\pi)^{-d}$ for each integral. Cut each integral off at Λ, including Θ-functions as in (8.27) if you feel fussy.
>   - Multiply by $\frac{1}{2}$ for each link whose two ends are joined to the same vertex.
>   - Multiply by $(l!)^{-1}$ for each set of l links joining the same two vertices.
>   - If the internal points of a diagram can be rearranged r ways and yet leave the diagram looking exactly the same, divide by r.
> - Add up the resulting expressions for all these diagrams.

We now show that $\beta \widetilde{G}_c^{(2)}(0)$ is equal to the susceptibility χ, the rate of change of the local magnetization $\langle \phi(\mathbf{x}) \rangle$ when a uniform magnetic field is applied. A small magnetic field $B(\mathbf{y})$ corresponds to a small source field $J(\mathbf{y}) = \beta B(\mathbf{y})$. The change in $\langle \phi(\mathbf{x}) \rangle$ produced by such a field can be represented by the following integral, which is analogous to the chain rule of ordinary differential calculus:

$$\delta \langle \phi(\mathbf{x}) \rangle = \int d^d \mathbf{y}\, J(\mathbf{y}) \left. \frac{\delta \langle \phi(\mathbf{x}) \rangle}{\delta J(\mathbf{y})} \right|_{J=0} \tag{8.33}$$

(see equation (L.12)). If this field is uniform and equal to $\mathcal{J}$, (8.33) becomes

$$\begin{aligned}\delta\langle\phi(\mathbf{x})\rangle &= \mathcal{J}\int d^d\mathbf{y}\,\frac{\delta\langle\phi(\mathbf{x})\rangle}{\delta J(\mathbf{y})}\\ &= \mathcal{J}\int d^d\mathbf{y}\,\frac{\delta^2\log Z[J]}{\delta J(\mathbf{x})\delta J(\mathbf{y})} \\ &= \mathcal{J}\int d^d\mathbf{y}\,G_c^{(2)}(\mathbf{x},\mathbf{y}).\end{aligned} \qquad (8.34)$$

Now

$$\begin{aligned}G_c^{(2)}(\mathbf{x},\mathbf{y}) &= \int d^d\check{\mathbf{k}}_1 d^d\check{\mathbf{k}}_2\,\widetilde{G}_c^{(2)}(\mathbf{k}_1,\mathbf{k}_2)e^{i(\mathbf{x}\cdot\mathbf{k}_1+\mathbf{y}\cdot\mathbf{k}_2)}\\ &= \int d^d\check{\mathbf{k}}_1\,\widetilde{G}_c^{(2)}(\mathbf{k}_1)e^{i\mathbf{k}_1\cdot(\mathbf{x}-\mathbf{y})}.\end{aligned} \qquad (8.35)$$

Substituting this into (8.34) gives

$$\chi = \frac{\delta\langle\phi\rangle}{\delta B} = \beta\widetilde{G}_c^{(2)}(0), \qquad (8.36)$$

as claimed. This is an example of the linear response theorem of §2.2.

The k-space correlation functions are somewhat easier to calculate than the position-space ones, and for this reason the subsequent chapters will work almost exclusively in k-space. The reason for beginning our study of the Landau–Ginzburg model with the position-space correlation functions is that they are conceptually simpler. Also it is easier to derive the Feynman rules and the symmetry factor in position-space and to obtain the k-space rules from these, rather than to start off in k-space directly (though the latter is perfectly possible).

When studying critical phenomena we are interested in the behaviour of the theory at large scales. For example, we might study the limit

$$\ell(\mathbf{x}_1,\ldots,\mathbf{x}_n) \equiv \lim_{b\to\infty} G_c^{(n)}(b\mathbf{x}_1,\ldots,b\mathbf{x}_n). \qquad (8.37)$$

Taking the Fourier transform of (8.37) gives us

$$\widetilde{\ell}(\mathbf{k}_1,\ldots,\mathbf{k}_n) = \lim_{b\to\infty} b^{-nd}\widetilde{G}_c^{(n)}(\mathbf{k}_1/b,\ldots,\mathbf{k}_n/b). \qquad (8.38)$$

So the information about the critical behaviour of a model is contained in the small-wavevector limits of its correlation functions. When we use the phrase **small-wavevector limit** later in this book we shall mean a limiting process similar to that in (8.38).

8.3 The Helmholtz free energy

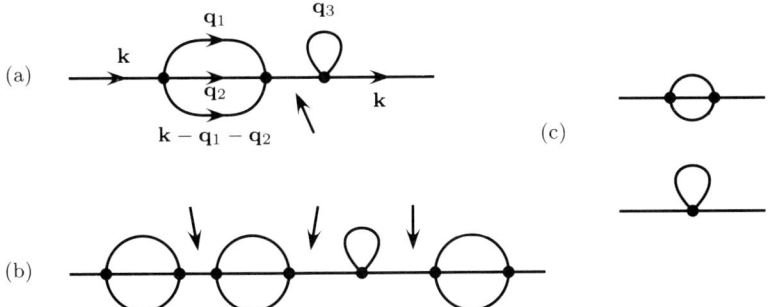

Figure 8.6 (a) A contribution to $\tilde{G}_c^{(2)}(\mathbf{k})$, discussed in the text. (b) A more complicated contribution to $\tilde{G}_c^{(2)}(\mathbf{k})$. (c) The two subdiagrams which make up the diagrams in (a) and (b).

8.3.2 Vertex functions

We have seen that to find $\log Z$ and the connected correlation functions we must calculate and sum all connected diagrams according to the above rules. This is much simpler than having to consider all diagrams, disconnected as well as connected. But the number of diagrams that need to be calculated can be reduced still further. This is possible because some connected diagrams are made up of two or more connected diagrams joined together in a simple way. If we know the expressions for these building blocks, we can easily find the expression for the complete diagram.

To illustrate this, consider the diagram shown in Figure 8.6(a). This diagram contributes to $\tilde{G}_c^{(2)}(\mathbf{k})$, and the corresponding expression is

$$D_1 = \frac{1}{\alpha^2 k^2 + \mu^2} \left(\frac{1}{2}\lambda^2 \int \frac{d^d \check{q}_1}{\alpha^2 q_1^2 + \mu^2} \frac{d^d \check{q}_2}{\alpha^2 q_2^2 + \mu^2} \frac{1}{\alpha^2(\mathbf{k} - \mathbf{q}_1 - \mathbf{q}_2)^2 + \mu^2} \right) \quad (8.39)$$

$$\times \frac{1}{\alpha^2 k^2 + \mu^2} \left(-\frac{1}{2}\lambda \int \frac{d^d \check{q}_3}{\alpha^2 q_3^2 + \mu^2} \right) \frac{1}{\alpha^2 k^2 + \mu^2}.$$

Notice the structure of this expression. First there is a bare propagator; then a complicated integral; then another propagator; then another integral, and finally the propagator again. All of these components are simply multiplied together. The expression for the diagram in Figure 8.6(b) has a similar structure and we can easily write it down too. Calling the two expressions in the brackets in (8.39) I_a and I_b respectively, it is

$$D_2 = \Delta(\mathbf{k}) I_a(\mathbf{k}) \Delta(\mathbf{k}) I_a(\mathbf{k}) \Delta(\mathbf{k}) I_b(\mathbf{k}) \Delta(\mathbf{k}) I_a(\mathbf{k}) \Delta(\mathbf{k}), \quad (8.40)$$

where $\Delta(\mathbf{k}) \equiv (\alpha^2 k^2 + \mu^2)^{-1}$ is the bare propagator in wavevector space. More and more complicated connected diagrams can be made by stringing together the two subdiagrams shown in Figure 8.6(c), and the corresponding expressions can be written down as a product of the bare propagator and the expressions I_a and I_b. The subdiagrams in Figure 8.6(c) act as building blocks for this set of diagrams.

We can formalize this idea of some diagrams being building blocks from which other diagrams are constructed as follows. A connected diagram is called **one-particle reducible**[4] (**1PR**) if, by cutting a single propagator not joined to an external point, it can be split into two separate connected subdiagrams. A connected diagram is called **one-particle irreducible** (**1PI**) if it cannot be split into two separate subdiagrams in this way. The diagrams in Figure 8.6(a) and (b) are 1PR. If cut at any of the places indicated by arrows they fall into two pieces. The diagrams in Figure 8.6(c) are 1PI. They cannot be split apart by a single cut. It is clear that all diagrams are either 1PR or 1PI. So the separate diagrams produced by the splitting of a 1PR diagram are either themselves 1PI or can be further reduced. We conclude that all 1PR diagrams can be written as two or more 1PI diagrams joined together. The 1PI diagrams are the 'building blocks' for connected diagrams.

It is worth stating explicitly that one is allowed to cut links joined to source-blobs, unless the other end of the link is attached to an external point. Thus the only 1PI diagram that involves J is ⟶● $\tilde{J}$. This rule allows source-blobs to act as building blocks.

So any connected diagram is either itself 1PI or is made up of 1PI diagrams joined together with bare propagators to form a tree structure. There can be no loops in this structure since a loop of 1PI diagrams would itself be 1PI. The wavevectors flowing in the bare propagators that link the 1PI subdiagrams are entirely determined by the wavevectors flowing into the diagram from its external legs, and from any source-blobs present. The wavevectors internal to each 1PI diagram are local to each diagram; they do not flow through the rest of the tree at all. In the example above the wavevectors $\mathbf{q}_1$, $\mathbf{q}_2$, and $\mathbf{q}_3$ appeared only in the expressions I_a and I_b. So it is enough to know the expressions corresponding to the 1PI diagrams. If these are known, then any connected diagram can be made up by multiplying them together.

The expressions I_a and I_b are described as **amputated**, because they do not contain propagators for their external legs. Normally the expressions corresponding to the diagrams in Figure 8.6(c) would each have two factors of $(\alpha^2 k^2 + \mu^2)^{-1}$, in addition to the integrals I_a and I_b. Amputated expressions are useful because they can be joined together directly with a propagator to form a valid diagram. Multiplying together two non-amputated expressions and a propagator would not be correct. It is the n-point amputated 1PI diagrams which are the building blocks for connected diagrams.

For $n > 2$, *minus* the sum of all n-point amputated 1PI diagrams is called the n-point **vertex function**, and is written $\Gamma^{(n)}(\mathbf{k}_1, \ldots, \mathbf{k}_n)$.[5] $\Gamma^{(2)}$ is *minus* the sum of all two-point amputated 1PI diagrams, *plus* the inverse

---

[4] The terms 'one-particle reducible' and 'one-particle irreducible' come from quantum field theory, in which each link in a Feynman diagram represents the propagation of a particle. The same terms are in common use in statistical physics, even though the interpretation of diagrams is completely different.

[5] Like connected correlation functions, vertex functions in zero field come in full and

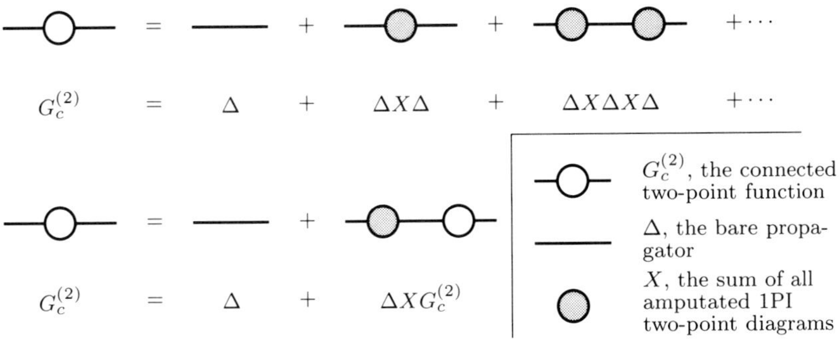

8.3 The Helmholtz free energy

Figure 8.7 The two-point connected correlation function expressed as an infinite series involving amputated two-point diagrams. The top line shows the first terms of this infinite series. The bottom line indicates how the series may be summed diagrammatically to yield $G_c^{(2)}(1 - \Delta X) = \Delta$ (cf. (8.43)).

Δ^{-1} of the bare propagator. The reasons behind these slightly eccentric definitions are revealed in the next section, and in Appendix M.

As an example of how connected diagrams can be constructed from 1PI subdiagrams, consider the connected two-point function $\widetilde{G}_c^{(2)}(\mathbf{k})$ for the case $J = 0$. All diagrammatic contributions to this, apart from the bare propagator, take the form of one or more 1PI subdiagrams strung together in a line, because as $J = 0$ there can be no branches. The same wavevector $\mathbf{k}$ flows through all the subdiagrams. Algebraically we can write

$$\widetilde{G}_c^{(2)}(\mathbf{k}) = \Delta(\mathbf{k}) + \Delta(\mathbf{k})X(\mathbf{k})\Delta(\mathbf{k}) + \Delta(\mathbf{k})X(\mathbf{k})\Delta(\mathbf{k})X(\mathbf{k})\Delta(\mathbf{k}) + \cdots, \quad (8.41)$$

where $X(\mathbf{k})$ is the sum of all amputated 1PI contributions to $\widetilde{G}_c^{(2)}(\mathbf{k})$ excluding the bare propagator. The integrals I_a and I_b of (8.40) appear in this sum, along with many others. From the definition of $\Gamma^{(2)}$ given above, it follows that

$$\Gamma^{(2)}(\mathbf{k}) = \Delta^{-1}(\mathbf{k}) - X(\mathbf{k}). \quad (8.42)$$

Equation (8.41) is illustrated in Figure 8.7. The series can be summed to give:

$$\widetilde{G}_c^{(2)}(\mathbf{k}) = \frac{1}{\Delta^{-1}(\mathbf{k}) - X(\mathbf{k})}$$
$$= \frac{1}{\Gamma^{(2)}(\mathbf{k})}, \quad (8.43)$$

---

reduced versions. The two are related by

$$\Gamma^{(n)}(\mathbf{k}_1, \ldots, \mathbf{k}_n) = (2\pi)^d \delta(\mathbf{k}_1 + \cdots + \mathbf{k}_n) \Gamma^{(n)}(\mathbf{k}_1, \ldots, \mathbf{k}_{n-1}).$$

> **Box 8.4: Feynman rules for vertex functions in k-space**
>
> Let $Z_G[J]$ denote the partition function of the Gaussian model, given in (8.5). Then $\Gamma^{(0)}$ is $-\log Z_G[0]$ minus the sum of all connected 1PI diagrams with no external legs constructed according to the rules of Box 8.3. $\Gamma^{(1)}$ is minus the sum of all amputated 1PI contributions to $\widetilde{G}_c^{(1)}$, excluding the diagram involving the source $\widetilde{J}$. For the Landau–Ginzburg model, $\Gamma^{(1)}$ vanishes. The only one of the $\Gamma^{(n)}$ which has a contribution zeroth-order in λ is $\Gamma^{(2)}$. This contribution is the inverse of the bare propagator:
>
> $$\Gamma^{(2)}(\mathbf{k}) = \alpha^2 k^2 + \mu^2 + \mathrm{O}(\lambda).$$
>
> The remaining contributions to $\Gamma^{(2)}$, and all of the contributions to $\Gamma^{(m)}$ for $m \geq 3$, are given by the rules below. They differ from the rules for the $\widetilde{G}_c^{(n)}$ in that:
> - Only 1PI diagrams are considered.
> - There are no propagators for links connected to external points (the diagrams are amputated).
> - There is an overall minus sign.
>
> All else is unchanged. In particular the symmetry factors are calculated in the same way as before.
>
> To find the $\mathrm{O}(\lambda^n)$ contribution to $\Gamma^{(m)}(\mathbf{k}_1, \ldots, \mathbf{k}_m)$:
> - Draw all distinct connected diagrams containing n four-link vertices, and m external points labelled $\mathbf{k}_1$, ..., $\mathbf{k}_m$, joining all these up with links (exactly four links per vertex, and one per point). Two diagrams are distinct if they cannot be deformed into each other by moving the vertices and points, without cutting any links. (Sliding links over other links is allowed.) For each of these diagrams:
>   - Assign to the diagram a factor $(-1) \times (-\lambda)^n$.
>   - Assign a wavevector $\mathbf{q}_i$ to flow along each link and a factor $(\alpha^2 q_i^2 + \mu^2)^{-1}$, subject to conservation of wavevector at each vertex. The wavevector flowing along a link connected to an external point labelled $\mathbf{k}_i$ is equal to $\mathbf{k}_i$. Do *not* assign a factor $(\alpha^2 k_i^2 + \mu^2)^{-1}$ to links connected to external points.
>   - Integrate over all free wavevectors, including a factor of $(2\pi)^{-d}$ for each integral and cutting integrals off at Λ.
>   - Multiply by $\frac{1}{2}$ for each link whose two ends are joined to the same vertex.
>   - Multiply by $(l!)^{-1}$ for each set of l links joining the same two vertices.
>   - If the internal points of a diagram can be rearranged r ways and yet leave the diagram looking exactly the same, divide by r.
> - Add up the resulting expressions for all these diagrams.

8.4 The Gibbs free energy

Table 8.2. Dimensions of quantities introduced in this chapter in powers of length

Variable	Dimension	Variable	Dimension
$\phi(\mathbf{x})$	$-\frac{1}{2}d$	$G_c^{(n)}(\mathbf{x}_1,\ldots,\mathbf{x}_n)$	$-\frac{n}{2}d$
$\widetilde{\phi}(\mathbf{k})$	$\frac{1}{2}d$	$\widetilde{G}_c^{(n)}(\mathbf{k}_1,\ldots,\mathbf{k}_n)$	$\frac{n}{2}d$
$J(\mathbf{x})$	$-\frac{1}{2}d$	$\widetilde{G}_c^{(n)}(\mathbf{k}_1,\ldots,\mathbf{k}_{n-1})$	$(\frac{n}{2}-1)d$
$\widetilde{J}(\mathbf{k})$	$\frac{1}{2}d$	$\Gamma^{(n)}(\mathbf{k}_1,\ldots,\mathbf{k}_n)$	$\frac{n}{2}d$
$G^{(n)}(\mathbf{x}_1,\ldots,\mathbf{x}_n)$	$-\frac{n}{2}d$	$\Gamma^{(n)}(\mathbf{k}_1,\ldots,\mathbf{k}_{n-1})$	$(\frac{n}{2}-1)d$

where the last line follows from (8.42). Thus the two-point vertex function is the inverse of the propagator, and hence from (8.36) $\Gamma^{(2)}(0) = \beta/\chi$.

From now on we shall deal mainly with vertex functions, since any correlation function can be constructed from them. The dimensions of all the quantities associated with the Landau–Ginzburg model that we have introduced in this chapter are given in Table 8.2.

Connected correlation functions are the derivatives of the Helmholtz free energy with respect to the source J (see equation (8.2)). It is natural to ask whether vertex functions have a physical interpretation. In the next section (and in Appendix M) we will see that, above T_c, the vertex functions are the derivatives of the Gibbs free energy with respect to the thermal average of the order parameter ϕ. So vertex functions are not only technically useful, but are physically interesting in their own right. The peculiarities in their definitions—the apparently gratuitous minus sign, and the anomalous definition of $\Gamma^{(2)}$—are present in order to simplify their relationship to the Gibbs free energy. The Feynman rules for calculating k-space vertex functions are given in Box 8.4. (This box also gives the rules for $\Gamma^{(0)}$ and $\Gamma^{(1)}$; these two vertex functions are defined in the next section.)

8.4 The Gibbs free energy of the Landau–Ginzburg model

In this section we calculate the Gibbs free energy G of the Landau–Ginzburg model. (Note that G is *not* a correlation function.) This calculation illustrates the techniques developed in the previous section, as well as being useful in following chapters. We have seen how to calculate the Helmholtz free energy as a functional of the source $J(\mathbf{x})$. G, the Legendre transform of the Helmholtz free energy, is given by

$$\beta G[\overline{\phi}] = \left[\int d^d\mathbf{x}\, J(\mathbf{x})\overline{\phi}(\mathbf{x})\right] - \log Z[J] \equiv \Gamma[\overline{\phi}], \qquad (8.44)$$

where

$$\overline{\phi}(\mathbf{x}) \equiv \frac{\delta \log Z[J]}{\delta J(\mathbf{x})} \qquad (8.45)$$
$$= \langle \phi(\mathbf{x}) \rangle$$

is the thermal average of the order parameter $\phi(\mathbf{x})$. Equation (8.44) defines the **dimensionless Gibbs free energy** Γ, which is the quantity we shall calculate. (We shall frequently refer to Γ itself as the Gibbs free energy.) The calculation involves working out Feynman diagrams, and we saw in the last section that these are easier in k-space. So we introduce the variable $\tilde{\varphi}$, of which $\overline{\phi}$ is the Fourier transform:

$$\overline{\phi}(\mathbf{x}) = \int d^d\check{\mathbf{k}}\, e^{i\mathbf{k}\cdot\mathbf{x}} \tilde{\varphi}(\mathbf{k});$$
$$\tilde{\varphi}(\mathbf{k}) = (2\pi)^d \frac{\delta \log Z}{\delta \tilde{J}(-\mathbf{k})} = \langle \tilde{\phi}(\mathbf{k}) \rangle. \qquad (8.46)$$

(It would be consistent with the rest of our notation to write $\tilde{\varphi}$ as $\overline{\tilde{\phi}}$, but this symbol doesn't look very nice.) Then Γ can be written in terms of $\tilde{\varphi}$ and $\tilde{J}$:

$$\Gamma[\tilde{\varphi}] = \left[\int d^d\check{\mathbf{k}}\, \tilde{\varphi}(\mathbf{k})\tilde{J}(-\mathbf{k})\right] - \log Z[\tilde{J}]. \qquad (8.47)$$

Remember that $\tilde{J}$ is to be eliminated in favour of $\tilde{\varphi}$ so that the left-hand side of (8.47) is a functional of $\tilde{\varphi}(\mathbf{k})$.

Γ is useful in at least two ways. It can locate the temperature at which spontaneous symmetry breaking begins: we saw in §2.3 that, for an infinite system, this is the temperature at which the second derivative of Γ at $\tilde{\varphi}=0$ vanishes. It will turn out that a full knowledge of Γ is not needed for this. Also, below the critical temperature we can use it to find the spontaneous magnetization. This is the largest value of $\tilde{\varphi}$ for which the gradient of Γ vanishes, as we saw in §2.3.

8.4.1 The rules for finding $\Gamma[\tilde{\varphi}]$

Above T_c, the vertex functions $\Gamma^{(n)}$ are defined to be the coefficients in the functional Taylor series expansion of $\Gamma[\tilde{\varphi}]$ about $\tilde{\varphi}=0$:

$$\Gamma[\tilde{\varphi}] = \sum_{n=0}^{\infty} \frac{1}{n!} \int d^d\check{\mathbf{q}}_1 \ldots d^d\check{\mathbf{q}}_n\, \tilde{\varphi}(-\mathbf{q}_1)\ldots\tilde{\varphi}(-\mathbf{q}_n)\,\Gamma^{(n)}(\mathbf{q}_1,\ldots,\mathbf{q}_n). \qquad (8.48)$$

Below T_c we would not expect Γ to have such an expansion (see Figure 2.6); this matter will be discussed later. The $\Gamma^{(n)}$ can be obtained by differentiating Γ, just as the $G_c^{(n)}$ can be obtained by differentiating $\log Z$ (compare (8.2) and (8.21)):

$$\Gamma^{(n)}(\mathbf{k}_1,\ldots,\mathbf{k}_n) = (2\pi)^{dn} \frac{\delta}{\delta\tilde{\varphi}(-\mathbf{k}_1)} \cdots \frac{\delta}{\delta\tilde{\varphi}(-\mathbf{k}_n)} \Gamma[\tilde{\varphi}]\bigg|_{\tilde{\varphi}=0}. \qquad (8.49)$$

8.4 The Gibbs free energy

We now go some way towards proving that the vertex functions as defined above are given by the rules in Box 8.4; the proof is completed in Appendix M.

Let us begin with $\Gamma^{(0)}$. On putting $\widetilde{\varphi} = 0$, equations (8.47) and (8.48) give directly
$$\Gamma^{(0)} = -\log Z[\widetilde{J}[0]]. \tag{8.50}$$
In this equation $\widetilde{J}[0]$ denotes the source field $\widetilde{J}$ that corresponds to $\widetilde{\varphi} = 0$. In the Landau–Ginzburg model, $\widetilde{J}[0] = 0$. Now all the diagrams contributing to $\log Z[0]$ are 1PI diagrams (can you see why?), so (8.50) tells us that $\Gamma^{(0)}$ is $-\log Z_G[0]$ (Z_G is the Gaussian partition function), minus the sum of all 1PI contributions to $\log Z$. This confirms the first rule of Box 8.4.

Next, putting $n = 1$ in (8.49) gives
$$\Gamma^{(1)}(\mathbf{k}) = (2\pi)^d \frac{\delta \Gamma[\widetilde{\varphi}]}{\delta \widetilde{\varphi}(-\mathbf{k}_1)}\bigg|_{\widetilde{\varphi}=0}. \tag{8.51}$$
On differentiating $\Gamma[\widetilde{\varphi}]$ as defined in (8.47) we get
$$(2\pi)^d \frac{\delta \Gamma[\widetilde{\varphi}]}{\delta \widetilde{\varphi}(-\mathbf{k}_1)} = \widetilde{J}(\mathbf{k}_1). \tag{8.52}$$
This vanishes when $\widetilde{\varphi} = 0$, as then $\widetilde{J} = 0$. So from the last two equations $\Gamma^{(1)} = 0$ in the Landau–Ginzburg model, again as in Box 8.4.

The $n = 2$ term is more complicated. Putting $n = 2$ in (8.49) gives
$$(2\pi)^d \delta(\mathbf{k}_1 + \mathbf{k}_2) \Gamma^{(2)}(\mathbf{k}_1) = (2\pi)^{2d} \frac{\delta^2 \Gamma[\widetilde{\varphi}]}{\delta \widetilde{\varphi}(-\mathbf{k}_2) \delta \widetilde{\varphi}(-\mathbf{k}_1)}\bigg|_{\widetilde{\varphi}=0}. \tag{8.53}$$
Differentiating (8.52) gives
$$(2\pi)^{2d} \frac{\delta^2 \Gamma[\widetilde{\varphi}]}{\delta \widetilde{\varphi}(-\mathbf{k}_2) \delta \widetilde{\varphi}(-\mathbf{k}_1)}\bigg|_{\widetilde{\varphi}=0} = (2\pi)^d \frac{\delta \widetilde{J}(\mathbf{k}_1)}{\delta \widetilde{\varphi}(-\mathbf{k}_2)}\bigg|_{\widetilde{\varphi}=0}. \tag{8.54}$$
To work out the functional derivative on the right-hand side of this last equation, note that
$$\frac{\delta \widetilde{\varphi}(\mathbf{k}_1)}{\delta \widetilde{\varphi}(-\mathbf{k}_2)} = \delta(\mathbf{k}_1 + \mathbf{k}_2). \tag{8.55}$$
But this derivative can also be written
$$\begin{aligned}\frac{\delta \widetilde{\varphi}(\mathbf{k}_1)}{\delta \widetilde{\varphi}(-\mathbf{k}_2)} &= (2\pi)^d \frac{\delta^2 \log Z[\widetilde{J}]}{\delta \widetilde{\varphi}(-\mathbf{k}_2) \delta \widetilde{J}(-\mathbf{k}_1)} \\ &= (2\pi)^d \int d^d \mathbf{q} \, \frac{\delta^2 \log Z[\widetilde{J}]}{\delta \widetilde{J}(\mathbf{q}) \delta \widetilde{J}(-\mathbf{k}_1)}\bigg|_{\widetilde{J}=0} \frac{\delta \widetilde{J}(\mathbf{q})}{\delta \widetilde{\varphi}(-\mathbf{k}_2)}\bigg|_{\widetilde{\varphi}=0} \quad \begin{bmatrix}\text{chain rule for func-} \\ \text{tional derivatives}\end{bmatrix} \\ &= (2\pi)^d \int d^d \mathbf{q} \, (2\pi)^{-2d} \widetilde{G}_c^{(2)}(-\mathbf{q})(2\pi)^d \delta(\mathbf{q} - \mathbf{k}_1) \frac{\delta \widetilde{J}(\mathbf{q})}{\delta \widetilde{\varphi}(-\mathbf{k}_2)}\bigg|_{\widetilde{\varphi}=0} \\ &= \widetilde{G}_c^{(2)}(-\mathbf{k}_1) \frac{\delta \widetilde{J}(\mathbf{k}_1)}{\delta \widetilde{\varphi}(-\mathbf{k}_2)}\bigg|_{\widetilde{\varphi}=0}.\end{aligned} \tag{8.56}$$

Together (8.55) and (8.56) imply that

$$\left.\frac{\delta \widetilde{J}(\mathbf{k}_1)}{\delta \widetilde{\varphi}(-\mathbf{k}_2)}\right|_{\widetilde{\varphi}=0} = \frac{\delta(\mathbf{k}_1 + \mathbf{k}_2)}{\widetilde{G}_c^{(2)}(-\mathbf{k}_1)}. \tag{8.57}$$

Combining this with (8.53) and (8.54) finally tells us

$$\Gamma^{(2)}(\mathbf{k}) = \frac{1}{\widetilde{G}_c^{(2)}(\mathbf{k})}. \tag{8.58}$$

The second derivative of the Gibbs free energy is the inverse of the connected two-point function. This result means that we can locate the critical temperature knowing only $\Gamma^{(2)}(\mathbf{k})$; at the critical temperature, $\Gamma^{(2)}(0)$ must vanish. In fact this is not a surprise, as we know that $\Gamma^{(2)}(0)$ is the inverse of the susceptibility, but it is nice to see different approaches giving the same result.

By a precisely similar method it can be shown that

$$\Gamma^{(3)}(\mathbf{k}_1, \mathbf{k}_2) = -\frac{\widetilde{G}_c^{(3)}(\mathbf{k}_1, \mathbf{k}_2)}{\widetilde{G}_c^{(2)}(\mathbf{k}_1)\widetilde{G}_c^{(2)}(\mathbf{k}_2)\widetilde{G}_c^{(2)}(-\mathbf{k}_1 - \mathbf{k}_2)}, \tag{8.59}$$

and also

$$\frac{\widetilde{G}_c^{(4)}(\mathbf{k}_1, \mathbf{k}_2, \mathbf{k}_3)}{\widetilde{G}_c^{(2)}(\mathbf{k}_1)\widetilde{G}_c^{(2)}(\mathbf{k}_2)\widetilde{G}_c^{(2)}(\mathbf{k}_3)\widetilde{G}_c^{(2)}(-\mathbf{k}_1 - \mathbf{k}_2 - \mathbf{k}_3)} = \tag{8.60}$$
$$-\Gamma^{(4)}(\mathbf{k}_1, \mathbf{k}_2, \mathbf{k}_3) + \Gamma^{(3)}(\mathbf{k}_1, \mathbf{k}_2)\widetilde{G}_c^{(2)}(\mathbf{k}_1 + \mathbf{k}_2)\Gamma^{(3)}(\mathbf{k}_1 + \mathbf{k}_2, \mathbf{k}_3)$$
$$+\Gamma^{(3)}(\mathbf{k}_1, \mathbf{k}_3)\widetilde{G}_c^{(2)}(\mathbf{k}_1 + \mathbf{k}_3)\Gamma^{(3)}(\mathbf{k}_1 + \mathbf{k}_3, \mathbf{k}_2).$$

Equations (8.59) and (8.60) are illustrated in Figure 8.8. For example, Figure 8.8(a) shows the connected three-point function written as a three-point vertex function, with connected two-point functions for its legs. The pictures make it plausible that $\Gamma^{(3)}$ and $\Gamma^{(4)}$ are indeed the sum of all amputated three-point and four-point 1PI diagrams respectively, since one would expect the relationships between the connected n-point functions and 1PI n-point functions to be precisely those illustrated in the figure. However, the algebra needed to derive the above equations, particularly (8.60), is rather tedious. It can be carried on further to find the higher-order vertex functions, and indeed a proof by induction that $\Gamma^{(n)}$ is the sum of all n-point amputated 1PI diagrams can be constructed in this way. We shall do it differently, trying to give an intuitive picture of this odd result—that the derivatives of the Gibbs free energy can be expressed in terms of 1PI diagrams only.

Let us return to the definition of $\Gamma[\widetilde{\varphi}]$, equation (8.47). The task of finding $\Gamma[\widetilde{\varphi}]$ directly from this might seem rather daunting. First, we have

8.4 The Gibbs free energy

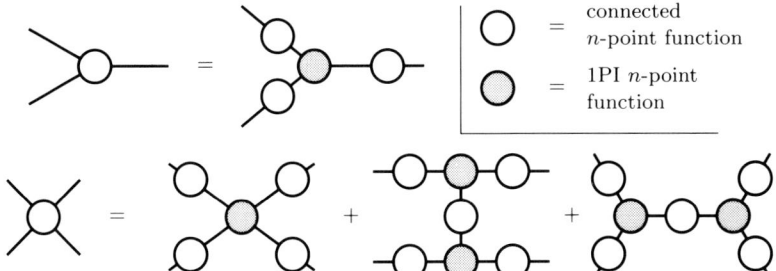

Figure 8.8 The top line illustrates the relation between $\Gamma^{(3)}$ and the connected correlation functions $\widetilde{G}_c^{(3)}$ and $\widetilde{G}_c^{(2)}$. The bottom line illustrates the relation between $\Gamma^{(4)}$ and $\widetilde{G}_c^{(4)}$, $\widetilde{G}_c^{(3)}$, and $\widetilde{G}_c^{(2)}$.

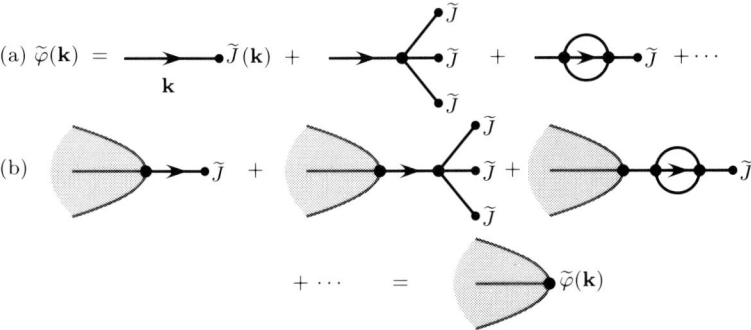

Figure 8.9 (a) The first terms in the diagrammatic representation of $\widetilde{\varphi}(\mathbf{k})$. (b) A set of diagrams contributing to $\log Z[J]$. The members of this set all have one part—the body, represented by the shaded area—in common, but they have different pieces attached to it. The possible pieces that may be attached to the body are the diagrams which appear in (a).

to find an expression for $\widetilde{\varphi}(\mathbf{k})$ in the presence of a source $\widetilde{J}(\mathbf{k})$. This can be done diagrammatically using the techniques of the previous section. Next, the resulting power series in λ must be inverted to give us $\widetilde{J}(\mathbf{k})$ as a functional of $\widetilde{\varphi}(\mathbf{k})$. Finally this expression must be substituted into the diagrammatic expression for $\log Z[\widetilde{J}]$ to eliminate $\widetilde{J}$, and $\Gamma[\widetilde{\varphi}]$ calculated using (8.47). This sounds horrendous, and you might be forgiven for thinking that it would be simpler to stick with the Helmholtz free energy! But amazingly the rules for finding $\Gamma[\widetilde{\varphi}]$ turn out to be simpler than those for $\log Z[\widetilde{J}]$.

The key to the problem lies in the representation of $\widetilde{\varphi} = \widetilde{G}_c^{(1)}$ as a sum of Feynman diagrams, according to the rules of Box 8.3. A few diagrams in this expansion are shown in Figure 8.9(a). The thermal average of $\widetilde{\varphi}(\mathbf{k})$ is equal to the sum of all diagrams which have wavevector **k** flowing in through their one external link. There are an infinite number of such diagrams. Let us call them $D_1(\mathbf{k})$, $D_2(\mathbf{k})$ ..., so that

$$\tilde{\varphi}(\mathbf{k}) = \sum_n D_n(\mathbf{k}). \tag{8.61}$$

Now consider for a moment Figure 8.9(b). This shows a set of diagrams contributing to $\log Z$, all with the same 'bodies' (not shown explicitly, but represented by the shaded part) but with different 'bits sticking out the side'. And these 'bits' are *precisely* the diagrams in Figure 8.9(a) which all add up to give $\tilde{\varphi}$. So the diagrams in Figure 8.9(b) all add up to the result shown—the 'body', with $\tilde{\varphi}(\mathbf{x})$ attached where the sticking-out bits used to be. Algebraically, if we let the value of the 'body', common to all the diagrams in Figure 8.9(b), be $B(\mathbf{k})$, then we can write the sum S of these diagrams as

$$\begin{aligned} S &= \int \mathrm{d}^d \check{\mathbf{k}}\, B(\mathbf{k}) \sum_n D_n(\mathbf{k}) \\ &= \int \mathrm{d}^d \check{\mathbf{k}}\, B(\mathbf{k}) \tilde{\varphi}(\mathbf{k}). \end{aligned} \tag{8.62}$$

The only diagrams in the expansion of $\log Z[J]$ which contribute to Γ are those that have 'no bits sticking out', because all the sticking-out pieces appear in sums such as (8.62), and so are replaced by $\tilde{\varphi}$s. These diagrams are precisely the 1PI diagrams of the previous section, because a 1PI diagram has no 'bits sticking out' by definition.

So we see that it is plausible that only 1PI diagrams contribute to Γ and therefore to its derivatives, the $\Gamma^{(n)}$. We have not yet shown that each 1PI diagram turns up the correct number of times, nor have we considered the term $\int \mathrm{d}^d \mathbf{k}\, \tilde{\varphi}(\mathbf{k}) \tilde{J}(-\mathbf{k})$ in the definition of Γ. But the argument can be tightened up to cover these points; this is done in Appendix M. The final result is that vertex functions, defined as functional derivatives of the Gibbs free energy, can be calculated by the rules of Box 8.4.

There are two related matters which must be cleared up before we go further. In our derivation we have assumed that the Gibbs free energy is an analytic function of $\tilde{\varphi}$ (equation (8.48)), and also that we can find a $\tilde{J}$ that makes the magnetization equal to any given value of $\tilde{\varphi}$. While these assumptions are unobjectionable above the critical temperature T_c, at T_c the first fails if the critical exponent δ is at all interesting, and below T_c it is clear from the graphs in §2.3 that both are untrue. Below T_c, the Gibbs free energy is flat over a finite range of $\tilde{\varphi}$ and is definitely not analytic. Also, an arbitrarily small applied field makes the magnetization take a finite value, so the second assumption fails. What happens?

At T_c the expansion does break down. When we calculate vertex functions in later chapters we shall see that at the critical temperature and at

8.4 The Gibbs free energy

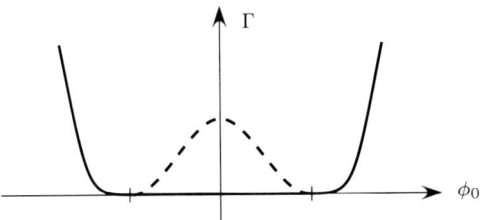

Figure 8.10 This graph shows the true Gibbs free energy (solid line), and the continuation of its analytic part (dashed line). Since the two coincide up to the points where the true Gibbs free energy becomes flat, the minima of the analytically-continued graph give the spontaneous magnetization of the system.

zero external wavevector they diverge. The quantity Γ cannot be expanded as a power series in $\tilde{\varphi}$. Below T_c the situation is somewhat more complicated. Γ as defined by equation (8.44) has a flat section around $\tilde{\varphi} = 0$ (the solid curve in Figure 8.10), its derivatives vanish and the expansion (8.48) fails. However the $\Gamma^{(n)}$ calculated according to the rules of Box 8.4 do not vanish, and neither do they diverge. This means that we are free to plug them into (8.48) and to treat this as the *definition* of a new Γ. The quantity calculated in this way is not related to the true Gibbs free energy by equation (8.47). Instead, $e^{-\Gamma}$ turns out to be proportional to the probability of the order-parameter field ϕ taking a particular value. It is equal to βF_3, where F_3 is the function discussed on page 25. It is sketched as the dashed line in Figure 8.10. Outside its flat region the Gibbs free energy is an analytic function, and both definitions of Γ coincide.

In practice the dashed curve is what is calculated since this can be found from the rules of Box 8.4, and so from now on the symbol Γ will denote the sum of the series (8.48), both above and below T_c. Γ has a stationary point at the edge of the flat region of the true Gibbs free energy, and we have seen in §2.3 that this determines the value of the magnetization, (This is also consistent with our interpretation of $e^{-\Gamma}$ as a probability.) If we want to know the true Gibbs free energy we can always draw the flat portion in by hand.

8.4.2 The loop expansion

We are now almost ready to calculate something interesting—the Gibbs free energy beyond the Landau approximation. But before we do this, there is one more matter to sort out: namely, the question of how to organize the stages of the calculation. When calculating an individual correlation function or vertex function, it makes sense to do the calculation order by order in λ. Thus, first one calculates the O(1) terms; then the O(λ) terms; then the O(λ^2) terms ... and so on. However, ordering the calculation of

Γ in this way is not so satisfactory. In the previous chapter we calculated the Helmholtz free energy in the Landau approximation (equation (7.17)). Combining this with (8.47) gives

$$\frac{\Gamma(\phi_0)}{V} = \tfrac{1}{2}\mu^2\phi_0^2 + \frac{\lambda}{4!}\phi_0^4, \tag{8.63}$$

where ϕ_0 is the position-independent value of the order parameter. This result, which one feels ought to be 'zeroth order', contains terms of both zeroth and first order in λ. If we were to arrange the calculation of Γ in powers of λ the zeroth-order result would not contain the ϕ^4 term, while the next order would contain both the ϕ^4 term and $O(\lambda)$ corrections to the other terms. Landau theory would not appear in our series of approximations to Γ, whereas physically one feels that it should.

An approximation scheme which gets around this is the **loop expansion**, which organizes diagrams according to the number of 'independent loops' they contain. The number of **independent loops** in an amputated 1PI diagram is the smallest number of loops (closed paths of links) having the property that each internal link of the diagram belongs to at least one loop. Thus, the diagram in Figure 8.5 has two independent loops. The number of independent loops is also equal to the number of independent internal wavevector integrations that appear when the diagram is interpreted according to the k-space Feynman rules. Let us introduce the partition function $Z(b)$, defined by

$$Z(b) \equiv \int \mathcal{D}\phi \, \exp\left(-\frac{H[\phi]}{b}\right). \tag{8.64}$$

This is the usual partition function Z when $b = 1$. If the parameter b is much less than one, the contribution to $Z(b)$ of field configurations away from the minimum of H is suppressed. For this reason b is analogous to a temperature, though it is not the physical temperature. (Recall that the physical temperature resides in the parameters α, μ, and λ; see equation (7.9).) Now Landau theory consists in approximating the partition function by its largest term, and neglecting the contributions from configurations away from the minimum of H. If b is very small only this largest term survives, and we get the Landau approximation to the free energy; as b grows towards one, the free energy should approach its true value.

Using the diagrammatic methods developed in this chapter we can calculate $\log Z(b)$. The theory described by the effective Hamiltonian H/b is a Landau–Ginzburg theory, but with a bare propagator

$$\Delta_b(\mathbf{x}-\mathbf{y}) \equiv b\Delta(\mathbf{x}-\mathbf{y}) \tag{8.65}$$

and a quartic coupling

$$\lambda_b \equiv \frac{\lambda}{b}. \tag{8.66}$$

8.4 The Gibbs free energy

The Feynman rules for the expansion of $\log Z(b)$ are thus exactly the same as those for $\log Z(b = 1)$, except that each diagram is multiplied by a factor of b^{I-V}, where I is the number of links and V is the number of internal four-point vertices. In the diagrams for the vertex functions, I and V satisfy

$$I - V = L - 1, \qquad (8.67)$$

where L is the number of independent loops in the diagram (see Problem 8.5), so an expansion in powers of b is also an expansion in the number of loops. The contributions to Γ of lowest order in b are thus the 'zero-loop' diagrams

$$\left(\text{———}\right)^{-1} \quad \text{and} \quad \times$$

of order b^{-1}. When multiplied by $\frac{1}{2!}$ and $\frac{1}{4!}$ respectively (see equation (8.48)) these two diagrams do indeed give the Landau approximation to the Gibbs free energy. Diagrams containing loops give corrections to this approximation.

8.4.3 The one-loop Gibbs free energy

Now we calculate the Gibbs free energy to one-loop order for a constant field, $\overline{\phi}(\mathbf{x}) = \phi_0$. This restriction simplifies the calculation. Also, we are interested in stationary points of the Gibbs free energy, and the translational invariance of the theory suggests that these occur for a spatially constant field. The Fourier transform of a constant field ϕ_0 is

$$\tilde{\varphi}(\mathbf{k}) = (2\pi)^d \phi_0 \delta(\mathbf{k}). \qquad (8.68)$$

Substituting (8.68) into the expression (8.48) for Γ gives

$$\Gamma(\phi_0) = (2\pi)^d \delta(0) \sum_{n=0}^{\infty} \frac{1}{n!} \phi_0^n \Gamma^{(n)}(0, \ldots, 0) \qquad (8.69)$$

(the $\Gamma^{(n)}$ here and below are reduced vertex functions). The term involving $\Gamma^{(0)}$ is different to the others since it does not involve ϕ_0. We will put it to one side for now, and return to it when we have calculated the others. The δ-function with zero argument requires comment. What can it mean? Well, the Gibbs free energy is an extensive quantity, so we would expect it to be proportional to the volume of the system. The δ-function is indeed proportional to V:

$$(2\pi)^d \delta(\mathbf{k}) = \int d^d\mathbf{x}\, e^{-i\mathbf{k}\cdot\mathbf{x}}, \quad \text{so} \quad (2\pi)^d \delta(0) = \int d^d\mathbf{x} \times 1 = V. \qquad (8.70)$$

So that's all right.

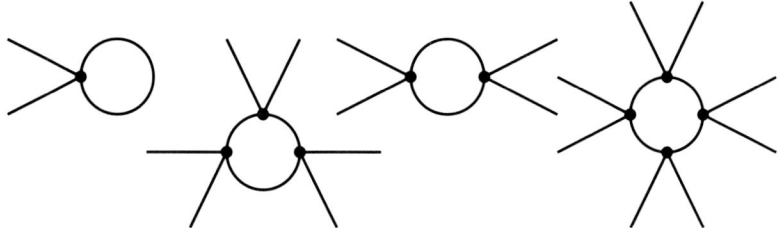

Figure 8.11 One-loop diagrams contributing to $\Gamma^{(2p)}$ for $p = 1, 2, 3,$ and 4.

To work out (8.69) we need to find the vertex functions for zero external wavevector. Let us denote by $\Gamma_{1\,\text{loop}}(\phi_0)$ the contribution to the Gibbs free energy from one-loop diagrams. To find $\Gamma^{(n)}(0,\ldots,0)$ at one loop, we must write down all 1PI diagrams with one loop and n external links, with zero wavevector coming through the external links. Some of these diagrams are shown in Figure 8.11. They all have the form of single loops, with external links attached in pairs around the outside. The diagrams with p pairs of links contribute to $\Gamma^{(2p)}$ an amount

$$-\left\{\begin{array}{c}\text{symmetry}\\ \text{factor}\end{array}\right\} \times (-\lambda)^p \int \frac{d^d\check{\mathbf{q}}}{(\alpha^2 q^2 + \mu^2)^p}. \tag{8.71}$$

There is a factor of $-\lambda$ from each of the p vertices, and of $(\alpha^2 q^2 + \mu^2)^{-1}$ from each of the p links between them.

The diagram for $\Gamma^{(2)}$ has one link starting and finishing at the same vertex; it therefore has a symmetry factor of $\frac{1}{2}$. The diagram for $\Gamma^{(4)}$ has a pair of links joining the same two vertices; it too has a symmetry factor of $\frac{1}{2}$. The symmetry factors for all the other diagrams are one. The only complication is that there is more than one diagram for each value of p. Consider the case of $\Gamma^{(4)}$. This has the three diagrams shown in Figure 8.12. These are all distinct cases. Although they have the same values if the external wavevectors are zero, it is easy to check that they are all different otherwise. In general, we need to count the number of different ways of assigning the $2p$ external links to the $2p$ connection points on the loop. There are $(2p)!$ ways of doing this, but some of these ways are the same. First we must divide by 2^p, since the links in each pair may be swapped without making any difference. Next we must divide by p; this takes account of link assignments which can be rotated into each other. Finally, if $p > 2$ there is a factor of $\frac{1}{2}$ to account for the symmetry of the diagram under reflection. If $p \leq 2$ reflection does not generate a different diagram and there is no factor of $\frac{1}{2}$. This last effect exactly cancels the different symmetry factors noted above, with the result that the one-loop contribution to $\Gamma^{(2p)}$ is

$$-\frac{(2p)!}{2^p\,2p}(-\lambda)^p \int \frac{d^d\check{\mathbf{q}}}{(\alpha^2 q^2 + \mu^2)^p}. \tag{8.72}$$

8.4 The Gibbs free energy

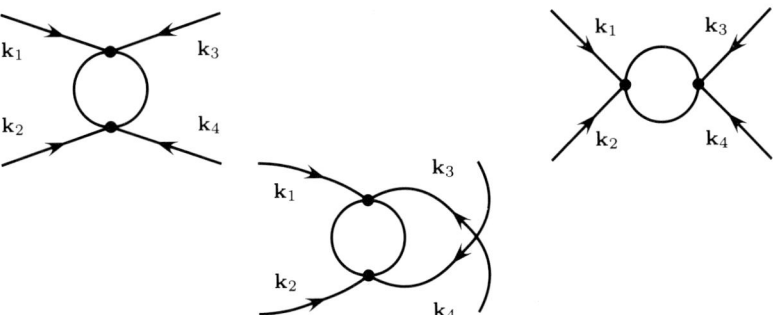

Figure 8.12 The three distinct diagrams which contribute to $\Gamma^{(4)}$ at one loop.

So the expression (8.69) for Γ gives us the following for $\Gamma_{1\,\mathrm{loop}}(\phi_0)$:

$$\Gamma_{1\,\mathrm{loop}}(\phi_0) = -V \sum_{p=1}^{\infty} \frac{\phi_0^{2p}}{(2p)!} \frac{(2p)!}{2^p\, 2p} (-\lambda)^p \int \frac{d^d\check{\mathbf{q}}}{(\alpha^2 q^2 + \mu^2)^p}$$

$$= -\frac{V}{2} \int d^d\check{\mathbf{q}} \sum_{p=1}^{\infty} \frac{1}{p} \left(-\frac{\lambda}{2} \frac{\phi_0^2}{\alpha^2 q^2 + \mu^2}\right)^p \qquad (8.73)$$

$$= \frac{V}{2} \int d^d\check{\mathbf{q}}\, \log\left(1 + \frac{\lambda}{2} \frac{\phi_0^2}{\alpha^2 q^2 + \mu^2}\right),$$

where we have exchanged the order of the sum and integration and recognized the series expansion of $\log(1+x)$. The integral over $\mathbf{q}$ can be done (remembering that the integration is cut off at $q = \Lambda$) by writing

$$\log\left(1 + \frac{\lambda}{2} \frac{\phi_0^2}{\alpha^2 q^2 + \mu^2}\right) \quad \text{as} \quad \log(\alpha^2 q^2 + \mu^2 + \tfrac{\lambda}{2}\phi_0^2) - \log(\alpha^2 q^2 + \mu^2). \quad (8.74)$$

For $d = 3$ the resulting expression is horrible. For $d = 2$ the result is

$$\Gamma_{1\,\mathrm{loop}}(\phi_0) = \frac{V}{8\pi\alpha^2} \left[(\alpha^2\Lambda^2 + \mu^2) \log\left(1 + \frac{\lambda}{2} \frac{\phi_0^2}{\alpha^2\Lambda^2 + \mu^2}\right) \right. \qquad (8.75)$$

$$\left. - \mu^2 \log\left(1 + \frac{\lambda}{2}\frac{\phi_0^2}{\mu^2}\right) + \frac{\lambda}{2}\phi_0^2 \log\left(1 + \frac{\alpha^2\Lambda^2}{\mu^2 + \tfrac{\lambda}{2}\phi_0^2}\right) \right].$$

This is zero when $\phi_0 = 0$. It also vanishes when λ is zero, as it should.

As promised, we now consider the term $\Gamma^{(0)}$. According to the rules of Box 8.4, this is $-\log Z_G[0]$ (where Z_G is the Gaussian partition function), minus the sum of all 1PI diagrams with no external legs. The first such diagram is

But this diagram has two loops, so it does not contribute to Γ at one loop. Only the Gaussian contribution remains. From equation (8.5):

$$\Gamma^{(0)} = \tfrac{1}{2}V \int^\Lambda d^d\check{k}\, \log(\alpha^2 k^2 + \mu^2). \tag{8.76}$$

The total of equations (8.76), (8.63), and (8.73) gives the Gibbs free energy to one loop. We have succeeded in our attempt to improve on Landau theory.

This has been a long chapter. Its main purpose has been to provide a tool for use in later chapters—the Feynman rules for calculating the Helmholtz free energy of the Landau–Ginzburg model. We are now able to calculate this free energy, or any of the correlation functions, order by order in the coupling λ.

Problems

8.1 Define $\Phi \equiv \int d^d\mathbf{x}\, \phi(\mathbf{x}) e^{-x^2/a^2}$. Show that

$$\langle \Phi^2 \rangle = (a\sqrt{\pi})^{2d} \int d^d\check{k}\, e^{-k^2 a^2/2} \widetilde{G}_c^{(2)}(k). \tag{8.77}$$

Evaluate this integral for the Gaussian model in $d = 3$, when $a\mu/\alpha \gg 1$.

8.2 Consider the one-dimensional integral

$$I(\lambda) = \int_{-\infty}^{\infty} dx\, e^{-(\mu x^2 + \lambda x^4)}. \tag{8.78}$$

By expanding $e^{-\lambda x^4}$, write the integral as a power series in λ. Evaluate the coefficients of the power series, and show that the series diverges for all $\lambda \neq 0$.

8.3 Draw the Feynman diagrams for the following correlation functions, and write down the corresponding algebraic expressions (not forgetting symmetry factors!):
(i) $\widetilde{G}_c^{(3)}$ to $O(\lambda)$ (with $J \neq 0$);
(ii) The $O(\lambda^2)$ contributions to $\widetilde{G}_c^{(2)}(\mathbf{k})$ (with $J = 0$);
(iii) The two-loop contributions to $\Gamma^{(4)}(\mathbf{k}_1, \mathbf{k}_2, \mathbf{k}_3)$.

8.4 There is a slight asymmetry between our treatments of the $\widetilde{G}_c^{(n)}$ and the $\Gamma^{(n)}$. We've allowed the $\widetilde{G}_c^{(n)}$ to be functions of J, but we've not taken the $\Gamma^{(n)}$ to be functions of $\overline{\phi}$. This is easily rectified, however. Let us define a **field-dependent vertex function** to be

$$\Gamma^{(n)}(\mathbf{k}_1, \ldots, \mathbf{k}_n; \widetilde{\varphi}_0) \equiv (2\pi)^{dn} \frac{\delta}{\delta\widetilde{\varphi}(-\mathbf{k}_1)} \cdots \frac{\delta}{\delta\widetilde{\varphi}(-\mathbf{k}_n)} \Gamma[\widetilde{\varphi}(\mathbf{k})] \bigg|_{\widetilde{\varphi} = \widetilde{\varphi}_0}. \tag{8.79}$$

Use (8.79) and (8.48) to express $\Gamma^{(3)}(\mathbf{k}_1, \mathbf{k}_2, \mathbf{k}_3; \widetilde{\varphi}_0)$ in terms of the ordinary vertex functions, and write down the Feynman diagrams (and corresponding algebraic expressions) that contribute to $\Gamma^{(3)}(\mathbf{k}_1, \mathbf{k}_2, \mathbf{k}_3; \widetilde{\varphi}_0)$ to $O(\lambda)$. Compare your answer to that of Problem 8.3(i), and comment on the similarities and differences.

8.5 Derive (8.67). [Hint: consider Euler's theorem for a polyhedron.]

8.6 We can caricature the partition function of the Landau–Ginzburg model in a finite volume by

$$Z = \int d\phi \, \exp\left[-V\left(\tfrac{1}{2}\mu^2\phi^2 + \tfrac{1}{4!}\lambda\phi^4 - J\phi\right)\right]. \tag{8.80}$$

If $\mu^2 < 0$ and J is sufficiently small, the exponent has two maxima. Approximating Z by the value of the integrand at these maxima (Landau theory), calculate $\langle\phi\rangle$ as a function of J for small values of J/μ^2, and sketch the forms of the Helmholtz and Gibbs free energies. Would this behaviour be observed in the perturbation theory developed in this chapter, if the Feynman rules were modified so that all integrals over position were restricted to a finite volume and wavevector integrals were made into sums?

8.7 Using the ideas of Appendix E, show that the probability $\rho(\phi_0)\mathrm{d}\phi_0$ of $\left(\int \mathrm{d}^d\mathbf{x}\,\phi\right)/V$ lying between ϕ_0 and $\phi_0 + \mathrm{d}\phi_0$ is

$$\rho(\phi_0)\mathrm{d}\phi_0 = \frac{1}{Z}\int' \mathcal{D}\phi\,\mathrm{e}^{-H}\,\mathrm{d}\phi_0, \tag{8.81}$$

where H is the Landau–Ginzburg Hamiltonian, Z is the Landau–Ginzburg partition function, and the primed functional integral is over the $\mathbf{k} \neq 0$ Fourier components of ϕ, the $\mathbf{k} = 0$ component of ϕ being constrained to satisfy

$$\widetilde{\phi}(0) = V\phi_0. \tag{8.82}$$

Construct the Feynman rules for the evaluation of (8.81) (Appendix N gives details of the Feynman rules when powers of ϕ other than ϕ^4 are present in the Hamiltonian), and thus show that $-\log\rho(\phi_0)$ is the dimensionless 'Gibbs free energy' $\Gamma[\widetilde{\varphi}]$ of §8.4, evaluated for a constant field ϕ_0.

9
Renormalization

In Chapter 7 we introduced the Landau–Ginzburg 'metamodel' of phase transitions. The model's partition function takes the form of a functional integral which cannot be done exactly. So in the last chapter we expressed this integral as an infinite series of functional integrals that can be reduced to ordinary multi-dimensional integrals. To keep track of these integrals we associated with each a Feynman diagram and used these diagrams to classify the integrals by the number of integrations or 'loops' each involves.

In the last chapter we saw that it is expedient to focus on the 'vertex functions' $\Gamma^{(n)}(\mathbf{k}_1, \ldots, \mathbf{k}_{n-1})$ since these are relatively easy to calculate, and once they are known the free energy or any correlation function can be readily evaluated. Hitherto we have considered the $\Gamma^{(n)}$ and everything that follows from them to be functions of the basic parameters of the Landau–Ginzburg model, namely α, μ, λ and Λ. In this chapter we re-express the $\Gamma^{(n)}$ as functions of the values assumed by $\Gamma^{(2)}$ and $\Gamma^{(4)}$ at some arbitrarily chosen 'renormalizing' wavevectors rather than as functions of α, μ, λ and Λ. The motivation for this process of 'renormalization' is two-fold:

(i) It expresses the predictions of the theory in terms of readily measurable quantities rather than the entirely theoretical quantities α, μ, λ and Λ. It is essential to do this since the Landau–Ginzburg model is a phenomenological one: recall from §7.1 that the value of μ involves the number of microstates compatible with a given value of the order-parameter field $\boldsymbol{\phi}(\mathbf{x})$. In general we do not know how to work out this number exactly, so we must regard it as something to be fitted to

Introduction

measurable quantities such as correlation functions.

(ii) We shall find that there is a one-parameter family of models, each having a different value of Λ, all of which yield the same correlation functions $\Gamma^{(n)}$ at macroscopic separations. So given that a system is described by the Landau–Ginzburg model, only three measurements are needed to predict unambiguously the value of *any* macroscopic observable $\Gamma^{(n)}$. From this fact it follows that any integrals over wavevector that may occur in relations between macroscopic observables can be written without reference to the cutoff parameter Λ. In fact, we shall see that we can express the $\Gamma^{(n)}$ in terms of convergent integrals over wavevectors that run all the way from zero to infinity.

Item (ii) is important in two respects. At an elementary level it is a great convenience because integrals to infinity are easier to evaluate than integrals to $k = \Lambda$. More fundamentally, it assures us that our predictions do not depend on exactly how we restrict the fields $\boldsymbol{\phi}$ that are summed over to 'smooth' ones—recall that Λ was introduced in §8.1 as a way of ensuring that the summed fields are all smooth. This assurance is of the utmost importance because there are many ways in which we could impose a smoothness condition on the $\boldsymbol{\phi}$ and there is no obvious way of choosing between them. For example, given an arbitrary field $\boldsymbol{\phi}$, we can smooth it by convolving it with a Gaussian, or a cosine-bell, or by filtering its Fourier transform with a Lorentzian or a Heaviside step function or whatever. In short, it is by no means clear that the correct way to exclude jagged fields is simply to set to zero the amplitudes of all Fourier components with wavevectors larger than Λ. Furthermore, we cannot feel any confidence in results that depend on the details of how we eliminate jagged fields. We can feel confidence in relations between measurable quantities only if they are substantially independent of the value of Λ used in their derivation; only in this case is it reasonable to hope to get the same relations between experimentally measurable quantities when we reconstruct the theory around a different scheme for ensuring that the summed fields are all smooth.

It is natural to ask whether there is any connection between the renormalization process discussed in this chapter and that of Chapter 5. Superficially there is no connection: here we are concerned with changing the parameters of the Landau–Ginzburg model from α, μ, λ and Λ to experimentally more accessible quantities, while Chapter 5 was concerned with the construction of new models from old by averaging the dynamical variables of the old model to form the 'block variables' of the new one. However, at a deeper level there is a close connection. In Chapter 7 we remarked on the connection between $\boldsymbol{\phi}$ and a set of block variables. The effective scale of the blocks is $1/\Lambda$ so halving Λ, is equivalent to doubling the size of the blocks. From Chapter 5 we know that this process leaves the correlation functions unchanged on scales much larger than $1/\Lambda$, but changes the coefficients in the effective Hamiltonian so that changing the size of the blocks is equivalent to a change in the system's parameters. Now, we noted above that there exists

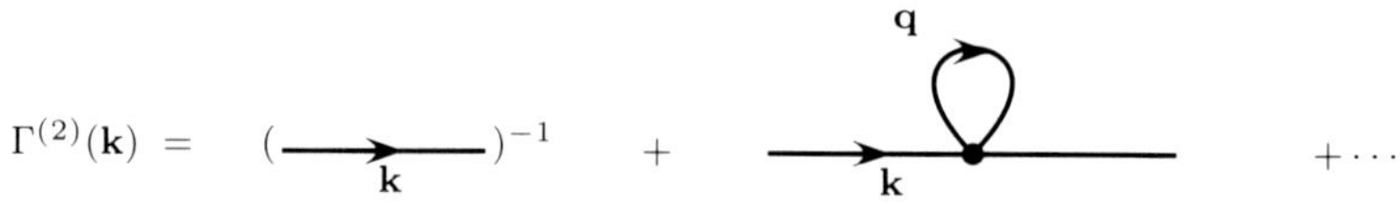

Figure 9.1 The two-point vertex function to one loop. The diagram at right is called the **bubble** diagram.

a one-parameter family of Landau–Ginzburg models, each member of which has a different value of Λ, that generate the same macroscopic physics. Thus changing the value of Λ and changing the values of α, μ and λ so as to remain within the same family, is entirely equivalent to the blocking transformations of Chapter 5.

We eliminate the parameters of the Landau–Ginzburg model one by one. The first to go is μ.

9.1 Mass renormalization

A quantity of primary physical interest is the two-point vertex function $\Gamma^{(2)}(k)$, a dimensionless function equal to one over the Fourier transform of the connected two-point correlation function $G_c^{(2)}(|\mathbf{x}_1 - \mathbf{x}_2|)$ (see equation (8.43)). Figure 9.1 and the rules of Box 8.4 yield for $\Gamma^{(2)}$

$$\Gamma^{(2)}(k) = \mu^2 + \alpha^2 k^2 + \frac{\lambda}{2} \int_0^\Lambda \frac{d^d \check{\mathbf{q}}}{\alpha^2 q^2 + \mu^2} + O(\lambda^2), \qquad (9.1)$$

where we have explicitly written out terms to one loop only. Although μ^2 is linearly related to temperature (see equation (7.9)), its actual value has no direct physical meaning. By contrast, the value taken by $\Gamma^{(2)}$ at $k = 0$ is of particular physical interest, being β divided by the susceptibility χ (see equation (8.36)). So let us investigate the possibility of trading the parameter μ^2 of the Landau–Ginzburg effective Hamiltonian, which physically is a measure of temperature, for the quantity

$$m^2 \equiv \Gamma^{(2)}(0) = \mu^2 + \frac{\lambda}{2} \int_0^\Lambda \frac{d^d \check{\mathbf{q}}}{\alpha^2 q^2 + \mu^2} + O(\lambda^2). \qquad (9.2)$$

In the new scheme the state of the system will be specified not by giving the value of μ^2, but by citing the value of its susceptibility. m^2 does not depend linearly on T, but we know that near T_c we have $m^2 \sim |T - T_c|^\gamma$. The exponent γ will be calculated in §10.2 from an equation having its origin in (9.2). Once it has been found, the temperature dependence of all m^2-dependent results will be known, and μ^2 will no longer be needed.

9.1 Mass renormalization

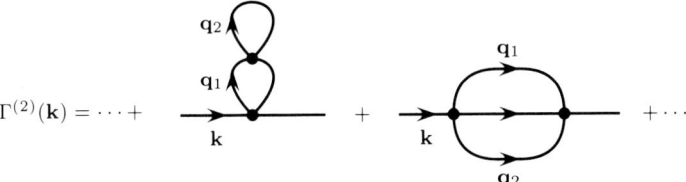

Figure 9.2 The two-point vertex function to two loops. The diagram on the extreme right is called the **Saturn** diagram and is associated with the integral A of equation (9.5).

When we use equations (9.1) and (9.2) to express $\Gamma^{(2)}(k)$ in terms of m, we find

$$\begin{aligned}
\Gamma^{(2)}(k) &= \Gamma^{(2)}(k) - \Gamma^{(2)}(0) + m^2 \quad \text{(from the definition of } m^2\text{)} \\
&= \left(\mu^2 + \alpha^2 k^2 + \frac{\lambda}{2}\int_0^\Lambda \frac{d^d\check{q}}{\alpha^2 q^2 + \mu^2} + \cdots \right) \\
&\quad - \left(\mu^2 + \frac{\lambda}{2}\int_0^\Lambda \frac{d^d\check{q}}{\alpha^2 q^2 + \mu^2} + \cdots \right) + m^2 \\
&= m^2 + \alpha^2 k^2 + O(\lambda^2).
\end{aligned} \tag{9.3}$$

Here the two integrals over $\mathbf{q}$ have cancelled to leave a spectacularly simple result.

Particle physicists call this process of eliminating μ in favour of m **mass renormalization** because for them both μ and m are masses and equation (9.2) is to be interpreted as expressing the physical, or **dressed**, mass m in terms of the (physically meaningless) **bare** mass μ. This jargon makes no sense in statistical physics, since here neither μ nor m is a mass, and the two quantities have entirely different physical interpretations from the outset. Nonetheless, we shall conform to the usual practice of statistical mechanics in speaking of 'mass renormalization'.

Now let's look at how this process works to two loops. To two loops equation (9.1) reads (see Figure 9.2)

$$\begin{aligned}
\Gamma^{(2)}(k) =& \mu^2 + \alpha^2 k^2 + \frac{\lambda}{2}\int_0^\Lambda \frac{d^d\check{q}}{\alpha^2 q^2 + \mu^2} \\
& -\lambda^2\left(\frac{1}{4}\int_0^\Lambda \frac{d^d\check{q}_1 d^d\check{q}_2}{(\alpha^2 q_1^2 + \mu^2)(\alpha^2 q_2^2 + \mu^2)^2} + \frac{1}{6}A(k)\right) + O(\lambda^3),
\end{aligned} \tag{9.4}$$

where

$$A(k) \equiv \int_0^\Lambda \frac{d^d\check{q}_1 d^d\check{q}_2}{(\alpha^2 q_1^2 + \mu^2)(\alpha^2 q_2^2 + \mu^2)[\alpha^2(\mathbf{k}-\mathbf{Q})^2 + \mu^2]} \quad (\mathbf{Q} \equiv \mathbf{q}_1 + \mathbf{q}_2). \tag{9.5}$$

We now subtract from the right-hand side of (9.4) the same expression evaluated at $k = 0$, and then add back the numerical value, m^2, of what we have just subtracted. This procedure yields

$$\begin{aligned}\Gamma^{(2)}(k) &= \Gamma^{(2)}(k) - \Gamma^{(2)}(0) + m^2 \\ &= m^2 + \alpha^2 k^2 - \tfrac{1}{6}\lambda^2 \Delta A(k) + O(\lambda^3),\end{aligned} \quad (9.6)$$

where $\Delta A \equiv A(k) - A(0)$. The first of the two-loop integrals in (9.4) has been eliminated by this renormalization process because it does not depend on k. However, the second two-loop integral of (9.4) appears in (9.6) twice because it does depend on k. So to two loops our new expression for $\Gamma^{(2)}(k)$ in terms of m^2 depends on the difference of two ghastly integrals. This is obviously bad news. But the difference of these two integrals is actually less bothersome than either taken alone. The key point is the nature of A's dependence on the cutoff parameter Λ.

A has $2d$ powers of q on top and 6 powers of q on the bottom. Hence for $d = 3$ we have to expect A to depend logarithmically on Λ. But for $Q \gg k$ the integrand of $A(k)$ depends only weakly on k. Hence it is plausible that the portion of A that increases like $\log \Lambda$ is essentially k-independent, with the result that the difference $\Delta A \equiv A(k) - A(0)$ involved in (9.6) is probably independent of Λ so long as $\Lambda \gg k$. Box 9.1 demonstrates that this conjecture is valid. Consequently, the results we obtain for $k \ll \Lambda$ will not be significantly in error if we replace Λ by ∞ in equation (1) of Box 9.1. This change makes evaluation of the integral much simpler. We have finally

$$\begin{aligned}\Gamma^{(2)}(k) = m^2 + \alpha^2 k^2 - \frac{\lambda^2}{6} \int_0^\infty \frac{d^d \check{q}_1 d^d \check{q}_2}{(\alpha^2 q_1^2 + \mu^2)(\alpha^2 q_2^2 + \mu^2)(\alpha^2 Q^2 + \mu^2)} \\ \times \frac{-\alpha^2 k^2 + 2\alpha^2 \mathbf{k} \cdot \mathbf{Q}}{[\alpha^2 (\mathbf{k} - \mathbf{Q})^2 + \mu^2]} + O(\lambda^3).\end{aligned} \quad (9.7)$$

Our elimination of μ in favour of m can be taken one stage further by replacing μ by m in all our integrals. By the renormalization condition (9.2), m differs from μ by a term of order λ. Hence one might suppose that replacing μ by m in a term of order λ^n would induce an error of order λ^{n+1}. In reality the replacement $\mu \to m$ increases the accuracy of our expressions, and at the same time reduces the number of diagrams we need to consider. This happy state of affairs arises as follows.

μ appears in any of our integrals, say the integral $A(k)$, through the zeroth-order propagator $[\alpha^2(\mathbf{q} - \mathbf{k})^2 + \mu^2]^{-1}$, which is represented in A's Feynman diagram by a link. At the level of one more loop there will be a diagram just like A's except that a bubble is attached to this link, and two levels of loops higher there will be a diagram in which this link becomes Saturn's equator. Similarly, at every stage in the hierarchy of diagrams we will

9.1 Mass renormalization

Box 9.1: Differencing the $A(k)$ defined by (9.5)

Arranging as a single integral the difference of the two integrals defined by (9.5), we have (with $\mathbf{Q} \equiv \mathbf{q}_1 + \mathbf{q}_2$)

$$\Delta A(k) \equiv A(k) - A(0)$$
$$= \int_0^\Lambda \frac{d^d\check{\mathbf{q}}_1 d^d\check{\mathbf{q}}_2}{(\alpha^2 q_1^2 + \mu^2)(\alpha^2 q_2^2 + \mu^2)} \frac{\alpha^2 Q^2 - \alpha^2(\mathbf{k}-\mathbf{Q})^2}{[\alpha^2(\mathbf{k}-\mathbf{Q})^2 + \mu^2](\alpha^2 Q^2 + \mu^2)}$$
$$= \int_0^\Lambda \frac{d^d\check{\mathbf{q}}_1 d^d\check{\mathbf{q}}_2}{(\alpha^2 q_1^2 + \mu^2)(\alpha^2 q_2^2 + \mu^2)(\alpha^2 Q^2 + \mu^2)} \frac{-\alpha^2 k^2 + 2\alpha^2 \mathbf{k}\cdot\mathbf{Q}}{[\alpha^2(\mathbf{k}-\mathbf{Q})^2 + \mu^2]}. \tag{1}$$

To understand how this integral depends on Λ, imagine first averaging the integrand over all directions of $\mathbf{k}$, which cannot affect the value of ΔA as a whole. When Q is large compared with k this angular averaging can be done by expanding the bottom in powers of $\cos\theta$, where θ is the angle between $\mathbf{Q}$ and $\mathbf{k}$. Specifically, we write the θ-dependent factor in (1) thus

$$\frac{-\alpha^2 k^2 + 2\alpha^2 \mathbf{k}\cdot\mathbf{Q}}{\alpha^2(\mathbf{k}-\mathbf{Q})^2 + \mu^2} = \frac{-k^2 + 2\mathbf{k}\cdot\mathbf{Q}}{\mathcal{Q}_k^2(1 - 2\mathbf{k}\cdot\mathbf{Q}/\mathcal{Q}_k^2)}$$
$$= \frac{-k^2 + 2\mathbf{k}\cdot\mathbf{Q}}{\mathcal{Q}_k^2}\left(1 + 2\frac{\mathbf{k}\cdot\mathbf{Q}}{\mathcal{Q}_k^2} + \cdots\right), \tag{2}$$

where $\mathcal{Q}_k^2 \equiv Q^2 + k^2 + \mu^2/\alpha^2$. On averaging over the directions of $\mathbf{k}$, two terms in this expansion dominate at large Q: the 1 of the binomial expansion multiplied by the k^2 from the top, and the binomial's $\mathbf{k}\cdot\mathbf{Q}$ term multiplied by the $\mathbf{k}\cdot\mathbf{Q}$ term from the top. The relevant angular average is

$$\left\langle\frac{-\alpha^2 k^2 + 2\alpha^2 \mathbf{k}\cdot\mathbf{Q}}{\alpha^2(\mathbf{k}-\mathbf{Q})^2 + \mu^2}\right\rangle_{\hat{\mathbf{k}}} = \frac{-k^2}{\mathcal{Q}_k^2} + \frac{2k^2 Q^2}{\mathcal{Q}_k^4} + O((k^2/\mathcal{Q}_k^2)^2)$$
$$= \frac{k^2}{\mathcal{Q}_k^2}\left(2\frac{Q^2}{\mathcal{Q}_k^2} - 1\right) + O((k^2/\mathcal{Q}_k^2)^2). \tag{3}$$

Hence the leading q-dependent term in the integrand of (1) is k^2/q^8, causing the overall integrand to scale as $q^{(2d-8)}$. Hence for $d < 4$, ΔA converges as $\Lambda \to \infty$. When $d = 4$, ΔA grows as $\log \Lambda$ for large Λ.

---

be able to identify diagrams identical to A's but with this link replaced by ever more complex subdiagrams. The nice thing is that these subdiagrams form precisely the Feynman series for the full connected two-point correlation function $G_c^{(2)}(|\mathbf{q}-\mathbf{k}|)$. Hence we can sum the whole infinite string of them by simply replacing $[\alpha^2(\mathbf{q}-\mathbf{k})^2 + \mu^2]^{-1}$ in the original integral A

by

$$G_c^{(2)}(|\mathbf{q}-\mathbf{k}|) = \frac{1}{\Gamma^{(2)}(|\mathbf{q}-\mathbf{k}|)} = \frac{1}{m^2 + \alpha^2(\mathbf{q}-\mathbf{k})^2 + O(\lambda^2)}, \quad (9.8)$$

where the first equality is assured by equation (8.43) and the second follows from (9.3). Box 9.2 illustrates this idea. This shows that replacing μ by m is equivalent to first summing a subset of diagrams to all orders in λ and then approximating the answer by an expression whose error is smaller than it by a factor $O(\lambda^2)$.

Since $\Gamma^{(2)}(0) = m^2$ by definition, the $O(\lambda^2)$ error in (9.8) vanishes with k, so the replacement $\mu \to m$ has the effect of exactly including the contributions of all k-independent diagrams. In particular, this substitution implicitly includes all diagrams that can be obtained from explicitly included diagrams by merely adding bubbles to links. Hence once we have replaced μ by m, we should exclude from further consideration all diagrams with bubbles. Diagrams that are not k-independent are only partially included and in some applications it may be necessary to add in by hand the k-dependent portions of them. This can be done by modifying the Feynman rules so that the integral associated with each two-point subdiagram is equal to the integral assigned to that diagram by the original rules but with the substitution $\mu \to m$, less the value of that integral at $k = 0$.

For future reference here are the results of making the substitution $\mu \to m$ as far as they affect $\Gamma^{(2)}$:

$$\Gamma^{(2)}(k) = m^2 + \alpha^2 k^2 - \tfrac{1}{6}\lambda^2 \Delta A(k) + O(\lambda^3) \quad (9.9)$$

$$\Delta A(k) = \int_0^\infty \frac{d^d \check{\mathbf{q}}_1 d^d \check{\mathbf{q}}_2}{(\alpha^2 q_1^2 + m^2)(\alpha^2 q_2^2 + m^2)(\alpha^2 Q^2 + m^2)}$$
$$\times \frac{-\alpha^2 k^2 + 2\alpha^2 \mathbf{k}\cdot\mathbf{Q}}{[\alpha^2(\mathbf{k}-\mathbf{Q})^2 + m^2]} \quad (9.10)$$

$$\mu^2 = m^2 - \frac{\lambda}{2}\int_0^\Lambda \frac{d^d \check{\mathbf{q}}}{\alpha^2 q^2 + m^2} + \tfrac{1}{6}\lambda^2 A(0) + O(\lambda^3), \quad (9.11)$$

where $\mathbf{Q} \equiv \mathbf{q}_1 + \mathbf{q}_2$. Notice that the first of the two-loop integrals of (9.4) does not appear in equation (9.11)—this integral is represented by a bubble within a bubble and is therefore automatically included in the first integral of (9.11).

Box 9.2: Replacing μ^2 by m^2 in all propagators

The Feynman series for $\widetilde{G}_c^{(2)}$ is the sum of all connected graphs with two external legs:

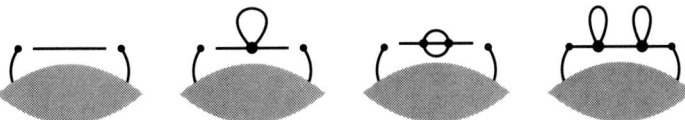

Now consider a graph that contributes to $\Gamma^{(n)}$, consisting entirely of 'plain' links, without bubbles or Saturn diagrams etc. That is, it cannot be split into two subdiagrams by cutting just two links (it is a 2PI graph). We choose one link from this graph and call it Link 1. Amongst all the graphs contributing to $\Gamma^{(n)}$ we are able to find a graph that differs only in having a bubble attached to Link 1. Also there is a graph in which Link 1 becomes Saturn's equator, and so on. Thus from $\Gamma^{(n)}$'s graphs we can pick out this subset:

where the shading indicates stuff that is precisely the same in all graphs. The sum of all these graphs is equal to the original graph with Link 1 replaced by the full propagator $\widetilde{G}_c^{(2)}(q)$. We make this replacement in the first graph and delete all the others. Then we find a graph like the original one but with a bubble on some other link, say Link 2, and assemble the graphs in which Link 2 always has one bubble and Link 1 is simple, has a bubble, becomes Saturn's equator and so forth. All these graphs add to a single graph in which Link 1 is replaced by $\widetilde{G}_c^{(2)}(q)$ and Link 2 has a bubble. Then we sum the set in which Link 2 is Saturn's equator, and so forth until we have assembled all the graphs needed to replace Link 2 by $\widetilde{G}_c^{(2)}(q')$. Repeating this process we eventually replace every link of the original 2PI graph by a full propagator. We repeat this process for every 2PI diagram contributing to $\Gamma^{(n)}$.

Notice that although this process condenses a great many graphs into one, most of $\Gamma^{(n)}$'s many-loop graphs are untouched by it.

9.2 Field renormalization

Box 9.1 demonstrates that for $d < 4$ the integral ΔA involved in equation (9.9) for $\Gamma^{(2)}$ is insensitive to the cutoff parameter Λ. So to two loops $\Gamma^{(2)}(k)$ depends on Λ only through the relation (9.11) between μ and m. In §10.2 we shall show that this enables us to derive the critical exponent γ from formulae which we already have in hand. However, to obtain most of the Landau–

Ginzburg model's other critical exponents and to demonstrate that its free energy F and correlation function $G_c^{(2)}$ satisfy the Widom and Kadanoff homogeneity hypotheses of §1.5.1, we shall find it necessary to approach the physically interesting dimensions $d = 2, 3$ by analytic continuation in d from $d = 4$. In preparation for this enterprise we now derive an expression for $\Gamma^{(2)}(k)$ which is insensitive to Λ even at $d = 4$.

In high-energy physics the procedure by which one eliminates from $\Gamma^{(2)}(k)$ the Λ-sensitivity introduced by ΔA is called **field renormalization**. (See §9.5 for an explanation of this name.)

It turns out that we can eliminate the residual Λ-sensitivity of $\Gamma^{(2)}$ as follows. We define a^2 to be the rate of change of $\Gamma^{(2)}$ with k^2 at some arbitrarily chosen wavevector $\boldsymbol{\kappa}$:

$$a^2 \equiv \left.\frac{\mathrm{d}\Gamma^{(2)}}{\mathrm{d}k^2}\right|_\kappa. \tag{9.12}$$

a is a quantity that can be readily measured and is finite even at a critical point; we choose to define a at $\mathbf{k} = \boldsymbol{\kappa} \neq 0$ because at a critical point $\mathrm{d}\Gamma^{(2)}/\mathrm{d}k^2$ diverges as $k \to 0$—see Problem 9.1. We now rewrite (9.9) in a form which introduces a:

$$\begin{aligned}
\Gamma^{(2)}(k) &= \Gamma^{(2)}(k) - k^2 \left.\frac{\mathrm{d}\Gamma^{(2)}}{\mathrm{d}k^2}\right|_\kappa + k^2 \left.\frac{\mathrm{d}\Gamma^{(2)}}{\mathrm{d}k^2}\right|_\kappa \\
&= m^2 + a^2 k^2 - \tfrac{1}{6}\lambda^2 \Delta A(k) - k^2 \left.\frac{\mathrm{d}\Gamma^{(2)}}{\mathrm{d}k^2}\right|_\kappa + k^2 a^2 + O(\lambda^3) \\
&= m^2 + k^2\left(a^2 - \tfrac{1}{6}\lambda^2 B(k,\kappa)\right) + O(\lambda^3),
\end{aligned} \tag{9.13}$$

where

$$B(k,\kappa) \equiv \frac{1}{k^2}\Delta A(k) + \frac{6}{\lambda^2}\left(\left.\frac{\mathrm{d}\Gamma^{(2)}}{\mathrm{d}k^2}\right|_\kappa - a^2\right). \tag{9.14}$$

With (9.9) and equation (1) of Box 9.1, and bearing in mind that $\mathrm{d}\Delta A/\mathrm{d}k^2 = \mathrm{d}A/\mathrm{d}k^2$, this becomes

$$\begin{aligned}
B(k,\kappa) &= \frac{1}{k^2}\Delta A(k) - \left.\frac{\mathrm{d}A}{\mathrm{d}k^2}\right|_\kappa \\
&= \int_0^\Lambda \frac{\mathrm{d}^d \check{q}_1 \mathrm{d}^d \check{q}_2}{(\alpha^2 q_1^2 + m^2)(\alpha^2 q_2^2 + m^2)} \\
&\quad \times \left(\frac{-\alpha^2 k^2 + 2\alpha^2 \mathbf{k}\cdot\mathbf{Q}}{[\alpha^2(\mathbf{k}-\mathbf{Q})^2 + m^2][\alpha^2 Q^2 + m^2]k^2} + \frac{\alpha^2 \kappa^2 - \alpha^2 \boldsymbol{\kappa}\cdot\mathbf{Q}}{[\alpha^2(\boldsymbol{\kappa}-\mathbf{Q})^2 + m^2]^2 \kappa^2}\right).
\end{aligned} \tag{9.15}$$

9.2 Field renormalization

Box 9.3: The integral B defined by equation (9.14)

Box 9.1 shows that at large q the k-dependent part of the integrand involved in $\Delta A(k)$ behaves as k^2/Q^2. Differentiating equation (9.9) for $\Gamma^{(2)}$ generates a similarly divergent integral:

$$a^2 = \left.\frac{d\Gamma^{(2)}}{dk^2}\right|_\kappa$$

$$= a^2 + \frac{\lambda^2}{6} \int_0^\Lambda \frac{d^d\check{q}_1 d^d\check{q}_2}{(\alpha^2 q_1^2 + m^2)(\alpha^2 q_2^2 + m^2)} \frac{2\alpha^2(\boldsymbol{\kappa} - \mathbf{Q}) \cdot (d\boldsymbol{\kappa}/d\kappa^2)}{[\alpha^2(\boldsymbol{\kappa} - \mathbf{Q})^2 + m^2]^2} \quad (1)$$

$$= a^2 + \frac{\lambda^2}{6} \int_0^\Lambda \frac{d^d\check{q}_1 d^d\check{q}_2}{(\alpha^2 q_1^2 + m^2)(\alpha^2 q_2^2 + m^2)\kappa^2} \frac{\alpha^2 \kappa^2 - \alpha^2 \boldsymbol{\kappa} \cdot \mathbf{Q}}{[\alpha^2(\boldsymbol{\kappa} - \mathbf{Q})^2 + m^2]^2}.$$

Taking angular averages and analysing the integrand's behaviour as in Box 9.1, we find

$$\left\langle \frac{\alpha^2 \kappa^2 - \alpha^2 \boldsymbol{\kappa} \cdot \mathbf{Q}}{[\alpha^2(\boldsymbol{\kappa} - \mathbf{Q})^2 + m^2]^2} \right\rangle_{\hat{\boldsymbol{\kappa}}} = \frac{1}{\alpha^2 \mathcal{Q}_\kappa^2} \left(\frac{\kappa^2}{\mathcal{Q}_\kappa^2} - \frac{2\kappa^2 Q^2}{\mathcal{Q}_\kappa^4} \right) + O((\kappa^2/Q^2)^2)$$

$$= -\frac{\kappa^2}{\alpha^2 \mathcal{Q}_\kappa^4} \left(2\frac{Q^2}{\mathcal{Q}_\kappa^2} - 1 \right) + O((\kappa^2/Q^2)^2), \quad (2)$$

where $\mathcal{Q}_\kappa^2 \equiv Q^2 + \kappa^2 + m^2/\alpha^2$. It follows that the leading q-dependent term in the integrand of (1) is $-1/Q^4$ so this integrand will at large Q cancel $1/k^2$ times the leading term in the integrand of equation (1) of Box 9.1. The difference of the integrands will be $O(Q^{-6})$, and thus the combined integral will be insensitive to Λ for $d < 6$. In other words, if we combine into one the integrals involved in the sum

$$B(k, \kappa) \equiv \frac{1}{k^2}\Delta A(k) + \frac{6}{\lambda^2}\left(\left.\frac{d\Gamma^{(2)}}{dk^2}\right|_\kappa - a^2\right), \quad (3)$$

we may approximate Λ by ∞ since at large q its integrand scales as $q^{(2d-10)}$.

Box 9.3 shows that when $d < 5$, B is insensitive to Λ, which may therefore be approximated by ∞.

The point of these manouevres is that once the measurable parameter a has been specified, we have in (9.13) an expression for $\Gamma^{(2)}(k)$, with k arbitrary, in terms of integrals in which Λ may be approximated by ∞ even when $d = 4$. Thus we have localized the Λ-sensitivity of $\Gamma^{(2)}$ into the integral in equation (1) of Box 9.3 that defines a. This is a real step forward, if only because integrals to ∞ are easier to evaluate than integrals to finite upper limits.

Equation (1) of Box 9.3 can be regarded from more than one point of view. We can consider it to define a as a function of α and Λ. Or we may regard it as defining α as a function of a and Λ. This latter point of view is closely analogous to our interpretation of (9.2) as specifying μ as a function of m and Λ.

Equation (9.13) shows that our scheme for localizing the Λ-sensitivity of the Saturn diagram has the consequence of casting the propagator into the form

$$\frac{1}{m^2 + q^2 a^2 + \mathrm{O}(\lambda^2)}. \tag{9.16}$$

Thus the change in propagator introduced by our renormalization scheme can be effected by simply replacing every α by a. We henceforth make this change.

A word should be said about the replacement of α by a in *all* integrals, including that in equation (9.15) for $B(k,\kappa)$. Two arguments at different levels of sophistication justify this change. First, in equation (9.13) B is multiplied by two powers of λ, so to achieve two-loop accuracy overall, B need be only accurate to zeroth order in λ. From this point of view the proposed replacement of α by $a = \alpha + \mathrm{O}(\lambda^2)$ is of no consequence. Second, and more fundamentally, the **k**'s, **q**'s etc. in (9.15) arise from differencing $\Gamma^{(2)}$ at different values of **k**. The accuracy of an expression such as (9.15) can only increase if we derive it from a more accurate expression for $\Gamma^{(2)}$. In §9.1 we improved the accuracy of all our integrals by replacing μ in every propagator by m. After making this change the wavevector-independent part of the propagator was exact. But the dependence of the propagator on wavevector remained inaccurate at the two-loop level. By replacing α by a we now ensure that the dependence of the propagator on **q** at the renormalization point is accurate to two loops. Thus this replacement is a natural extension of our replacement of μ by m.

In this book we shall not require diagrams accurate to more than two loops. Should such a diagram be required, it can be constructed in the normal way but with the Feynman rules modified so that each two-point subdiagram contributes its full value with certain subtractions. This is investigated in Problem 9.3.

9.3 Renormalizing the coupling constant

We now eliminate the Landau–Ginzburg parameter λ in favour of the value $g \equiv \Gamma^{(4)}(0,0,0)$ of the four-point vertex function. This change of variables is very much in the spirit of our elimination of μ in favour of m in our expressions for $\Gamma^{(n)}$; in each case we are replacing one of the undetermined parameters of the Landau–Ginzburg model by a readily measurable quantity. Just as the replacement of μ eliminated many Λ-sensitive integrals from the Feynman series for $\Gamma^{(2)}$, so elimination of λ in favour of g removes the remaining Λ-sensitivity from the series for $\Gamma^{(4)}$. Better than this, we

9.3 Renormalizing the coupling constant

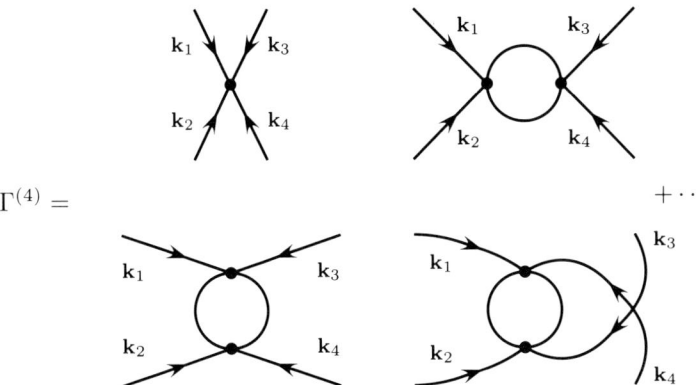

$\Gamma^{(4)} =$

Figure 9.3 The series for $\Gamma^{(4)}$ to one loop. The second diagram in the top row is called the **fish** diagram. The diagrams of the bottom row are trivial modifications of the fish diagram.

shall see that this replacement eliminates *all* remaining Λ-sensitivity from the Feynman series for *all* vertex functions.

By Box 8.4 and Figure 9.3 we have

$$\Gamma^{(4)}(\mathbf{k}_1, \mathbf{k}_2, \mathbf{k}_3) = \lambda\left[1 - \frac{\lambda}{2}\int_0^\Lambda \frac{d^d\check{\mathbf{q}}}{(a^2q^2 + m^2)}\left(\frac{1}{a^2(\mathbf{k}_1 + \mathbf{k}_2 - \mathbf{q})^2 + m^2}\right.\right.$$
$$\left.\left. + \frac{1}{a^2(\mathbf{k}_3 + \mathbf{k}_1 - \mathbf{q})^2 + m^2} + \frac{1}{a^2(\mathbf{k}_2 + \mathbf{k}_3 - \mathbf{q})^2 + m^2}\right)\right]$$
$$+ O(\lambda^3). \tag{9.17}$$

The integral in this expression has d powers of q on top and 4 below. Hence at the dimension $d = 4$ around which we shall find it expedient to expand, the integral grows with Λ as $\log \Lambda$. By analogy with the case of mass renormalization we can localize the inconvenience of this Λ-sensitivity by adding and subtracting

$$g \equiv \Gamma^{(4)}(0,0,0)$$
$$= \lambda\left(1 - \frac{3\lambda}{2}\int_0^\Lambda \frac{d^d\check{\mathbf{q}}}{(a^2q^2 + m^2)^2}\right) + O(\lambda^3). \tag{9.18}$$

Specifically,

$$\Gamma^{(4)}(\mathbf{k}_1, \mathbf{k}_2, \mathbf{k}_3) = \Gamma^{(4)}(\mathbf{k}_1, \mathbf{k}_2, \mathbf{k}_3) - \Gamma^{(4)}(0,0,0) + g$$
$$= g - \frac{\lambda^2}{2}\int_0^\Lambda \frac{d^d\check{\mathbf{q}}}{(a^2q^2 + m^2)^2}\left(\frac{2a^2\mathbf{q}\cdot\mathbf{K} - a^2K^2}{a^2(\mathbf{K} - \mathbf{q})^2 + m^2}\right. \tag{9.19}$$
$$\left. + 2 \text{ permutations}\right) + O(\lambda^3),$$

where $\mathbf{K} \equiv \mathbf{k}_1 + \mathbf{k}_2$. It is straightforward to show that the integral in (9.19) is insensitive to Λ for $d < 6$ (see Problem 9.4).

On inverting (9.18) we find

$$\lambda = g\left(1 + \frac{3g}{2}\int_0^\Lambda \frac{d^d\check{q}}{(a^2q^2 + m^2)^2}\right) + O(g^3). \tag{9.20}$$

Thus so long as we are working only to one loop, we may approximate λ in the $O(\lambda^2)$ term of (9.19) by g, to obtain

$$\Gamma^{(4)}(\mathbf{k}_1, \mathbf{k}_2, \mathbf{k}_3) = g - \frac{g^2}{2}\int_0^\Lambda \frac{d^d\check{q}}{(a^2q^2 + m^2)^2}\left(\frac{2a^2\mathbf{q}\cdot\mathbf{K} - a^2K^2}{a^2(\mathbf{K}-\mathbf{q})^2 + m^2}\right. \\ \left. + 2\text{ permutations}\right) + O(g^3). \tag{9.21}$$

When we use (9.20) to eliminate λ in favour of g in equations (9.9) and (9.13) we obtain the results summarized in Box 9.4.

It is not hard to see that localization of Λ-sensitivity can be achieved by adding and subtracting $\Gamma^{(4)}$ evaluated at any three wavevectors—the arguments of $\Gamma^{(4)}$ do not have to vanish as in our treatment. Thus the right-hand sides of (9.18) could be replaced by $\Gamma^{(4)}(\boldsymbol{\kappa}_1, \boldsymbol{\kappa}_2, \boldsymbol{\kappa}_3)$, where the $\boldsymbol{\kappa}_i$ are any three fixed wavevectors. Though the algebra is then more cumbersome (see Problem 9.5), from §10.3 onwards we shall find it expedient to proceed with non-zero $\boldsymbol{\kappa}_i$. Usually the $\boldsymbol{\kappa}_i$ are chosen to be symmetrically distributed vectors of magnitude $\sqrt{3/4}\kappa$, where κ is the magnitude of the wavevector at which the field renormalization constant a of §9.2 is defined—see §10.3. One refers to the set of vectors $(\boldsymbol{\kappa}_1, \boldsymbol{\kappa}_2, \boldsymbol{\kappa}_3)$ as the **renormalization point**.

We argued above that wholesale replacement of μ by m in all integrals actually increases the accuracy of our expressions by implicitly including terms of arbitrary power in λ. Unfortunately the replacement of λ by g has no comparably satisfactory side-effect. The root of this disappointment is that whereas the difference between $\mu^2 + a^2k^2$ and the one-loop approximation to $\Gamma^{(2)}$ is independent of k, the difference between λ and the one-loop approximation to $\Gamma^{(4)}(\mathbf{k}_1, \mathbf{k}_2, \mathbf{k}_3)$ depends on the $\mathbf{k}_i$. So when we subtract $\Gamma^{(4)}(0,0,0)$ or, more generally, $\Gamma^{(4)}(\boldsymbol{\kappa}_1, \boldsymbol{\kappa}_2, \boldsymbol{\kappa}_3)$ from $\Gamma^{(4)}(\mathbf{k}_1, \mathbf{k}_2, \mathbf{k}_3)$, we do not eliminate the one-loop contribution to the series. We *do* diminish its coefficient, however, at least for $\{\mathbf{k}_i\}$ in the neighbourhood of the renormalization point, since as $\{\mathbf{k}_i\}$ tends to that point the coefficient vanishes by construction. Hence we will in the following replace λ by g and modify the Feynman rules where necessary to take into account the difference between $\Gamma^{(4)}$ evaluated at $(\mathbf{k}_1, \mathbf{k}_2, \mathbf{k}_3)$ and at the renormalization point $(\boldsymbol{\kappa}_1, \boldsymbol{\kappa}_2, \boldsymbol{\kappa}_3)$.

9.4 Renormalization at higher orders

> **Box 9.4: Summary of renormalization formulae**
>
> The two- and four-point vertex functions are
>
> $$\Gamma^{(2)}(k) = m^2 + k^2\left(a^2 - \frac{g^2}{6}B(k,\kappa)\right) + O(g^3)$$
>
> $$\Gamma^{(4)}(\{\mathbf{k}_i\}) = g - \frac{g^2}{2}\int_0^\infty \frac{d^d\check{\mathbf{q}}}{(a^2q^2+m^2)^2}\left(\frac{2a^2\mathbf{q}\cdot\mathbf{K} - a^2K^2}{a^2(\mathbf{K}-\mathbf{q})^2+m^2}\right. \quad (1)$$
>
> $$\left. + \text{ 2 permutations}\right) + O(g^3),$$
>
> where $\mathbf{K} \equiv \mathbf{k}_1 + \mathbf{k}_2$. Should they be required, the 'bare' parameters μ^2, α^2 and λ are obtained from the 'dressed' ones $m^2 \equiv \Gamma^{(2)}(0)$, $a^2 \equiv d\Gamma^{(2)}/dk^2\big|_\kappa$ and $g \equiv \Gamma^{(4)}(0)$ through
>
> $$\lambda = g\left(1 + \frac{3g}{2}\int_0^\Lambda \frac{d^d\check{\mathbf{q}}}{(a^2q^2+m^2)^2}\right) + O(g^3)$$
>
> $$\mu^2 = m^2 - \frac{\lambda}{2}\int_0^\Lambda \frac{d^d\check{\mathbf{q}}}{a^2q^2+m^2} + \tfrac{1}{6}\lambda^2 A(0) + O(\lambda^3)$$
>
> $$\alpha^2 = a^2 - \frac{g^2}{6}\int_0^\Lambda \frac{d^d\check{\mathbf{q}}_1 d^d\check{\mathbf{q}}_2}{(a^2q_1^2+m^2)(a^2q_2^2+m^2)\kappa^2}\frac{a^2\kappa^2 - a^2\boldsymbol{\kappa}\cdot\mathbf{Q}}{[a^2(\mathbf{Q}-\boldsymbol{\kappa})^2+m^2]^2} + O(g^3). \quad (2)$$
>
> In these formulae $\mathbf{Q} \equiv \mathbf{q}_1 + \mathbf{q}_2$ and
>
> $$A(k) = \int_0^\Lambda \frac{d^d\check{\mathbf{q}}_1 d^d\check{\mathbf{q}}_2}{(a^2q_1^2+m^2)(a^2q_2^2+m^2)[a^2(\mathbf{k}-\mathbf{Q})^2+m^2]}$$
>
> $$B(k,\kappa) = \int_0^\infty \frac{d^d\check{\mathbf{q}}_1 d^d\check{\mathbf{q}}_2}{(a^2q_1^2+m^2)(a^2q_2^2+m^2)}$$
>
> $$\times \left(\frac{-a^2k^2 + 2a^2\mathbf{k}\cdot\mathbf{Q}}{[a^2(\mathbf{k}-\mathbf{Q})^2+m^2][a^2Q^2+m^2]k^2} + \frac{a^2\kappa^2 - a^2\boldsymbol{\kappa}\cdot\mathbf{Q}}{[a^2(\boldsymbol{\kappa}-\mathbf{Q})^2+m^2]^2\kappa^2}\right). \quad (3)$$
>
> For $d < 4$ the first and third integrals in equations (2) are Λ-insensitive.

9.4 Renormalization at higher orders

The results of the foregoing sections suffice for the calculation of the critical exponents of the Landau–Ginzburg model to the order to which they will be evaluated in this book. However, it is natural to enquire how the renormalization programme—that is, the confinement of Λ-sensitivity to a few integrals—may be extended to higher-order vertex functions and calculations involving many loops. The central question is the following: what is the minimum number of Λ-sensitive integrals in the theory? So far we have got by with only the three Λ-sensitive integrals that occur in equations (2) of Box 9.4. It turns out that *these are the only Λ-sensitive integrals that need be considered when $d \leq 4$*. Thus the minimum number of Λ-sensitive integrals is equal to the number of free parameters, α, μ and λ in the effective Hamiltonian. This equality of the number of Λ-sensitive integrals and parameters is important since it ensures that if a set of experimental results is consistent with the theory for one value of Λ, then that set is consistent with the theory for an infinite number of other values of Λ: changing Λ changes the values of only three integrals, and these changes can be compensated by corresponding changes in the values of α, μ and λ, in such a way that the values of all vertex functions, and thus all correlation functions, remain the same.

Thus the Landau–Ginzburg theory admits one-parameter families of quadruples of values for α, μ, λ and Λ which all generate the same macroscopic quantities—families are labelled by the corresponding triple (a, m, g) of values of the renormalized variables. A theory which is degenerate in this way is said to be **renormalizable**. Each quadruple $(\alpha, \mu, \lambda, \Lambda)$ in a given family corresponds to a particular smallest scale $1/\Lambda$ on which ϕ is permitted to fluctuate. Different families describe systems that genuinely differ, for example because they are at different temperatures.

Obviously the minimum scale of fluctuations, $1/\Lambda$, has to exceed the scale of atoms, but it is not otherwise strongly constrained by physical considerations since ϕ is an inherently macroscopic quantity. So it is important that at the end of the day the results we obtain from the Landau–Ginzburg theory are independent of the value we choose for this smallest scale of fluctuation. From §7.1 it is clear that by varying $1/\Lambda$ we must vary μ because we are then changing the number of microstates associated with a particular value of ϕ. The other two parameters α and λ vary with $1/\Lambda$ for similar though less transparent reasons in the same way that all the parameters in the effective Hamiltonians of Chapter 5 varied when we formed new block variables.

In particle physics, renormalizability allows the elimination of all Λ-sensitive integrals, and thus Λ itself, from the theory. The necessity for this is that in particle physics one hasn't the least idea what the bare quantities α, μ, and λ are. So one has absolutely no interest in relations such as equations (2) of Box 9.4 between bare and dressed quantities. Our situation is rather different. We *are* interested in the connection between μ^2 and m^2, since the first is a measure of the temperature and the second the inverse of the susceptibility. Nonetheless, we are anxious to work with a renormalizable theory, both because we don't want to have to calculate a significant number of integrals with finite upper limits, and because our results would not be credible if they depended on such an ill-determined quantity as Λ.

So let us investigate the Λ-sensitivity of a graph of $\Gamma^{(n)}$ that contains V vertices. Our first step is simply to determine how the integrand (including the volume elements in wavevector space) scales with the magnitude q of the

9.4 Renormalization at higher orders

integration variables $\mathbf{q}_i$. If the integrand scales as a non-negative power of q, it is likely that the integral is sensitive to the value of Λ, and conversely if the integrand scales as a negative power of q. We shall see that it is essential to include the word 'likely' in the last sentence because no guarantee is provided either way. But naïve dimensional analysis gives us a valuable guide to the Λ-sensitivity of integrals.

Of the $4V$ links that emerge from the V vertices of a graph of $\Gamma^{(n)}$, n are required for external connections and the rest are joined in pairs to form internal links. Thus the graph has

$$I = \tfrac{1}{2}(4V - n) \qquad (9.22)$$

internal links. The wavevectors associated with these links and with the external waves are coupled by V equations, since the wavevectors flowing into each vertex must sum to zero. Thus of the $n + (4V - n)/2$ wavevectors associated with the graph, $[n + (4V - n)/2] - V = \tfrac{1}{2}n + V$ may be freely chosen. $n - 1$ of these choices are associated with the external wavevectors, leaving

$$L = V - \tfrac{1}{2}n + 1 \qquad (9.23)$$

free choices for the internal wavevectors. Consequently, the graph involves integration over $V - \tfrac{1}{2}n + 1$ d-dimensional wavevectors, and the top of the integral will look something like this

$$\mathrm{d}^d \tilde{\mathbf{q}}_1 \ldots \mathrm{d}^d \tilde{\mathbf{q}}_{V+1-n/2}. \qquad (9.24)$$

This product of volume elements will be multiplied by $(4V-n)/2$ propagators such as $(a^2 q_i^2 + m^2)^{-1}$, one for each internal link. So at large q the integrand scales as

$$\frac{q^{d(V+1-n/2)}}{q^{4V-n}} = q^{d+V(d-4)+n(1-d/2)}. \qquad (9.25)$$

For example, when $d = 4$ all graphs of $\Gamma^{(n)}$ have integrands which scale as the same power of q, independent of the number of vertices V. In particular, the integrands of $\Gamma^{(2)}$ scale as q^2 with the consequence that the corresponding graphs are all likely to be highly sensitive to the value of Λ. The integrands of $\Gamma^{(4)}$ scale as q^0, so we expect $\Gamma^{(4)}$'s graphs to grow as $\log \Lambda$. Similarly naïve power counting predicts that when $d = 4$ the vertex functions $\Gamma^{(n)}$ with $n > 4$ are Λ-insensitive. These results are summarized in Table 9.1.

Obviously, any graph which has embedded within it a Λ-sensitive graph will itself be Λ-sensitive. Graphs which are Λ-sensitive without containing a Λ-sensitive subgraph are said to be **primitively divergent**. The only primitively divergent graphs of the Landau–Ginzburg model at $d \leq 4$ are the bubble and the Saturn graphs that contribute to the relation between μ and

Table 9.1. Naïve scaling of the integrands of graphs of $\Gamma^{(n)}$

		\multicolumn{3}{c}{V}		
		1	2	3
$\Gamma^{(2)}$	$d = 3$	1	0	-1
	$d = 4$	2	2	2
$\Gamma^{(4)}$	$d = 3$	—	-1	-2
	$d = 4$	—	0	0
$\Gamma^{(n)}$		\multicolumn{3}{c}{$d + V(d-4) + n(1 - d/2)$}		

m, plus the fish graph that contributes to the relation between λ and g. We do not prove this statement, but Table 9.1 makes it plausible by suggesting that primitive divergences should be confined to the graphs of $\Gamma^{(2)}$ and $\Gamma^{(4)}$. Furthermore, it is natural to encounter primitively divergent diagrams at a small number of loops, since diagrams with many loops are 'more of the same'—they consist principally of pieces which have already appeared lower in the hierarchy as diagrams in their own right.

The continuations to arbitrary order in the loop expansion of our expressions for $\Gamma^{(2)}(k)$ and $\Gamma^{(4)}(\mathbf{k}_1, \mathbf{k}_2, \mathbf{k}_3)$ in terms of the dressed quantities m and g contain no Λ-sensitive integrals because the parts of $\Gamma^{(2)}$ and $\Gamma^{(4)}$ that grow with Λ depend only very weakly on $\mathbf{k}$, so they are cleanly subtracted when we subtract from the general Feynman series the series for $\Gamma^{(2)}(0) + k^2 (d\Gamma^{(2)}/dk^2)_\kappa$ and $\Gamma^{(4)}(0)$ in the two cases, respectively. Hence we should not expect to find new Λ-sensitive integrals at more than two loops in the series for $\Gamma^{(2)}$ and $\Gamma^{(4)}$.

The absence of new Λ-sensitive integrals in the Feynman series for higher-order vertex functions such as $\Gamma^{(6)}$ is also natural because, as Table 9.1 shows, these series should contain no primitively divergent integrals. Consequently, whatever divergent subdiagrams they involve will have been already encountered and mastered in connection with $\Gamma^{(2)}$ and $\Gamma^{(4)}$. Another way of expressing this idea is to say that every diagram is built of links ($\Gamma^{(2)}$) and vertices ($\Gamma^{(4)}$). Once one has expressed the series for these elements in terms of dressed quantities and Λ-insensitive integrals, it is reasonable to be able to express any diagram that is built out of these elements in terms of the same, acceptable quantities. We say 'reasonable' because it is logically possible for an integral over Λ-insensitive vertices to be Λ-sensitive. Indeed, in many field theories this happens—the theory is then said to be 'unrenormalizable'. It does not happen in the Landau–Ginzburg model, but in Chapter 14 we shall discuss theories in which it does.

9.5 More on field renormalization

In this chapter we have expressed the vertex functions in terms of the three dressed parameters, (a, m, g), which should be thought of as phenomenological parameters chosen to ensure that theory and experiment agree as well as possible at one particular wavevector, κ. In later chapters we shall make extensive use of our ability to choose κ at will, and in particular to change it continuously. So it is important to have the clearest possible understanding of the way in which the vertex functions vary as a, m and g are varied. Much of this information could be gleaned from formulae already in hand, but there is a more elegant way to proceed, which has the advantage of being valid to all orders in the coupling constant.

We return to the Landau–Ginzburg Hamiltonian density (7.10):

$$\mathcal{H}_{LG} \equiv \tfrac{1}{2}\alpha^2|\nabla\phi|^2 + \tfrac{1}{2}\mu^2\phi^2 + \tfrac{1}{4!}\lambda\phi^4. \tag{9.26}$$

This is manifestly invariant under the transformation

$$\begin{aligned}\alpha^2 \to \alpha_R^2 \equiv Z\alpha^2, \quad &\mu^2 \to \mu_R^2 \equiv Z\mu^2, \\ \lambda \to \lambda_R \equiv Z^2\lambda, \quad &\phi \to \phi_R \equiv Z^{-1/2}\phi.\end{aligned} \tag{9.27}$$

So the probability of realizing the field $\phi(\mathbf{x})$ in a system characterized by (α, λ, μ) is equal to the probability of realizing the field $\phi_R(\mathbf{x})$ in a system characterized by $(\alpha_R, \lambda_R, \mu_R)$. Hence if we use the same functional measure[1] $\mathcal{D}\phi$ to sum over all possible fields $\phi_R(\mathbf{x})$ as we used to sum over all fields $\phi(\mathbf{x})$, it follows that under this transformation $G^{(n)}$ transforms like the simple product $|\phi|^n$. That is

$$\widetilde{G}^{(n)}(\mathbf{k}_1,\ldots,\mathbf{k}_{n-1}) \to \widetilde{G}_R^{(n)}(\mathbf{k}_1,\ldots,\mathbf{k}_{n-1}) \equiv Z^{-n/2}\widetilde{G}^{(n)}(\mathbf{k}_1,\ldots,\mathbf{k}_{n-1}). \tag{9.28}$$

From the relations such as equations (8.58) and (8.59) between $\Gamma^{(n)}$ and $\widetilde{G}_c^{(n)}$ it follows that the former scales as[2]

$$\Gamma^{(n)}(\mathbf{k}_1,\ldots,\mathbf{k}_{n-1}) \to \Gamma_R^{(n)}(\mathbf{k}_1,\ldots,\mathbf{k}_{n-1}) = Z^{n/2}\Gamma^{(n)}(\mathbf{k}_1,\ldots,\mathbf{k}_{n-1}). \tag{9.29}$$

From this and the definition (9.12) of a it is now apparent that a scales in the same way as α:

$$a^2 = \frac{\mathrm{d}\Gamma^{(2)}}{\mathrm{d}k^2} \to a_R^2 = Za^2. \tag{9.30}$$

---

[1] In practice this means retaining the same upper cutoff, Λ, on the wavevectors considered.

[2] Connoisseurs of quantum field theory (QFT) may be misled by equation (9.29). $\Gamma_R^{(n)}$ plays a similar rôle in the present theory and in QFT, but no close parallel exists between the quantities denoted here and in QFT by $\Gamma^{(n)}$ and Z; in QFT these quantities are Λ-sensitive and here they are not. Indeed, here $\Gamma^{(n)}$ is essentially the Fourier transform of the n-point correlation function and thus an eminently physical quantity, while $\Gamma_R^{(n)}$ has no direct physical significance. In QFT $\Gamma^{(n)}$, being Λ-sensitive, has no clear physical interpretation.

Therefore, if we set $Z = a^{-2}$, we will have $a_R = 1$, and our Feynman integrals will be usefully simplified. So we make this choice of Z, evaluate all vertex functions for the model whose objects carry subscripts R, and then recover the vertex functions of interest by application of the scaling (9.29):

$$\Gamma^{(n)}(\mathbf{k}_1, \ldots, \mathbf{k}_{n-1}) = a^n \Gamma_R^{(n)}(\mathbf{k}_1, \ldots, \mathbf{k}_{n-1}). \tag{9.31}$$

Since $m^2 = \Gamma^{(2)}(0)$ and $g = \Gamma^{(4)}(\{\kappa_i\})$, it follows that the values m_R^2 and g_R which should be used in evaluating the $\Gamma_R^{(n)}$ are

$$m_R^2 = m^2/a^2; \quad g_R = g/a^4. \tag{9.32}$$

In other words, $\Gamma_R^{(n)}$ involves a only in the combinations m^2/a^2 and g/a^4. In later chapters we shall make extensive use of this knowledge.

Somewhat confusingly, it is customary to refer to ϕ_R as the **renormalized field** and $\Gamma_R^{(n)}$ as a **renormalized vertex function**. ϕ_R and $\Gamma_R^{(n)}$ certainly are renormalized versions of ϕ and $\Gamma^{(n)}$ inasmuch as they are the latter rescaled by an appropriate number of powers of the constant a. Thus the word 'renormalized' is here being used in the literal sense in which it was used in Chapter 5, wherein the block variables were arbitrarily renormalized to ensure that they retained unit variance at every stage of the blocking process. But beware the danger of confusing this straightforward sense of 'renormalized' with the sense in which 'renormalization' is used to describe the process which gives this chapter its name. For example, mass 'renormalization' involves no mere rescaling, but a change of independent variable from μ^2 to m^2, a quantity with a totally different physical interpretation.

9.6 The Ginzburg criterion

In Chapter 7 we introduced the approximation upon which Landau theory is based, namely that one field ϕ_0 dominates the partition function. Phenomena such as critical opalescence tell us that sufficiently near T_c this approximation cannot be a good one, since they show that near T_c many macroscopically distinguishable fields occur with non-negligible probabilities. What has our freshly renormalized theory to say about how near to T_c one must go before Landau theory becomes invalid?

From equations (8.36) and (8.43) we know that $m^2 = \Gamma^{(2)}(0)$ is β divided by the susceptibility χ, and in §7.2 we saw that Landau theory predicts that $\chi \propto |T - T_c|^{-1}$ near T_c. So Landau theory predicts that $m^2 \propto |T - T_c| \propto |\mu^2 - \mu_c^2|$, where μ_c^2 is the value of μ^2 at which m^2 vanishes. The second of equations (2) in Box 9.4 relates μ^2 to m^2. Subtracting from this equation its value with $m^2 = 0$ we have

$$\Delta \mu^2 \equiv \mu^2 - \mu_c^2 = m^2 \left(1 + \frac{\lambda}{2} \int_0^\Lambda \frac{d^d \check{q}}{a^2 q^2 (a^2 q^2 + m^2)} + O(\lambda^2) \right). \tag{9.33}$$

9.6 The Ginzburg criterion

Since the integral depends on m^2, Landau's proportionality $m^2 \propto |\mu^2 - \mu_c^2|$ will break down if the integral becomes comparable with $1/\lambda$. Suppose we try to evaluate the integral with $m^2 = 0$. Then everything will go fine provided $d > 4$, since there are then sufficient powers of q on the top to cancel most of the four powers of q on the bottom. So when $d > 4$ the second term in the big bracket of (9.33) is negligible for all sufficiently small λ and Landau's proportionality can hold all the way to T_c.

When $d \leq 4$, by contrast, the integral in (9.33) diverges at small q when $m^2 = 0$. So in this case the integral becomes arbitrarily large as $m^2 \to 0$, and, no matter how small λ is, Landau's prediction will fail sufficiently near T_c. The temperature range around T_c within which the second term in the big bracket of (9.33) is non-negligible, and $m^2 \propto |T - T_c|^2$ does not hold, is called the **critical region**.

Let us derive an estimate of the width Δ_G of the critical region for $d < 4$. The smallest value of m^2 for which Landau's proportionality holds, m_G^2, is given by

$$1 \approx \frac{\lambda \Omega_{d-1}}{(2\pi)^d a^2} \int_0^\Lambda \frac{q^{d-3} dq}{a^2 q^2 + m_G^2}$$

$$= \frac{\lambda \Omega_{d-1}}{(2\pi a)^d m_G^{4-d}} \int_0^{a\Lambda/m_G} \frac{x^{d-3} dx}{x^2 + 1}, \qquad (9.34)$$

where $\Omega_{d-1} = 2\pi^{d/2}/[\frac{1}{2}(d-2)]!$ is the area of the unit $(d-1)$-sphere and $x \equiv aq/m_G$. When $d < 4$ the integral in (9.34) is insensitive to its upper limit $a\Lambda/m_G$, and we shall ignore the associated m_G-dependence. Then the value of the susceptibility at which the critical region is entered is $m_G^{-2} \propto (\lambda/a^d)^{-2/(4-d)}$. The difference between the temperature at this point and the critical temperature scales as $\Delta_G \propto m_G^2 \propto (\lambda/a^d)^{2/(4-d)}$. This relation between the width of the critical region and λ is known as the **Ginzburg criterion** for the breakdown of Landau theory. The smaller d is, the wider is the range of temperatures within which, for a given value of λ/a^d, mean-field theory cannot be relied upon.

A simple argument helps explain the physical content of the Ginzburg criterion. Consider the situation just below T_c. Then Landau theory predicts that the order parameter will have a small mean value (equation (7.18))

$$\phi_0 = 6|\mu^2|/\lambda. \qquad (9.35)$$

How does this mean value compare with the predicted amplitude of fluctuations in ϕ on the scale of a correlation length? Fluctuations are measured by the connected two-point correlation function $G_c^{(2)}(r)$, whose average value is proportional to the susceptibility χ (see (8.36)). If we make the approximation that $G_c^{(2)}(r)$ is roughly constant within a sphere of radius one correlation length a/m and then negligible, we have

$$\chi = \beta \widetilde{G}_c^{(2)}(0) \simeq \beta G_c^{(2)}(0)(a/m)^d. \qquad (9.36)$$

By equation (7.26) Landau theory predicts $\chi = \frac{1}{2}\beta/|\mu^2|$. So the RMS fluctuations are of order

$$\phi_f \equiv \sqrt{G_c^{(2)}(0)} \simeq \frac{m^d}{2|\mu^2|a^d}, \qquad (9.37)$$

and the ratio of the fluctuating to the mean components is

$$\frac{\phi_f}{\phi_0} \simeq \frac{\lambda m^d}{12|\mu^2|^2 a^d}. \qquad (9.38)$$

If we now replace both m^2 and μ^2 by $\Delta_G = |T - T_c|$ and require $\phi_f < \phi_0$, we again obtain the Ginzburg criterion, $\Delta_G \propto (\lambda/a^d)^{2/(4-d)}$.

Problems

9.1 Show that at a critical point $d\Gamma^{(2)}/dk^2$ diverges as $k \to 0$ whenever $\eta > 0$.

9.2 By explicitly differentiating equation (9.13), show that $(d\Gamma^{(2)}/dk^2)_\kappa = a^2$.

9.3 Imagine labelling the 1PI diagrams contributing to $\Gamma^{(2)}(q)$ with an index i, and let $D_i(q)$ be the value assigned to the ith such diagram by the Feynman rules of Chapter 8. Then

$$\Gamma^{(2)}(q) = \alpha^2 q^2 + \mu^2 + \sum_i D_i(q). \qquad (9.39)$$

If instead we write $\Gamma^{(2)}$ as $a^2 q^2 + m^2 + \sum_i D'_i(q)$, where $m^2 \equiv \Gamma^{(2)}(0)$ and $a^2 \equiv (d\Gamma^{(2)}/dq^2)|_{q=\kappa}$, what value must our modified Feynman rules assign to the ith 1PI diagram?

9.4 Show that the integral of (9.19) is insensitive to Λ for $d < 6$.

9.5 Obtain the analogue of equation (9.21) for the case in which the coupling constant is renormalized at non-zero wavevectors κ_1, κ_2 and κ_3.

10
The calculation of critical exponents for $T \geq T_c$

For $T \geq T_c$ five critical exponents are defined. These are α, γ, δ, η and ν. The remaining exponent β is only defined below the critical temperature (see §1.1.1). In this chapter we will calculate two exponents—γ (which is defined for $T > T_c$) and η (which is defined at $T = T_c$). Since the six exponents are connected by the four scaling laws of §1.5.1, the calculation of these two is in fact sufficient to tell us the values of all of them provided we are willing to assume that these scaling laws apply to the Landau–Ginzburg model. In Chapters 11 and 12 we shall show that these laws do indeed apply to the Landau–Ginzburg model, as one might hope in view of the experimental evidence for them and the theoretical arguments of §5.7.

10.1 Ultraviolet and infrared divergences

Before we dive into the calculation of the first of our critical exponents, it will be helpful for us to consider a little more closely some of the problems that arise from the perturbation expansion we have developed for our correlation functions. This expansion is an expansion in powers of the coupling constant λ, whose coefficients are integrals given by the Feynman rules. In §9.4, we were concerned with the behaviour of some of these integrals as we let the wavevector cutoff Λ tend to infinity. In general we found that, in sufficiently high dimension, all the integrals diverge as we let $\Lambda \to \infty$. These divergences are known as **ultraviolet divergences**, because they are the result of the behaviour of the integrands at large wavevectors. Ultraviolet divergences are not a mathematical problem with the theory; they are a genuine physical effect, which arises because the correlation functions in the theory we have constructed are sensitive to the value we choose for Λ. We showed however, that, at least below a certain dimension, this Λ-sensitivity can be confined to a small number of quantities appearing in the theory—m, g and the field renormalization parameter a—leaving all the other integrals finite in the limit $\Lambda \to \infty$. This was the important achievement of renormalization. Apart from anything else, it allows us to set the upper limit of our integrals to infinity, which makes them much simpler to perform.

However, there is another kind of divergence which can arise in our integrals, which we have not mentioned yet: this is the **infrared divergence**. Infrared divergences occur when we let $T \to T_c$, so that $m \to 0$, *and* we let all the external wavevectors $\mathbf{k}_1, \mathbf{k}_2 \ldots$ tend to zero. To see why this can produce a divergence, recall that in §9.1 we replaced every bare mass μ appearing in the denominators of our integrals with the renormalized mass m, thereby actually increasing the accuracy of our expansion. However, after performing this replacement, our denominators are entirely made up of factors like $a^2(\mathbf{q} - \mathbf{k})^2 + m^2$, and all these factors will become simply $a^2 q^2$ if we let both $\mathbf{k}$ and m^2 go to zero. This is bad news, because given a sufficiently large number of such factors and a sufficiently small dimension d, the integrand will acquire a (non-integrable) pole at $q = 0$, and we will not be able to perform the integral. This problem only arises in the particular limit $k, m \to 0$, but unfortunately this is precisely the limit we are interested in if we are going to calculate critical exponents. The exponent γ for example is calculated from the behaviour of $\Gamma^{(2)}(k)$ at $k = 0$ as $m \to 0$, and the exponent η is calculated from the behaviour of $\Gamma^{(2)}(k)$ at $m = 0$ as $k \to 0$. In both these cases, the coefficients in our perturbation expansion will, for sufficiently small d, diverge as we take the limit, and the expansion will become invalid.

In fact, it turns out that in the limit $m \to 0$ with $k = 0$, these infrared divergences are not a problem. As we shall see in the next section, they can be removed by the renormalization of the coupling constant. We replace the bare coupling constant λ by the renormalized coupling g, which (as we shall demonstrate) tends to zero in the limit $m, k \to 0$, and kills the divergence of the integrals. In the interesting cases (at $d = 4$ and below), it does so just fast enough that the resulting terms in the perturbation series are finite, giving rise to corrections to the leading-order (Gaussian) results for the critical exponents. However, when we look at $k \to 0$ with $m = 0$, we will find that infrared divergences persist, even after we renormalize the coupling constant. In order to get around this problem, we must resort to the 'ϵ-expansion' (§10.5).

10.1 Ultraviolet and infrared divergences

To simplify the calculation of the critical exponents, it will be helpful to study first the general form of the integrals derived from the Feynman graphs, and to extract their dependence on m and k. To do this we make use of an argument very similar to that of §9.4 for the Λ-sensitivity of graphs. Strictly we should give the development separately for the two cases $k = 0$, $m \to 0$ and $m = 0$, $k \to 0$. However, these cases are very similar and to run through them both now would be tedious, so we will only do the former. The reader can generalize the argument we will give to the latter, though the idea should become clear in §10.3 anyway when we calculate η.

So, imagine we are calculating the n-point vertex function $\Gamma^{(n)}$, with all the external momenta $\mathbf{k}_1, \mathbf{k}_2 \ldots$ set to zero. The Feynman rules set out in Box 8.4 allow us to write this vertex function as a function of the bare coupling constant λ to any order we desire. We have seen this done for a few special cases in Chapter 9. Then we can replace the bare mass μ where it occurs in the denominators of the integrands with the renormalized mass m and, as we have argued in §9.1, this will improve the accuracy of our expansion by adding in a subset of the higher-order diagrams that we would otherwise have missed out. Now consider a general term in this series. Suppose this term arises from a diagram which has I internal lines, each giving rise to a factor like $a^2 q^2 + m^2$ in the denominator of the integrand, and L closed loops, each giving rise to an integration $\mathrm{d}^d \check{\mathbf{q}}$. Let us make a change of integration variables from $\mathbf{q}$ to the dimensionless variable $\mathbf{x}$, defined by $a\mathbf{q} \equiv m\mathbf{x}$. This change enables us to express the integral as a product of some power of m and an integral that is essentially independent of m. It is the power of m in this expression which tells us how the integral will diverge as $m \to 0$. As an example, let us look at the integral in equation (9.33), which will be important for the calculation of the exponent γ in the next section. We have

$$\int_0^\Lambda \frac{\mathrm{d}^d \check{\mathbf{q}}}{a^2 q^2 (a^2 q^2 + m^2)} = m^{d-4} \int_0^{a\Lambda/m} \frac{\mathrm{d}^d \check{\mathbf{x}}}{x^2 (x^2 + 1)}, \tag{10.1}$$

where

$$\mathrm{d}^d \check{\mathbf{x}} \equiv \frac{\mathrm{d}^d \mathbf{x}}{(2\pi a)^d}. \tag{10.2}$$

This leaves the integrand well-behaved in the limit $m \to 0$ for dimensions $d < 4$, and instead we have a factor of m^{d-4} out the front. It is this factor which will give us an infrared divergence. The reason we can isolate the infrared divergence in this way is that, for this, and all other physical quantities we will be considering, we can arrange things so that the integrals are insensitive to their upper limits. This is exactly because we have renormalized them to remove such sensitivity—see Chapter 9. For integrals written in terms of dimensionless variables like this, the only dependence on m is in the upper limit, so this insensitivity is precisely an insensitivity to m.

In general then, we can see that every integration will produce a factor of m^d in front of the integrand, and every internal line (i.e., every $a^2 q^2 + m^2$ in the denominator) a factor of m^{-2}. Thus for our general diagram with L loops and I internal lines, the overall power P of m will be

$$P = dL - 2I. \tag{10.3}$$

Now suppose the diagram we are considering is of order V in λ. This means that the diagram has V vertices. In §9.4 it was demonstrated that I and L can be expressed in terms of V and the number n of external legs thus:

$$I = \tfrac{1}{2}(4V - n), \tag{10.4}$$

and[1]

$$L = V - \tfrac{1}{2}n + 1. \tag{10.5}$$

(See equations (9.22) and (9.23).)

Substituting from (10.4) and (10.5) into (10.3), the overall power of m multiplying the integrand in our general diagram is

$$P = (d-4)V + d + n - \tfrac{1}{2}dn = -\epsilon V + \delta_n, \tag{10.6}$$

where

$$\epsilon \equiv 4 - d \quad \text{and} \quad \delta_n \equiv d + n - \tfrac{1}{2}dn. \tag{10.7}$$

The quantity δ_n in equation (10.6) for P is independent of V, the order in λ of the graph we are looking at. In other words, there is a factor of m^{δ_n} in every graph appearing in the expansion of our vertex function, which we can take out as a multiplier for the whole expansion. The remaining multiplier is $m^{-\epsilon V}$, or one factor of $m^{-\epsilon}$ for every power of λ. It is standard practice therefore to regard the expansion as being in powers of $\lambda m^{-\epsilon}$, rather than λ itself. If we are below $d = 4$, ϵ is negative and this expansion parameter will diverge as we approach the critical point $m = 0$. It will turn out that this divergence can be removed by the renormalization of the coupling constant. The same is not true of the leading factor of m^{δ_n} on the other hand. Below $d = 4$, for example, this factor ensures that $\Gamma^{(4)}(0) \to 0$ with m as m^{δ_4}, a result which will prove important in §10.2.2. We might also expect that our correlation functions would diverge as we went to the critical temperature whenever $\delta_n < 0$, or in other words when

$$d > \frac{2n}{n-2}, \quad (n \neq 2). \tag{10.8}$$

---

[1] Note that (10.4) implies that for any particular vertex function (i.e., for any particular n) the number of loops goes up in step with the number of vertices. In other words, an expansion in numbers of loops is equivalent to one in powers of the coupling constant, as we argued in §8.4.2.

10.1 Ultraviolet and infrared divergences

(In the case $n = 2$, $\delta_n = 2$ independent of d, so the leading power of m would never diverge.) However, for large d the powers of m in the coupling constant $\lambda m^{-\epsilon}$ converge to zero as $m \to 0$, and it turns out that this effect is exactly large enough to cancel both the leading power of m and any ultraviolet divergence in the diagram. Thus, happily, our correlation functions remain finite. This does mean however, that we have to treat the cases of ϵ above and below zero separately, and, as we shall see in §§10.2 and 10.3, this gives rise to an important qualitative difference in the behaviour of the theory in these two cases.

Actually, whilst we are on the subject of powers and expansion parameters and things, there is one other messy little detail we ought to clear up. Referring to Table 7.1, we see that the coupling constant λ has dimensions L^d. This is unsatisfactory. Expansions are never made in powers of dimensionful quantities because there can be no real sense in which a dimensionful quantity is small. Its size depends on the scale on which you measure it. The coefficients in the expansion also have dimensions, since the integration measure $d^d\tilde{x}$ contains a factor of $(2\pi a)^{-d}$. We can therefore tidy things up considerably by shifting some factors of a (which has dimensions of length) out of the integrals and into the expansion parameter. Each diagram involves $L = V - \frac{1}{2}n + 1$ integrations, so let us shift V powers of $(2\pi a)^{-d}$ out of each expansion coefficient and into the λ^V term in front of it, giving us a dimensionless expansion parameter

$$\hat{\lambda} \equiv \frac{\lambda m^{-\epsilon}}{(2\pi a)^d}. \tag{10.9}$$

The remaining $1 - \frac{1}{2}n$ powers of $(2\pi a)^{-d}$ are the same for every term in the expansion, so we can factor them out and put them up front with the m^{δ_n}. This leaves our integrals entirely in terms of dimensionless variables, which has the added bonus of making them easier to perform.

So, the final picture we have at the end of this section is of an expansion in powers of the dimensionless coupling constant $\hat{\lambda}$ defined in equation (10.9). The expansion of $\Gamma^{(n)}$ with all the external wavevectors set to zero has a leading factor of $m^n (m/2\pi a)^{d(1-n/2)}$, giving it the correct dimension of $L^{d(n/2-1)}$ (see Table 8.2), and then a completely dimensionless expansion in terms of the dimensionless coupling constant, with dimensionless coefficients made up of dimensionless integrals over dimensionless variables.

10.2 The calculation of γ

The exponent γ governs the divergence of the susceptibility χ with the temperature, as we approach T_c from above. When talking about the Landau–Ginzburg model, it is convenient to work with the variables m and μ, rather than χ and T. By equation (8.36)

$$m^2 \propto \chi^{-1}. \tag{10.10}$$

And by equation (7.12)

$$\Delta\mu^2 \equiv \mu^2 - \mu_c^2 \propto \frac{T - T_c}{T_c}, \quad \text{as } T \to T_c. \tag{10.11}$$

Thus the defining relation for the exponent γ (see Table 1.1) becomes

$$\Delta\mu^2 \sim (m^2)^{1/\gamma}, \quad m \to 0, \tag{10.12}$$

and hence, if we know how $\Delta\mu^2$ behaves as a function of m^2 in the vicinity of the phase transition, we can calculate γ from

$$\frac{1}{\gamma} = \lim_{m \to 0} \frac{\partial \log \Delta\mu^2}{\partial \log m^2}. \tag{10.13}$$

The relationship between the bare and renormalized 'masses' μ and m can be found directly using perturbation theory. In fact, we have already done this to order two loops in Chapter 9. To keep things simple, let us for the moment content ourselves with a one-loop calculation. The one-loop result for $\Delta\mu^2$ appears in equation (9.33). For convenience we give it again here:

$$\begin{aligned}\Delta\mu^2 &= m^2 + m^2 \frac{\lambda}{2} \int_0^{\Lambda} \frac{d^d\check{\mathbf{q}}}{a^2 q^2(a^2 q^2 + m^2)} + O(\lambda^2) \\ &= m^2 \left\{ 1 + \frac{\hat\lambda}{2} \int_0^{a\Lambda/m} \frac{d^d\mathbf{x}}{x^2(x^2 + 1)} + O(\hat\lambda^2) \right\},\end{aligned} \tag{10.14}$$

where we have taken out factors of m to leave an expression of exactly the form described at the end of the last section, and written the integral in terms of the new variable $\mathbf{x} \equiv a\mathbf{q}/m$. Taking logarithms, this becomes

$$\begin{aligned}\log \Delta\mu^2 &= \log m^2 + \log\left\{ 1 + \frac{\hat\lambda}{2} \int_0^{a\Lambda/m} \frac{d^d\mathbf{x}}{x^2(x^2+1)} + O(\hat\lambda^2) \right\} \\ &= \log m^2 + \frac{\hat\lambda}{2} \int_0^{a\Lambda/m} \frac{d^d\mathbf{x}}{x^2(x^2+1)} + O(\hat\lambda^2).\end{aligned} \tag{10.15}$$

10.2 The calculation of γ

10.2.1 $d = 4$ and above

The behaviour of equation (10.15) below four dimensions differs greatly from its behaviour at $d = 4$ and above. For large values of the integration variable, the integrand goes as x^{-4}, which means that the integral must depend on its upper limit as $(a\Lambda/m)^{d-4}$. For dimensions greater than or equal to four then, the integral is strongly dependent on the mass m, whereas for $d < 4$ it is virtually independent of it. We will treat the former case first.

Let us call the integral in (10.15) I. If I depends on m as $(a\Lambda/m)^{d-4}$, then we must be able to write it in the form

$$I = A + B\left(\frac{a\Lambda}{m}\right)^{d-4} = A + B\left(\frac{a\Lambda}{m}\right)^{-\epsilon}, \qquad (10.16)$$

plus terms that become negligible as $a\Lambda/m \to \infty$. A and B are constants independent of m. We now substitute this into equation (10.15) and perform the derivative, equation (10.13), not forgetting the m-dependence of $\hat{\lambda}$. A useful formula to have to hand when performing the derivative is

$$m^2 \frac{\partial \hat{\lambda}^n}{\partial m^2} = -\tfrac{1}{2} n \epsilon \hat{\lambda}^n, \qquad (10.17)$$

which is easily proven starting from (10.9). Using it we find that

$$\frac{1}{\gamma} = 1 - \tfrac{1}{4}\epsilon A \lim_{m \to 0} \hat{\lambda}. \qquad (10.18)$$

But with ϵ negative in dimensions greater than four, $\hat{\lambda} \to 0$ when we take the limit and so the second term vanishes. Actually at $d = 4$, $\hat{\lambda}$ does not tend to zero when we take the limit, but on the other hand $\epsilon = 0$ at $d = 4$, so we still get the second term in (10.18) vanishing. Either way then we find that

$$\gamma = 1. \qquad (10.19)$$

Thus in the Landau–Ginzburg model at four dimensions and above we find the same value for γ to this order as we would in mean-field theory (see §6.7) or Landau theory (see §7.2).[2]

Below four dimensions however, we find a very different story. The calculation becomes considerably more complicated, and, as we shall see, the higher-order corrections to the value of γ no longer vanish. Because $d = 4$ divides two regimes of behaviour in this way, it is given the name **critical dimension**, which we denote d_c. (In fact $d = 4$ is strictly the **upper critical dimension**. The **lower critical dimension** is the largest value of d at which long-range order is not possible for $T > 0$—see Chapter 13. The

---

[2] Actually this result extends to all orders; the exact result for γ is 1 for all $d > 4$.

lower critical dimension for this model is 2.) The value of 4 for the critical dimension is not common to all theories. For instance, a theory with only a ϕ^6 term after the $\frac{1}{2}\mu^2\phi^2$ term in the Hamiltonian would have an upper critical dimension of 3.[3] For the moment however, we are studying the ϕ^4 theory, and we need not bother with such complications.

10.2.2 Below four dimensions

Below the critical dimension, the integral in (10.15) depends only weakly on its upper limit, so we can approximate the integral by its value as $\Lambda \to \infty$, which renders it independent of m altogether. (This approximation will become better and better as we approach the critical point $m = 0$ at which $a\Lambda/m$ genuinely is infinite. Thus no approximation is being made at all as far as the calculation of the critical exponent is concerned.) The derivative in equation (10.13) is then simple to perform, the only dependence on m being in the definition (10.9) of $\hat{\lambda}$. Using (10.17) we then have

$$\frac{1}{\gamma} = 1 - \tfrac{1}{4}\epsilon \lim_{m \to 0} \hat{\lambda} \int_0^\infty \frac{d^d\mathbf{x}}{x^2(x^2+1)} + O(\hat{\lambda}^2). \qquad (10.20)$$

The integral here can be performed. In three dimensions, for example, it is

$$4\pi \int_0^\infty \frac{dx}{x^2+1} = 2\pi^2. \qquad (10.21)$$

Unfortunately this does not mean that we can calculate γ, because below d_c, ϵ is positive and so the factor $m^{-\epsilon}$ in the definition of $\hat{\lambda}$ diverges as we approach the critical temperature. This is a serious problem since it is not just confined to the first term in the series for $1/\gamma$. As we have shown, our whole expansion is in powers of $\hat{\lambda}$. There is unlikely to be any argument we can make to support the validity of an expansion in powers of a quantity that diverges in the very limit we are interested in exploring.

As we hinted in §10.1 however, there is a way out of this, which is to transform our expansion in powers of $\hat{\lambda}$ into one in powers of a different dimensionless quantity, which *is* well behaved in the limit that $m \to 0$. It is here that the renormalization of the coupling constant, introduced in

---

[3] As we shall see in the next section, the crucial difference in the behaviour of a theory above and below d_c is that below d_c the renormalized coupling constant g tends to zero as $T \to T_c$. In a theory in which the lowest power of ϕ appearing after the $\frac{1}{2}\mu^2\phi^2$ term is ϕ^r, the renormalized coupling constant is equal to $\Gamma^{(r)}(0)$. From equation (10.8) it can then be shown that, in general, the upper critical dimension is given by

$$d_c = \frac{2r}{r-2}.$$

10.2 The calculation of γ

§9.3, finds its use. The quantity we now expand in is the renormalized dimensionless coupling constant

$$\hat{g} \equiv \frac{gm^{-\epsilon}}{(2\pi a)^d}. \tag{10.22}$$

We use exactly the techniques introduced in §9.3 to eliminate $\hat{\lambda}$ in favour of $\hat{g}$, and so produce a power series expansion for the critical exponent which (as we shall shortly see) is valid below the critical dimension, near to T_c. In particular, we make use of the equation from Box 9.4 which says

$$\lambda = g + \tfrac{3}{2}g^2 \int_0^\Lambda \frac{d^d\check{q}}{(a^2q^2 + m^2)^2} + O(g^3). \tag{10.23}$$

Making the substitution $a\mathbf{q} = m\mathbf{x}$ and dividing both sides of the formula by $m^\epsilon (2\pi a)^d$, we get

$$\hat{\lambda} = \hat{g} + \tfrac{3}{2}\hat{g}^2 \int_0^{a\Lambda/m} \frac{d^dx}{(x^2 + 1)^2} + O(\hat{g}^3). \tag{10.24}$$

In fact all this tells us is that, to the order to which we are calculating γ here (which is $O(\hat{g})$, equivalent to one loop—see equation (10.5)), the bare and renormalized couplings are just the same. Let us assume for the moment that $\hat{g}$ will tend to a finite limit as $m \to 0$. Then we can simply replace $\hat{\lambda}$ by $\hat{g}$ in (10.20), to get

$$\frac{1}{\gamma} = 1 - \lim_{m\to 0} \tfrac{1}{4}\epsilon\hat{g} \int_0^\infty \frac{d^dx}{x^2(x^2 + 1)} + O(\hat{g}^2). \tag{10.25}$$

However, as we showed in §10.1, below $d = 4$, $g \to 0$ as m^{δ_4} in the limit of small m. But $\delta_4 = \epsilon$, so $\hat{g}$ will tend to a finite limit as $m \to 0$. So we can take the limit in (10.25) and get

$$\frac{1}{\gamma} = 1 - \tfrac{1}{4}\epsilon\hat{g}_c \int_0^\infty \frac{d^dx}{x^2(x^2 + 1)} + O(\hat{g}_c^2), \tag{10.26}$$

where $\hat{g}_c$ is the limiting value of $\hat{g}$ at the critical point.

The only task remaining to us before we get our answer for the exponent γ at this order then is the calculation of $\hat{g}_c$. The way we calculate it is to consider the β-**function** $\beta(\hat{g})$, which is defined by

$$\beta(\hat{g}) \equiv \left(\frac{\partial \hat{g}}{\partial \log m^2}\right)_{\lambda,\Lambda}. \tag{10.27}$$

As we have implied by our notation, it will turn out that the β-function can be expressed as a function of only one variable $\hat{g}$. This is rather an odd state of affairs. We start off with a quantity $\hat{g}$, which may be expressed as a function of m, λ and Λ. Then we calculate its derivative with respect to $\log m^2$ (keeping λ and Λ constant) and presumably we get another function of m, λ and Λ. What we find however, is that β can be expressed far more simply than this as a function of $\hat{g}$, the thing we started off with. This means that all sets of values $\{m, \lambda, \Lambda\}$ which give rise to the same value of $\hat{g}$ will give rise to the same value of β. Strange as it may seem, we can easily see that this must be the case. Our plan for calculating the β-function will be to take the inverse of equation (10.23), which first appeared in §9.3 as equation (9.18):

$$g = \lambda - \tfrac{3}{2}\lambda^2 \int_0^{\Lambda} \frac{d^d\check{\mathbf{q}}}{(a^2 q^2 + m^2)^2} + O(\lambda^3). \tag{10.28}$$

Making the substitution $a\mathbf{q} = m\mathbf{x}$ and dividing both sides by $m^\epsilon(2\pi a)^d$ as we did with (10.23), we get

$$\hat{g} = \hat{\lambda} - \tfrac{3}{2}\hat{\lambda}^2 \int_0^{a\Lambda/m} \frac{d^d\mathbf{x}}{(x^2+1)^2} + O(\hat{\lambda}^3). \tag{10.29}$$

But below the critical dimension $d_c = 4$, the integral will be insensitive to its upper limit, so the only dependence on m in this expression is that implicit in the definition (10.9) of $\hat{\lambda}$. But now the formula (10.17) tells us that when we perform the logarithmic derivative to find $\beta(\hat{g})$, we will just get back another expansion in $\hat{\lambda}$. So the β-function is a function of $\hat{\lambda}$ only. We can then make use of equation (10.24) to eliminate $\hat{\lambda}$ in favour of $\hat{g}$ and so derive an expression for the β-function which is a function of $\hat{g}$ only. The extension of this argument to all orders depends only on our being able to renormalize our expansion for $\hat{g}$ in powers of $\hat{\lambda}$ so that none of the integrals depend strongly on their limits of integration. If we can do this (and we assume that we can), then β is a function of $\hat{g}$ only, to all orders.

Our interest in the β-function stems from the fact that, as can be seen from its definition as a derivative, if $\hat{g}$ tends to a finite limit $\hat{g}_c$ as $m \to 0$ (and therefore $\log m^2 \to -\infty$), then β will have a zero at this value of $\hat{g}$. It will turn out that (to this order at least) β has one and only one non-trivial zero, which is at a finite value of $\hat{g}$. The problem of finding $\hat{g}_c$ is thus reduced to that of solving the equation $\beta(\hat{g}_c) = 0$. Note in addition, that this means that the value of $\hat{g}_c$ must be independent of λ and all the other bare parameters. This fact, along with our observation that γ can be written as an expansion in powers of $\hat{g}_c$ with coefficients also independent of the bare parameters, means that to all orders the value of γ must be independent of the bare parameters in the theory. A quantity with these properties is

10.2 The calculation of γ

called **universal**. As we shall see in Chapter 14, the values of *all* the critical exponents are universal. In fact, not only are they the same for different values of the Landau–Ginzburg bare parameters, but they are the same for many other Hamiltonians as well. This is what gives the Landau–Ginzburg model its importance.

So, let us calculate $\beta(\hat{g})$ in the way that we have described. Since the integral in (10.29) is insensitive to its upper limit in dimensions less than four, we can set Λ to infinity. Using equation (10.17) we can then write the β-function as

$$\beta(\hat{g}) = -\tfrac{1}{2}\epsilon\hat{\lambda} + \tfrac{3}{2}\epsilon\hat{\lambda}^2 \int_0^\infty \frac{d^d\mathbf{x}}{(x^2+1)^2} + O(\hat{\lambda}^3). \tag{10.30}$$

Now we use (10.24) to eliminate $\hat{\lambda}$ from this equation in favour of $\hat{g}$. Note that in contrast to the elimination of $\hat{\lambda}$ from equation (10.20), we are in this case working to $O(\hat{\lambda}^2)$, which means that we must keep the $O(\hat{\lambda}^2)$ term in (10.24). The result is that

$$\beta(\hat{g}) = -\tfrac{1}{2}\epsilon\hat{g} + \tfrac{3}{4}\epsilon\hat{g}^2 \int_0^\infty \frac{d^d\mathbf{x}}{(x^2+1)^2} + O(\hat{g}^3). \tag{10.31}$$

As expected, this gives the β-function as a function of $\hat{g}$ only. At this order $\beta(\hat{g})$ has two zeros. One is the trivial or **Gaussian** zero at $\hat{g} = 0$, which will obviously not give us any non-Gaussian contributions to the critical exponents given that we are calculating them as expansions in powers of $\hat{g}$. It is the other zero which gives us the critical value $\hat{g}_c$ of the renormalized dimensionless coupling constant:[4]

$$\hat{g}_c = \tfrac{2}{3}\left[\int_0^\infty \frac{d^d\mathbf{x}}{(x^2+1)^2}\right]^{-1}. \tag{10.32}$$

The integral is one that can be done for general $d < 4$. When $d = 3$ we get

$$\hat{g}_c = \frac{2}{3\pi^2} = 0.0675\ldots \tag{10.33}$$

Substituting this result into equation (10.26) and making use of (10.21), we arrive at the result

$$\gamma = \tfrac{3}{2} \tag{10.34}$$

---

[4] We can regard this choice between the two roots as being justified by experiments, which clearly show that the critical exponents can assume non-Gaussian values for many systems. In §11.1 however, we will show that within the Landau–Ginzburg theory there are also good mathematical reasons for choosing the non-trivial root, rather than the trivial one.

in three dimensions to one loop order. This then is conclusively different from the $\gamma = 1$ found in Landau theory and mean-field theory.

This all seems very nice, but you may well be asking yourself by now what on earth is going on here. Consider. We have developed, in a moderately rigorous fashion, an expansion for $\Delta\mu^2$ as a function of m^2 in powers of $\hat\lambda$; an expansion that, having carefully considered every line in its derivation, one can feel some faith in. But, it transpires, this expansion does not converge in the vicinity of the critical point, and so fails to give us the very information that we have been searching for through all these chapters. This is in turn because of the divergence of the expansion parameter $\hat\lambda$ near the critical point. And how have we got around this shortcoming? We have argued that, to the order we are interested in, $\hat\lambda$ and $\hat g$ are equal so we can replace $\hat\lambda$ everywhere by $\hat g$ which is finite and small at the critical point, and so render our expansion convergent and extract an answer.

"Well, come on," you will be saying. "Who you trying to fool? This argument is clearly potty. For surely, if $\hat\lambda$ diverges near the critical point and $\hat g$ does not, there is no way it can be legitimate to make the substitution $\hat\lambda \to \hat g$ in the vicinity of T_c? The expansion is divergent, and no amount of shuffling of variables is going to change this."

There is an answer to this criticism. It runs like this. We have made an expansion of the derivative $\partial\log\Delta\mu^2/\partial\log m^2$ in powers of the dimensionless coupling constant $\hat\lambda$, and we believe that this derivative will give us the exponent γ when we evaluate it at the critical temperature. Near the critical temperature however, we have little hope that the series will converge, because $\hat\lambda$ becomes large. On the other hand, $\hat\lambda$ is small when we are well away from the critical point; it varies with temperature because of the factor of $m^{-\epsilon}$ in its definition (see equation (10.9)). So we can make it as small as we like by going sufficiently far from the critical temperature. In this regime we expect our series for the derivative to converge quite well. However, this argument can also be applied to the expansion of $\hat g$ in powers of $\hat\lambda$. In the regime far from the critical temperature then, it is legitimate to eliminate $\hat\lambda$ in favour of $\hat g$, giving us two different expansions for the derivative as a function of m^2, both valid far from the critical point. As we approach the critical point, $\hat g$ remains finite, so there is reasonable hope of obtaining worthwhile results by evaluating the series for $\partial\log\Delta\mu^2/\partial\log m^2$ in powers of $\hat g$ at $\hat g = \hat g_c$. On the other hand, the expansion in powers of $\hat\lambda$ is definitely invalid in this regime, since, no matter how large its radius of convergence may be, $\hat\lambda$ will eventually exceed that radius as $m \to 0$.

But now you may ask, why did we not replace $\hat\lambda$ by $\hat g$ at an earlier stage? It might seem more natural to eliminate $\hat\lambda$ from equation (10.14) for $\Delta\mu^2$ so that we have a series expressed entirely in terms of renormalized quantities. If we had done this, we would have found a very different answer for γ. The bracket in (10.14) would have been entirely independent of m in the limit $m \to 0$ (since the integrals in the expansion are insensitive to their upper limits and $\hat{g}$ is independent of m in this limit) and so on differentiating with respect to $\log m^2$, we would have found $\gamma = 1$ to all orders.

The problem with this argument is that γ can only differ from unity if $\Delta\mu^2$ is a non-analytic function of $m^2 \equiv \Gamma^{(2)}(0)$. And if $\Delta\mu^2$ is such a function, it would be surprising if it were not also a non-analytic function of $\hat{g} \equiv \Gamma^{(4)}(0)/m^\epsilon$. Moreover, the curly bracket in (10.14) must diverge as $m \to 0$ in order that $\Delta\mu^2$ can tend to zero more slowly than m^2, and this behaviour is incompatible with its being an analytic function of $\hat{g}$ which tends to a well-defined limit as $m \to 0$. Now $\Delta\mu^2$ can be expanded in integer powers of $\hat{g}$ only if it is an analytic function of $\hat{g}$. So we should not be surprised if strange results emerge from a calculation which simply *assumes* that $\Delta\mu^2$ can be expanded in powers of $\hat{g}$. The key to our derivation of a non-trivial value of γ was to stick with power series in $\hat{\lambda}$ until we had an expression for $\partial \log \Delta\mu^2 / \partial \log m^2$, which doesn't do anything funny in the limit $m \to 0$ and so can plausibly be expanded in powers of $\hat{g}$.

Another interesting result of the replacement of $\hat{\lambda}$ by $\hat{g}$ is that our expansion of the logarithmic derivative which gives us γ in powers of $\hat{g}$ now contains contributions from all orders in $\hat{\lambda}$ as we take the limit $m \to 0$. It is typical of the perturbative theory of phase transitions that we can only get results for quantities defined actually at the critical temperature by including terms to all orders in the bare coupling constant. This does not mean that our calculation incorporates all the terms in the expansion in powers of $\hat{\lambda}$. Far from it. We have here performed only the simplest one-loop calculation, truncating our series at first order in $\hat{g}$. In effect the calculation corresponds to a partial summation to all orders of the series in $\hat{\lambda}$, taking in the dominant terms but leaving others out.

The best value for γ is 1.239 ± 0.003, so our new estimate, $\gamma = 1.5$, brings us no nearer to the true result than did Landau theory. However, the shift is in the right direction and the value $0.0675\ldots$ of the expansion parameter $\hat{g}_c$ is small enough that we might expect the series to converge quite fast, the next term being a good deal smaller than the first correction. In Chapter 12 we develop some powerful tools which reduce the tedium of these calculations. For now, let us press on to another exponent.

10.3 The calculation of η

In §6.7.1, we showed that the behaviour of $\widetilde{G}_c^{(2)}(k)$ at small wavevectors was related to the exponent η thus (cf. equation (6.81)):

$$\widetilde{G}_c^{(2)}(k) \sim k^{-2+\eta}, \qquad k \to 0,\, T = T_c. \tag{10.35}$$

In fact, for the Landau–Ginzburg model, it is more convenient to work with the vertex function $\Gamma^{(2)}(k)$ which is the reciprocal of $\widetilde{G}_c^{(2)}(k)$ (see equation (8.43)), so that

$$\Gamma^{(2)}(k) \sim k^{2-\eta}, \qquad k \to 0,\, T = T_c. \tag{10.36}$$

Taking logarithms of both sides of this equation and differentiating, we find that

$$2 - \eta = \lim_{k \to 0} \frac{\partial \log \Gamma^{(2)}}{\partial \log k}. \tag{10.37}$$

To one-loop order, $\Gamma^{(2)}(k)$ is given by equation (9.3), which says

$$\Gamma^{(2)}(k) = m^2 + a^2 k^2 + O(\lambda^2). \tag{10.38}$$

At the critical temperature $m = 0$ and so, comparing with (10.36), we have

$$\eta = 0, \tag{10.39}$$

to this order for all values of d. This is not a particularly interesting result—it is just the same as in Landau theory and shows no anomalous behaviour at the critical temperature. The reason is that the $O(\lambda)$ term in the expansion of $\Gamma^{(2)}$ (the bubble diagram) is independent of k, so it is not going to add any interesting k behaviour to this vertex function at small k or indeed anywhere. To get a non-Gaussian result for the exponent η, we must include the next term in the expansion, the two-loop term. This we also calculated in Chapter 9. Equation (9.9) gives us

$$\Gamma^{(2)}(k) = m^2 + a^2 k^2 - \tfrac{1}{6}\lambda^2 \Delta A(k, m) + O(\lambda^3), \tag{10.40}$$

where

$$\Delta A(k, m) = \int_0^\Lambda \frac{d^d \check{q}_1 d^d \check{q}_2}{(a^2 q_1^2 + m^2)(a^2 q_2^2 + m^2)} \left[\frac{1}{a^2 (\mathbf{Q} - \mathbf{k})^2 + m^2} - \frac{1}{a^2 Q^2 + m^2} \right] \tag{10.41}$$

with $\mathbf{Q} \equiv \mathbf{q}_1 + \mathbf{q}_2$. We are interested in the behaviour of $\Gamma^{(2)}$ at the critical temperature, so we can straight away set $m = 0$. The result we write as

$$\Gamma^{(2)}(k) = a^2 k^2 \left[1 - \frac{\lambda^2}{6 a^2 k^2} \Delta A(k, 0) \right] + O(\lambda^3), \tag{10.42}$$

and take logarithms to get

$$\log \Gamma^{(2)}(k) = 2 \log(ak) - \frac{\lambda^2}{6 a^2 k^2} \Delta A(k, 0) + O(\lambda^3). \tag{10.43}$$

Changing to the dimensionless variables $\mathbf{x}_1$, $\mathbf{x}_2$ and $\mathbf{X}$, defined by

$$\mathbf{x}_1 \equiv \frac{\mathbf{q}_1}{k}, \quad \mathbf{x}_2 \equiv \frac{\mathbf{q}_2}{k}, \quad \mathbf{X} \equiv \frac{\mathbf{Q}}{k}, \tag{10.44}$$

10.3 The calculation of η

we can write this as

$$\log \Gamma^{(2)}(k) = 2\log(ak) - \tfrac{1}{6}\hat{\lambda}^2 \int_0^{\Lambda/k} \frac{d^d\mathbf{x}_1 d^d\mathbf{x}_2}{x_1^2 x_2^2}\left[\frac{1}{(\mathbf{X}-\hat{\mathbf{k}})^2} - \frac{1}{X^2}\right] + \mathrm{O}(\hat{\lambda}^3). \tag{10.45}$$

Here $\hat{\mathbf{k}}$ is the unit vector in the direction of $\mathbf{k}$ and $\hat{\lambda}$ has changed its definition since the last time we used the symbol; it is now

$$\hat{\lambda} \equiv \frac{\lambda(ak)^{-\epsilon}}{(2\pi a)^d}. \tag{10.46}$$

Similar dimensional arguments apply to this expansion as did to the one we used for calculating γ. Here we have expressed all our wavevectors as multiples of the magnitude of the external wavevector k, rather than as multiples of the mass m, but we still, by the same arguments, end up with an expansion in powers of the new dimensionless coupling constant $\hat{\lambda}$, with coefficients given by integrals over dimensionless variables.

10.3.1 $d = 4$ and above

As with the calculation of γ the argument here takes different directions, depending on whether we are below the critical dimension or not. And as with γ, let us look first at the case $d \geq 4$.

It was demonstrated in Box 9.1 that the integral in equation (10.45) depends on its upper limit as $(\Lambda/k)^{-2\epsilon}$. Thus, we can write it in the form

$$I = A + B\left(\frac{\Lambda}{k}\right)^{-2\epsilon}, \tag{10.47}$$

plus terms that become negligible as $\Lambda/k \to \infty$. A and B are constants independent of k. We now substitute this into (10.45) and perform the derivative (10.37) to find η. We employ the formula

$$k\frac{\partial \hat{\lambda}^n}{\partial k} = -n\epsilon \hat{\lambda}^n, \tag{10.48}$$

which is derived from (10.46) in exactly the same way as equation (10.17) was from (10.9), and we find

$$\eta = \tfrac{1}{3}\epsilon A \lim_{k\to 0} \hat{\lambda}^2. \tag{10.49}$$

As before, with ϵ being positive for $d > 4$, we get $\hat{\lambda} \to 0$ when we take the limit, and η vanishes. This argument does not apply actually at $d = 4$, but at $d = 4$, $\epsilon = 0$, so η vanishes anyway. Either way then, we find that

$$\eta = 0 \tag{10.50}$$

at four dimensions and above, to this order.[5] Again, this is the same result that we found using both mean-field theory (§6.7) and Landau theory (§7.2).

---

[5] Actually this result is known to extend to all orders; η is exactly 0 for all $d > 4$.

10.3.2 Below four dimensions

Below the critical dimension, the integral in equation (10.45) is only weakly dependent on its upper bound and we can approximate it by setting Λ to infinity. (Again this becomes exact in the limit $k \to 0$, so there is no approximation involved in the calculation of the critical exponent.) The derivative in equation (10.37) is then easily performed with the aid of (10.48), giving

$$\eta = \tfrac{1}{3}\epsilon \lim_{k \to 0} \hat{\lambda}^2 \int_0^\infty \frac{d^d\mathbf{x}_1 d^d\mathbf{x}_2}{x_1^2 x_2^2} \left\{ \frac{1}{X^2} - \frac{1}{(\mathbf{X} - \hat{\mathbf{k}})^2} \right\} + O(\hat{\lambda}^3). \qquad (10.51)$$

The parallel with the calculation of γ should by now be apparent, and it should be clear that we have exactly the same problem now as we did then; we cannot take the limit in equation (10.51), because with ϵ positive below $d = 4$, the factor of $k^{-\epsilon}$ in the definition of $\hat{\lambda}$, equation (10.46), will diverge. As before, our whole expansion is in powers of $\hat{\lambda}$, so it is the very quantity that we are expanding in that tends to infinity as we take the limit $k \to 0$. What we do here is also exactly analogous to the one we used in §10.2—we change variables to the renormalized coupling g, though we now shift the renormalization point away from the origin thus:

$$g \equiv \Gamma^{(4)}(\boldsymbol{\kappa}_1, \boldsymbol{\kappa}_2, \boldsymbol{\kappa}_3)\big|_{\mathrm{SP}}. \qquad (10.52)$$

The reason for shifting the renormalization point is that we wish to set $m = 0$ in our calculation, and that would give infrared-divergent integrals in the calculation of g below if all the $\boldsymbol{\kappa}_i$ were also zero.

The notation 'SP' in equation (10.52) means that the wavevectors $\boldsymbol{\kappa}_1$, $\boldsymbol{\kappa}_2$, $\boldsymbol{\kappa}_3$ are chosen to satisfy the condition[6]

$$\boldsymbol{\kappa}_i \cdot \boldsymbol{\kappa}_j = (4\delta_{ij} - 1)\frac{\kappa^2}{4}, \qquad (10.53)$$

which in turn implies that

$$(\boldsymbol{\kappa}_i + \boldsymbol{\kappa}_j)^2 = \kappa^2. \qquad (10.54)$$

A point satisfying the condition (10.53) is called a **symmetry point**.

We are in theory free to choose any value we like for the scalar quantity κ, but it turns out that in order to get a well-behaved series relating the bare and renormalized coupling constants, we must choose κ proportional to k. The simplest and commonest choice is $\kappa = k$.

Equation (10.53) leaves considerable latitude in the vectors' directions. But this freedom is not a problem; any set of wavevectors satisfying (10.53)

---

[6] It is not possible to fulfil this condition in two dimensions, for which we have to make another choice of renormalization point.

10.3 The calculation of η

will do for our purposes. In three dimensions for example, (10.53) is satisfied by choosing the vectors all to have length $\kappa\sqrt{3/4}$ and to point towards three of the corners of a regular tetrahedron, though this still leaves the vectors a good deal of rotational freedom in their values.

Choosing the wavevectors to be at a symmetry point like this is to a certain extent arbitrary, but most writers make the same choice as we have because it makes the integrals easy.

We can express λ in terms of this new coupling constant by inverting equation (9.17). Because we have chosen the wavevectors at a symmetry point, the three terms in the integrand in equation (9.17) all contribute the same amount to the integral. It then follows that

$$\lambda = g + \tfrac{3}{2}g^2 \int_0^\Lambda \frac{d^d \check{q}}{a^4 q^2 (\kappa_1 + \kappa_2 - \mathbf{q})^2} + O(g^3). \tag{10.55}$$

Changing to dimensionless variables in the usual way and dividing throughout by $(2\pi a)^d$, we then get

$$\hat{\lambda} = \hat{g} + \tfrac{3}{2}\hat{g}^2 \int_0^{\Lambda/\kappa} \frac{d^d \mathbf{x}}{x^2 (\hat{\kappa}_1 + \hat{\kappa}_2 - \mathbf{x})^2} + O(\hat{g}^3), \tag{10.56}$$

where

$$\hat{g} \equiv \frac{g(a\kappa)^{-\epsilon}}{(2\pi a)^d}, \tag{10.57}$$

and

$$\mathbf{x} \equiv \frac{\mathbf{q}}{\kappa}; \quad \hat{\kappa}_1 \equiv \frac{\kappa_1}{\kappa}; \quad \hat{\kappa}_2 \equiv \frac{\kappa_2}{\kappa}. \tag{10.58}$$

Note that $\hat{\kappa}_1$ and $\hat{\kappa}_2$ are not unit vectors. In three dimensions, for example, they have length $\sqrt{3/4}$.

Below $d = 4$, the integral in equation (10.56) is insensitive to its upper limit, so we can approximate it by its value as $\Lambda/\kappa \to \infty$. The equation then tells us that to the order we are interested in (which is two loops, equivalent to $O(\hat{g}^2)$—see equation (10.5)) we can, as before, simply replace $\hat{\lambda}$ by $\hat{g}$ in equation (10.51). Taking the limit, we then have

$$\eta = \tfrac{1}{3}\epsilon \hat{g}_c^2 \int_0^\infty \frac{d^d \mathbf{x}_1 d^d \mathbf{x}_2}{x_1^2 x_2^2} \left\{ \frac{1}{X^2} - \frac{1}{(\mathbf{X} - \hat{\mathbf{k}})^2} \right\} + O(\hat{g}_c^3), \tag{10.59}$$

where $\hat{g}_c$ is the limiting value of $\hat{g}$ as κ goes to zero. We find this limiting value in a way exactly analogous to the method we employed in §10.2. We define a new β-function thus:[7]

$$\beta(\hat{g}) \equiv \frac{\partial \hat{g}}{\partial \log \kappa}. \tag{10.60}$$

---

[7] This function will play an important rôle in the developments of the next three chapters—it appears as one of the coefficients in the 'renormalization group' equation of Chapter 11 which in turns leads to some of the most important and general results in the theory of critical phenomena.

266   Chapter 10: The calculation of critical exponents for $T \geq T_c$

As before this function will have a zero at the critical value of $\hat{g}$. We work out $\beta(\hat{g})$ exactly as we did in §10.2 as well. We write down the inverse of equation (10.56) thus:

$$\hat{g} = \hat{\lambda} - \tfrac{3}{2}\hat{\lambda}^2 \int_0^\infty \frac{d^d \mathbf{x}}{x^2(\hat{\boldsymbol{\kappa}}_1 + \hat{\boldsymbol{\kappa}}_2 - \mathbf{x})^2} + O(\hat{\lambda}^3), \tag{10.61}$$

and differentiate it using (10.48) to get

$$\beta(\hat{g}) = -\epsilon\hat{\lambda} + 3\epsilon\hat{\lambda}^2 \int_0^\infty \frac{d^d \mathbf{x}}{x^2(\hat{\boldsymbol{\kappa}}_1 + \hat{\boldsymbol{\kappa}}_2 - \mathbf{x})^2} + O(\hat{\lambda}^3). \tag{10.62}$$

Using equation (10.56) to eliminate $\hat{\lambda}$ in favour of $\hat{g}$ we find

$$\beta(\hat{g}) = -\epsilon\hat{g} + \tfrac{3}{2}\epsilon\hat{g}^2 \int_0^\infty \frac{d^d \mathbf{x}}{x^2(\hat{\boldsymbol{\kappa}}_1 + \hat{\boldsymbol{\kappa}}_2 - \mathbf{x})^2} + O(\hat{g}^3). \tag{10.63}$$

As before, there is one trivial (Gaussian) zero at $\hat{g} = 0$ and another, which is the one we are after, at

$$\hat{g}_c = \tfrac{2}{3} \left[\int_0^\infty \frac{d^d \mathbf{x}}{x^2(\hat{\boldsymbol{\kappa}}_1 + \hat{\boldsymbol{\kappa}}_2 - \mathbf{x})^2} \right]^{-1}. \tag{10.64}$$

The integral here is one that can be done. In three dimensions, for example, we can perform it as follows. We write the bracket in the denominator as

$$(\hat{\boldsymbol{\kappa}}_1 + \hat{\boldsymbol{\kappa}}_2 - \mathbf{x})^2 = (\hat{\boldsymbol{\kappa}}_1 + \hat{\boldsymbol{\kappa}}_2)^2 - 2(\hat{\boldsymbol{\kappa}}_1 + \hat{\boldsymbol{\kappa}}_2) \cdot \mathbf{x} + x^2. \tag{10.65}$$

We work in spherical polar coordinates and choose the polar axis to lie in the direction of $\hat{\boldsymbol{\kappa}}_1 + \hat{\boldsymbol{\kappa}}_2$. Noting that equation (10.54) implies $|\hat{\boldsymbol{\kappa}}_1 + \hat{\boldsymbol{\kappa}}_2| = 1$, we then have

$$(\hat{\boldsymbol{\kappa}}_1 + \hat{\boldsymbol{\kappa}}_2 - \mathbf{x})^2 = 1 - 2x\cos\theta + x^2. \tag{10.66}$$

Substituting back into (10.64), we get an integral that can readily be performed and gives

$$\hat{g}_c = \frac{2}{3\pi^3} = 0.0215\ldots \tag{10.67}$$

This is not the same value as the $\hat{g}_c$ in (10.33), but neither should we expect it to be; though we have used the same symbol for the critical value of the renormalized dimensionless coupling constant as we did in the last section, it is defined completely differently. In the former case we were examining

10.3 The calculation of η

Box 10.1: The integral of equation (10.59) for $d = 3$

We can perform the first half of the inner integral of (10.59) with $d = 3$ by choosing axes such that $\mathbf{x}_1 = x_1 \hat{\mathbf{z}}$ and by defining $\mathbf{y} \equiv \mathbf{x}_2/x_1$:

$$\int \frac{\mathrm{d}^3 \mathbf{x}_2}{x_2^2 |\mathbf{x}_1 + \mathbf{x}_2|^2} = \frac{2\pi}{x_1} \int_0^\infty \mathrm{d}y \int_{-1}^1 \frac{\mathrm{d}\cos\theta_2}{1 + 2y\cos\theta_2 + y^2} \quad (1)$$

$$= \frac{2\pi}{x_1} \int_0^\infty \frac{\mathrm{d}y}{y} \log\left(\frac{|1+y|}{|1-y|}\right).$$

By Taylor-expanding the logarithm in powers of y, it is straightforward to show that the integrand tends to a constant as $y \to 0$. Making the substitution $y \to u = 1/y$ demonstrates that the integral also converges for large y. In fact, it has value $\pi^2/2$. Integrating the entire first integral of (10.59) over the region $a \leq x_1 \leq b$, we therefore find

$$\int_{a \leq x_1 \leq b} \frac{\mathrm{d}^3 \mathbf{x}_1 \mathrm{d}^3 \mathbf{x}_2}{x_1^2 x_2^2 X^2} = 4\pi^4 \int_a^b \frac{\mathrm{d}x_1}{x_1}. \quad (2)$$

The other half of the inner integral of (10.59) is the same as (1) with the substitution $\mathbf{x}_1 \to \mathbf{x}_1 - \hat{\mathbf{k}}$. Thus we have to evaluate

$$\int_{a \leq x_1 \leq b} \frac{\mathrm{d}^3 \mathbf{x}_1 \mathrm{d}^3 \mathbf{x}_2}{x_1^2 x_2^2 |\mathbf{X} - \hat{\mathbf{k}}|^2} = 2\pi^4 \int_a^b \mathrm{d}x_1 \int_{-1}^1 \frac{\mathrm{d}\cos\theta_1}{|\mathbf{x}_1 - \hat{\mathbf{k}}|}$$

$$= 2\pi^4 \int_a^b \mathrm{d}x_1 \int_{-1}^1 \frac{\mathrm{d}\cos\theta_1}{\sqrt{x_1^2 - 2x_1\cos\theta_1 + 1}} \quad (3)$$

$$= 2\pi^4 \int_a^b \frac{\mathrm{d}x_1}{x_1} \big(|x_1 + 1| - |x_1 - 1|\big).$$

Subtracting (3) from (2) yields

$$\int \frac{\mathrm{d}^3 \mathbf{x}_1 \mathrm{d}^3 \mathbf{x}_2}{x_1^2 x_2^2} \left(\frac{1}{X^2} - \frac{1}{|\mathbf{X} - \hat{\mathbf{k}}|^2}\right) = 2\pi^4 \int_a^b \frac{\mathrm{d}x}{x}(2 - |x+1| + |x-1|)$$

$$= 4\pi^4 \big[\ln x - x\big]_a^1 \quad \text{for } b > 1.$$

Since it is independent of b, this expression is well-behaved as we let $b \to \infty$, but it diverges logarithmically in the limit $a \to 0$.

---

the limiting behaviour as $m \to 0$ with $k = 0$ and in the latter, as $k \to 0$ with $m = 0$.

We now have all that we need to calculate η to order two loops. The only remaining task is to evaluate the integral appearing in equation (10.59). In Box 10.1 we do this for the case $d = 3$. But here a disaster occurs! The integral diverges. After all this work, we do not actually get an answer for η. This seems incomprehensible. Surely we have not put a foot wrong in our derivation? Can there be any reason why this integral should diverge, unless η itself is divergent? And yet we know experimentally that the value of η is perfectly finite, and indeed quite small. As we will show in the following section, this problem is solved by switching once more to a different expansion parameter. It turns out that the best way to calculate the properties of the Landau–Ginzburg model at the critical temperature is to make an 'ϵ-expansion'.

10.4 The ϵ-expansion

Feynman series are expansions of vertex functions in integral powers of a (bare or renormalized) coupling constant. Unfortunately, the essence of critical phenomena is that near T_c quantities of physical interest are non-analytic functions of one another—this non-analyticity is essential if the relations between them are to involve non-integer exponents. In §10.2, where $\Delta\mu^2$ was evaluated as a function of m^2, we saw how easily one can be seduced into obtaining erroneous results by assuming that one quantity of interest can be expanded in integral powers of another, i.e., by assuming that two quantities are analytic functions of one another. In §10.2 a non-trivial value of γ was obtained by postponing the change of expansion parameter from λ to g until the logarithmic derivative which tends to γ was in hand. The key point was to introduce an expansion in powers of $\hat{g}$ only when dealing with a quantity which is well-behaved in the limit $T \to T_c$, and in the closely related limit $\hat{g} \to \hat{g}_c$.

Now it is not clear that every critical exponent can be expressed as the limit as $T \to T_c$ of some well-behaved quantity. Consider, for example, the exponent δ, defined by the relation $m \propto B^{1/\delta}$ at $T = T_c$. (m is the magnetization here, of course, not the reciprocal of the susceptibility.) Precisely at $T = T_c$, $\delta = \mathrm{d}\log B/\mathrm{d}\log m$. But it is easy to see that δ cannot be obtained as the limit as $T \to T_c$ of any smoothly evolving slope of the curve $m(B)$. For $T > T_c$, B and m are proportional, so the required logarithmic derivative is fixed at unity, while for $T < T_c$ a glance at Figure 1.7 shows that the curve $m(B)$ has infinite slope at $B = 0$, and unit slope for B slightly above zero, no matter how close T is to T_c. There simply isn't any location where we can evaluate $\mathrm{d}\log B/\mathrm{d}\log m$ and watch it tend smoothly to δ as T tends to T_c.

The ϵ-**expansion** is a cunning device for making critical exponents into analytic functions of a variable that can be gradually increased from a value at which they and all their derivatives with respect to this variable can be evaluated. The magic variable is the dimensionality of the system, and the idea behind the expansion is as follows.

So far we have sought to expand everything as a function of the coupling constant. The experiment we imagine is one in which the coupling is

10.4 The ϵ-expansion

increased from zero, when the system will be above its critical temperature, until it reaches its critical value. Our attention was attracted to the problem of critical phenomena in the first place because weird, non-analytic things happen as the system passes through criticality, so it should not surprise us that many quantities of interest cannot be Taylor expanded in powers of $\hat{g}$ near $\hat{g}_c$. There is another expansion however, which meets with more success.

In this expansion, instead of simply varying the coupling constant $\hat{g}$, we consider what happens as we vary the system's dimensionality d continuously through both integer and non-integer values. Of course, this is not an experiment that we can actually perform—we cannot construct a system in which $d = 3.14159\ldots$, for example—but mathematically it is fairly straightforward to formulate the theory in non-integer dimensions, as we will shortly demonstrate. For each different value of d there will be a different critical coupling $\hat{g}_c$, and as we vary d, we also vary $\hat{g}$ so that our system is always at the critical point, $\hat{g} = \hat{g}_c$.

When we do this, we find that the deviations in the critical exponents from their Landau-theory values are zero for $d > 4$, then as d is reduced below 4, they increase smoothly towards the values we are really interested in, namely those at $d = 3$ and $d = 2$. The emphasis here is on the word smoothly. Whatever horrid non-analytic behaviour we would observe if we allowed T to stray from T_c, all quantities evolve as smooth functions of d so long as we hold the coupling at its critical value. And smooth functions can usually be Taylor expanded over a decent interval[8] so there is a good prospect that the Taylor series expansions of the critical exponents in d about $d = 4$ remain valid at $d = 3$, and possibly even at $d = 2$.

So the plan is this. We evaluate all Feynman integrals in arbitrary dimension d. This step makes every integral into a function of $\epsilon \equiv 4 - d$. Now we use this technology to express $\hat{g}_c$ as a Taylor series in ϵ and replace $\hat{g}$ in all Feynman series with this Taylor series. At this stage every vertex function has been expressed as a Taylor series in ϵ, which sums (we trust) to the value taken by that vertex function at criticality in dimension $d = 4 - \epsilon$. Taylor series for the critical exponents η and δ defined at $T = T_c$ can now be obtained as power series in ϵ by differentiating these vertex functions with respect to their other arguments.

The evaluation of integrals in non-integer dimensions also has another very desirable effect. It makes them easier to perform. This may appear strange at first, for surely it must be easier to perform an integral at one particular value of d than it is at general non-integer d (whatever that means). This is true, but nonetheless, via a clever line of argument, it turns out that this continuation of the integrals in d actually allows us to perform ones which hitherto would have defeated us. The technique is known as **dimensional regularization** and was first developed in the context of quantum field

---

[8] A counter-example is $f(x) = 1/(1+x^2)$, which is wonderfully smooth over the whole real line but can be Taylor expanded about the origin only for $x < 1$.

theory by 't Hooft and Veltman (1972).[9] We will examine this technique first, since the results we derive will lead us directly to the ϵ-expansion.

10.4.1 Dimensional regularization

In Chapter 9 we expressed most Λ-sensitive integrals as sums of some Λ-sensitive quantity such as m^2 and the difference of two integrals whose Λ-sensitivity cancels. The integral ΔA of equation (9.9), which is essentially the integral of equation (10.59), is a typical specimen of the type of integral this process confronts us with. It is a much harder integral to do than either of the integrals whose difference it is, though it does have the advantage of Λ-insensitivity, which allows us to set the upper limit to infinity. Dimensional regularization renders the individual, previously Λ-sensitive integrals which we are subtracting from one another also insensitive to Λ. These regularized integrals are comparatively simple to evaluate, and when we subtract them, one from the other, we get the value of the original difficult integral over the differenced integrands.

In essence the procedure is this. Our integral is of the form

$$I_0(d) \equiv \int \mathrm{d}^d\mathbf{k} \big[f_1(\mathbf{k}) - f_2(\mathbf{k}) \big]. \tag{10.68}$$

For the value of d that we are interested in, we assume that the integral converges at large k because in this limit f_1 and f_2 cancel more and more exactly, making the integrand tend rapidly to zero. However, the algebra involved in subtracting $f_2(\mathbf{k})$ from $f_1(\mathbf{k})$ can be heavy, and the resulting expression exceedingly hard to integrate.

For the dimension of interest, d_p, the integrals $I_1(d) \equiv \int \mathrm{d}^d\mathbf{k} f_1(\mathbf{k})$ and $I_2(d) \equiv \int \mathrm{d}^d\mathbf{k} f_2(\mathbf{k})$ make no sense because $f_1(\mathbf{k})$ and $f_2(\mathbf{k})$ do not vanish sufficiently fast for large k. But these individual integrals may well be evaluable for small d. Dimensional regularization involves:

(i) defining the integrals $I_i(d)$ for a range of values of the real variable d,
(ii) analytically continuing the resulting functions $I_i(d)$ to give functions which are defined at d_p, and finally
(iii) calculating the quantity $I_0'(d) \equiv I_1(d) - I_2(d)$.

$I_0'(d)$ is an analytic function, as was the original integral $I_0(d)$. Moreover, for all those values of d for which the integrals underlying I_1 and I_2

---

[9] To a field-theorist, dimensional regularization is much more than just a trick for evaluating hard integrals. In field theory, where there is no known wavevector cut-off like Λ, there is no mechanism that renders otherwise divergent integrals finite *ab initio* and so one is forced to invent one. One could regularize the integrals as in statistical mechanics by introducing a wavevector cut-off, but for gauge field theories this approach is unsatisfactory since it breaks the gauge symmetry. In this situation dimensional regularization is one of the simplest techniques for rendering the integrals finite in a controlled way, preserving the gauge symmetry and allowing the calculation to proceed. For statistical physicists however, none of this matters.

10.4 The ϵ-expansion

are defined, we clearly have $I_0' = I_0$. But two analytic functions can be equal over a non-zero range only if they are everywhere equal. So we have $I_0'(d_p) = I_0(d_p)$ also, and thus the integral of physical interest can be evaluated by evaluating the simpler integrals I_1 and I_2 for small, but arbitrary d.

10.4.2 Calculating γ by dimensional regularization

To show how dimensional regularization works in practice, let us take first a problem for which we already know the answer, and see how it may be tackled using the new technique. The problem we take is the calculation of γ to one loop, which we performed directly in §10.2.

Consider then equation (10.26) for the critical exponent γ to one-loop order. Looking back over equations (9.33) and (10.14), you will see that the integral here arises from the difference of two Λ-sensitive integrals, which we have run together into one to make an integral which is insensitive to Λ for $d < 4$. Undoing this process,[10] we now write $1/\gamma$ in terms of the two separate original integrals thus:

$$\frac{1}{\gamma} = 1 + \tfrac{1}{4}\epsilon \hat{g}_c \lim_{u \to 0} \left[\int_0^\infty \frac{\mathrm{d}^d \mathbf{x}}{x^2 + 1} - \int_0^\infty \frac{\mathrm{d}^d \mathbf{x}}{x^2 + u^2} \right] + \mathrm{O}(\hat{g}_c^2)$$

$$= 1 + \tfrac{1}{4}\epsilon \hat{g}_c \lim_{u \to 0} [I_1(1) - I_1(u)] + \mathrm{O}(\hat{g}_c^2), \qquad (10.69)$$

where

$$I_1(u) \equiv \int_0^\infty \frac{\mathrm{d}^d \mathbf{x}}{x^2 + u^2}. \qquad (10.70)$$

Notice the upper limit here. We have set it to infinity. Above $d = 2$ this is only a formal notation. We cannot actually set the limit to infinity like this because $I_1(u)$ would diverge for $d \geq 2$. Nonetheless, the difference of the two integrals in equation (10.69) will converge if the limits $\Lambda \to \infty$ are taken simultaneously in both. This is tantamount to the procedure of §10.2. Here though, we take a different course. The integral $I_1(u)$ converges if the dimension is less than two and if $u > 0$. For a real physical system of course this means that the dimension is one. But we can generalize the idea of the dimension to non-integer values quite simply. We can write $I_1(u)$ as

$$I_1(u) = \Omega_{d-1} \int_0^\infty \frac{x^{d-1} \mathrm{d}x}{x^2 + u^2}. \qquad (10.71)$$

---

[10] It may occur to the reader that there is a certain similarity between this splitting of the integral into several simpler ones, and the trick of 'partial fractions' familiar from elementary calculus courses. Dimensional regularization is in a sense just a slightly more sophisticated version of this old trick, which allows us to go ahead and form partial fractions, even though the integrals to which they give rise do not converge.

Everything in this expression generalizes straight away to the case of non-integral d, except for Ω_{d-1}, which, if you recall, is the surface area of the unit sphere in $d-1$ dimensions. In Problem 3.4 it was shown that for d a positive integer, Ω_{d-1} is given by

$$\Omega_{d-1} = \frac{2\pi^{d/2}}{(d/2-1)!}. \tag{10.72}$$

The factorial function can be generalized to non-integral d by defining it to be[11]

$$x! \equiv \int_0^\infty dt\, e^{-t} t^x, \tag{10.73}$$

which takes the accepted values when x is a non-negative integer and is an analytic function at all positive values of x. For $x < 0$, the integral diverges, so we define the factorial function in this regime by analytic continuation from positive x instead. This gives a finite function for all x except for simple poles at the points $-1, -2, -3\ldots$ on the real line. If we make this generalization, we have a complete prescription for writing the integral $I_1(u)$ as a function of arbitrary real d. The remaining integral in (10.71) can be performed, and we find that

$$I_1(u) = u^{d-2} \pi^{d/2} (-\tfrac{1}{2}d)!. \tag{10.74}$$

Substituting this expression into equation (10.69), we have

$$\frac{1}{\gamma} = 1 + \tfrac{1}{4}\epsilon \hat{g}_c \pi^{d/2}(-\tfrac{1}{2}d)! \lim_{u \to 0}[1 - u^{d-2}] + O(\hat{g}_c^2). \tag{10.75}$$

Finally, we set $d = 3$, $\epsilon = 1$ and proceed to the limit $u = 0$:

$$\frac{1}{\gamma} = 1 - \tfrac{1}{2}\hat{g}_c \pi^2 + O(\hat{g}_c^2). \tag{10.76}$$

Using equation (10.33) for $\hat{g}_c$, we then find once more that $\gamma = \tfrac{3}{2}$ to this order.

What we have gained by this procedure is that, instead of having to evaluate an integral of the form (10.21), we only need do one like (10.70), although we do have to be able to do it at arbitrary, non-integral dimension d. In this case, both integrals were possible, but as we get onto higher-order calculations, such as the calculation of η to two loops of §10.3, the dimensionally regularized integrals turn out to be simpler, and using the technique makes all the difference between being able to calculate the quantity of interest and getting stuck.

---

[11] Conventionally the notation $x!$ is reserved for integer x and its generalization is written in terms of **Euler's Γ-function**

$$\Gamma(x+1) \equiv \int_0^\infty dt\, e^{-t} t^x = x!.$$

The authors find the extra '+1' in this definition confusing and in the interests of clarity and the elimination of errors have resolved not to use the Γ-function in this book.

10.4 The ε-expansion

10.4.3 Calculating η by dimensional regularization

Let us now see how dimensional regularization can be used in the calculation of η to evaluate the integral of equation (10.59). This calculation is much more complicated than that of the previous section, and what is more we do not expect it to give an answer, since, as we showed in Box 10.1, the integral diverges. It will however lead us directly to the ε-expansion, which *will* give us an answer.

As with the calculation of γ in the last section, we split the integral we want to do into two parts, and write it as

$$\int_0^\infty \frac{d^d\mathbf{x}_1 d^d\mathbf{x}_2}{x_1^2 x_2^2} \left\{ \frac{1}{X^2} - \frac{1}{(\mathbf{X} - \hat{\mathbf{k}})^2} \right\} = I_2(0) - I_2(1), \tag{10.77}$$

where

$$I_2(k) \equiv \int_0^\infty \frac{d^d\mathbf{x}_1 d^d\mathbf{x}_2}{x_1^2 x_2^2 (\mathbf{X} - \mathbf{k})^2}. \tag{10.78}$$

As in the last example, we are not really entitled to set the upper limit of the integral to infinity in this way at $d = 3$, but we can do it at lower dimensions, and we define I_2 at $d = 3$ and above by analytic continuation.

The integral (10.78) is much harder to do than the previous example, equation (10.71). The reasons for this are that it is a double integral over two wavevectors, and that the integrand is not a function only of the magnitudes of the wavevectors, but also of the angles between them and between each of them and the external wavevector $\mathbf{k}$. To perform this integral, we are going to need some tricks. Luckily, we have a few up our sleeve. The most useful of these is the 'Feynman parameterization', due to R. P. Feynman who was something of a whiz with integrals if his autobiography is to be believed (Feynman 1985).

10.4.4 Feynman parameters

One quite general integral which we can perform in arbitrary non-integer dimension is

$$\int_0^\infty \frac{d^d\mathbf{x}}{(x^2 + 2\mathbf{x} \cdot \mathbf{p} + c^2)^\alpha} = \pi^{d/2} \frac{(\alpha - \tfrac{1}{2}d - 1)!}{(\alpha - 1)!} (c^2 - p^2)^{d/2 - \alpha}. \tag{10.79}$$

To see how this result was arrived at, we rewrite the denominator of the integrand using

$$x^2 + 2\mathbf{x} \cdot \mathbf{p} + c^2 = (\mathbf{x} + \mathbf{p})^2 - p^2 + c^2. \tag{10.80}$$

When we make the substitution $\mathbf{x}' = \mathbf{x} + \mathbf{p}$, the integral becomes simply

$$\int_0^\infty \frac{d^d\mathbf{x}'}{(x'^2 + c^2 - p^2)^\alpha} = \Omega_{d-1} \int_0^\infty \frac{dx' \, x'^{d-1}}{(x'^2 + c^2 - p^2)^\alpha}, \tag{10.81}$$

which is a simple, easily-performed scalar integral giving the result quoted above.

Feynman parameterization is a trick for putting any Feynman integral—the integral over k-space of the product of a bunch of propagators—in the form of the integral (10.79), which we can then perform. The trick works like this. From the definition (10.73) of the factorial function, we have, for any real number α

$$(\alpha - 1)! = \int_0^\infty dt\, e^{-t} t^{\alpha-1}. \tag{10.82}$$

Making the replacement $t \to at$ we can then show that

$$\frac{1}{a^\alpha} = \frac{1}{(\alpha-1)!} \int_0^\infty dt\, e^{-at} t^{\alpha-1}, \tag{10.83}$$

where a is another real number. a will be our propagator when we come to use the Feynman parameterization on our Feynman integrals.

Now suppose we have two different pairs of numbers (a_1, α_1) and (a_2, α_2). Then we can write

$$\frac{1}{a_1^{\alpha_1} a_2^{\alpha_2}} = \frac{1}{(\alpha_1-1)!(\alpha_2-1)!} \int_0^\infty dt_1\, dt_2\, e^{-(a_1 t_1 + a_2 t_2)} t_1^{\alpha_1-1} t_2^{\alpha_2-1}. \tag{10.84}$$

Now let us make the substitutions

$$t_1 = s u_1, \qquad t_2 = s u_2. \tag{10.85}$$

u_1 and u_2 are the eponymous Feynman parameters and s is another variable which we will eventually eliminate. We have expressed our two t variables in terms of three new ones, so we have not yet tied down the values of the new variables. To do this we also impose the condition

$$u_1 + u_2 = 1. \tag{10.86}$$

Thus we could have written $t_2 = s(1 - u_1)$ and eliminated u_2 altogether. However, we will keep the notation as it is for the moment because it helps to emphasize the symmetry between the Feynman parameters, and it will make the later generalization of our formulae to more than two pairs of variables (a, α) simple.

Since the integrals in (10.84) are over non-negative values of t_1 and t_2 only, we can cover the entire domain of integration by allowing only non-negative values of s and the Feynman parameters u_1 and u_2. u_1 and u_2 are then constrained by (10.86) to lie in the range 0 to 1. The Jacobian for the change of variables turns out to be simply s, so that (10.84) becomes

$$\frac{1}{a_1^{\alpha_1} a_2^{\alpha_2}} = \frac{1}{(\alpha_1-1)!(\alpha_2-1)!} \int_0^1 du_1\, du_2\, u_1^{\alpha_1-1} u_2^{\alpha_2-1} \delta(u_1 + u_2 - 1)$$
$$\times \int_0^\infty ds\, e^{-(a_1 u_1 + a_2 u_2) s} s^{\alpha_1+\alpha_2-1}. \tag{10.87}$$

10.4 The ϵ-expansion

The integral over s can now be performed using equation (10.83) to give

$$\frac{1}{a_1^{\alpha_1} a_2^{\alpha_2}} = \frac{(\alpha_1 + \alpha_2 - 1)!}{(\alpha_1 - 1)!(\alpha_2 - 1)!} \\ \times \int_0^1 du_1\, du_2\, \delta(u_1 + u_2 - 1) \frac{u_1^{\alpha_1 - 1} u_2^{\alpha_2 - 1}}{(a_1 u_1 + a_2 u_2)^{\alpha_1 + \alpha_2}}. \tag{10.88}$$

This is Feynman's result as we shall use it in this book, although it is in fact a straightforward matter to generalize it to more than two pairs of variables. The general result is that

$$\prod_{i=1}^n \frac{1}{a_i^{\alpha_i}} = \frac{(\sum_i \alpha_i - 1)!}{\prod_i (\alpha_i - 1)!} \int_0^1 du_1 \ldots du_n\, \delta\!\left(\sum_i u_i - 1\right) \frac{\prod_i u_i^{\alpha_i - 1}}{(\sum_i a_i u_i)^{\sum_i \alpha_i}}. \tag{10.89}$$

The proof is left as an exercise for the reader.

One other result we will need in order to perform the integrals over the Feynman parameters u_i is

$$\int_0^1 du\, u^\mu (1-u)^\nu = \frac{\mu!\, \nu!}{(\mu + \nu + 1)!}. \tag{10.90}$$

10.4.5 The calculation of η again

So here is how we do the integral $I_2(k)$ defined in (10.78) for arbitrary non-integer dimension d. First we treat the simpler integral (which forms a part of $I_2(k)$)

$$J(k) \equiv \int_0^\infty \frac{d^d \mathbf{x}}{x^2 (\mathbf{x} - \mathbf{k})^2}, \tag{10.91}$$

using the Feynman parameterization, equation (10.88), with

$$a_1 = x^2, \quad a_2 = (\mathbf{x} - \mathbf{k})^2, \quad \alpha_1 = \alpha_2 = 1, \tag{10.92}$$

to give

$$J(k) = \int_0^1 du_1\, du_2\, \delta(u_1 + u_2 - 1) \int_0^\infty \frac{d^d \mathbf{x}}{(x^2 u_1 + (\mathbf{x} - \mathbf{k})^2 u_2)^2} \\ = \int_0^1 du_2 \int_0^\infty \frac{d^d \mathbf{x}}{(x^2 - 2\mathbf{x} \cdot \mathbf{k} u_2 + k^2 u_2)^2}. \tag{10.93}$$

Now we use equation (10.79) with $\mathbf{p} = -\mathbf{k} u_2$, $c^2 = k^2 u_2$ and $\alpha = 2$ to get

$$J(k) = \pi^{d/2} k^{-\epsilon} (\tfrac{1}{2}\epsilon - 1)! \int_0^1 du_2 [u_2(1 - u_2)]^{-\epsilon/2} \\ = \pi^{d/2} k^{-\epsilon} \frac{(\tfrac{1}{2}\epsilon - 1)! [(-\tfrac{1}{2}\epsilon)!]^2}{(1 - \epsilon)!}, \tag{10.94}$$

where we have in the second line employed (10.90).

We make use of this result to perform the integral over $\mathbf{x}_2$ in equation (10.78) by putting $\mathbf{k} \to \mathbf{k} - \mathbf{x}_1$ and $\mathbf{x} \to \mathbf{x}_2$ to get

$$I_2(k) = \pi^{d/2} \frac{(\tfrac{1}{2}\epsilon - 1)![(-\tfrac{1}{2}\epsilon)!]^2}{(1-\epsilon)!} \int_0^\infty \frac{d^d \mathbf{x}_1}{x_1^2[(\mathbf{x}_1 - \mathbf{k})^2]^{\epsilon/2}}. \qquad (10.95)$$

The last integral here we can perform by exactly the same technique as we used for $J(k)$, employing first the Feynman parameterization and then equation (10.79). Making use of equation (10.90) once more, the result is

$$\begin{aligned} I_2(k) &= (k^2)^{1-\epsilon} \pi^d \frac{(\epsilon - 2)![(-\tfrac{1}{2}\epsilon)!]^2}{(1-\epsilon)!} \int_0^1 du_2\, u_2^{-\epsilon/2}(1-u_2)^{1-\epsilon} \\ &= (k^2)^{1-\epsilon} \pi^d \frac{[(-\tfrac{1}{2}\epsilon)!]^3(\epsilon - 2)!}{(2 - \tfrac{3}{2}\epsilon)!}. \end{aligned} \qquad (10.96)$$

We must be rather careful how we treat this expression when $d = 3$, because the factorial $(\epsilon - 2)!$ diverges as $\epsilon \to 1$. The quantity we want to calculate is the difference $I_2(0) - I_2(1)$, which in the absence of this divergence we might expect to vanish at $d = 3$ because $I_2(k) \sim k^{2(1-\epsilon)}$ would then be k-independent in the limit $\epsilon \to 1$. To see which of these two effects dominates, we consider the more general difference $I_2(k) - I_2(k')$ and expand about $\epsilon = 1$. We write

$$(k^2)^{1-\epsilon} = e^{(1-\epsilon)\log k^2} = 1 + (1-\epsilon)\log k^2 + O[(1-\epsilon)^2]. \qquad (10.97)$$

The factorial function $(\epsilon - 2)!$ can be Laurent expanded about the pole at $\epsilon = 1$ using equation (8) of Box 10.2 with $x = \epsilon - 1$:

$$(\epsilon - 2)! = \frac{1}{\epsilon - 1} + O(1). \qquad (10.98)$$

Close to $\epsilon = 1$, we then have

$$I_2(k) - I_2(k') = 2\pi^d \frac{[(-\tfrac{1}{2}\epsilon)!]^3}{(2 - \tfrac{3}{2}\epsilon)!}\left[\log \frac{k}{k'} + O(1-\epsilon)\right]. \qquad (10.99)$$

Taking the limit $\epsilon \to 1$, all the higher terms in this expansion vanish, and we get

$$I_2(k) - I_2(k') = 4\pi^4 \log \frac{k}{k'}, \qquad (10.100)$$

which is finite as long as $k' \neq 0$, but diverges as $k' \to 0$. At first sight this may appear to conflict with the result of Box 10.1 for the same integral, but the quantities calculated in the two cases are not the same. In this calculation

10.4 The ϵ-expansion

Box 10.2: Expansions of $x!$ about integers and half-integers

Below we give the Taylor and Laurent expansions of the factorial function $x!$ about the integers and half-integers from $+2$ to -2. Following each expansion we give its range of validity. These expansions are useful when dimensionally regularizing integrals, and when making ϵ-expansions. In these expressions, γ is Euler's constant

$$\gamma \equiv \lim_{n \to \infty} \left[\sum_{m=1}^{n} \frac{1}{m} - \log n \right] = 0.577\ldots \quad (1)$$

We find

$$(x+2)! = 2 + (3 - 2\gamma)x + O(x^2)$$
where $-3 < x < 3$, (2)

$$(x + \tfrac{3}{2})! = \tfrac{3}{4}\sqrt{\pi}[1 + (\tfrac{8}{3} - \gamma - 2\log 2)x + O(x^2)]$$
where $-2\tfrac{1}{2} < x < 2\tfrac{1}{2}$, (3)

$$(x+1)! = 1 + (1 - \gamma)x + O(x^2)$$
where $-2 < x < 2$, (4)

$$(x + \tfrac{1}{2})! = \tfrac{1}{2}\sqrt{\pi}[1 + (2 - \gamma - 2\log 2)x + O(x^2)]$$
where $-1\tfrac{1}{2} < x < 1\tfrac{1}{2}$, (5)

$$x! = 1 - \gamma x + O(x^2)$$
where $-1 < x < 1$, (6)

$$(x - \tfrac{1}{2})! = \sqrt{\pi}[1 - (\gamma + 2\log 2)x + O(x^2)]$$
where $-\tfrac{1}{2} < x < \tfrac{1}{2}$, (7)

$$(x-1)! = \frac{1}{x} - \gamma + \tfrac{1}{2}[\gamma^2 + \tfrac{1}{6}\pi^2]x + O(x^2)$$
where $-1 < x < 1$, (8)

$$(x - \tfrac{3}{2})! = -2\sqrt{\pi}[1 + (2 - \gamma - 2\log 2)x + O(x^2)]$$
where $-\tfrac{1}{2} < x < \tfrac{1}{2}$, (9)

$$(x-2)! = -\frac{1}{x} - (1 - \gamma) - \tfrac{1}{2}[(1-\gamma)^2 + \tfrac{1}{6}\pi^2 + 1]x + O(x^2)$$
where $-1 < x < 1$. (10)

---

we have worked out the difference $I_2(k) - I_2(k')$, which we find to diverge as $k' \to 0$, whereas in Box 10.1 we calculated the difference $I_2(k) - I_2(0)$ with the integrals evaluated from a minimum wavevector a up to infinity, rather than from zero to infinity. The difference then diverged as $a \to 0$. We have therefore not calculated the same quantity in the two cases. In both cases however, there is an infrared divergence, which persists even though we have renormalized the coupling constant, a procedure which for $T > T_c$ removed all such divergences. Dimensional regularization has not freed us of this problem and we are still exactly where we were at the end of §10.3. How then are we to proceed? The answer, as we explained before, is to make an ϵ-expansion.

10.4.6 Calculation of η by the ϵ-expansion

The idea behind the ϵ-expansion is to expand both the coefficients in the power series for η and the coupling constant $\hat{g}_c$ itself in powers of $\epsilon \equiv 4 - d$, and so change our power series from one in $\hat{g}_c$ to one in ϵ. The power series in ϵ turns out to have better convergence properties than the one in $\hat{g}_c$, and will ultimately give us an answer for η. For simplicity we include only the leading-order terms in the expansion for the moment.

So, to begin, let us see what we get when we expand $\hat{g}_c$ about $d = 4$. To one-loop order, $\hat{g}_c$ is given by equation (10.64). The integral in this equation is of the same form as the integral $J(k)$ of §10.4.5, so it can immediately be performed in arbitrary non-integral dimension, giving

$$\hat{g}_c = \frac{2(1-\epsilon)!}{3\pi^{d/2}[(-\frac{1}{2}\epsilon)!]^2(\frac{1}{2}\epsilon - 1)!} = \frac{1}{3\pi^2}\epsilon + O(\epsilon^2). \tag{10.101}$$

Here we have made use of the fact that at the symmetry point $|\boldsymbol{\kappa}_1 + \boldsymbol{\kappa}_2|$ is always equal to one. When we set $\epsilon = 1$ in the middle expression of (10.101), we get $\hat{g}_c = \frac{2}{3}\pi^{-3} = 0.0215\ldots$, which agrees with equation (10.67), and the $O(\epsilon)$ approximation, $\hat{g}_c = \frac{1}{3}\pi^{-2} = 0.0337\ldots$, is not too far off the mark. Moreover, it turns out that since $\hat{g}_c$ is of leading order ϵ, if we calculate $\beta(\hat{g})$ to higher order and examine the zeros of the function, we will get no more contributions to $\hat{g}_c$ at this order. This point is covered in Problem 10.4, so for the moment, let us take it as read. Since we are only working to leading order here, equation (10.101) then contains all the contributions to $\hat{g}_c$ that we will need.

The fact that $\hat{g}_c$ is of leading order ϵ also means that, in the expansion of any critical exponent in powers of $\hat{g}_c$, all contributions to terms of order $\hat{g}_c^m$ are of order ϵ^m or higher. Thus to calculate the exponent to order ϵ^n, we need only work to order ϵ^{n-m} in any integral that multiplies $\hat{g}_c^m$. As an obvious corollary of this result, we can altogether ignore all diagrams which contribute to our exponent at order higher than $\hat{g}_c^n$.[12] Thus, to leading order in ϵ, equation (10.59) gives us all we need to calculate η.

---

[12] In fact things can be even simpler than this if the expression for the exponent contains an explicit overall factor of some power of ϵ, as in equation (10.59) for η.

10.4 The ϵ-expansion

Let us write (10.59) as

$$\eta = \tfrac{1}{3}\epsilon \hat{g}_c^2 [I_2(0) - I_2(1)], \tag{10.102}$$

where $I_2(k)$ is the integral defined in equation (10.78), whose value is given in equation (10.96). If we are going to make use of this expression at $d = 4$, we really ought also to include the 'field renormalization' term given in equation (9.15) (i.e., the term in the integrand involving κ) in our calculation of η. However, this term is k-independent and so vanishes when we take the logarithmic derivative to calculate the critical exponent. It is left for the reader to verify that the eventual result for η is just (10.102), unchanged from the case $d < 4$ (Problem 10.2).

Let us then ϵ-expand the value (10.96) of the integral $I_2(k)$. We write

$$(k^2)^{1-\epsilon} = k^2 e^{-\epsilon \log k^2} = k^2 (1 - \epsilon \log k^2 + O(\epsilon^2)), \tag{10.103}$$

and, from equation (9) of Box 10.2

$$(\epsilon - 2)! = -\frac{1}{\epsilon} + O(1). \tag{10.104}$$

The first of these expansions immediately tells us that, to any finite order in ϵ, no matter what the value of ϵ, the limit of $I_2(k)$ as $k \to 0$ is zero, because the logarithmic divergences in the series are always dominated by the leading power of k^2. This means that, within the ϵ-expansion, we can ignore the $I_2(0)$ term. This was the very term that caused the infrared divergence when we tried to evaluate η in a straightforward manner in §10.3.2. Thus the ϵ-expansion has already solved our infrared problems. Now we multiply our two series together to get the ϵ-expansion of $I_2(k)$:

$$I_2(k) = -\tfrac{1}{2} \pi^4 k^2 \frac{1}{\epsilon} + O(1), \tag{10.105}$$

to leading order in ϵ. Setting $k = 1$, substituting this result into equation (10.102), and making use of the expansion of $\hat{g}_c$ in equation (10.101), we then get

$$\eta = \frac{\epsilon^2}{54} + O(\epsilon^3). \tag{10.106}$$

Thus in three dimensions ($\epsilon = 1$) we find

$$\eta = \frac{1}{54} = 0.0185\ldots \tag{10.107}$$

And this is our estimate of the exponent η for the Landau–Ginzburg model in three dimensions to leading order. It must be admitted that it's not a very accurate result—the true value is thought to be nearly a factor of

Table 10.1. Crude values for γ and η from the ϵ-expansion with $\epsilon = 1$ ($d = 3$)

Order n	0	1	2	3	4	5
γ	1	1.167	1.244	1.195	1.375	0.96
η	0	0	0.0185	0.0372	0.0289	0.0649

After Zinn-Justin (1989).

two larger: 0.032 ± 0.005. Given the considerable labour required to get this one number, and the poor value we found for γ to leading order in §10.2, one might be forgiven for despairing of this field and turning to something easier. However, when remorselessly applied the techniques we have introduced in this chapter do yield excellent values for the critical exponents. In fact, over time the values deduced from numerical simulations and experiments have tended to converge towards the values predicted by the ϵ-expansion.

Getting credible results from the ϵ-expansion does, however, require a deal of cunning, as Table 10.1 demonstrates: It shows the values of γ and η obtained by calculating the ϵ-expansion through terms of order ϵ^n, and a glance along the bottom two rows of numbers shows that the series show no sign of converging! Indeed it is thought that the ϵ-expansion is divergent for all ϵ (Zinn-Justin 1989). Sophisticated summation techniques such as Borel summation (Eckmann et al. 1975, Zinn-Justin 1989), make it possible nonetheless to deduce credible values of the exponents, and these may be checked against exact results in special cases. For example, the Borel-summed ϵ-expansion yields $\gamma = 1.73 \pm 0.06$ and $\eta = 0.26 \pm 0.05$ at $\epsilon = 2$ ($d = 2$), compared with the exact results $\gamma = 1.75$ and $\eta = 0.25$ yielded by the Ising model. The Borel-summed values for $\epsilon = 1$ ($d = 3$) are $\gamma = 1.239 \pm 0.0025$ and $\eta = 0.0375 \pm 0.0025$ (Zinn-Justin 1989). For comparison, Table 3.1 reports $\gamma = 1.239 \pm 0.003$ and $\eta = 0.024 \pm 0.007$ for the $d = 3$ Ising model. Thus the ϵ-expansion really does work.[13]

The ϵ-expansion is only strictly necessary when calculating quantities with $m = 0$, i.e., at the critical temperature. However, the convenience of working with $m = 0$ is such that in Chapter 12 we shall develop a formalism which enables us to set $m = 0$ even when evaluating quantities at $T \neq T_c$. We could not extract the quantities of interest from these calculations if the ϵ-expansion were not available. Thus the ϵ-expansion, and its hand-maiden dimensional regularization, will be central to subsequent developments.

---

[13] It is not at all obvious why the critical exponents for the Landau–Ginzburg model and the Ising model should agree, since we show in Appendix K that their Hamiltonians are quite different. Nor is is evident why their critical exponents coincide with experimental values for the $D = 1$ systems presented in Table 1.2. In Chapter 14 we endeavour to explain these coincidences.

Problems

10.1 Show that with $\epsilon = 1$, equation (10.94) agrees with the direct evaluation of the integral in equation (10.64).

10.2 Using equations (9.13) and (9.15), calculate $\partial \log \Gamma^{(2)} / \partial \log k$, and thus show that equation (10.59) is still true even when we employ field renormalization.

10.3 By analogy with the developments of §10.4.3, write down an expression for a product $\prod_i a_i^{-\alpha_i}$ of an arbitrary number of terms, and then, defining Feynman parameters $t_i = su_i$, show that equation (10.89) is the correct generalization to many variables of (10.88).

10.4 Using equation (10.63) and the integral $J(k)$ of §10.4.5, show that

$$\beta(\hat{g}) = -\epsilon\hat{g} + 3\pi^2 \hat{g}^2 + \mathrm{O}(\epsilon \hat{g}^2, \hat{g}^3). \tag{10.108}$$

By writing

$$\beta(\hat{g}) = \sum_{n=0}^{\infty} \sum_{m=1}^{\infty} a_{nm} \epsilon^n \hat{g}^m \quad \text{and} \quad \hat{g}_c = \sum_{r=1}^{\infty} b_r \epsilon^r, \tag{10.109}$$

where $\beta(\hat{g}_c) = 0$ and the coefficients a_{nm} and b_r are independent of ϵ and $\hat{g}$, show that in the Landau–Ginzburg model

$$\frac{\partial b_r}{\partial a_{pq}} = 0 \quad \text{if} \quad p + q > r + 1. \tag{10.110}$$

Deduce that to calculate $\hat{g}_c$ to $\mathrm{O}(\epsilon)$, we need only know β to the accuracy of (10.108).

11
The renormalization group

We are part of the way towards understanding the Landau–Ginzburg model in the critical regime. We have been able to calculate two critical exponents, one of which describes the system at $T = T_c$ and the other at $T \neq T_c$. If we are prepared to accept the scaling laws for which we found experimental evidence in Chapter 1, and theoretical evidence in Chapter 5, we can stop there; all four remaining critical exponents follow. However, this is rather unsatisfactory. In Chapter 5 we saw that the scaling relations can be deduced from the properties of the critical fixed point that is reached by repeatedly reblocking and rescaling the system; it would be good to provide a similarly general derivation in the context of the Landau–Ginzburg model.

In this chapter we introduce the last component of the formal edifice needed to support such a derivation: the 'renormalization group'. We then go half-way towards our goal, by calculating the critical exponents η and δ that describe the system at $T = T_c$. We have already found η in the last chapter, but the methods we shall use this time, as well as giving the value of δ, show very naturally that the two exponents are connected by a scaling relation. In the next chapter we shall finish the job by using these same methods to extend the calculation to $T \neq T_c$.

11.1 The renormalization group at $T = T_c$

The renormalization process that we applied to the Landau–Ginzburg model in 9 involved isolating the dependence of correlation functions on small length scales into the values of just three parameters, m, g and a. In the renormalization scheme we have adopted, these parameters are:

$$m^2 \equiv \Gamma^{(2)}(\mathbf{k} = 0; \alpha, \mu, \lambda, \Lambda),$$
$$a^2 \equiv \left.\frac{\partial}{\partial k^2}\Gamma^{(2)}(\mathbf{k}; \alpha, \mu, \lambda, \Lambda)\right|_{k=\kappa}, \quad (11.1)$$
$$g \equiv \left.\Gamma^{(4)}(\boldsymbol{\kappa}_1, \boldsymbol{\kappa}_2, \boldsymbol{\kappa}_3; \alpha, \mu, \lambda, \Lambda)\right|_{\text{SP}}.$$

At the critical temperature, when $\mu = \mu_c$, m^2 vanishes by definition and equations (11.1) become:

$$0 = \Gamma^{(2)}(\mathbf{k} = 0; \alpha, \mu_c, \lambda, \Lambda),$$
$$a^2 \equiv \left.\frac{\partial}{\partial k^2}\Gamma^{(2)}(\mathbf{k}; \alpha, \mu_c, \lambda, \Lambda)\right|_{k=\kappa}, \quad (11.2)$$
$$g \equiv \left.\Gamma^{(4)}(\boldsymbol{\kappa}_1, \boldsymbol{\kappa}_2, \boldsymbol{\kappa}_3; \alpha, \mu_c, \lambda, \Lambda)\right|_{\text{SP}}.$$

For the rest of this chapter we shall be working at the critical temperature, and it is to be understood that m and μ take the values 0 and μ_c respectively. The notation SP has the same meaning as in equations (10.53) and (10.54). The $\Gamma^{(n)}$ can now be written in terms of a and g. Equation (9.31) shows that powers of a may be factored out of the expression for $\Gamma^{(n)}$ in the form

$$\Gamma^{(n)}(\mathbf{k}_1, \ldots, \mathbf{k}_{n-1}; \alpha, \lambda, \Lambda) = a^n \Gamma_R^{(n)}(\mathbf{k}_1, \ldots, \mathbf{k}_{n-1}; g_R, \kappa), \quad (11.3)$$

where we have suppressed the arguments m and μ. $\Gamma_R^{(n)}$ is called the renormalized vertex function and

$$g_R \equiv \frac{g}{a^4}. \quad (11.4)$$

This quantity has dimensions $L^{-\epsilon}$ (where $\epsilon \equiv 4 - d$), so it is possible to make a dimensionless combination of g_R and κ in the form

$$\hat{g} \equiv \frac{g_R}{(2\pi)^d}\kappa^{-\epsilon}, \quad (11.5)$$

as in equation (10.57). $\hat{g}$ plays the rôle of the natural dimensionless expansion parameter for perturbation theory. From now on we shall write all vertex functions depending on renormalized parameters as functions of $\hat{g}$, rather than g.

Once the cut-off parameter Λ and the renormalization point κ have been specified, equations (11.2) can be regarded as defining g and a in terms of λ and α or *vice versa*. As we stressed in Chapter 9, there exists a one-parameter family of values of $(\alpha, \lambda, \Lambda)$ corresponding to each set of values (a, g, κ). However, the reverse is also true. This is because we are free to choose any finite and non-zero value for κ. Different choices simply correspond to choosing different parameters (the values of $(\partial \Gamma^{(2)}/\partial k^2)$ and $\Gamma^{(4)}$ at the new value of κ) in terms of which to describe the macroscopic physics, without affecting that physics in any way. In other words, there is also a one-parameter family of values for (a, g, κ) given by equations (11.2) for each set of values of the bare parameters.

The set of transformations between possible values of κ, and the associated values of $\hat{g}$ and a, form a group which is known as the **renormalization group**. In some ways this nomenclature is unhelpful, because it leads one to expect more richness than is present in its simple Abelian structure. However, the terminology is standard and we shall use it. As frequently happens with continuous groups, it turns out to be advantageous to study the infinitesimal transformations of the group. These express the invariance of the $\Gamma^{(n)}$ when the value of κ is changed by an infinitesimal amount:

$$\left(\kappa \frac{\partial}{\partial \kappa} [a^n \Gamma_R^{(n)}(\mathbf{k}_1, \ldots, \mathbf{k}_{n-1}; \hat{g}, \kappa)]\right)_{\alpha, \lambda, \Lambda} = 0, \tag{11.6}$$

where the partial derivative keeps the bare parameters $(\alpha, \lambda, \Lambda)$ constant but includes the changes of the renormalized parameters $(a, \hat{g})$ as κ is changed. The κ-dependence of the left-hand side comes from three sources:

(i) the explicit dependence of $\Gamma_R^{(n)}$ on κ;
(ii) the implicit dependence of $\Gamma_R^{(n)}$ on κ through the value of $\hat{g}$;
(iii) the dependence of a on κ.

Rewriting (11.6) to reflect this, we obtain

$$\left[\kappa \frac{\partial}{\partial \kappa} + \beta \frac{\partial}{\partial \hat{g}} - \frac{n}{2} \gamma_1\right] \Gamma_R^{(n)}(\mathbf{k}_1, \ldots, \mathbf{k}_{n-1}; \hat{g}, \kappa) = 0, \tag{11.7}$$

where the partial derivative with respect to κ is now taken with $\hat{g}$ and a constant as well as the bare parameters, and β and γ_1 reflect respectively the dependence of $\hat{g}$ and a on κ:

$$\beta \equiv \kappa \left(\frac{\partial \hat{g}}{\partial \kappa}\right)_{\alpha, \lambda, \Lambda},$$
$$\gamma_1 \equiv -2\kappa \left(\frac{\partial \log a}{\partial \kappa}\right)_{\alpha, \lambda, \Lambda}. \tag{11.8}$$

11.1 Renormalization at $T = T_c$

We have already come across β in equation (10.60). The factor of -2 in the definition of γ_1 is purely conventional. The quantities β and γ_1 are known as **flow functions**, since they describe the flow of the quantities $\hat{g}$ and a as κ is changed.

Equation (11.7) is known as the **renormalization group equation**. We can look at it in two ways: if we had made a complete calculation of $\Gamma_R^{(n)}$ and a for all values of κ, we could use it to find the functions β and γ_1. Alternatively, we could invert the procedure; if we could somehow find the functions β and γ_1, we could use (11.7) to extract information about $\Gamma_R^{(n)}$ and hence about $\Gamma^{(n)}$.

We shall pursue the second route in this chapter. But what sort of information can it give us? As we have pointed out, κ is a parameter without direct physical significance. A solution of (11.7) which tells us only how the $\Gamma_R^{(n)}$ depend on κ and nothing else, will not therefore be very helpful. It will not, for example, immediately lead us to the value of any of the critical exponents. Fortunately, we are rescued by dimensional analysis; we know the dimensions of $\Gamma_R^{(n)}$ and of its arguments, so we know how they all change if we simply change the length units by a given factor. This will enable us to transform our knowledge of how the vertex functions depend on κ into knowledge of how they depend on wavevector.

It follows from equation (11.3) and Table 8.2 that the dimensions of $\Gamma_R^{(n)}(\mathbf{k}_1, \ldots, \mathbf{k}_{n-1})$ are $L^{-\delta_n}$, where

$$\delta_n \equiv n - (\tfrac{1}{2}n - 1)d \qquad (11.9)$$

as in Chapter 10. Of the quantities on which $\Gamma^{(n)}$ depends, $\hat{g}$ is dimensionless while κ and k both have dimensions L^{-1}. Therefore, $\Gamma_R^{(n)}$ must be a **homogeneous function** of degree δ_n in κ and k, and so must satisfy an **Euler equation** (see Box 11.1) of the form[1]

$$\left(\kappa\frac{\partial}{\partial \kappa} - b\frac{\partial}{\partial b}\right)\Gamma_R^{(n)}(\mathbf{k}_i/b; \hat{g}, \kappa) = \delta_n \Gamma_R^{(n)}(\mathbf{k}_i/b; \hat{g}, \kappa), \qquad (11.10)$$

where we have scaled all the wavevector arguments of $\Gamma_R^{(n)}$ by the same factor $1/b$.

We can now eliminate the derivatives with respect to κ between equations (11.7) and (11.10) to get an equation which tells us something of direct physical interest, namely the behaviour of $\Gamma_R^{(n)}$ as its wavevector arguments are scaled at fixed κ:

$$\left[b\frac{\partial}{\partial b} + \beta\frac{\partial}{\partial \hat{g}} - \frac{n}{2}\gamma_1 + \delta_n\right]\Gamma_R^{(n)}(\mathbf{k}_i/b; \hat{g}, \kappa) = 0 \qquad (11.11)$$

---

[1] In (11.10) and subsequent equations we write $\Gamma_R^{(n)}(\mathbf{k}_i; \hat{g}, \kappa)$ as an abbreviation for $\Gamma_R^{(n)}(\mathbf{k}_1, \ldots, \mathbf{k}_{n-1}; \hat{g}, \kappa)$.

> **Box 11.1: Euler's equation and homogeneous functions**
>
> Suppose that a function $f(x_1, \ldots, x_n)$ is homogeneous of degree D in the variables $x_1 \ldots x_n$. By definition this means that, on multiplying each x_i by an arbitrary factor ρ, the value of f is multiplied by ρ^D:
>
> $$f(\rho x_1, \ldots, \rho x_n) = \rho^D f(x_1, \ldots, x_n).$$
>
> Now let us differentiate both sides with respect to ρ and multiply by ρ. We find that
>
> $$\rho \left[\sum_{i=1}^n x_i \frac{\partial}{\partial x_i} \right] f(\rho x_1, \ldots, \rho x_n) = D \rho^D f(x_1, \ldots, x_n).$$
>
> Setting $\rho = 1$ gives **Euler's equation**:
>
> $$\left[\sum_{i=1}^n x_i \frac{\partial}{\partial x_i} - D \right] f(x_1, \ldots, x_n) = 0.$$

Before we can solve this equation, we need to know more about the functions β and γ_1. First, let us specify precisely what we mean by the function β. If κ takes the value κ_A, the values of g and a are determined by

$$\begin{aligned} g_A &= \Gamma^{(4)}(\kappa_{A1}, \kappa_{A2}, \kappa_{A3}; \alpha, \lambda, \Lambda) \Big|_{\text{SP } A} \\ a_A^2 &= \frac{\partial}{\partial k^2} \Gamma^{(2)}(\mathbf{k}; \alpha, \lambda, \Lambda) \Big|_{k=\kappa_A} . \end{aligned} \qquad (11.12)$$

Now let us use (11.3) to write $\Gamma^{(2)}$ and $\Gamma^{(4)}$ in terms of $\Gamma_R^{(2)}$ and $\Gamma_R^{(4)}$ defined at a *different* renormalization point $\kappa = \kappa_B$:

$$\begin{aligned} g_A &= a_B^4 \Gamma_R^{(4)}(\kappa_{A1}, \kappa_{A2}, \kappa_{A3}; \hat{g}_B, \kappa_B) \Big|_{\text{SP } A} \\ a_A^2 &= a_B^2 \frac{\partial}{\partial k^2} \Gamma_R^{(2)}(\mathbf{k}; \hat{g}_B, \kappa_B) \Big|_{k=\kappa_A} . \end{aligned} \qquad (11.13)$$

The first of these yields

$$\hat{g}_A = \frac{\kappa_A^{-\epsilon}}{(2\pi)^d} \Gamma_R^{(4)}(\kappa_{A1}, \kappa_{A2}, \kappa_{A3}; \hat{g}_B, \kappa_B) \Big|_{\text{SP } A} . \qquad (11.14)$$

Therefore the value of β when $\kappa = \kappa_B$ is

$$\beta = \kappa_A \frac{\partial \hat{g}_A}{\partial \kappa_A} \Big|_{\kappa_A = \kappa_B} . \qquad (11.15)$$

11.1 Renormalization at $T = T_c$

Box 11.2: The method of characteristics

The method of characteristics enables one to solve a partial differential equation by integrating a series of coupled ordinary differential equations. Consider, for example, the general, first-order quasi-linear equation

$$\psi_t + c(x,t,\psi)\psi_x = f(x,t,\psi), \qquad (1)$$

where the subscripts t and x indicate partial derivatives. Let $(x(\lambda), t(\lambda))$ be some curve in (x,t) space. Then the change in ψ along this path is

$$\frac{d\psi}{d\lambda} = \psi_t \frac{dt}{d\lambda} + \psi_x \frac{dx}{d\lambda}. \qquad (2)$$

We choose the curve such that $(dx/d\lambda)/(dt/d\lambda) = c(x,t,\psi)$; such a curve is called a **characteristic**. Then (1) implies that

$$\frac{d\psi}{d\lambda} = f\frac{dt}{d\lambda}. \qquad (3)$$

By solving this differential equation along all the characteristics which intersect the line along which our initial data are specified, we can obtain ψ at any point. It is usually convenient to set λ equal to one of the independent variables, for example t.

Problem 11.1 generalizes this theory to completely non-linear equations.

From (11.14) we see that β can depend only on κ_B and $\hat{g}_B$, since κ_A has been set equal to κ_B. However, β is dimensionless, and a dimensionless quantity can depend only on other dimensionless quantities because it has to remain invariant when the units in which everything is measured are changed. So the value of β when $\kappa = \kappa_B$ depends only on $\hat{g}_B$. A similar argument shows that γ_1 at $\kappa = \kappa_B$ also depends only on $\hat{g}_B$. This is a very important result, because it enables us to solve (11.15) easily by the 'method of characteristics' (see Box 11.2).

To apply the results of the box to the problem in hand, we regard $\Gamma_R^{(n)}(\mathbf{k}_i/b; \hat{g}, \kappa)$ as a function of the two variables b and $\hat{g}$. Our initial conditions tell us the value of $\Gamma_R^{(n)}(\mathbf{k}_i/b; \hat{g}, \kappa)$ for $b = 1$; we are interested in finding the value of this function for an arbitrary value of b. To be precise, let us suppose that we are interested in the value of $\Gamma_R^{(n)}(\mathbf{k}_i/b_0; \hat{g}_0, \kappa)$.

Equation (11.11) is of exactly the form of (1) in Box 11.2, with

$$t = \log b, \quad x = \hat{g}, \quad c = \beta, \quad \psi = \Gamma_R^{(n)},$$
$$\text{and} \quad f = \left[\tfrac{1}{2}n\gamma_1 - \delta_n\right]\Gamma_R^{(n)}. \qquad (11.16)$$

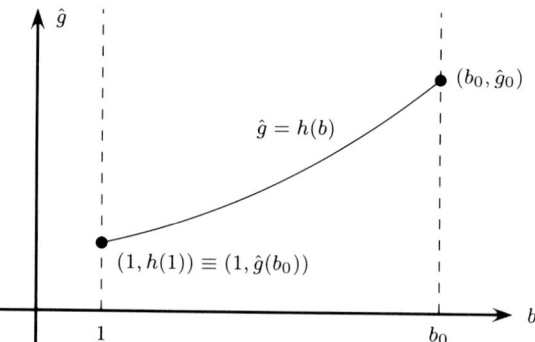

Figure 11.1 The application of the method of characteristics to the solution of (11.11). Along the characteristic curve shown (it is one of infinitely many such curves), $\Gamma_R^{(n)}$ satisfies an ordinary differential equation. Integrating this equation relates the values of $\Gamma_R^{(n)}$ at different points along the curve. In particular we can relate its value at $(b_0, \hat{g}_0)$ to its value at $(1, h(1)) \equiv (1, \hat{g}(b_0))$. These constraints on the behaviour of $\Gamma_R^{(n)}$ when its wavevectors are scaled will allow us to determine the critical behaviour of the system.

The characteristics are the curves $\hat{g} = h(b)$, where $h(b)$ satisfies

$$b\frac{dh(b)}{db} = \beta(h(b)), \quad \text{with} \quad h(b_0) = \hat{g}_0. \tag{11.17}$$

One of these curves is sketched in Figure 11.1 (for the case of a positive β-function). Along each such curve, $\Gamma_R^{(n)}$ satisfies an ordinary differential equation analogous to equation (3) of Box 11.2. Integrating this equation along a characteristic relates the values of $\Gamma_R^{(n)}$ at different points on the characteristic. This is the reason for the choice of boundary condition in (11.17); we wish to relate the value of $\Gamma_R^{(n)}$ at $(b, \hat{g}) = (b_0, \hat{g}_0)$ to some quantity evaluated at $b = 1$.

Equation (3) of Box 11.2 now becomes

$$b\frac{d}{db}\Gamma_R^{(n)}(\mathbf{k}_i/b; h(b), \kappa) = [\tfrac{1}{2}n\gamma_1(h(b)) - \delta_n]\Gamma_R^{(n)}(\mathbf{k}_i/b; h(b), \kappa). \tag{11.18}$$

This is a separable equation that can be integrated from $b = 1$ to $b = b_0$ to give

$$\Gamma_R^{(n)}(\mathbf{k}_i/b_0; \hat{g}_0, \kappa) = b_0^{-\delta_n} \alpha^n(1) a^{-n} \Gamma_R^{(n)}(\mathbf{k}_i; h(1), \kappa), \tag{11.19}$$

where the function $\alpha(b)$ satisfies

$$b\frac{d\log\alpha(b)}{db} = -\tfrac{1}{2}\gamma_1(h(b)) \quad \text{with} \quad \alpha(b_0) = a. \tag{11.20}$$

The quantity $h(1)$, the value of h where the characteristic intersects the line $b = 1$, is a function both of $\hat{g}_0$ and b_0. Because it appears in $\Gamma_R^{(n)}$ in the

11.1 Renormalization at $T = T_c$

rôle of a coupling, it is usual to write $h(1) \equiv \hat{g}(b_0)$, where the function $\hat{g}(b)$ satisfies the equation

$$b\frac{d\hat{g}(b)}{db} = -\beta(\hat{g}(b)) \quad \text{with} \quad \hat{g}(1) = \hat{g}_0. \tag{11.21}$$

In the same way we may define a function $a(b_0) \equiv \alpha(1)$ satisfying

$$b\frac{d\log a(b)}{db} = \tfrac{1}{2}\gamma_1(\hat{g}(b)) \quad \text{with} \quad a(1) = a. \tag{11.22}$$

Dropping the subscripts on b_0 and $\hat{g}_0$, (11.19) becomes

$$\Gamma_R^{(n)}(\mathbf{k}_i/b; \hat{g}, \kappa) = b^{-\delta_n} a^n(b) a^{-n} \Gamma_R^{(n)}(\mathbf{k}_i; \hat{g}(b), \kappa). \tag{11.23}$$

In words, dividing all the wavevectors by b has three effects:

(i) $\Gamma_R^{(n)}$ is multiplied by a factor $b^{-\delta_n}$ corresponding to its dimension (recall that $\Gamma_R^{(n)}$ has dimension $L^{-\delta_n}$);
(ii) $\Gamma_R^{(n)}$ is multiplied by an additional factor $a^n(b)$;
(iii) The coupling is changed from $\hat{g}$ to $\hat{g}(b)$.

$\hat{g}(b)$ is referred to as the **running coupling constant**.

It is now trivial to write down the solution for the original vertex functions $\Gamma^{(n)} = a^n \Gamma_R^{(n)}$:

$$\Gamma^{(n)}(\mathbf{k}_i/b; \hat{g}, \kappa) = b^{-\delta_n} a^n(b) \Gamma_R^{(n)}(\mathbf{k}_i; \hat{g}(b), \kappa). \tag{11.24}$$

Equation (11.24) gives us just what we want—an expression for the vertex functions with the physical value of $\hat{g}$ but different values of the wavevector. Notice that the behaviour of $\hat{g}(b)$ for large b controls the long-distance bahaviour of the correlation functions.

The fact that β depends only on $\hat{g}$ strongly restricts the possible behaviour of $\hat{g}$ as a function of b. A priori we might imagine the following possibilities as b decreases:

(i) $\hat{g}$ might increase without limit;
(ii) $\hat{g}$ might tend to a finite limiting value;
(iii) $\hat{g}$ might oscillate periodically;
(iv) $\hat{g}$ might oscillate aperiodically or behave chaotically.

However, we can now discount possibilities (iii) and (iv), since they cannot be realized while $d\hat{g}/d\log b$ remains a single-valued function of $\hat{g}$ alone.

Let us consider some possible shapes for the function $\beta(\hat{g})$ and their consequences for the dependence of $\hat{g}$ on b. Suppose first that β is everywhere positive; this implies that $\hat{g}$ must become very large and positive as $b \to 0$

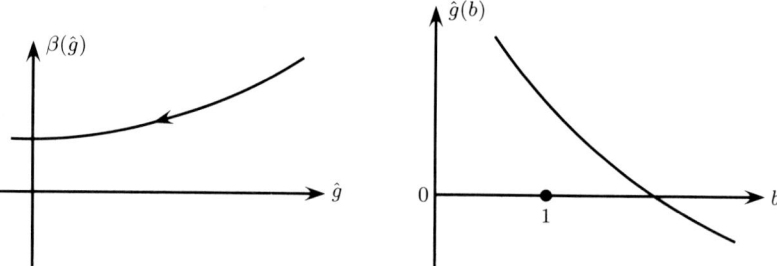

Figure 11.2 A β-function that is everywhere positive. The arrow on the graph of $\beta(\hat{g})$ indicates the direction of flow of $\hat{g}(b)$ as b increases.

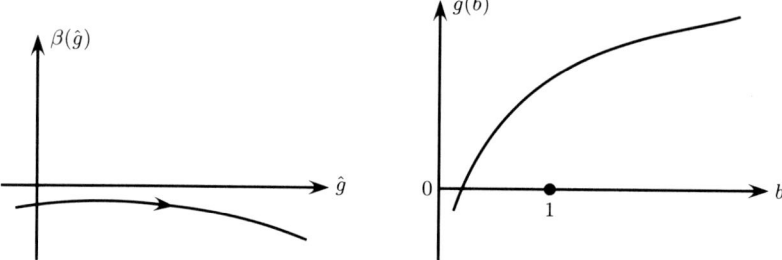

Figure 11.3 A β-function that is everywhere negative. The arrow on the graph of $\beta(\hat{g})$ indicates the direction of flow of $\hat{g}(b)$ as b increases.

(i.e., as $\log b \to -\infty$) and very large and negative as $b \to \infty$, no matter what its value at intermediate values of b (see Figure 11.2).

Similarly if β is everywhere negative, then $\hat{g}$ will become large and negative as $b \to 0$ and large and positive as $b \to \infty$, again regardless of its value at intermediate points (see Figure 11.3).

Now let us consider cases in which β changes sign just once, at $\hat{g} = \hat{g}_c$. This could happen in two ways: β might pass through zero, or it might have a pole of odd order. We shall consider the case of a zero here, and leave the physically less interesting case of the pole to Problem 11.2. First suppose that β crosses zero with a negative slope (see Figure 11.4); then if $\hat{g} > \hat{g}_c$, we have $d\hat{g}/db = -\beta/b > 0$ and $\hat{g}$ will increase without limit as $b \to \infty$. Conversely, if $\hat{g} < \hat{g}_c$ it will decrease without limit. In either case, however, $\hat{g} \to \hat{g}_c$ as $b \to 0$. Such a zero is said to be an **ultraviolet stable fixed point**, since then the ultraviolet (large-wavevector) behaviour of the vertex functions is controlled by the fixed value of $\hat{g}_c$.

Finally, suppose that β has a positive slope at $\hat{g} = \hat{g}_c$ (see Figure 11.5). The argument we have just given is now reversed; regardless of its value at finite b, $\hat{g}$ is attracted to $\hat{g}_c$ as $b \to \infty$. Similarly, $\hat{g}$ becomes large and

11.1 Renormalization at $T = T_c$

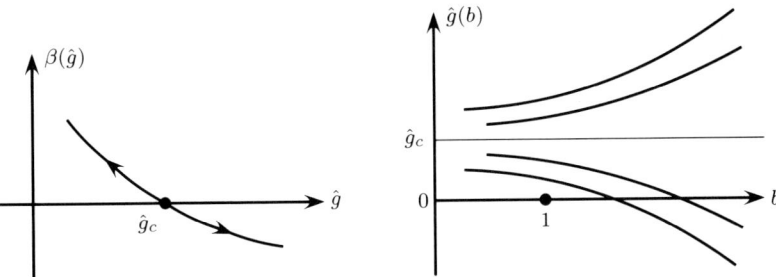

Figure 11.4 A β-function that has a zero with negative slope. The arrows on the graph of $\beta(\hat{g})$ indicate the direction of flow of $\hat{g}(b)$ as b increases.

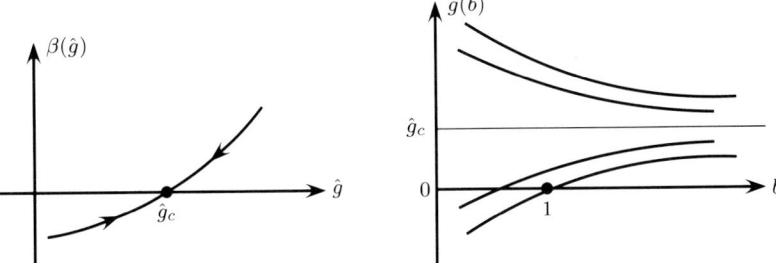

Figure 11.5 A β-function that has a zero with positive slope. The arrows on the graph of $\beta(\hat{g})$ indicate the direction of flow of $\hat{g}(b)$ as b increases.

positive or large and negative as $b \to 0$, depending on whether it starts from a value above or below $\hat{g}_c$. This type of zero is said to be an **infrared stable fixed point**, since the infrared (small-wavevector) behaviour of the vertex functions is then controlled by the fixed value of $\hat{g}_c$.

It is not hard to see that if there are several zeros of the β-function, any model with a value of $\hat{g}$ that starts between a zero with positive slope and a zero with negative slope will move towards the zero with positive slope as $b \to \infty$, and towards the zero with negative slope as $b \to 0$. The set of starting values of $\hat{g}$ which flow to a particular infrared or ultraviolet stable fixed point as $b \to \infty$ or as $b \to 0$ are said to form the **domain of attraction** of the fixed point. Notice that an infrared stable fixed point is inevitably ultraviolet unstable and vice versa.

What have fixed points to do with critical phenomena? Critical exponents and other macroscopic properties of a phase transition result from the long-range (small-wavevector) behaviour of the correlation functions, and therefore also of the vertex functions. Equation (11.24) tells us that if an infrared stable fixed point exists, its properties control the long-range behaviour; the macroscopic properties depend *only* on the fixed point, not on the parameters in the microscopic Hamiltonian.

We know from the experiments presented in Chapter 1 that correlation functions display power-law behaviour at the critical temperature. Furthermore, the Widom homogeneity hypothesis (§1.5.1) suggests that the vertex functions will depend only on particular special combinations of the basic variables. Now consider the form of equation (11.24) when $\hat{g}(b)$ comes close to a value $\hat{g}_c$ at which the β-function is zero. In that case, $\beta(\hat{g}(b)) \sim 0$ and $\gamma_1(\hat{g}(b)) \sim \gamma_1^* \equiv \gamma_1(\hat{g}_c)$, so for large b we find

$$a(b) \sim \exp(\tfrac{1}{2}\gamma_1^* \log b) = b^{\gamma_1^*/2}. \tag{11.25}$$

The solution (11.24) becomes

$$\Gamma^{(n)}(\mathbf{k}_i/b; \hat{g}, \kappa) \sim b^{-\delta_n + \frac{n}{2}\gamma_1^*} \Gamma_R^{(n)}(\mathbf{k}_i; \hat{g}_c, \kappa). \tag{11.26}$$

This tells us that $\Gamma^{(n)}$ depends on b as a power law when, and only when, $\hat{g}$ takes the special value $\hat{g}_c$. We can therefore identify zeros of the β-function with critical behaviour; they tell us the special values of the dimensionless coupling constant for which our power-series expansions of the vertex functions sum up to give power laws.

Let us study the β-function of (10.60) in the light of this. We found in §10.3 (equation (10.63), plus the integral $J(k)$ of §10.4.5) that, to order $\hat{g}^2$,

$$\beta(\hat{g}) = -\epsilon \hat{g} + 3\pi^2 \hat{g}^2 + O(\epsilon \hat{g}^2, \hat{g}^3). \tag{11.27}$$

This always has two zeros, one at $\hat{g} = 0$ and one at

$$\hat{g} = \frac{1}{3\pi^2} \epsilon. \tag{11.28}$$

For $d > 4$ ($\epsilon < 0$), the zero at the origin has positive slope. It is therefore infrared stable; the correlation functions at large distances are given by setting $\hat{g}(b) = 0$ on the right of (11.24) and the critical behaviour of the Landau–Ginzburg model is indistinguishable from that of the Gaussian model, in spite of the presence of the ϕ^4 term in the Hamiltonian. However, for $d < 4$ ($\epsilon > 0$) the zero at the origin develops a negative slope and becomes instead ultraviolet stable. The other zero, now at a positive value of $\hat{g}$, becomes infrared stable and controls the long-distance behaviour of the theory. This is why we used the non-zero root of the β-function in our calculations in the last chapter; it is infrared stable for the values of d of physical interest.

By this time you have probably noticed the similarity, both in nomenclature and in procedure, to the real-space renormalization techniques of Chapter 5. There we systematically removed sites from our lattice, eliminating large-wavevector fluctuations and effectively reducing the value of the cut-off parameter Λ. By itself, this would just have told us the dependence of the correlation functions on the microscopic cut-off, which is not of direct interest. However, by combining the removal of the sites with a rescaling of the lattice parameter, we were able to convert the dependence on the cut-off to information about the dependence of the correlation functions on distance for a fixed value of the cut-off. Finally, if the system moved towards a fixed point in parameter space under a large number of rescalings, we were able to deduce simple scaling forms for the correlation functions at large distances. This is just the situation that pertains for the Landau–Ginzburg model in the event of an infrared stable zero of the β-function; the renormalized parameters tend to constants or scale, and the large-distance behaviour of the correlation functions becomes simple. It is for this reason that we use the same term 'fixed point' here and in Chapter 5. The value of $\hat{g}$, which is particularly important since it determines the values of β and γ_1, does indeed become fixed. The other renormalized parameter, a, has a simple power-law dependence on b at such a fixed point.

11.2 The exponents η and δ

Equation (11.11) in the limit $b \to \infty$ gives us the small-wavevector (large-distance) behaviour of the vertex functions. Now let us go further and calculate the relationship between an assumed infrared stable fixed point and the critical exponents at $T = T_c$.

There are two exponents defined at $T = T_c$; we shall start with the exponent η, which describes the large-distance behaviour of the two-point correlation function.

11.2.1 The exponent η

The dependence on b in (11.26) is of the form

$$\Gamma^{(n)}(\mathbf{k}_i/b; \hat{g}, \kappa) \sim b^{-\delta_n + \frac{n}{2}\gamma_1^*}. \tag{11.29}$$

Putting $n = 2$ we obtain (see equation (11.9))

$$\Gamma_R^{(2)}(\mathbf{k}/b; \hat{g}, \kappa) \sim b^{\gamma_1^* - 2}. \tag{11.30}$$

Comparing with equation (10.36) and noting that b appears in (11.30) in the combination $\mathbf{k}/b$, we find that

$$\eta = \gamma_1^*. \tag{11.31}$$

The exponent η is the value of the flow function γ_1 at the fixed point $\hat{g} = \hat{g}_c$. It was to obtain this particularly simple relationship that we introduced the factor of -2 in the definition (11.8) of γ_1.

11.2.2 The exponent δ

To calculate δ, which controls the magnetization when a small magnetic field is applied exactly at the critical temperature, we need to extend our calculation a little. Within the Landau–Ginzburg model, we must add a spatially uniform value of the source field

$$\beta B(\mathbf{x}) = J(\mathbf{x}) = J_0. \tag{11.32}$$

We can find the relationship between the source field J_0 and the uniform thermal average ϕ_0 of the order parameter which it induces from an expansion of the dimensionless Gibbs free energy Γ (defined in equation (8.44)) in terms of the vertex functions and powers of ϕ_0. Equation (8.69) gives such an expansion for the case of a spatially constant field: [2]

$$\begin{aligned} J_0 &= \frac{1}{V} \frac{\partial \Gamma}{\partial \phi_0} \\ &= \frac{\partial}{\partial \phi_0} \sum_{n=0}^{\infty} \frac{\phi_0^n}{n!} \Gamma^{(n)}(\mathbf{k}_i = 0; \lambda, \Lambda) \\ &= \sum_{n=0}^{\infty} \frac{\phi_0^n}{n!} \Gamma^{(n+1)}(\mathbf{k}_i = 0; \lambda, \Lambda) \\ &= a \sum_{n=0}^{\infty} \frac{\phi_{0R}^n}{n!} \Gamma_R^{(n+1)}(\mathbf{k}_i = 0; \hat{g}, \kappa), \end{aligned} \tag{11.33}$$

where we have used equation (11.3) and defined a renormalized value of the uniform field

$$\phi_{0R} \equiv a\phi_0. \tag{11.34}$$

Let us similarly define a renormalized source field

$$J_{0R} \equiv J_0/a; \tag{11.35}$$

then, using the fact that each of the vertex functions on the right-hand side separately satisfies an equation of the form (11.7), we see that J_{0R} must satisfy

$$\left[\kappa \frac{\partial}{\partial \kappa} + \beta \frac{\partial}{\partial \hat{g}} - \tfrac{1}{2}\gamma_1 - \tfrac{1}{2}\gamma_1 \phi_{0R} \frac{\partial}{\partial \phi_{0R}} \right] J_{0R}(\phi_{0R}, \hat{g}, \kappa) = 0. \tag{11.36}$$

We can now combine this with dimensional analysis to eliminate κ from the equation. We know that $\phi(\mathbf{x})$ and $J(\mathbf{x})$ have dimensions $L^{-d/2}$. Therefore,

---

[2] You may wonder how it is that we can calculate the thermal average of ϕ, which we expect to behave as a fractional power of J_0 at T_c, using a power-series expansion. We discuss this problem in §12.1.7.

11.2 Solving the equation

ϕ_{0R} and J_{0R} have dimensions $L^{1-d/2}$ and $L^{-1-d/2}$, respectively. From this information, we can deduce the Euler equation

$$\left[\kappa\frac{\partial}{\partial\kappa} - b\frac{\partial}{\partial b}\right]J_{0R}(b^{1-d/2}\phi_{0R},\hat{g},\kappa) = (1+\tfrac{1}{2}d)J_{0R}(b^{1-d/2}\phi_{0R},\hat{g},\kappa) \quad (11.37)$$

and use this to eliminate from (11.36) the derivative with respect to κ. We obtain

$$\left[(1+\frac{\gamma_1}{d-2})b\frac{\partial}{\partial b} + \beta\frac{\partial}{\partial\hat{g}} - \tfrac{1}{2}\gamma_1 + 1 + \tfrac{1}{2}d\right]J_{0R}(b^{1-d/2}\phi_{0R},\hat{g},\kappa) = 0. \quad (11.38)$$

We are interested in the what happens as $\phi_0 \to 0$. Equation (11.38) shows that, for $d > 2$, we can reach this limit by letting b tend to infinity. We can solve (11.38) by the method of characteristics just as we did equation (11.11). Again, in the limit $b \to \infty$ the solution is determined by an infrared stable zero in the β-function (assuming that one exists). Since for our purposes we are not interested in the approach to the $b \to \infty$ limit but in the limit itself, we can save work by putting $\hat{g} = \hat{g}_c$ at $b = 1$ to get the limiting form of the solution straight away. If we put $\hat{g} = \hat{g}_c$ in (11.38) then β vanishes, and $\gamma_1 = \gamma_1^*$. The resulting equation gives

$$J_{0R}(b^{1-d/2}\phi_{0R},\hat{g},\kappa) = b^{\delta(1-d/2)}J_{0R}(\phi_{0R},\hat{g}_c,\kappa), \quad (11.39)$$

with

$$\delta = \frac{d+2-\gamma_1^*}{d-2+\gamma_1^*}. \quad (11.40)$$

Therefore, as $\phi_{0R} \to 0$, we have

$$\lim_{\phi_{0R}\to 0} J_{0R} \propto \phi_{0R}^\delta. \quad (11.41)$$

It follows that $\phi_0 \propto J_0^{1/\delta}$, since ϕ_0 and J_0 are related to ϕ_{0R} and J_{0R} by factors of a which are independent of the applied field. The quantity δ we have defined is therefore exactly the critical exponent of the same name.

This connects the critical exponents δ and η at the critical temperature to the flow functions β and γ_1^* that emerged from the renormalization group. In doing so, it shows that η and δ are not independent, but are related by the expression

$$\delta = \frac{d+2-\eta}{d-2+\eta}. \quad (11.42)$$

This is exactly the equation (5.56) which we derived from real-space renormalization arguments in §5.5 and which we showed in §5.7 may be derived from the experimental scaling relations of §1.5.1.

Note that so far in this chapter all results have been obtained without the use of any perturbation theory. We have used general properties of the Feynman graph expansion, but we have not needed to evaluate it to any particular order. You will not be surprised to learn that this state of affairs is too good to last; we now turn to a perturbative calculation of the critical exponents in powers of ϵ.

11.3 The calculation of β and γ_1

We have reduced the problem of calculating the properties of the Landau–Ginzburg model at the critical temperature to that of determining the form of two functions of $\hat{g}$. These are $\beta(\hat{g})$, the positions of whose zeros tell us the fixed-point values $\hat{g}_c$ of $\hat{g}$, and $\gamma_1(\hat{g})$, whose value at $\hat{g} = \hat{g}_c$ tells us the exponents η and δ.

In §§10.3 and 10.4 we have seen that infrared divergences of Feynman integrals with $m = 0$ oblige us to calculate quantities of interest as power series in $\epsilon \equiv 4 - d$ rather than as a series in powers of $\hat{g}$. As in §10.5, we shall find that to calculate η to $O(\epsilon^2)$ it suffices to know $\hat{g}_c$ to $O(\epsilon)$. We have already calculated $\hat{g}_c$ to this order in equation (10.101): $\hat{g}_c = \epsilon/(3\pi^2) + O(\epsilon^2)$.

11.3.1 The calculation of γ_1 to order ϵ^2

The derivative in the definition of γ_1 is best evaluated using the chain rule:

$$\gamma_1 \equiv -2\frac{\partial \log a}{\partial \log \kappa} = -2\beta(\hat{g})\frac{\partial \log a}{\partial \hat{g}}, \tag{11.43}$$

the differentiations being taken with the bare parameters held fixed. Equation (11.43) is helpful because a^2 is equal to α^2 multiplied by a power series in $\hat{g}$ whose coefficients are κ-independent integrals. However, this must be done with care because we know that β has a zero at the fixed point; consequently, if γ_1 is to have a non-zero value at the fixed point, the second factor on the right of (11.43) must diverge. We must be careful, therefore, to work to a definite order in $\hat{g}$ in the *finite* quantity $\beta \partial \log a/\partial \hat{g}$.

The first k-dependent correction to $\Gamma^{(2)}$ (i.e., the first term making a^2 different from α^2) comes from the Saturn diagram, Figure 11.6. We have already evaluated this in §10.4 by dimensional regularization. There, we found (see equations (10.40), (10.77) and (10.78)) that, for $m = 0$,

$$\Gamma^{(2)}(k) = \alpha^2 k^2 \left\{ 1 - \tfrac{1}{6} \frac{g^2(\alpha k)^{-2\epsilon}}{(2\pi a)^{2d}} [I_2(1) - I_2(0)] \right\}, \tag{11.44}$$

where $I_2(k)$ (defined by equation (10.78)) is the value of Figure 11.6. We were able to show that

$$I_2(k) = (k^2)^{1-\epsilon} \pi^d \frac{[(-\tfrac{1}{2}\epsilon)!]^3 (\epsilon - 2)!}{(2 - \tfrac{3}{2}\epsilon)!}, \tag{11.45}$$

and that within the ϵ-expansion, $I_2(0) = 0$. So we have

$$a^2 = \left.\frac{\partial \Gamma^{(2)}}{\partial k^2}\right|_{k=\kappa} = \alpha^2 \left[1 - \tfrac{1}{6}\hat{g}^2 \pi^d (1 - \epsilon) \frac{[(-\tfrac{1}{2}\epsilon)!]^3 (\epsilon - 2)!}{(2 - \tfrac{3}{2}\epsilon)!} + O(\hat{g}^3) \right]. \tag{11.46}$$

11.3 The calculation of β and γ_1

Figure 11.6 The Saturn diagram provides the second-order contribution to a^2.

Making an ϵ-expansion of the factorials, we obtain

$$a^2 = \alpha^2 \left[1 + \tfrac{1}{6}\hat{g}^2 \pi^4 \left(-\frac{1}{2\epsilon} + O(1) \right) + O(\hat{g}^3) \right]. \tag{11.47}$$

Keeping the terms to order $\hat{g}^2$ in γ_1 gives us

$$\begin{aligned}\gamma_1(\hat{g}) &= \beta(\hat{g}) \left[\tfrac{1}{3}\hat{g}\pi^4 \left(-\frac{1}{2\epsilon} + O(1) \right) + O(\hat{g}^2) \right] \\ &= \hat{g}^2 \left[\tfrac{1}{6}\pi^4 + O(\epsilon) \right] + O(\hat{g}^3),\end{aligned} \tag{11.48}$$

Substituting for $\hat{g}_c$ from equation (10.101) and using equation (11.31), we find

$$\begin{aligned}\eta = \gamma_1^* &= \frac{\epsilon^2}{54} + O(\epsilon^3) \\ &= 0.0185\ldots \quad \text{for } d = 3.\end{aligned} \tag{11.49}$$

This value of η agrees with that obtained in §10.4.6. We may now calculate δ from the scaling relation (11.40). Using the above value for γ_1^* we find

$$\delta = 4.89\ldots \tag{11.50}$$

in three dimensions.

As we explained at the end of the last chapter, credible results can be extracted from the ϵ-expansion only with a sophisticated summation technique such as Borel summation. So the calculation presented here is really only a sample of the larger calculation required to obtain several terms of the ϵ-expansion, which can be Borel summed to produce a final answer. Nothing conceptually new is involved in such a larger calculation, however. The best available bottom line is $\delta = 4.814 \pm 0.015$. For comparison Table 3.1 reports $\delta = 4.80 \pm .05$ for the $d = 3$ Ising model, while Table 1.2 reports $\gamma = 4.85 \pm 0.03$ from experiments on binary fluids. Notice again the surprising agreement between the critical exponents of the Landau–Ginzburg model and those of other systems.

Besides taking the first step towards a remarkably satisfactory value of the critical exponent δ, the main achievement of this chapter must be counted the demonstration that η and δ are connected by a scaling relation, because they are both derived from the value of the flow function γ_1 at the infrared stable fixed point. In the next chapter we extend this way of looking at critical exponents to those that are only defined away from the critical temperature.

Problems

11.1 Let $\psi(\mathbf{x})$ be a function of n coordinates $(x_1, \ldots, x_n)$ which satisfies the partial differential equation $H(\psi, \mathbf{x}, \mathbf{p}) = 0$, where $p_i \equiv \partial\psi/\partial x_i$. Given that the curve $\mathbf{x}(\lambda)$ is chosen such that

$$\frac{\mathrm{d}x_i}{\mathrm{d}\lambda} = \frac{\partial H}{\partial p_i},$$

show that these o.d.es can be complemented with

$$\frac{\mathrm{d}p_i}{\mathrm{d}\lambda} = -p_i \frac{\partial H}{\partial \psi} - \frac{\partial H}{\partial x_i}$$
$$\frac{\mathrm{d}\psi}{\mathrm{d}\lambda} = p_i \frac{\partial H}{\partial p_i}$$

to form a complete set of coupled o.d.es for the curves $(\mathbf{x}(\lambda), \mathbf{p}(\lambda))$. The latter are called the characteristics of $H = 0$. [Hint: express the derivative $\partial H/\partial x_i$ with $\mathbf{p}$ and ψ replaced by their values from the solution $\psi(\mathbf{x})$, in terms of partial derivatives of H in which only one argument varies at a time.]

11.2 How does the running coupling constant defined by (11.21) behave when the β-function has the following forms: (a) $\beta(\hat{g}) = A(\hat{g} - \hat{g}_c)$; (b) $\beta(\hat{g}) = B \operatorname{sgn}(\hat{g} - \hat{g}_c)|\hat{g} - \hat{g}_c|^p$ with $p > 1$; (c) $\beta(\hat{g}) = C/(\hat{g} - \hat{g}_c)$?

11.3 Use equations (11.31), (11.40) and (11.49) to obtain an expansion of δ in powers of ϵ. How does the value of δ obtained in this way agree with that derived in the text? Which do you consider to be the more reliable?

12
The renormalization group at $T \neq T_c$

To calculate the remaining critical exponents and to complete our understanding of the Landau–Ginzburg model in the critical regime, we have to calculate quantities in the limit $m^2 \to 0$, rather than at $m^2 = 0$. How can we extend our theory of the renormalization group to cover this more general case?

There are two possible ways to go about it. The first, and most obvious, is simply to repeat the whole renormalization of the theory with $m^2 \neq 0$. However, this method has several disadvantages. It involves evaluating integrals with both m^2 and k non-zero; this can be very difficult. It also involves appealing to some rather delicate physics to show that the vertex functions do indeed scale as $m^2 \to 0$. Finally, it is somewhat inelegant to adopt an approach for $T \neq T_c$ which is completely different from that adopted for $T = T_c$.

What is the alternative? Instead of treating the case $m^2 \neq 0$ as distinct, we can expand the vertex functions $\Gamma^{(n)}$ in powers of

$$\tau \equiv -(\mu^2 - \mu_c^2), \tag{12.1}$$

which is proportional to the deviation from the critical temperature.[1] This has the great advantage that all the coefficients in the expansion are computed at $m^2 = 0$. So, when we come to write down the finite-mass analogue

---
[1] The apparently perverse minus sign in (12.1) is present in order to avoid a host of other minus signs later on.

of the renormalization group equation (11.7), we shall find that we have already calculated most of the terms appearing in it.

12.1 Expansion about the critical temperature

We propose to carry out an expansion of the vertex functions about the critical temperature $T = T_c$. We could achieve this quite simply by replacing μ^2 in the Landau–Ginzburg Hamiltonian with $\mu_c^2 - \tau$, and then calculating the derivatives of the vertex functions with respect to τ at $\tau = 0$. However, it is useful to make a more general expansion. We shall allow τ to be a function of position, so that the Landau–Ginzburg Hamiltonian becomes

$$H_{LG} = \int d^d\mathbf{x} \left[\tfrac{1}{2}\alpha^2(\nabla\phi)^2 + \tfrac{1}{2}(\mu_c^2 - \tau(\mathbf{x}))\phi^2 + \tfrac{1}{4!}\lambda\phi^4 - J(\mathbf{x})\phi\right]. \quad (12.2)$$

We shall make a functional Taylor series expansion of the correlation functions of the theory about $\tau(\mathbf{x}) = 0$, similar to the expansions about $J(\mathbf{x}) = 0$ that we made in Chapter 8.

Let us consider the functional derivatives of $Z[J,\tau]$ with respect to $\tau(\mathbf{x})$, evaluated at $\tau(\mathbf{x}) = 0$. These are the thermal averages of powers of the quantity $\phi^2(\mathbf{x})$ at $T = T_c$, in just the same way as the functional derivatives of Z with respect to $J(\mathbf{x})$ are the thermal averages of powers of $\phi(\mathbf{x})$ at $T = T_c$. Specifically, we have

$$\frac{1}{Z} \frac{\delta}{\delta\tau(\mathbf{y}_1)} \frac{\delta}{\delta\tau(\mathbf{y}_2)} \cdots \frac{\delta}{\delta\tau(\mathbf{y}_l)} Z[J,\tau] = \left(\tfrac{1}{2}\right)^l \langle \phi^2(\mathbf{y}_1)\phi^2(\mathbf{y}_2)\ldots\phi^2(\mathbf{y}_l)\rangle. \quad (12.3)$$

We can also produce thermal averages of products of n powers of ϕ and l powers of ϕ^2, by differentiating n times with respect to J and l times with respect to τ:

$$\begin{aligned}
G^{(n,l)}(\mathbf{x}_1,\ldots,\mathbf{x}_n;\mathbf{y}_1,\ldots,\mathbf{y}_l) \\
\equiv \left(\tfrac{1}{2}\right)^l \langle \phi(\mathbf{x}_1)\ldots\phi(\mathbf{x}_n)\phi^2(\mathbf{y}_1)\ldots\phi^2(\mathbf{y}_l)\rangle \\
= \frac{1}{Z} \frac{\delta}{\delta J(\mathbf{x}_1)} \cdots \frac{\delta}{\delta J(\mathbf{x}_n)} \frac{\delta}{\delta\tau(\mathbf{y}_1)} \cdots \frac{\delta}{\delta\tau(\mathbf{y}_l)} Z[J,\tau].
\end{aligned} \quad (12.4)$$

Similarly, connected correlation functions involving ϕ^2 can be found by differentiating $\log Z$. For example, one of the simplest non-trivial connected correlation functions involving $\phi^2(\mathbf{y})$ is

$$\begin{aligned}
G_c^{(0,2)}(\mathbf{y}_1,\mathbf{y}_2) &\equiv \left(\tfrac{1}{2}\right)^2 \left[\langle\phi^2(\mathbf{y}_1)\phi^2(\mathbf{y}_2)\rangle - \langle\phi^2(\mathbf{y}_1)\rangle\langle\phi^2(\mathbf{y}_2)\rangle\right] \\
&= \frac{\delta}{\delta\tau(\mathbf{y}_1)}\frac{\delta}{\delta\tau(\mathbf{y}_2)} \log Z.
\end{aligned} \quad (12.5)$$

12.1 Expansion about the critical temperature

Notice that the derivatives with respect to J and τ commute, so we can write $G_c^{(n,l)}$ as a functional derivative either of $G_c^{(n)}$, or of $G_c^{(0,l)}$:

$$G_c^{(n,l)}(\mathbf{x}_1,\ldots,\mathbf{x}_n;\mathbf{y}_1,\ldots,\mathbf{y}_l) = \frac{\delta}{\delta\tau(\mathbf{y}_1)}\cdots\frac{\delta}{\delta\tau(\mathbf{y}_l)} G_c^{(n)}(\mathbf{x}_1,\ldots,\mathbf{x}_n) \\ = \frac{\delta}{\delta J(\mathbf{x}_1)}\cdots\frac{\delta}{\delta J(\mathbf{x}_n)} G_c^{(0,l)}(\mathbf{y}_1,\ldots,\mathbf{y}_l). \tag{12.6}$$

12.1.1 Functional Taylor expansions

The reason we are introducing these new correlation functions is as follows. Since the $G_c^{(n,l)}$ are the functional derivatives of the $G_c^{(n)}$ with respect to $\tau(\mathbf{x})$, we can reconstruct the correlation functions for an arbitrary distribution $\tau(\mathbf{x})$ by a functional Taylor expansion in powers of τ in which the coefficients are precisely the $G_c^{(n,l)}$ evaluated at some reference field $\tau_0(\mathbf{x})$. (See Appendix L for more about functional Taylor series.) In practice, we shall always choose the reference field to be zero everywhere, so that the expansion is about the critical temperature. For example, the functional Taylor series for $G_c^{(n)}$ is

$$G_c^{(n)}(\mathbf{x}_1,\ldots,\mathbf{x}_n;\tau) = \sum_{l=0}^{\infty} \frac{1}{l!} \int d^d\mathbf{y}_1 \ldots d^d\mathbf{y}_l\, \tau(\mathbf{y}_1)\ldots\tau(\mathbf{y}_l) \\ \times G_c^{(n,l)}(\mathbf{x}_1,\ldots,\mathbf{x}_n;\mathbf{y}_1,\ldots,\mathbf{y}_l). \tag{12.7}$$

(From here on, $G_c^{(n,l)}$ in which τ does not appear explicitly are understood to be evaluated with $\tau = 0$.) This series enables us to write the connected correlation functions $G_c^{(n)}$ for an arbitrary temperature distribution $\mu_c^2 - \tau(\mathbf{x})$ in terms of our more general correlation functions $G_c^{(n,l)}$ at the critical temperature.

12.1.2 Diagrammatic representation of the ϕ^2 correlation functions

How can we represent these new correlation functions diagrammatically? This is straightforward, once we remember that the quantity $\phi^2(\mathbf{x})$ is just the product of two factors of $\phi(\mathbf{x})$. The correlation function $G_c^{(n,l)}$ in position space is therefore a special case of the correlation function $G_c^{(n+2l)}$, in which l pairs of position arguments have been set equal to each other.

We can symbolize the fact that l pairs of position arguments are to be set equal by introducing a new symbol, a square block. When we make diagrams, we regard this block as having two 'hooks' on which to hang links. The block represents a pair of external points whose positions have been made to coincide; as such, the position of the block is not integrated over when

Figure 12.1 Contributions to $G_c^{(2,1)}(\mathbf{x}_1, \mathbf{x}_2; \mathbf{y}_1)$ at zeroth and first order in λ.

Figure 12.2 A square block is the limit of two external points as their separation δ tends to zero, but there are two contributions to this limit.

evaluating the contribution of the graph. For example, the two diagrams shown in Figure 12.1 contribute to $G_c^{(2,1)}$ up to first order in λ.

As always, the thing to be careful about when evaluating the diagrams is their symmetry factors. Since the diagrams for the $G_c^{(n,l)}$ are just those for $G_c^{(n+2l)}$ with l pairs of legs coinciding, we can obtain the correct symmetry factor by treating the end points of each pair of joined legs as distinct and then applying the rules of §8.2.2 and Box 8.2. However, one thing deserves comment: each time we join two legs with a square block, there are two possible ways we can arrange the two position arguments as they approach each other (see Figure 12.2). This results in a factor of 2^l for a diagram containing l blocks, and this conveniently cancels the factor of $\left(\frac{1}{2}\right)^l$ that occurred in the relation (12.4) between the $G^{(n,l)}$ and the correlation function of l powers of ϕ^2. The net result is that we use the rules of Chapter 8 with no additional factors of 2 or $\frac{1}{2}$.

12.1.3 Wavevector space

In Chapter 8 it proved much more convenient to work in k-space when we actually carry out calculations. So let us define quantities $\widetilde{G}_c^{(n,l)}$ by the equations

$$G_c^{(n,l)}(\mathbf{x}_1, \ldots, \mathbf{x}_n; \mathbf{y}_1, \ldots, \mathbf{y}_l) = \int d^d\check{\mathbf{k}}_1 e^{i\mathbf{k}_1 \cdot \mathbf{x}_1} \ldots d^d\check{\mathbf{k}}_n e^{i\mathbf{k}_n \cdot \mathbf{x}_n}$$
$$\times d^d\check{\mathbf{p}}_1 e^{i\mathbf{p}_1 \cdot \mathbf{y}_1} \ldots d^d\check{\mathbf{p}}_l e^{i\mathbf{p}_l \cdot \mathbf{y}_l} \quad (12.8)$$
$$\times \widetilde{G}_c^{(n,l)}(\mathbf{k}_1, \ldots, \mathbf{k}_n; \mathbf{p}_1, \ldots, \mathbf{p}_l)$$

(compare equation (8.22)). In words, $G_c^{(n,l)}$ is the $(n+l)$-dimensional Fourier transform of $\widetilde{G}_c^{(n,l)}$. We know, however, that $G_c^{(n,l)}$ must be numerically equal to $G_c^{(n+2l)}$ with its arguments set equal in pairs. When we write $G_c^{(n+2l)}$ in

12.1 Expansion about the critical temperature

terms of $\widetilde{G}_c^{(n+2l)}$ and substitute this in (12.8), l of the wavevector integrations just give δ-functions. This implies a relationship between $\widetilde{G}_c^{(n,l)}$ and $\widetilde{G}_c^{(n+2l)}$:

$$\widetilde{G}_c^{(n,l)}(\mathbf{k}_1,\ldots,\mathbf{k}_n;\mathbf{p}_1,\ldots,\mathbf{p}_l) =$$
$$\int d^d\check{\mathbf{q}}_1\ldots\int d^d\check{\mathbf{q}}_l\, \widetilde{G}_c^{(n+2l)}(\mathbf{k}_1,\ldots,\mathbf{k}_n;\mathbf{p}_1+\mathbf{q}_1,-\mathbf{q}_1,\ldots,\mathbf{p}_l+\mathbf{q}_l,-\mathbf{q}_l). \tag{12.9}$$

From §8.3 we know that $\widetilde{G}_c^{(n+2l)}$ is the connected thermal average of $(2n+l)$ powers of $\widetilde{\phi}$. Also, we have

$$\phi^2(\mathbf{x}) = \int d^d\check{\mathbf{p}}\, e^{i\mathbf{p}\cdot\mathbf{x}} \int d^d\check{\mathbf{q}}\, \widetilde{\phi}(\mathbf{p}+\mathbf{q})\widetilde{\phi}(-\mathbf{q}) = \int d^d\check{\mathbf{p}}\, e^{i\mathbf{p}\cdot\mathbf{x}} \widetilde{\phi^2}(\mathbf{p}), \tag{12.10}$$

with

$$\widetilde{\phi^2}(\mathbf{p}) \equiv \int d^d\check{\mathbf{q}}\, \widetilde{\phi}(\mathbf{p}+\mathbf{q})\widetilde{\phi}(-\mathbf{q}). \tag{12.11}$$

Comparing the arguments of $\widetilde{\phi}$ in (12.11) with the arguments of $\widetilde{G}_c^{(n+2l)}$ in (12.9), we see that the functions $\widetilde{G}_c^{(n,l)}$ must be the connected correlation functions for n powers of $\widetilde{\phi}$ and l powers of $\widetilde{\phi^2}$:

$$\widetilde{G}_c^{(n,l)}(\mathbf{k}_1,\ldots,\mathbf{k}_n;\mathbf{p}_1,\ldots,\mathbf{p}_l)$$
$$= \left(\tfrac{1}{2}\right)^l \langle \widetilde{\phi}(\mathbf{k}_1)\ldots\widetilde{\phi}(\mathbf{k}_n)\widetilde{\phi^2}(\mathbf{p}_1)\ldots\widetilde{\phi^2}(\mathbf{p}_l)\rangle_{\text{connected}}$$
$$= (2\pi)^{d(n+l)} \frac{\delta}{\delta \widetilde{J}(-\mathbf{k}_1)}\ldots\frac{\delta}{\delta \widetilde{J}(-\mathbf{k}_n)}\frac{\delta}{\delta \widetilde{\tau}(-\mathbf{p}_1)}\ldots\frac{\delta}{\delta \widetilde{\tau}(-\mathbf{p}_l)} \log Z. \tag{12.12}$$

This shows that $\widetilde{G}_c^{(n,l)}$ is the lth functional derivative of $\widetilde{G}_c^{(n)}$ with respect to $\widetilde{\tau}$, as expected from (12.6). It follows that

$$\widetilde{G}_c^{(n)}(\mathbf{k}_1,\ldots,\mathbf{k}_n;\widetilde{\tau}) = \sum_{l=0}^{\infty}\frac{1}{l!}\int d^d\check{\mathbf{p}}_1\ldots d^d\check{\mathbf{p}}_l\, \widetilde{\tau}(-\mathbf{p}_1)\ldots\widetilde{\tau}(-\mathbf{p}_l) \\ \times \widetilde{G}_c^{(n,l)}(\mathbf{k}_1,\ldots,\mathbf{k}_n;\mathbf{p}_1,\ldots,\mathbf{p}_l;\widetilde{\tau}=0) \tag{12.13}$$

(compare (12.7)).

What effect does all this have on the diagram rules for the $\widetilde{G}_c^{(n,l)}$? Well, introducing the square blocks into our set of diagram rules effectively creates l additional loops in the diagrams for $\widetilde{G}_c^{(n,l)}$ relative to those for $\widetilde{G}_c^{(n+2l)}$, because external points which were separate get joined together. Equation (12.9) tells us to treat these additional loops just as we would loops appearing elsewhere in the diagram, but with a wavevector $\mathbf{p}_i$ entering the diagram through the ith square block. We shall not display the wavevector entering the diagram at a square block explicitly, but it can easily be inferred from

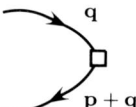

Figure 12.3 The symbol we shall use for a $\widetilde{\phi^2}$ vertex. A wavevector **p** is entering the diagram at the square block; such a square block appears when the thermal average represented by the diagram includes a factor of $\widetilde{\phi^2}(\mathbf{p})$. The wavevector **q** comes from the remainder of the diagram (not shown); in general **q** is integrated over. Although each block looks like a vertex with two legs, there is no associated coupling constant. The factor associated with each block is unity.

wavevector conservation and the wavevectors on the two propagator lines attached to it. Figure 12.3 shows a square block at which wavevector **p** is entering the diagram; if the wavevector **q** is not fixed uniquely by the wavevectors in the rest of the diagram, it is integrated over just as for any other undetermined wavevector. This is just about the simplest set of rules for such a graph that one could imagine!

12.1.4 Vertex functions

In §8.4 we showed that the vertex functions, defined in terms of amputated 1PI diagrams, are the coefficients in the functional Taylor expansion of the dimensionless Gibbs free energy $\Gamma[\widetilde{\varphi}]$ in powers of $\widetilde{\varphi} \equiv \langle \widetilde{\phi} \rangle$. The obvious way to proceed here is to define $\Gamma^{(n,l)}$ as the lth functional derivative of the vertex function $\Gamma^{(n)}$ with respect to $\widetilde{\tau}(\mathbf{p})$:

$$\Gamma^{(n,l)}(\mathbf{k}_1, \ldots, \mathbf{k}_n; \mathbf{p}_1, \ldots, \mathbf{p}_l) \equiv \frac{\delta}{\delta \widetilde{\tau}(-\mathbf{p}_1)} \cdots \frac{\delta}{\delta \widetilde{\tau}(-\mathbf{p}_l)} \Gamma^{(n)}(\mathbf{k}_1, \ldots, \mathbf{k}_n). \tag{12.14}$$

Then

$$\Gamma^{(n)}(\mathbf{k}_1, \ldots, \mathbf{k}_n; \widetilde{\tau}) = \sum_{l=0}^{\infty} \frac{1}{l!} \int d^d \check{\mathbf{p}}_1 \ldots d^d \check{\mathbf{p}}_l \, \widetilde{\tau}(-\mathbf{p}_1) \ldots \widetilde{\tau}(-\mathbf{p}_l)$$
$$\times \Gamma^{(n,l)}(\mathbf{k}_1, \ldots, \mathbf{k}_n; \mathbf{p}_1, \ldots, \mathbf{p}_l; \widetilde{\tau} = 0). \tag{12.15}$$

The definition (12.14) implies that the $\Gamma^{(n,l)}$ are made out of the 1PI graphs contributing to $\widetilde{G}_c^{(n,l)}$ in just the same way as the $\Gamma^{(n)}$ are made out of the 1PI graphs contributing to $\widetilde{G}_c^{(n)}$.[2] This strongly suggests that the diagrams

---

[2] Notice that this implies that the $\Gamma^{(n,l)}$ are the Legendre transforms of the $\widetilde{G}_c^{(n,l)}$ with respect to the pair of variables $(\widetilde{\varphi}, \widetilde{J})$ but not with respect to the pair $(\langle \widetilde{\phi^2} \rangle, \widetilde{\tau})$. They are therefore functions of the variables $\widetilde{\varphi}$ and $\widetilde{\tau}$. It would also be possible to perform an additional Legendre transform with respect to the pair $(\langle \widetilde{\phi^2} \rangle, \widetilde{\tau})$ to produce a function of $\widetilde{\varphi}$ and $\langle \widetilde{\phi^2} \rangle$, but we shall not consider these functions here.

12.1 Expansion about the critical temperature

contributing to $\Gamma^{(n,l)}$ are the 1PI diagrams contributing to $\widetilde{G}_c^{(n,l)}$ with the external legs that have not already been tied together, amputated. And, you will no doubt be relieved to hear, this suggestion is absolutely correct! (The only tricky bit in the proof of this statement concerns the symmetry factors.)

One consequence of these rules is that there is overall wavevector conservation throughout a diagram, and it is helpful to factorize out a δ-function factor and to define the reduced vertex function through the equation

$$\Gamma^{(n,l)}(\mathbf{k}_1, \ldots, \mathbf{k}_n; \mathbf{p}_1, \ldots, \mathbf{p}_l) =$$
$$(2\pi)^d \delta(\mathbf{k}_1 + \cdots + \mathbf{p}_l) \Gamma^{(n,l)}(\mathbf{k}_1, \ldots, \mathbf{k}_n; \mathbf{p}_1, \ldots, \mathbf{p}_{l-1}). \quad (12.16)$$

The reduced and full versions of $\Gamma^{(n,l)}$ are in precisely the same relationship to each other as the reduced and full versions of $\Gamma^{(n)}$—see §8.3—and we again distinguish the full and reduced forms by the number of their arguments. Notice that in (12.16) we have chosen to use the δ-function to eliminate one of the **p** arguments, rather than one of the **k** arguments; this was an arbitrary choice, and in the rest of the chapter we shall sometimes find it convenient to eliminate a **k** argument instead. It is only whether the *total* number of arguments is $n+l$ or $n+l-1$ that is important in determining whether a vertex function is of the full or the reduced type.

12.1.5 Renormalization

You might think that, since the correlation functions $G_c^{(n,l)}$ are related so simply in real space to the correlation functions $G_c^{(n+2l)}$, we would need to do no more work than we have already done in absorbing all Λ-sensitivity in functions of the latter type into a small number of renormalized parameters. Unfortunately, this is not the case; there is additional Λ-sensitivity in the $\Gamma^{(n,l)}$ over and above that present in the $\Gamma^{(n)}$.

This may seem surprising. However, examine equation (7.12), where we showed that μ^2 is linearly related to the temperature. There is no reason to believe that the constant of proportionality in this equation should be independent of the cutoff. The $\Gamma^{(n,l)}$ are therefore derivatives of the Gibbs free energy with respect to a potentially Λ-sensitive quantity. This Λ-sensitivity also manifests itself in the algebra of §§12.1.3 and 12.1.4; equation (12.9) shows that the expressions for $\Gamma^{(n,l)}$ with $l \geq 1$ involve l additional wavevector integrations relative to those for $\Gamma^{(n)}$. Our argument that we can write all our integrals in a form where they do not depend on their upper limits by expressing the theory in terms of (a, m, g, κ) instead of $(\alpha, \mu, \lambda, \Lambda)$ relies on a delicate balance between the number of links and the number of loops appearing in each graph. This balance has now been disturbed, so we need to reconsider the argument.

Let us start with a simple example. Consider the graphs for the function $\Gamma^{(4)}$ to one loop shown in Figure 12.4. After replacing μ^2 by $m^2 = 0$ in all

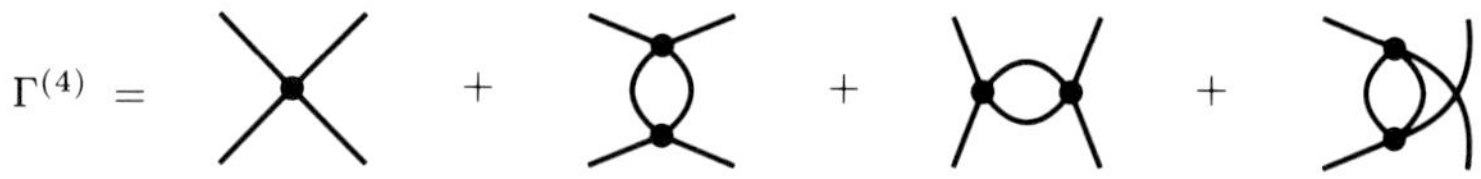

Figure 12.4 Contributions to $\Gamma^{(4)}$ up to one loop.

Figure 12.5 Contributions to $\Gamma^{(2,1)}$ up to one loop.

the propagators, they give rise to equation (9.17):

$$\Gamma^{(4)}(\mathbf{k}_1, \mathbf{k}_2, \mathbf{k}_3) = \lambda \left[1 - \frac{\lambda}{2} \int_0^\Lambda \frac{d^d \check{\mathbf{q}}}{a^2 q^2} \left(\frac{1}{a^2 (\mathbf{k}_1 + \mathbf{k}_2 - \mathbf{q})^2} \right. \right.$$
$$\left. \left. + 2 \text{ permutations} \right) \right] + O(\lambda^3). \quad (12.17)$$

We saw in Chapter 9 that we can absorb the Λ-sensitivity into the parameter $g \equiv \Gamma^{(4)}(\kappa_1, \kappa_2, \kappa_3)|_{\text{SP}}$. Compare this, however, with the diagrams for $\Gamma^{(2,1)}(\mathbf{k}, \mathbf{p})$ to one loop shown in Figure 12.5. The result for $\Gamma^{(2,1)}$ is

$$\Gamma^{(2,1)}(\mathbf{k}, \mathbf{p}) = 1 - \tfrac{1}{2}\lambda \int_0^\Lambda \frac{d^d \check{\mathbf{q}}}{a^4 q^2 (\mathbf{p} + \mathbf{q})^2} + O(\lambda^2). \quad (12.18)$$

As far as its dependence on Λ is concerned, (12.18) is similar to equation (12.17)—at $d = 4$, both are proportional to $\log \Lambda$ for large Λ. However, the powers of λ occurring in each are different: the first term in (12.18) is independent of λ, and the second term contains only a single power of λ. As a result, we cannot eliminate both λ and the Λ-sensitive integral simultaneously from our expression for $\Gamma^{(2,1)}$.

The physical origin of this Λ-sensitivity also suggests a simple way to deal with it. Let us multiply and divide $\Gamma^{(2,1)}(\mathbf{k}, \mathbf{p})$ by $\Gamma^{(2,1)}(\kappa, \kappa)$:

$$\Gamma^{(2,1)}(\mathbf{k}, \mathbf{p}) = \Gamma^{(2,1)}(\kappa, \kappa) \frac{\Gamma^{(2,1)}(\mathbf{k}, \mathbf{p})}{\Gamma^{(2,1)}(\kappa, \kappa)}. \quad (12.19)$$

This works because the coefficient of $\log \Lambda$ is the same in both $\Gamma^{(2,1)}(\mathbf{k}, \mathbf{p})$ and $\Gamma^{(2,1)}(\kappa, \kappa)$, so the ratio in the last equation is independent of Λ (to

12.1 Expansion about the critical temperature

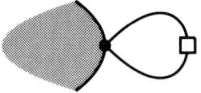

Figure 12.6 The primitively divergent sub-graph associated with a square block. The shaded portion represents the rest of the diagram.

$O(\lambda)$). This ratio is the derivative of the Gibbs free energy with respect to a temperature that has been rescaled by a factor $\Gamma^{(2,1)}(\kappa,\kappa)$ so as to remove its Λ-sensitivity. So we have isolated the Λ-sensitivity of $\Gamma^{(2,1)}(\mathbf{k},\mathbf{p})$ into $\Gamma^{(2,1)}(\kappa,\kappa)$, which we shall call $1/Z_2$.[3] To the order of the graphs in Figure 12.5, we have

$$Z_2 = 1 + \tfrac{1}{2}\lambda \int_0^\Lambda \frac{d^d\check{q}}{a^4 q^2(\mathbf{q}+\kappa)^2} + O(\lambda^2). \tag{12.20}$$

Now consider the renormalization of vertex functions $\Gamma^{(n,l)}$ more complex than $\Gamma^{(2,1)}$. In the language of §9.4, the square blocks have introduced a new class of primitively divergent sub-graphs (see Figure 12.6). Assuming that the 'body' of the diagram in Figure 12.6 is not Λ-sensitive, each such sub-graph will contribute exactly one Λ-sensitive integral of the same form as in (12.18). This Λ-sensitivity can therefore be absorbed by l powers of Z_2; it is convenient to factor these out and to define renormalized vertex functions $\Gamma_R^{(n,l)}$ by

$$\Gamma^{(n,l)}(\mathbf{k}_1,\ldots,\mathbf{k}_n;\mathbf{p}_1,\ldots,\mathbf{p}_{l-1};\lambda,\Lambda) \equiv \\ a^n Z_2^{-l} \Gamma_R^{(n,l)}(\mathbf{k}_1,\ldots,\mathbf{k}_n;\mathbf{p}_1,\ldots,\mathbf{p}_{l-1};\hat{g},\kappa), \tag{12.21}$$

(compare equation (9.31)). The quantities $\Gamma_R^{(n,l)}$ are then (almost) all Λ-independent for $d < 6$, since it turns out in the Landau–Ginzburg model that additional Λ-sensitivity does not arise when integrating over Λ-insensitive sub-graphs.

We say almost, because there are in fact two more primitive divergences that appear at *zeroth* order in λ, i.e., in the Gaussian model. One appears in the diagram for $\Gamma^{(0,1)}$ containing only a square block and a single propagator (see Figure 12.7(a)); this is, however, not a very important diagram, since it is just a contribution to the internal energy of the model. We would expect such a quantity to be Λ-sensitive. Much more important, however, is the Λ-sensitivity of the graph in which two square blocks are directly joined by two

---

[3] As so often, this apparently perverse notation has its origins in quantum field theory.

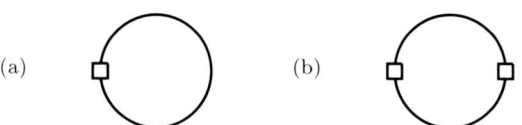

Figure 12.7 The lowest-order diagrams for (a) $\Gamma^{(0,1)}$ and (b) $\Gamma^{(0,2)}$.

propagators. Such a graph contributes only to $\Gamma^{(0,2)}$ (see Figure 12.7(b)), and is important because, as we shall see later, this vertex function gives the heat capacity of our model. So it is vital that we arrange for it to give a result independent of our cut-off procedure.

The graph in Figure 12.7(b) gives

$$\Gamma^{(0,2)}(\mathbf{p}) = \int_0^\Lambda \frac{d^d\check{\mathbf{q}}}{a^4 q^2 (\mathbf{q}+\mathbf{p})^2} + O(\lambda). \tag{12.22}$$

At $d = 4$ this integral goes as $\log \Lambda$. It is independent of λ, so this Λ-sensitivity occurs even in the Gaussian model. There is therefore no hope of making it Λ-independent by multiplying by Z_2 which differs from unity only at first order in the coupling—see equation (12.20). Instead we have to treat this diagram as a special case, and express it in terms of its value at $\mathbf{p} = \boldsymbol{\kappa}$, which we write as A. So

$$\Gamma^{(0,2)}(\mathbf{p}) = [\Gamma^{(0,2)}(\mathbf{p}) - \Gamma^{(0,2)}(\boldsymbol{\kappa})] + \Gamma^{(0,2)}(\boldsymbol{\kappa}) \equiv Z_2^{-2}\Gamma_R^{(0,2)}(\mathbf{p}) + A. \tag{12.23}$$

We have isolated the Λ-sensitivity of $\Gamma^{(0,2)}$ into Z_2 and A.

Our results can be summarized as

$$\Gamma^{(n,l)}(\mathbf{k}_1,\ldots,\mathbf{p}_{l-1};\alpha,\lambda,\Lambda) = a^n Z_2^{-l}\Gamma_R^{(n,l)}(\mathbf{k}_1,\ldots,\mathbf{p}_{l-1};\hat{g},\kappa) + A\delta_{n,0}\delta_{l,2}. \tag{12.24}$$

Notice that the second term on the right is required only for $\Gamma^{(0,2)}$.

12.1.6 Expanding the renormalized vertex functions

Now we can make an expansion for our vertex functions about the critical temperature. First, we consider the expansion in the presence of a uniform field $\tau(\mathbf{x}) = \tau$. Then we can write the vertex function $\Gamma^{(n)}$ as

$$\Gamma^{(n)}(\mathbf{k}_i;\mu_c - \tau,\lambda,\Lambda) = \sum_{l=0}^{\infty} \frac{\tau^l}{l!}\Gamma^{(n,l)}(\mathbf{k}_i,0;\mu_c,\lambda,\Lambda), \tag{12.25}$$

12.1 Expansion about the critical temperature

since the Fourier transform of $\tau(\mathbf{x})$ is $(2\pi)^d \tau \delta(\mathbf{k})$. For the time being we shall assume we are dealing with $(n,l) \neq (0,2)$. Then we can convert (12.25) into an equation expressed entirely in terms of renormalized vertex functions:

$$\Gamma_R^{(n)}(\mathbf{k}_i; t, \hat{g}, \kappa) = \sum_{l=0}^{\infty} \frac{t_R^l}{l!} \Gamma_R^{(n,l)}(\mathbf{k}_i, 0; t_R = 0, \hat{g}, \kappa), \qquad (12.26)$$

where $t = \tau/Z_2$ and we have made a change of variable so that the renormalized vertex functions depend on $t_R = t/a^2$, rather than m^2/a^2.

12.1.7 The validity of the expansion

This expansion is going to be extremely useful to us. In particular, we are going to use it to derive a renormalization group equation that will enable us to derive long-range properties of the vertex functions, and hence find critical exponents away from T_c. So, it is appropriate to ask whether the expansion is actually valid.

At first sight, this seems to be extremely doubtful. This is because we know that the properties of the system behave as non-integral powers of $|T - T_c|$. Such powers do not have Taylor series expansions about $T = T_c$, so our task appears hopeless. This manifests itself in the expansion (12.26) through infrared divergences of the $\Gamma_R^{(n,l)}$ when the $\mathbf{p}_i$ go to zero. We noted a similar problem when expanding $\langle \phi \rangle$ in powers of J in §11.2.2.

This is why we have been working with a *functional* Taylor expansion. It allows us to calculate the response of the system to an arbitrary pattern of temperature deviation from T_c. Suppose we choose a pattern $t_R f(\mathbf{x})$, where $f(\mathbf{x})$ is close to one over a volume of linear dimensions much greater than the correlation length, but at large distances is such that $\tilde{f}(0) = 0$. Removing the zero-wavevector component of the temperature distribution removes the infrared divergences from the $\Gamma_R^{(n,l)}$, giving the series (12.26) a finite radius of convergence about $t_R = 0$. This radius is very small, and it gets smaller as the volume over which $f(\mathbf{x}) \approx 1$ is increased. For values of t_R inside the radius of convergence the behaviour of the $\Gamma_R^{(n)}$ is quite different to the critical behaviour that interests us, because an analytic function cannot behave as a fractional power. However, this doesn't matter. The essential point is that we have moved the singularity that was present at $T = T_c$ off the real t_R axis. We shall assume that this leaves the $\Gamma_R^{(n)}$ analytic in a strip of non-zero width on either side of this axis.

For values of t_R inside the radius of convergence of (12.26) we can safely derive the renormalization group equation (12.30) used in the next section. The differential operator on the left of this equation is an analytic function of t_R, and so the entire left-hand side is analytic in the narrow strip where the $\Gamma_R^{(n)}$ are analytic. We know that the left-hand side of (12.30) is zero in a small but finite part of this region; therefore, it is zero everywhere in the region.

In particular, it is zero for values of t_R outside the radius of convergence of (12.26) and so equation (12.30) has a wider validity than might be expected, so long as the assumption at the end of the last paragraph holds. From here on we shall use the expansion (12.26) without comment.

12.2 The renormalization group equations

Since they correspond to real physical things, namely local derivatives of the $\Gamma^{(n)}$ with respect to a quantity proportional to temperature, the vertex functions $\Gamma^{(n,l)}$ must be independent of κ in just the same way as the $\Gamma^{(n)}$ in Chapter 11. Equation (12.24) therefore implies that

$$\kappa \frac{\partial}{\partial \kappa} \left[a^n Z_2^{-l} \Gamma_R^{(n,l)}(\mathbf{k}_i; t_R = 0, \hat{g}, \kappa) \right] = 0. \tag{12.27}$$

If we isolate the κ-dependence of the different terms and neglect the extra term on the right of (12.24), which is only present for $(n,l) = (0,2)$, we find

$$\left[\kappa \frac{\partial}{\partial \kappa} + \beta \frac{\partial}{\partial \hat{g}} - \frac{n}{2} \gamma_1 + l \gamma_2 \right] \Gamma_R^{(n,l)}(\mathbf{k}_i; t_R = 0, \hat{g}, \kappa) = 0 \quad (n,l) \neq (0,2), \tag{12.28}$$

where β and γ_1 are still defined by equation (11.8) and

$$\gamma_2 \equiv \kappa \frac{\partial \log t_R}{\partial \kappa} = -\kappa \frac{\partial \log Z_2}{\partial \kappa} \tag{12.29}$$

describes how t, or equivalently Z_2, changes with κ.

Since we are interested in deriving an equation satisfied by $\Gamma^{(n)}$ rather than the $\Gamma^{(n,l)}$, we multiply each of equations (12.28) by $t_R{}^l$ and add. Then, since $t_R dt_R{}^l/dt_R = l t_R{}^l$, we have with (12.26)

$$\left[\kappa \frac{\partial}{\partial \kappa} + \beta \frac{\partial}{\partial \hat{g}} - \tfrac{1}{2} n \gamma_1 + \gamma_2 t_R \frac{\partial}{\partial t_R} \right] \Gamma_R^{(n)}(\mathbf{k}_i; t_R, \hat{g}, \kappa) = 0 \quad (n \neq 0). \tag{12.30}$$

This is the desired renormalization group equation for $t \neq 0$. As promised, we find that we have already calculated the quantities β and γ_1 appearing on the left-hand side. The only additional work required to study the system away from the critical point is therefore the calculation of γ_2. Note that γ_2, like β and γ_1, can depend only on $\hat{g}$, since it too is evaluated at $m^2 = 0$, where $\hat{g}$ is the only dimensionless variable available.

However, just as was the case in §11.1, we must combine (12.30) with dimensional analysis if we are to find out anything more interesting than the dependence of $\Gamma^{(n)}$ on the arbitrary wavenumber κ. Since $t_R = t/a^2$, dimensional analysis implies that

$$\left[\kappa \frac{\partial}{\partial \kappa} - b \frac{\partial}{\partial b} + 2 t_R \frac{\partial}{\partial t_R} \right] \Gamma_R^{(n)}(\mathbf{k}_i/b; t_R, \hat{g}, \kappa) = [n - (\tfrac{1}{2}n - 1)d] \Gamma_R^{(n)}(\mathbf{k}_i/b; t_R, \hat{g}, \kappa). \tag{12.31}$$

12.2 The renormalization group equations

Eliminating κ between (12.30) and (12.31), we obtain

$$\left[b\frac{\partial}{\partial b} + \beta\frac{\partial}{\partial \hat{g}} - \frac{n}{2}\gamma_1 + (\gamma_2 - 2)t_R\frac{\partial}{\partial t_R} + n - (\tfrac{1}{2}n - 1)d\right]\Gamma_R^{(n)}(\mathbf{k}_i/b; t_R, \hat{g}, \kappa) = 0. \tag{12.32}$$

We can solve the equation by the method of characteristics (see Box 11.2). By analogy with equations (11.21) and (11.22), let us define a running value of t_R by the equation

$$b\frac{\mathrm{d}\log t_R(b)}{\mathrm{d}b} = 2 - \gamma_2(\hat{g}(b)) \qquad \text{with} \qquad t_R(1) = t_R. \tag{12.33}$$

Then the solution to (12.32) is

$$\Gamma_R^{(n)}(\mathbf{k}_i/b; t_R, \hat{g}, \kappa) = b^{(n/2-1)d-n}a^n(b)a^{-n}\Gamma_R^{(n)}(\mathbf{k}_i; t_R(b), \hat{g}(b), \kappa). \tag{12.34}$$

By equation (9.31), the corresponding solution for $\Gamma^{(n)}$ is

$$\Gamma^{(n)}(\mathbf{k}_i/b; t_R, \hat{g}, \kappa) = b^{(n/2-1)d-n}a^n(b)\Gamma_R^{(n)}(\mathbf{k}_i; t_R(b), \hat{g}(b), \kappa). \tag{12.35}$$

As $\hat{g}(b)$ tends to a fixed point $\hat{g}_c$ this solution becomes

$$\Gamma^{(n)}(\mathbf{k}_i/b; t_R, \hat{g}, \kappa) = b^{(n/2-1)d-n}b^{n\gamma_1^*/2}\Gamma_R^{(n)}(\mathbf{k}_i; t_R(b), \hat{g}_c, \kappa), \tag{12.36}$$

with

$$t_R(b) \sim t_R(1)b^{2-\gamma_2^*}. \tag{12.37}$$

We can now derive all the classical relationships between the critical exponents.

12.2.1 The exponent ν

This exponent controls the divergence of the correlation length with temperature:

$$\xi \sim (T - T_c)^{-\nu} \sim t_R^{-\nu}. \tag{12.38}$$

How can we recognize the correlation length from our expressions for the vertex functions? We expect that (see equation (6.81))

$$\Gamma^{(n)}(\mathbf{k}_i/b) \sim b^p f(b/\xi), \tag{12.39}$$

where p is some power and f some function. However, examining the right-hand side of (12.36) shows that it is of exactly this form: the function $\Gamma_R^{(n)}$ depends on b only through the value of $t_R(b)$. Using (12.37), we see that

$$b/\xi \sim \left(t_R(1)b^{2-\gamma_2^*}\right)^{1/(2-\gamma_2^*)}, \tag{12.40}$$

so

$$\xi \sim t_R^{-1/(2-\gamma_2^*)}. \tag{12.41}$$

Comparing with (12.38) shows that

$$\frac{1}{\nu} = 2 - \gamma_2^*. \tag{12.42}$$

The exponent ν is determined by the value of the flow function γ_2 at the fixed point.

12.2.2 The exponent γ

It is now very straightforward to extract the susceptibility exponent γ, which reflects the temperature dependence of $\Gamma^{(2)}(\mathbf{k} = 0) \sim \chi^{-1}$ on t_R as $t_R \to 0$ in the form

$$\Gamma^{(2)}(\mathbf{k} = 0; t_R, \hat{g}, \kappa) \sim t_R^\gamma. \tag{12.43}$$

However, (12.36) gives

$$\Gamma^{(2)}(\mathbf{k} = 0; t_R, \hat{g}, \kappa) \sim b^{\gamma_1^* - 2} \Gamma_R^{(2)}(\mathbf{k} = 0; t_R(b), \hat{g}_c, \kappa). \tag{12.44}$$

Using (12.37) and (12.43), we can deduce that

$$t_R^\gamma \sim b^{\gamma_1^* - 2} t_R^\gamma b^{\gamma(2 - \gamma_2^*)}. \tag{12.45}$$

Dividing through by t_R^γ yields

$$\gamma_1^* - 2 + \gamma(2 - \gamma_2^*) = 0, \tag{12.46}$$

or

$$\gamma = \frac{2 - \gamma_1^*}{2 - \gamma_2^*}.$$

From this and equation (11.31) we can immediately obtain the Fisher scaling law of equation (1.29)

$$\gamma = \nu(2 - \eta). \tag{12.47}$$

12.2.3 The exponent α

Consider the vertex function $\Gamma^{(0,2)}$. It is proportional to the second derivative of the Gibbs free energy with respect to the temperature:

$$\left(\frac{\partial^2 G}{\partial T^2}\right)_M = -\left(\frac{\partial S}{\partial T}\right)_M. \tag{12.48}$$

This in turn is proportional to the specific heat at constant magnetization.[4] Now, α is defined in terms of the specific heat at constant magnetic field; however, it is straightforward to show (see Problem 12.2) that the two specific heats diverge in the same way as $T \to T_c$. So, to calculate α, it suffices to find the behaviour of $\Gamma^{(0,2)}$:

$$\Gamma^{(0,2)}(\mathbf{p} = 0; t_R, \hat{g}, \kappa) \sim t_R^{-\alpha}. \tag{12.49}$$

---

[4] There is a subtlety here. Below T_c the system has a non-zero magnetization, but $\Gamma^{(2,0)}$ is the derivative of the Gibbs free energy evaluated at zero magnetization. However, because the derivative is being evaluated at the critical temperature the two derivatives are equal (see Problem 12.1).

12.2 The renormalization group equations

However, we saw in §12.1.5 that this vertex function is rather a special case; just as in §5.6.3, the specific heat exponent therefore has to be calculated in a slightly different way from the others. We can expand $\Gamma_R^{(0,2)}$ in the form

$$\Gamma_R^{(0,2)}(\mathbf{p}; t_R, \hat{g}, \kappa) = \sum_{l=0}^{\infty} \frac{t_R^l}{l!} \Gamma_R^{(0,2+l)}(\mathbf{p}, 0, \ldots, 0; t_R = 0, \hat{g}, \kappa). \tag{12.50}$$

Now all the terms on the right except the first satisfy an equation like (12.28). The first, however, satisfies an equation obtained by taking the logarithmic derivative of (12.24) for the case $n = 0$ and $l = 2$ (see Problem 12.3):

$$\left[\kappa \frac{\partial}{\partial \kappa} + \beta \frac{\partial}{\partial \hat{g}} + 2\gamma_2\right] \Gamma_R^{(0,2)}(\mathbf{p}; t_R = 0, \hat{g}, \kappa) = B(\kappa), \tag{12.51}$$

where

$$B(\kappa) \equiv -\kappa \frac{\partial A}{\partial \kappa}. \tag{12.52}$$

Combining these results, we find that away from the critical temperature, $\Gamma_R^{(0,2)}$ satisfies an equation of the form

$$\left[\kappa \frac{\partial}{\partial \kappa} + \beta \frac{\partial}{\partial \hat{g}} + \gamma_2\left(2 + t_R \frac{\partial}{\partial t_R}\right)\right] \Gamma_R^{(0,2)}(\mathbf{p}; t_R, \hat{g}, \kappa) = B(\kappa). \tag{12.53}$$

The dimensions of $\Gamma_R^{(0,2)}$ are L^{4-d}, so dimensional analysis gives

$$\left[\kappa \frac{\partial}{\partial \kappa} - b \frac{\partial}{\partial b} + 2 t_R \frac{\partial}{\partial t_R}\right] \Gamma_R^{(0,2)}(\mathbf{p}/b; t_R, \hat{g}, \kappa) = (d-4) \Gamma_R^{(0,2)}(\mathbf{p}/b; t_R, \hat{g}, \kappa). \tag{12.54}$$

Using this to eliminate the κ derivative from (12.53), we obtain

$$\left[b \frac{\partial}{\partial b} + \beta \frac{\partial}{\partial \hat{g}} + (\gamma_2 - 2) t_R \frac{\partial}{\partial t_R} + 2\gamma_2 + d - 4\right] \Gamma_R^{(0,2)}(\mathbf{p}/b; t_R, \hat{g}, \kappa) = B(\kappa), \tag{12.55}$$

which is very like (12.32) except for the inhomogeneous term on the right.

How can we solve this equation? The differential operator on the left is linear, so we know from the elementary theory of differential equations that the solution is given by the general solution with $B = 0$, plus any particular solution with $B \neq 0$. Since B is independent of b and t_R, such a solution is easy to construct; $\Gamma_R^{(0,2)}$ is simply a constant. We can therefore write the general solution in the form

$$\Gamma_R^{(0,2)}(\mathbf{p}/b; t_R, \hat{g}, \kappa) = b^{4-d-2\gamma_2} \Gamma_R^{(0,2)}(\mathbf{p}; t_R(b), \hat{g}(b), \kappa) + C, \tag{12.56}$$

where $\hat{g}(b)$ and $t_R(b)$ are given by (11.21) and (12.33). Now consider the limit $b \to \infty$, so that the wavenumber becomes small; if we reach an infrared stable fixed point, (12.56) becomes

$$\Gamma_R^{(0,2)}(\mathbf{p}/b; t_R, \hat{g}, \kappa) = b^{4-d-2\gamma_2^*}\Gamma_R^{(0,2)}(\mathbf{p}; t_R(b), \hat{g}_c, \kappa) + C, \qquad (12.57)$$

with $t_R(b)$ satisfying (12.37). As $t_R \to 0$, we know that $\Gamma_R^{(0,2)}$ diverges; the first term on the right must therefore dominate. Note that the singularity in the specific heat is coming from the part of $\Gamma_R^{(0,2)}$ which is not Λ-sensitive. The Λ-sensitive part of $\Gamma_R^{(0,2)}$ is not singular as $T \to T_c$. Using (12.49), we find

$$t_R^{-\alpha} \sim b^{4-d-2\gamma_2^*} t_R^{-\alpha} b^{-\alpha(2-\gamma_2^*)}. \qquad (12.58)$$

Equating the powers of b on each side, we get

$$\alpha = 2 - \frac{d}{2-\gamma_2^*}. \qquad (12.59)$$

The exponent α, like ν, is therefore determined solely by the fixed-point value of γ_2. Substituting for ν from (12.42), we see that

$$\alpha = 2 - \nu d. \qquad (12.60)$$

This is the Josephson scaling relation of equation (1.29). We initially derived this from the assumption equation (1.28); now we have proved it.

12.3 The renormalization group below T_c

It is a great advantage of our formulation of the theory as an expansion about the critical point, that we can at once extend it below T_c.[5] Nowhere in the last section did we have to specify the sign of the variable t_R, so we might equally well have been working below or above the transition. Of course the function $\Gamma_R^{(n)}$ will behave very differently for positive and negative arguments, but none of our arguments about critical exponents depended in any way on its form. We can therefore at once deduce that

$$\alpha' = \alpha, \qquad \gamma' = \gamma, \qquad \nu' = \nu. \qquad (12.61)$$

There is, however, one critical exponent defined below T_c that has no counterpart above T_c; this is β, which describes the growth of the expectation of the order parameter below the critical point. Some extra work is needed to calculate this.

---

[5] To derive the renormalization group equation, we must still make the assumption, discussed in §12.1.7, that a small region around the real t_R-axis can be cleared of singularities.

12.3 The renormalization group below T_c

12.3.1 The exponent β

We shall find β by generalizing the renormalization group equation (11.36) that we derived for the magnetic field at the critical point. Writing the expansion of equation (11.33) for $t \neq 0$ and noting that each vertex function on the right satisfies (12.30), we see that

$$\left[\kappa\frac{\partial}{\partial\kappa}+\beta\frac{\partial}{\partial\hat{g}}-\tfrac{1}{2}\gamma_1-\tfrac{1}{2}\gamma_1\phi_{0R}\frac{\partial}{\partial\phi_{0R}}+\gamma_2 t_R\frac{\partial}{\partial t_R}\right]J_{0R}(\phi_{0R},t_R,\hat{g},\kappa) = 0. \quad (12.62)$$

Dimensional analysis gives

$$\left[\kappa\frac{\partial}{\partial\kappa}-b\frac{\partial}{\partial b}+2t_R\frac{\partial}{\partial t_R}\right]J_{0R}(b^{1-d/2}\phi_{0R},t_R,\hat{g},\kappa)$$
$$= (1+\tfrac{1}{2}d)J_{0R}(b^{1-d/2}\phi_{0R},t_R,\hat{g},\kappa), \quad (12.63)$$

so on eliminating the κ derivative we find

$$\left[\left(1+\frac{\gamma_1}{d-2}\right)b\frac{\partial}{\partial b}+\beta\frac{\partial}{\partial\hat{g}}-\tfrac{1}{2}\gamma_1+(\gamma_2-2)t_R\frac{\partial}{\partial t_R}+1+\tfrac{1}{2}d\right]$$
$$\times J_{0R}(b^{1-d/2}\phi_{0R},t_R,\hat{g},\kappa) = 0. \quad (12.64)$$

We can solve this by using our now-familiar friend, the method of characteristics. It is convenient to redefine $t_R(b)$ so that it obeys the equation

$$b\frac{d\log t_R(b)}{db}=\frac{(d-2)[2-\gamma_2(\hat{g})]}{d-2+\gamma_1} \quad \text{with} \quad t_R(1)=t_R \quad (12.65)$$

instead of (12.33). With this definition we obtain

$$J_{0R}(b^{1-d/2}\phi_{0R},t_R,\hat{g},\kappa) = [b^{-1-d/2}a(b)a^{-1}]^{(d-2)/(d-2+\gamma_1^*)} \\ \times J_{0R}(\phi_{0R},t_R(b),\hat{g}(b),\kappa). \quad (12.66)$$

At a fixed point where $b \to \infty$ and therefore (for $d > 2$) $b^{1-d/2}\phi_{0R} \to 0$, this becomes

$$J_{0R}(b^{1-d/2}\phi_{0R},t_R,\hat{g},\kappa) \sim [b^{-1-d/2+\gamma_1^*/2}]^{(d-2)/(d-2+\gamma_1^*)} J_{0R}(\phi_{0R},t_R(b),\hat{g}_c,\kappa), \quad (12.67)$$

with

$$t_R(b) \sim t_R(1)b^{(d-2)(2-\gamma_2^*)/(d-2+\gamma_1^*)}. \quad (12.68)$$

If we study this at $t_R = 0$, i.e., at the critical temperature, and ask how J_{0R} depends on ϕ_{0R}, we find that

$$J_{0R} \sim \phi_{0R}^{1/\delta} \quad (12.69)$$

with δ given by equation (11.40). This is just as well! However, we can get new information by setting $J_{0R} = 0$ on the left-hand side of equation (12.67) and asking how ϕ_{0R} has to change with t_R to bring this about. The right-hand side will remain zero only if $t_R(b)$ remains constant, which implies that

$$b \sim t_R^{(d-2+\gamma_1^*)/(2-d)(2-\gamma_2^*)}. \tag{12.70}$$

The value of the order parameter is then proportional to

$$b^{1-d/2} \sim t_R^{(d-2+\gamma_1^*)/2(2-\gamma_2^*)}, \tag{12.71}$$

so

$$\beta = \frac{d-2+\gamma_1^*}{2(2-\gamma_2^*)}. \tag{12.72}$$

Equivalently, we can write β in terms of η and ν as

$$\beta = \tfrac{1}{2}\nu(d-2+\eta). \tag{12.73}$$

This concludes our calculation of the critical exponents in terms of the flow functions.

12.4 Calculating γ_2 to one loop

The only new piece of information that we need in order to perform an ϵ-expansion for all the critical exponents is the form of the function $\gamma_2(\hat{g})$ and in particular its value at $\hat{g} = \hat{g}_c$. Now, we have already written down the start of the series for Z_2, in equation (12.20). What is more, we have already done the integral appearing on the right-hand side, since it is exactly the integral $J(\kappa)$ which formed part of the calculation of η in §10.4. In dimensional regularization, the result is

$$J(\kappa) = \pi^{d/2} \kappa^{-\epsilon} \frac{(\tfrac{1}{2}\epsilon - 1)![(-\tfrac{1}{2}\epsilon)!]^2}{(1-\epsilon)!}, \tag{12.74}$$

which has an ϵ-expansion

$$J(\kappa) = \pi^2 \kappa^{-\epsilon} \left[\frac{2}{\epsilon} + O(1)\right]. \tag{12.75}$$

Therefore

$$\begin{aligned} Z_2 &= 1 + \tfrac{1}{2} \frac{1}{a^4 (2\pi)^d} J(\kappa) \lambda + O(\lambda^2) \\ &= 1 + \tfrac{1}{2}\pi^2 \left[\frac{2}{\epsilon} + O(1)\right]\hat{g} + O(\hat{g}^2), \end{aligned} \tag{12.76}$$

12.4 γ_2 to one loop

where we have replaced λ by g to first order and then written the right-hand side in terms of the dimensionless variable $\hat{g}$ to obtain the second line. From this we can calculate

$$\begin{aligned}\gamma_2(\hat{g}) &= -\kappa \frac{\partial Z_2}{\partial \kappa} \\ &= -\beta(\hat{g}) \frac{\partial Z_2}{\partial \hat{g}} \\ &= -\tfrac{1}{2}\beta(\hat{g})\pi^2 \left[\frac{2}{\epsilon} + O(1)\right] \\ &= \pi^2 \hat{g} + O(\hat{g}^2).\end{aligned} \qquad (12.77)$$

Substituting $\hat{g} = \hat{g}_c$, we find

$$\gamma_2^* = \frac{\epsilon}{3} + O(\epsilon^2). \qquad (12.78)$$

This gives for ν in three dimensions:

$$\nu = \frac{3}{5} + O(\epsilon^2). \qquad (12.79)$$

Inserting this into the scaling relations (12.47), (12.60) and (12.73), and including the results for η and δ from (11.49) and (11.50), we get the following values for the critical exponents of the Landau–Ginzburg model in three dimensions:

$$\begin{aligned}\nu &= 0.600\ldots, & (12.80a) \\ \gamma &= 1.189\ldots, & (12.80b) \\ \alpha &= 0.200\ldots, & (12.80c) \\ \beta &= 0.306\ldots, & (12.80d) \\ \eta &= 0.019\ldots, & (12.80e) \\ \delta &= 4.891\ldots. & (12.80f)\end{aligned}$$

The errors in the first four exponents are $O(\epsilon^2)$; the errors in η and δ are $O(\epsilon^3)$. So, at last, we have finite and sensible values for all the critical exponents. They differ from the mean-field values, and we can understand the relationships between them. What is more, even at first order in ϵ, the agreement with the exact values for real systems with $D = 1$ (see Table 1.2) is quite respectable.

It is natural to ask how we can improve on this approximation. With considerable labour, the series for β, γ_1 and γ_2 can be computed to higher order in $\hat{g}$; this in turn allows the values of $\hat{g}_c$ and of the critical exponents to be found to higher order in ϵ. To order ϵ^3, for example, it is found (Wallace 1976) that

$$\nu = 0.634\ldots, \tag{12.81a}$$
$$\gamma = 1.244\ldots, \tag{12.81b}$$
$$\alpha = 0.099\ldots, \tag{12.81c}$$
$$\beta = 0.329\ldots, \tag{12.81d}$$
$$\eta = 0.037\ldots, \tag{12.81e}$$
$$\delta = 4.786\ldots. \tag{12.81f}$$

However, as we explained at the end of Chapter 10, the likely divergence of the ϵ-expansion is believed to make it essential to play some additional tricks if good results are to be obtained from a finite number of terms. The best results, after resumming the expansions, are (Zinn-Justin 1989):

$$\nu = 0.6310 \pm 0.0015, \tag{12.82a}$$
$$\gamma = 1.2390 \pm 0.0025, \tag{12.82b}$$
$$\alpha = 0.1070 \pm 0.0045, \tag{12.82c}$$
$$\beta = 0.3270 \pm 0.0025, \tag{12.82d}$$
$$\eta = 0.0375 \pm 0.0025, \tag{12.82e}$$
$$\delta = 4.814 \pm 0.015. \tag{12.82f}$$

These results are in impressive agreement with those obtained from numerical studies of the three-dimensional Ising model (Ferrenberg and Landau 1991), which give $\nu = 0.6289 \pm 0.0008$ and $\gamma = 1.2390 \pm 0.0071$. They also lie within all the experimental error bounds for the $D = 1$ systems given in Table 1.2.

Problems

12.1 By expanding the dimensionless Gibbs free energy Γ as a power series in $\langle\phi\rangle$, show that its second derivative with respect to temperature evaluated at $\langle\phi\rangle \sim (T_c - T)^\beta$ is equal to the same derivative evaluated at $\langle\phi\rangle = 0$, so long as $T = T_c$. Hence justify equation (12.49).

12.2 Show that the heat capacities at constant field and constant magnetization are related by

$$C_B - C_M = \frac{T\alpha_B^2}{\chi_T},$$

where χ_T is the isothermal susceptibility and

$$\alpha_B \equiv \left(\frac{\partial M}{\partial T}\right)_B.$$

Deduce that the difference between the two heat capacities is zero above the critical temperature, and that the two heat capacities diverge in the same way below T_c.

Problems

12.3 Derive the renormalization group equation (12.51) obeyed by the vertex function $\Gamma_R^{(0,2)}$ at the critical temperature.

13
The lower critical dimension

In §§3.2 and 3.3 we saw that both the Ising model and the spherical model have a lower critical dimension, d_{LC}: the highest dimension at and below which the model is incapable of displaying long-range order. In §3.2 we understood physically why this was so for the Ising model: the size and thus the energy of an interface between two domains of n mutually aligned spins decreases with the dimension d, and when d is diminished to d_{LC}, the entropy increase associated with increasing the number of such domains outweighs the energetic cost of extra interfaces, no matter how low the temperature is (providing it exceeds zero). In this chapter we study this phenomenon, especially the dependence of d_{LC} on the dimensionality D of the order parameter $\boldsymbol{\phi}$, in the context of the Landau–Ginzburg model.

In §13.1 we develop a simple argument which indicates that no system with $d \leq 2$ and $D \geq 2$ in which the effective Hamiltonian is invariant under rotations of the order parameter $\boldsymbol{\phi}$, should be capable of displaying long-range order. In other words, we conclude that systems with rotationally invariant effective Hamiltonians have $d_{\text{LC}} \geq 2$. In §13.2 we study the 'non-linear σ-model', which provides a natural approximation to the Landau–Ginzburg model at low T, and from it we infer that $d_{\text{LC}} = 2$ for $D \neq 2$.

The case in which both d and D are equal to two is delicate and requires further analysis. In §3.1 we introduced the $d = 2$ XY model, which is the lattice system which the $d = D = 2$ Landau–Ginzburg and non-linear σ-models seek to approximate. Here we examine this system more closely and show that in passing from a discrete system to a continuous one, the

13.1 Order below T_c

non-linear σ-model has excluded the contributions of 'spin vortices' from the partition function. When we remedy this oversight we find that the presence of vortices gives rise to a phase transition of an entirely new kind; as the system is cooled through the critical temperature, it is not long-range order which arises, but short-range order. Specifically, the vortices become bound in pairs rather than drifting freely through the system.

Besides being both intrinsically fascinating and relevant to experiments on low-dimensional systems, this 'Kosterlitz–Thouless' transition provides a timely caution on the dangers inherent in approximating one model by another. Moreover, if it were permissible to use the Landau–Ginzburg model as an approximation to the Ising chain, the argument we develop in §13.1 would contradict the main result of §3.2, namely that the Ising ring is incapable of exhibiting long-range order. In the next and final chapter we investigate when systems can and cannot be safely approximated by the Landau–Ginzburg model.

13.1 Order below T_c

We investigate whether long-range order is possible by first assuming that this is so, and, indeed, that the system under study is so cold that the order is nearly total. Of course, if the system is not at absolute zero, the underlying order will be disturbed by fluctuations. We can calculate the effects of these by assuming that they are small—this procedure is known as making a 'low-temperature expansion'. If long-range order *is* possible, as we have assumed, the system will not be seriously disturbed by these fluctuations for T sufficiently far below T_c, and we shall obtain a consistent picture of thermal fluctuations below T_c. But if the system is incapable of long-range order no matter how cold it is, we shall find that fluctuations are non-negligible even near $T = 0$, and we will infer from this failure of our low-temperature expansion that long-range order is not, in fact, possible.

13.1.1 The case $D = 1$

Consider first the $D = 1$ Landau–Ginzburg model at a very low temperature—a temperature so low that we are very likely to find the system at or near to its configuration of lowest energy. The latter is just the one with a uniform order-parameter field $\phi(\mathbf{x}) = \phi_0$ studied in §7.2 on Landau theory. By equation (7.18) we have that

$$|\phi_0|^2 = -6\mu^2/\lambda, \qquad (13.1)$$

where we have taken $\mu^2 < 0$ since we are, by hypothesis, in the regime $T \ll T_c$.

What are the energies of the excitations of the ordered system? Let us expand

$$\phi(\mathbf{x}) = \phi_0 + \delta\phi(\mathbf{x}). \qquad (13.2)$$

Then the energy relative to that of the ground state is

$$\Delta H = \int d^d\mathbf{x} \left[\tfrac{1}{2}\alpha^2|\boldsymbol{\nabla}\delta\phi|^2 + |\mu^2|(\delta\phi)^2 + O\big((\delta\phi)^3\big)\right]. \tag{13.3}$$

Provided that the fluctuations in ϕ are small, so that we can ignore terms higher than quadratic in $\delta\phi$, this is just like the Hamiltonian for the Gaussian model that we studied in §8.1; we therefore know how to find its correlation functions. As usual, they are most conveniently expressed in terms of $\delta\tilde\phi(\mathbf{k})$, the field of which $\delta\phi(\mathbf{x})$ is the Fourier transform. Specifically,

$$\langle \delta\tilde\phi(\mathbf{k})\delta\tilde\phi(-\mathbf{k})\rangle = (2\pi)^d \delta(0) \frac{1}{\alpha^2 k^2 + 2|\mu^2|}. \tag{13.4}$$

The δ-function factor, $\delta(0)$, indicates that the correlation function is proportional to the volume of the system—see equation (8.70). So, for example, we can calculate the mean-squared value of the fluctuation $\delta\phi$ at a given position:

$$\langle \delta\phi(\mathbf{x})\delta\phi(\mathbf{x})\rangle = \int^\Lambda \frac{d^d \tilde{\mathbf{k}}}{\alpha^2 k^2 + 2|\mu^2|}. \tag{13.5}$$

The integral on the right of (13.5) is finite for all finite Λ and all $|\mu^2| \neq 0$. In fact, as the temperature becomes very low, $|\mu^2|$ becomes large and the fluctuations about the ordered state become small. This accords with our intuition about the energetic stability of order at low temperatures.[1]

13.1.2 Systems with more than one component

Now consider the same problem, but for a Landau–Ginzburg model with more than one component. The effective Hamiltonian is minimized, as before, by a uniform field $\boldsymbol{\phi}(\mathbf{x}) = \boldsymbol{\phi}_0$ of constant magnitude:

$$|\boldsymbol{\phi}_0|^2 = -6\mu^2/\lambda. \tag{13.6}$$

Consider the possible field configurations $\boldsymbol{\phi}(\mathbf{x}) = \boldsymbol{\phi}_0 + \delta\boldsymbol{\phi}(\mathbf{x})$, and resolve $\delta\boldsymbol{\phi}$ into components:

(i) Displacements $\delta\phi_\parallel$ which are parallel to the direction of the uniform field $\boldsymbol{\phi}_0$. These displacements change the length of the vector $\boldsymbol{\phi}$.
(ii) Displacements $\delta\phi_\perp$ which are perpendicular to the direction of $\boldsymbol{\phi}_0$. These displacements do not change the length of the vector $\boldsymbol{\phi}$ (at least, not to first order).

---

[1] You may wonder why this argument predicts that the one-component Landau–Ginzburg model orders at low temperatures even in one dimension, whereas we know that the Ising model does not. In fact, as we shall see in the next chapter, the Landau–Ginzburg model can no longer be used as a 'metamodel' below $d = 2$, in the sense that its behaviour is no longer expected to be typical of other models with the same symmetry and number of components.

13.1 Order below T_c

What is the resulting energy change relative to the uniform state? Rotational invariance guarantees that the terms in the effective Hamiltonian that do not involve any derivatives are functions only of $|\boldsymbol{\phi}|$. As before, there is no energy change proportional to $\delta|\boldsymbol{\phi}|$, since the effective Hamiltonian has already been minimized with respect to $|\boldsymbol{\phi}_0|$; the first contribution to the energy change is therefore proportional to

$$(\delta|\boldsymbol{\phi}(\mathbf{x})|)^2 = (|\boldsymbol{\phi}_0 + \delta\boldsymbol{\phi}| - |\boldsymbol{\phi}_0|)^2$$
$$= \left(\sqrt{|\boldsymbol{\phi}_0|^2 + 2\boldsymbol{\phi}_0 \cdot \delta\boldsymbol{\phi} + |\delta\boldsymbol{\phi}|^2} - |\boldsymbol{\phi}_0|\right)^2 \quad (13.7)$$
$$= (\delta\phi_\parallel)^2 + O(|\delta\boldsymbol{\phi}|^3).$$

From the terms that do contain derivatives, we get (see the definition of $|\boldsymbol{\nabla}\boldsymbol{\phi}|^2$ for a multicomponent field in Chapter 7)

$$|\boldsymbol{\nabla}\delta\boldsymbol{\phi}|^2 = |\boldsymbol{\nabla}\delta\phi_\parallel|^2 + |\boldsymbol{\nabla}\delta\boldsymbol{\phi}_\perp|^2.$$

Putting this all together, we find

$$\Delta H = \int d^d\mathbf{x} \left[\tfrac{1}{2}\alpha^2(|\boldsymbol{\nabla}\delta\phi_\parallel|^2 + |\boldsymbol{\nabla}\delta\boldsymbol{\phi}_\perp|^2) + |\mu^2|(\delta\phi_\parallel)^2 + O|\delta\boldsymbol{\phi}|^3\right], \quad (13.8)$$

where again we have assumed that the fluctuations in $\boldsymbol{\phi}$ are small so we can stop the expansion at second order. Once again, the effective Hamiltonian we have obtained looks like that of the Gaussian model. As far as the fluctuations $\delta\phi_\parallel$ in the direction of $\boldsymbol{\phi}_0$ are concerned, it is exactly the same as the effective Hamiltonian for the one-component model. So, for example, the mean-squared value of the fluctuation $\delta\phi_\parallel$ is also given by equation (13.5). However, for the perpendicular fluctuations $\delta\boldsymbol{\phi}_\perp$, there is a crucial difference: there is no term in (13.8) proportional to $|\mu^2||\delta\boldsymbol{\phi}_\perp|^2$. The mean-squared value of the perpendicular fluctuations is therefore:

$$\langle\delta\boldsymbol{\phi}_\perp(\mathbf{x}) \cdot \delta\boldsymbol{\phi}_\perp(\mathbf{x})\rangle = \int^\Lambda \frac{d^d\check{\mathbf{k}}}{\alpha^2 k^2}. \quad (13.9)$$

This time, the integrand contains a singularity at $k = 0$. For $d \geq 3$, this singularity is integrable; the perpendicular fluctuations in the order parameter are finite. However, for $d \leq 2$, the singularity is not integrable; the mean squared deviation from our assumed ordered state is infinite! This strongly suggests that our initial assumption of an ordered state was incorrect. You may object that we have shown only that the perpendicular fluctuations do not alter the magnitude of the field vector to first order. Surely higher-order corrections will become important when the fluctuations are big and will produce an energy cost even for very long-wavelength fluctuations $\delta\boldsymbol{\phi}_\perp$? However, at the price of significantly heavier algebra we could have written the field in terms of radial and angular components. The magnitude of the field would then have been independent of the angular variables to all orders, and the RMS fluctuations in these would be of order their full range, $0 - 2\pi$, or whatever. It is *conceivable* that the field could retain a non-zero expectation in these circumstances, but on physical grounds this must be judged very unlikely: it is hard to understand how the field could remember in what direction it should on average point once it starts exploring all directions with high probability.

In any event, in 1965 Mermin and Wagner conclusively demonstrated that the perpendicular fluctuations *do* completely destroy the long-range order by showing that the order parameter in a symmetric, multicomponent model with $d \leq 2$ is proportional to the applied external field, with a finite constant of proportionality. In other words, the order parameter does not develop a non-zero value spontaneously at low temperatures in zero applied field, and the type of phase transition we have been discussing throughout this book does not occur. In honour of this demonstration, the result that models with rotationally symmetric order-parameters are incapable of long-range order at $d \leq 2$ is known as the **Mermin–Wagner theorem**. Notice that the theorem implies that $d_{\mathrm{LC}} \geq 2$ for models with $D \geq 2$, whereas we showed in §3.4.2 that the Ising model ($D = 1$) has $d_{\mathrm{LC}} = 1$.

13.1.3 Goldstone modes

The physics of what we saw in the last subsection is simple. A fluctuation in the order parameter perpendicular to the direction of $\boldsymbol{\phi}_0$ is equivalent to a local change in the direction, but not the magnitude, of $\boldsymbol{\phi}_0$. If the fluctuation is of sufficiently long wavelength, it corresponds to a rotation of $\boldsymbol{\phi}_0$ over the whole sample. If the model is rotationally invariant to begin with, this cannot cost any energy; it simply corresponds to a different choice among the many possible lowest-energy ground states which are related to each other by the symmetry of the system. This means that deformation patterns exist for which the stiffness of the system goes to zero in the long-wavelength limit, and these destroy the long-range order for $d \leq 2$. Their existence is a reminder that the system is fully symmetric in the ordered phase above T_c and that this symmetry is retained in the effective Hamiltonian, even though it may have been hidden by the development of a non-zero macroscopic average for the order parameter.

In quantum field theory, the counterpart of these low-energy displacements is the existence of massless particles. Goldstone (1961) was the first to realize that massless particles are present whenever the Hamiltonian of a system has a continuous global symmetry, and for this reason one speaks of Goldstone bosons in quantum field theory and **Goldstone modes** in statistical physics.

13.2 The non-linear σ-model

The disappearance of long-range order in two and fewer dimensions is an interesting and important phenomenon, which we shall now study further. However, the Landau–Ginzburg model is not the most suitable vehicle for this study. There are a number of reasons for this. One is aesthetic: we have seen that the fluctuations $\delta\phi_\parallel$ are essentially irrelevant to this process, because they cost a finite amount of energy no matter how long their wavelength. It would be good to work with a minimal model that did not include unnecessary $\delta\phi_\parallel$ components. The other reasons are more technical. First, we have seen that we can make sense of the predictions of the Landau–Ginzburg model only by making an ϵ-expansion about the upper critical dimension $d = 4$. This expansion is not ideal for the study of what happens near $d = 2$. Second, the Landau–Ginzburg model is set up to deal with what happens at temperatures close to T_c, whereas we want to investigate T_c itself. We hope to see T_c going to zero as d approaches 2 from above, and the transition vanishing for $d < 2$. But in the Landau–Ginzburg model the absolute temperature does not appear (see equation (7.10)), so it is not really suitable for our purposes.

For all these reasons it is usual to use a different model to investigate this problem. The Heisenberg ferromagnet would be a suitable model, but it's easier to work with a model whose order parameter is a continuous field. We therefore start from the Hamiltonian of the Heisenberg ferromagnet, and derive a continuous field theory from it, taking care not to lose track of the absolute zero of temperature. The continuous field model we shall arrive at is the 'non-linear σ-model'.

We consider the Heisenberg model for N spins, each having D components. Let the spin at the ith site of the lattice be $\mathbf{s}_i$. The Hamiltonian for this system is

$$H = \tfrac{1}{2} \sum_{ij} \mathcal{J}_{ij} \mathbf{s}_i \cdot \mathbf{s}_j - \mathbf{B} \cdot \sum_k \mathbf{s}_k. \tag{13.10}$$

Each spin is of fixed length S and therefore satisfies

$$s_i^2 = S^2. \tag{13.11}$$

The partition function for this system is

$$Z = \prod_{i=1}^{N} \left[\int d^D \mathbf{s}_i \, \delta(s_i^2 - S^2) \right] e^{-\beta H}, \tag{13.12}$$

where the δ-function imposes the constraint (13.11) on each spin. When $\mathbf{B} = 0$, this system possesses $O(D)$ symmetry (see Appendix N).

We can use (13.11) to get rid of one component of each spin. The components we choose to eliminate are those parallel to the magnetic field

B. Let us call these components σ_i, and let $\boldsymbol{\pi}_i$ be the $(D-1)$ components of the ith spin perpendicular to **B**. In terms of $\boldsymbol{\pi}$ and σ the partition function becomes

$$Z = \prod_{i=1}^{N} \left[\int \mathrm{d}^{D-1}\boldsymbol{\pi}_i \, \mathrm{d}\sigma_i \, \delta(|\boldsymbol{\pi}_i|^2 + \sigma_i^2 - S^2) \right] \mathrm{e}^{-\beta H}. \qquad (13.13)$$

Because the δ-function involves σ^2 and not σ, some care is needed when doing the σ-integral. We use the result

$$\int \mathrm{d}x \, f(x) \delta\big(g(x)\big) = \sum_i \frac{f(x_i)}{|g'(x_i)|}, \qquad (13.14)$$

where the x_i are the roots of $g(x) = 0$. Here

$$|\boldsymbol{\pi}_i|^2 + \sigma_i^2 - S^2 = 0 \qquad (13.15)$$

has two real roots for σ_i when $|\boldsymbol{\pi}_i|^2 < S^2$:

$$\sigma_i = \pm\sqrt{S^2 - |\boldsymbol{\pi}_i|^2}. \qquad (13.16)$$

However, only one of these is relevant to us. The negative root for σ_i corresponds to the ith spin pointing against the direction of the magnetic field. At low temperatures the Boltzmann factor for this state is very small, and we can neglect its contribution to Z. With this approximation Z becomes

$$Z = \int \frac{\mathrm{d}^{D-1}\boldsymbol{\pi}_1}{2\sqrt{S^2 - |\boldsymbol{\pi}_1|^2}} \cdots \frac{\mathrm{d}^{D-1}\boldsymbol{\pi}_N}{2\sqrt{S^2 - |\boldsymbol{\pi}_N|^2}} \times \qquad (13.17)$$
$$\exp\left[-\frac{1}{2}\sum_{ij}\left(\frac{J_{ij}}{k_B T}\left(\boldsymbol{\pi}_i \cdot \boldsymbol{\pi}_j + \sqrt{S^2 - |\boldsymbol{\pi}_i|^2}\sqrt{S^2 - |\boldsymbol{\pi}_j|^2}\right)\right.\right.$$
$$\left.\left. - \frac{B}{k_B T}\sum_k \sqrt{S^2 - |\boldsymbol{\pi}_k|^2}\right)\right],$$

where we've used T rather than β to emphasize that we shall be making an expansion in powers of T. The integral over each set of $\boldsymbol{\pi}_i$ is constrained so that $|\boldsymbol{\pi}_i|^2 < S^2$. We ignore this tricky point, since at low temperatures values of $\boldsymbol{\pi}_i$ which violate this condition give rise to very small Boltzmann factors. Notice that the partition function (13.17) is not quite what we would have obtained if we had eliminated σ in favour of $\boldsymbol{\pi}$ in the Hamiltonian (13.10), and then integrated over the values of the $\boldsymbol{\pi}_i$. There is an additional factor present in the measure of the integral. As we shall see, this factor is important in maintaining the $O(D)$ symmetry of the model.

13.2 The non-linear σ-model

The next step is to write the factor in the measure as an exponential:

$$\prod_{i=1}^{N} \frac{1}{\sqrt{S^2 - |\boldsymbol{\pi}_i|^2}} = \exp\left[-\frac{1}{2}\sum_{i=1}^{N} \log(S^2 - |\boldsymbol{\pi}_i|^2)\right]. \quad (13.18)$$

This allows us to write

$$Z = 2^{-N} \int d^{D-1}\boldsymbol{\pi}_1 \ldots d^{D-1}\boldsymbol{\pi}_N \, e^{-H_{\text{eff}}/k_B T}, \quad (13.19)$$

where the effective Hamiltonian H_{eff} is

$$H_{\text{eff}} = -\frac{1}{2}\sum_{ij} \mathcal{J}_{ij}\left(\boldsymbol{\pi}_i \cdot \boldsymbol{\pi}_j + \sqrt{S^2 - |\boldsymbol{\pi}_i|^2}\sqrt{S^2 - |\boldsymbol{\pi}_j|^2}\right)$$
$$+ \frac{1}{2}k_B T \sum_k \log(S^2 - |\boldsymbol{\pi}_k|^2) - B\sum_k \sqrt{S^2 - |\boldsymbol{\pi}_k|^2}. \quad (13.20)$$

If we were to follow the route of Chapter 7 at this point, we would introduce a D-component order parameter $\boldsymbol{\phi}(\mathbf{x})$, representing the average spin in a small volume around $\mathbf{x}$. Although the individual spins have a fixed length this is not true of the spin averaged over a finite volume, and so the value of $|\boldsymbol{\phi}|^2$ would not be fixed. We would, in fact, end up with the D-component Landau–Ginzburg model. This is not what we do here. We want a model with a continuous order parameter that mimics the structure of (13.20) as closely as possible. In particular we want to avoid any averaging process that obscures the physical temperature T (compare equation (7.10)). So let us introduce a $(D-1)$-component field $\boldsymbol{\pi}(\mathbf{x})$, and replace the sums in (13.20) by integrals. Now if $f(\mathbf{x})$ is a slowly-varying function that takes the values f_i at the points $\mathbf{x}_i$, then

$$\sum_{i=1}^{N} f_i \approx \frac{N}{V}\int d^d\mathbf{x}\, f(x), \quad (13.21)$$

where V is the volume of the space being integrated over. So

$$B\sum_{k=1}^{N} \sqrt{S^2 - |\boldsymbol{\pi}_k|^2} \quad \text{becomes} \quad B\frac{N}{V}\int d^d\mathbf{x}\, \sqrt{S^2 - \boldsymbol{\pi}^2(\mathbf{x})}. \quad (13.22)$$

The term involving $\mathcal{J}_{ij}$ in H_{eff} is a little more complicated. It is left as a problem to show that, for slowly-varying fields, it can be written as

$$\text{const.} + \frac{1}{2}J\int d^d\mathbf{x}\left(|\boldsymbol{\nabla\pi}|^2 + |\boldsymbol{\nabla}\sqrt{S^2 - \pi^2}|^2 + \text{terms with higher derivatives}\right). \quad (13.23)$$

J is related to the $\mathcal{J}_{ij}$ and the separation between adjacent lattice points. The terms with higher derivatives can be shown to be irrelevant, in the sense that including them does not change the critical behaviour. Finally, the term from the measure becomes

$$\text{const.} + \tfrac{1}{2}k_B T \frac{N}{V} \int d^d\mathbf{x}\, \log(S^2 - \pi^2). \tag{13.24}$$

As in the Landau–Ginzburg model a smoothness condition must be imposed on $\boldsymbol{\pi}(\mathbf{x})$ if the continuum theory is to make any sense. We suppose that $\boldsymbol{\pi}(\mathbf{x})$ can be written[2]

$$\boldsymbol{\pi}(\mathbf{x}) = \int^\Lambda d^d\check{\mathbf{k}}\, \tilde{\boldsymbol{\pi}}(\mathbf{k}) e^{i\mathbf{k}\cdot\mathbf{x}}. \tag{13.25}$$

The ratio N/V appearing in the measure term (13.24) can be expressed in terms of the cutoff parameter Λ. Imagine putting the system in a d-dimensional rectangular box of volume V. Then the total number N_k of modes in k-space with $k < \Lambda$ is

$$N_k = \frac{V}{(2\pi)^d} \int^\Lambda d^d\mathbf{k} = \frac{V\Omega_{d-1}}{d}\left(\frac{\Lambda}{2\pi}\right)^d, \tag{13.26}$$

where Ω_{d-1} is the 'area' of the unit $(d-1)$-sphere (see Problem 3.4). Equating N_k to N, which amounts to saying that the number of degrees of freedom in position space and k-space should be equal, gives

$$N/V = \frac{\Omega_{d-1}}{d}\left(\frac{\Lambda}{2\pi}\right)^d \equiv \Lambda_0. \tag{13.27}$$

The partition function can now be written as

$$Z = \int \mathcal{D}\boldsymbol{\pi}\, \exp\left[\int d^d\mathbf{x}\left(-\frac{1}{T}\left(\tfrac{1}{2}|\boldsymbol{\nabla}\boldsymbol{\pi}|^2 + \tfrac{1}{2}|\boldsymbol{\nabla}\sqrt{S^2-\pi^2}|^2\right)\right.\right. \tag{13.28}$$

$$\left.\left. + \frac{B}{T}\sqrt{S^2-\pi^2} - \frac{\Lambda_0}{2}\log(S^2-\pi^2)\right)\right],$$

---

[2] This regularization breaks the $O(D)$ symmetry of the original model, because if $\boldsymbol{\pi}(\mathbf{x})$ takes the form (13.25) then $\sigma(\mathbf{x}) = (S^2 - \pi^2)^{1/2}$ in general does not. We take the point of view that having eliminated σ the $O(D)$ symmetry no longer exists and our job is merely to evaluate the functional integral over $\boldsymbol{\pi}$ by any convenient method. Indeed our subsequent calculations show no sign of this regularization causing problems. The alternatives are to work with the lattice model (13.20), or to use dimensional regularization. The first is hard to do analytically, while the second involves setting integrals of the form $\int (d^d\mathbf{k}/k^2)$ equal to zero, which is physically unclear.

13.2 The non-linear σ-model

where we have redefined T and B:

$$T \equiv \frac{k_B T_{\text{old}}}{J} \quad \text{and} \quad B \equiv \frac{\Lambda_0 B_{\text{old}}}{J}. \tag{13.29}$$

T and B are no longer the true temperature and magnetic field, but they are proportional to their former values. This is important, as it means that we can study what happens to the model at low temperatures by letting T go to zero. In the Landau–Ginzburg model, the absolute scale of temperature was lost.

The partition function (13.28) defines the non-linear σ-model. Four parameters appear explicitly in it: T, S, B, and Λ_0. In addition the model depends on the dimension and the number of spin components. The effective Hamiltonian in the exponent of (13.29) is unusual in that it contains the cutoff parameter Λ_0 explicitly as the coefficient of the term from the measure. Notice that this term is suppressed by one power of T compared with the other two terms.

13.2.1 The two-point vertex function

The partition function (13.28) and the Feynman rules derived from it will be the basis for our analysis of this model. What we are aiming for is a renormalization group equation similar to those of Chapters 11 and 12, which will tell us how the vertex functions behave at small wavevector. To find the Feynman rules for this model we must expand the exponential in (13.28) as a power series in π/S. The terms up to $\mathrm{O}(\pi^4)$ in this Hamiltonian are

$$H = \int d^d x \left[\frac{1}{2T} |\nabla \pi|^2 + \frac{S^2}{2T} \left| \nabla \frac{\pi^2}{2S^2} \right|^2 + \frac{BS}{T} \left(\frac{\pi^2}{2S^2} + \frac{\pi^4}{8S^4} \right) \right. \\ \left. - \frac{\Lambda_0}{2} \left(\frac{\pi^2}{S^2} + \frac{\pi^4}{S^4} \right) + \mathrm{O}(\pi^6) \right]. \tag{13.30}$$

When deriving the Landau–Ginzburg model in Chapter 7 we also dropped terms of higher order than ϕ^4, because such terms do not affect the critical behaviour (as we will see in Chapter 14). This is *not* the situation here. We have only dropped the terms of order π^6 and higher because they contribute to higher orders in T than the terms we have kept, and we are studying the system at low temperatures.

Since the system is $\mathrm{O}(D)$-symmetric in the absence of a magnetic field, its energy should depend only on how π varies from place to place, and not on its actual value. However the term proportional to Λ_0 in (13.28) appears to break this symmetry. It turns out that the effect of this term is cancelled at each order by diagrams involving loops. We shall see this explicitly when we calculate $\Gamma^{(2)}$ to one loop.

> **Box 13.1: The low-order Feynman rules for the non-linear σ-model**
>
> $\dfrac{T}{k^2 + (B/S)}$
>
> $\dfrac{\Lambda_0}{S^2}$
>
> $\dfrac{3B}{TS^3} - \dfrac{12\Lambda_0}{S^4} - \dfrac{1}{TS^2}\Big[(\mathbf{k}_1 + \mathbf{k}_2)\cdot(\mathbf{k}_3 + \mathbf{k}_4)$
> $+ (\mathbf{k}_1 + \mathbf{k}_3)\cdot(\mathbf{k}_2 + \mathbf{k}_4) + (\mathbf{k}_1 + \mathbf{k}_4)\cdot(\mathbf{k}_2 + \mathbf{k}_3)\Big]$
>
> $\dfrac{B}{TS^3} - \dfrac{4\Lambda_0}{S^4} - \dfrac{1}{TS^2}(\mathbf{k}_1 + \mathbf{k}_2)\cdot(\mathbf{k}_3 + \mathbf{k}_4)$
>
> The letters a and b on the legs of the above vertices represent different components of the order parameter. Thus, when four identical components appear together at a four-leg vertex, the third expression should be used; if two pairs of different components are present, the fourth expression should be used. No other combinations of components are coupled by the quartic terms in the Hamiltonian (13.30), so all four-leg vertices are one of these two types.

The Feynman rules for this model are considerably more complicated than those for the Landau–Ginzburg model. First, the order parameter field $\boldsymbol{\pi}$ has $(D-1)$ components, and second there are complicated terms such as $|\nabla \pi^2|^2$ present. The Feynman rules for terms such as these are explained in Appendix N, and summarized in Box 13.1. The main points to bear in mind are the following:

- Each link must bear a label, telling which component of $\boldsymbol{\pi}$ it represents. Those labels on the internal links of a diagram which are not fixed by

13.2 The non-linear σ-model

the labels on the external links must be summed over all of their possible values (just as for wavevectors).
- The presence of '∇' in an interaction term leads to factors of wavevector being associated with the corresponding vertex.

A word about dimensions: we are free to choose the dimension of the field $\pi(x)$, so we choose it to have dimension $L^{-d/2}$ as in the Landau–Ginzburg model. All the dimensions given in Tables 7.1 and 8.2 then apply here also. In addition it follows that T has dimension L^{-2}, B has dimension $L^{-(d/2)-2}$, and S has dimension $L^{-d/2}$.

It turns out that we can learn all we need from the two-point vertex function $\Gamma^{(2)}_{ij}$. The two indices i and j run from 1 to $(D-1)$; they label the components of π on each leg. Since the non-linear σ-model has an $O(D-1)$ symmetry that is unbroken by the magnetic field (and is not broken spontaneously), its vertex functions must display this symmetry. The unit matrix is the only $O(D-1)$-invariant object with two indices, so $\Gamma^{(2)}_{ij}$ must have the form

$$\Gamma^{(2)}_{ij} = \delta_{ij}\Gamma^{(2)}(\mathbf{k}_1, \mathbf{k}_2). \tag{13.31}$$

To lowest order in T the rules in Box 13.1 give[3]

$$\Gamma^{(2)}(\mathbf{k}_1, \mathbf{k}_2) = \frac{k_1^2 + (B/S)}{T}(2\pi)^d \delta(\mathbf{k}_1 + \mathbf{k}_2). \tag{13.32}$$

This is similar to the vertex function for the Gaussian model above its critical temperature; the correlation length in the present case is $\sqrt{S/B}$.

Even at this early stage we can extract some interesting information about the infrared behaviour of the model. We have

$$\begin{aligned}\Gamma^{(2)}(\mathbf{k}_1/b, \mathbf{k}_2/b) &= \frac{(k_1/b)^2 + (B/S)}{T}(2\pi)^d \delta((\mathbf{k}_1 + \mathbf{k}_2)/b) \\ &= \frac{k_1^2 + (B/S)b^2}{Tb^{2-d}}(2\pi)^d \delta(\mathbf{k}_1 + \mathbf{k}_2).\end{aligned} \tag{13.33}$$

An increase in scale by a factor of b is equivalent to multiplying (B/S) by b^2, and multiplying the temperature by b^{2-d}. The effective increase in (B/S) implies a decrease in the correlation length when the system is viewed on larger scales; this is reasonable. The effective change in T is more interesting. First, consider the case of $d > 2$. The effect of viewing the system on a larger scale is to cause the apparent temperature to decrease; the limit of the apparent temperature as $b \to \infty$ is zero. To understand the importance of

---
[3] The reader might be wondering why the magnetic field B appears in $\Gamma^{(2)}$, which is a derivative of the Gibbs free energy. The answer is that we find it convenient to regard B as a parameter of the model, on the same footing as T and S. The Gibbs free energy is the Legendre transform of the Helmholtz free energy not with respect to B, but with respect to a source term which we have not bothered to include because we never use it.

this, consider the case of zero B. Several of the steps in our derivation of the Hamiltonian (13.28) relied on π being small compared to S; having B non-zero ensured that this would be true at low enough temperature. But when $B = 0$ there is a risk that this won't work any more. However, the behaviour of the system under scaling shows that things are all right. We know that, at zero temperature, all the spins *must* be aligned since this is the lowest energy state of the system. The fact that the apparent temperature tends to zero at large scales means that the deviation from the state of total alignment also goes to zero in this limit. At sufficiently low temperatures there is long-range order in the system even when $B = 0$, and our approximations are justified.

But when $d < 2$ the situation is quite different. On moving to larger scales the apparent temperature *increases* (it remains constant when $d = 2$), which means that the observed fluctuations do not tend to zero. Now if the system really were fluctuating about an ordered state, the fluctuations would have to vanish as $b \to \infty$, since in this limit the $\tilde{\pi}(\mathbf{k})$ become averages over increasing numbers of spins which, by assumption, are all pointing more-or-less in the same direction. The fact that the fluctuations do not vanish is therefore inconsistent with the idea that the system is undergoing small fluctuations about an ordered state, and we conclude that for $d \leq 2$ (and $B = 0$) there is no long-range order at any non-zero temperature. We would expect this result not to change if $\Gamma^{(2)}$ is calculated to higher orders, since the corrections are of higher order in T than (13.32), and so should not affect the behaviour of the system at sufficiently low temperature.

All we've done so far is to reproduce the results of §13.1, though we've done it in a framework in which the rôle of the temperature is clear. What we are going to do next is to find the temperature at which the ordering transition takes place in $d > 2$; we shall see that this temperature goes to zero as $d \to 2^+$. This involves a considerable amount of work for what may seem like a small return, and you may prefer to look at the results—equation (13.58), the following paragraph, and Figure 13.2—and then skip to §13.3. If you want to see the details, read on.

To proceed we need to find $\Gamma^{(2)}$ to the next order in T. The diagrams and the corresponding expressions are shown in Figure 13.1. The final result is

$$\Gamma^{(2)}(\mathbf{k}) = \frac{k^2}{T}\left(1 + \frac{I}{S^2}T + O(T^2)\right) + \frac{B}{ST}\left(1 + \frac{I(D-1)}{2S^2}T + O(T^2)\right), \tag{13.34}$$

where

$$I = \int^{\Lambda} \frac{d^d\check{\mathbf{q}}}{q^2 + (B/S)}. \tag{13.35}$$

The contribution from the measure term, which looked as though it would break the $O(D)$ symmetry, has been cancelled by part of the one-loop diagrams in Figure 13.1. $\Gamma^{(2)}(\mathbf{k})$ vanishes when $k = 0$ and $B = 0$, showing that

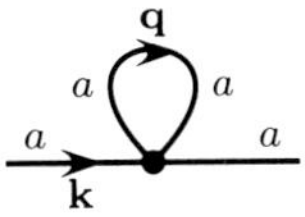

13.2 The non-linear σ-model

$$\frac{1}{2}\int \frac{d^d\check{q}}{q^2+(B/S)}\left(\frac{3B}{S^3}+\frac{2(k^2+q^2)}{S^2}\right)$$

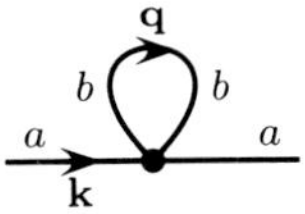

$$\frac{(D-2)}{2}\frac{B}{S^3}\int \frac{d^d\check{q}}{q^2+(B/S)}$$

$$\frac{\Lambda_0}{S^2}$$

Figure 13.1 These three diagrams are the O(1) contributions to $\Gamma^{(2)}(\mathbf{k})$. When they are added to the $O(T^{-1})$ piece in (13.32) the total is the expression (13.34). The above expressions follow from the rules in Box 13.1; we have neglected terms of order T. The factors of $\frac{1}{2}$ in front of the first two expressions come from the symmetry factor associated with the two ends of one link being joined to the same vertex; the factor of $(D-2)$ in front of the second expression is there because there are $(D-2)$ components of $\boldsymbol{\pi}$ different to a.

the symmetry is respected. Equation (13.35) depends on the cutoff parameter Λ through the integral I. This cutoff dependence is entirely natural, just as it was for the Landau–Ginzburg model.

13.2.2 The renormalization group equation

The first step in deriving the renormalization group equation for this model is knowing how the vertex functions depend on the cutoff. We shall *assume* that the non-linear σ-model is renormalizable; as in Chapter 9, this means that there are many different values of the parameters $\{\Lambda, S, T, B\}$ that give rise to the same vertex functions at wavevectors much less than the cutoff. To see that this is not unreasonable, consider the alternative. If the model were not renormalizable it would mean that changing Λ would cause new terms to appear in the Hamiltonian to keep the large-scale physics the same. But the model's $O(D)$ symmetry permits only certain terms with this symmetry to appear in the Hamiltonian, and as it happens all such terms that affect long-distance physics are already present. This argument can be upgraded into a proof of the renormalizability of the model; however, we shall not do this here.

Renormalizability means that we can find functions $S(\Lambda)$, $T(\Lambda)$, and $B(\Lambda)$ such that

$$\frac{d}{d\Lambda}\Gamma^{(n)}(\Lambda, S(\Lambda), T(\Lambda), B(\Lambda)) = 0. \tag{13.36}$$

These three functions are not independent. Let us expand a connected correlation function as a power series in B:[4]

$$\langle \tilde{\pi}_{i_1}(\mathbf{k}_1) \ldots \tilde{\pi}_{i_n}(\mathbf{k}_n) \rangle_c \Big|_B = \sum_{j=0}^{\infty} \frac{(B-B_0)^j}{j!} \frac{\partial^j}{\partial B^j} \langle \tilde{\pi}_{i_1}(\mathbf{k}_1) \ldots \tilde{\pi}_{i_n}(\mathbf{k}_n) \rangle_c \Big|_{B_0}. \tag{13.37}$$

The subscripts on $\tilde{\pi}$ denote different components of $\tilde{\pi}$, rather than different sites. It follows from (13.28) that

$$\frac{\partial^j}{\partial B^j} \langle \tilde{\pi}_{i_1}(\mathbf{k}_1) \ldots \tilde{\pi}_{i_n}(\mathbf{k}_n) \rangle_c \Big|_{B_0} = \frac{1}{T^j} \langle \tilde{\pi}_{i_1}(\mathbf{k}_1) \ldots \tilde{\pi}_{i_n}(\mathbf{k}_n) \tilde{\sigma}^j(0) \rangle_c. \tag{13.38}$$

Substituting (13.38) into (13.37) and taking a total derivative with respect to Λ gives

$$\frac{d}{d\Lambda} \left(\frac{B(\Lambda)}{T(\Lambda)} \right) = 0, \tag{13.39}$$

since by definition the total derivatives of the vertex functions vanish.

If we measure the vertex functions of the field $\boldsymbol{\pi}_R \equiv a\boldsymbol{\pi}$ instead of those of $\boldsymbol{\pi}$, the apparent values of S, T, and B change. It is easy to check that

$$T \mapsto T/a^2, \quad S \mapsto S/a, \quad \text{and} \quad B \mapsto B/a. \tag{13.40}$$

Because of this it is more convenient to use

$$t \equiv T/S^2, \quad h \equiv B/S \tag{13.41}$$

as measures of the temperature and magnetic field, because these do not depend on how the field is normalized[5]. If we choose a to have the particular value S, it follows that

$$\Gamma^{(n)}(\mathbf{k}_i; \Lambda, S, T, B) = S^{-n} \Gamma^{(n)}(\mathbf{k}_i; \Lambda, 1, t, h). \tag{13.42}$$

As in Chapter 9, we define

$$\Gamma_R^{(n)}(\mathbf{k}_i; \Lambda, t, h) \equiv \Gamma^{(n)}(\mathbf{k}_i; \Lambda, 1, t, h). \tag{13.43}$$

The quantities t and h have dimensions L^{d-2} and L^{-2} respectively. We define dimensionless Λ-independent quantities $\hat{s}$, $\hat{t}$, and $\hat{h}$ as follows:

$$S = Z_s \kappa^{d/2} \hat{s}; \quad t = Z_t \kappa^{2-d} \hat{t}; \quad h = Z_h \kappa^2 \hat{h}. \tag{13.44}$$

---

[4] We expand about a non-zero field B_0 because, if there is long-range order, the derivative in zero field are not well-defined.

[5] We use h for B/S rather than b, since we're using b to represent the factor by which wavevectors are scaled.

13.2 The non-linear σ-model

Here κ is a completely arbitrary quantity with the dimension of inverse length, introduced purely to soak up the dimensions. The Λ-dependence of S, t, and b is all to be contained in the dimensionless functions Z_s, Z_t, and Z_h. On dimensional grounds these can be functions of $\hat{s}$, $\hat{t}$, $\hat{h}$, and Λ/κ only. They are not independent: the condition (13.39) implies that

$$Z_h \propto Z_s Z_t. \tag{13.45}$$

We can now write equation (13.36) as

$$\Lambda \frac{d}{d\Lambda} \left[S^{-n} \Gamma_R^{(n)}(\mathbf{k}_i; \Lambda, t, h) \right] = 0, \tag{13.46}$$

or

$$\left[\Lambda \left(\frac{\partial}{\partial \Lambda} \right)_{t,h} - n\Lambda \left(\frac{\partial \log Z_s}{\partial \Lambda} \right)_{\Lambda,t,h} + \Lambda \left(\frac{\partial t}{\partial \Lambda} \right)_{\hat{s},\hat{t},\hat{h}} \left(\frac{\partial}{\partial t} \right)_{\Lambda,h} \right. \tag{13.47}$$
$$\left. + \Lambda \left(\frac{\partial b}{\partial \Lambda} \right)_{\hat{s},\hat{t},\hat{h}} \left(\frac{\partial}{\partial h} \right)_{\Lambda,t} \right] \Gamma_R^{(n)}(\mathbf{k}_i; \Lambda, t, h) = 0.$$

Dimensional analysis implies that

$$\left[\Lambda \left(\frac{\partial}{\partial \Lambda} \right)_{b,t,h} - b \left(\frac{\partial}{\partial b} \right)_{\Lambda,t,h} + (2-d)t \left(\frac{\partial}{\partial t} \right)_{\Lambda,b,h} \right. \tag{13.48}$$
$$\left. + 2h \left(\frac{\partial}{\partial h} \right)_{\Lambda,b,t} - d \right] \Gamma_R^{(n)}(\mathbf{k}_i/b; \Lambda, t, h) = 0.$$

Combining these two equations gives us a renormalization group equation for the vertex functions of the non-linear σ-model:

$$\left[b \left(\frac{\partial}{\partial b} \right)_{\Lambda,t,h} - \frac{n}{2} \gamma_s + \beta_t \left(\frac{\partial}{\partial t} \right)_{\Lambda,h} + \beta_h \left(\frac{\partial}{\partial h} \right)_{\Lambda,t} + d \right] \Gamma_R^{(n)}(\mathbf{k}_i/b; \Lambda, t, h) = 0, \tag{13.49}$$

where

$$\beta_t \equiv \Lambda \left(\frac{\partial t}{\partial \Lambda} \right)_{\hat{s},\hat{t},\hat{h}} - (2-d)t$$

$$\beta_h \equiv \Lambda \left(\frac{\partial h}{\partial \Lambda} \right)_{\hat{s},\hat{t},\hat{h}} - 2h \tag{13.50}$$

$$\gamma_s \equiv 2\Lambda \left(\frac{\partial \log Z_s}{\partial \Lambda} \right)_{\Lambda,t,h}.$$

Equation (13.49) is precisely analogous to the renormalization group equations in Chapters 11 and 12, and can be solved in the same way. We define the scale-dependent quantities

$$b\frac{dt(b)}{db} = -\beta_t\bigl(t(b), h(b)\bigr)$$
$$b\frac{dh(b)}{db} = -\beta_h\bigl(t(b), h(b)\bigr) \quad (13.51)$$
$$b\frac{d\log S(b)}{db} = -\tfrac{1}{2}\gamma_s\bigl(t(b), h(b)\bigr).$$

Then the solution is

$$\Gamma_R^{(n)}(\mathbf{k}/b; \Lambda, t, h) = b^{-d}\left(\frac{S}{S(b)}\right)^n \Gamma_R^{(n)}\bigl(\mathbf{k}; \Lambda, t(b), h(b)\bigr). \quad (13.52)$$

This will tell us how the vertex functions scale with wavevector, once we have calculated β_t, β_s, and γ_s. Using (13.44) and (13.45) we can write them as

$$\beta_t = t\left(\Lambda\frac{\partial \log Z_t}{\partial \Lambda} + (d-2)\right)$$
$$\gamma_s = 2\Lambda\frac{\partial \log Z_s}{\partial \Lambda} \quad (13.53)$$
$$\beta_h = h\left(\tfrac{1}{2}\gamma_s + \frac{\beta_t}{t} - d\right).$$

To get any further we must calculate Z_t and Z_s. We shall see that a knowledge of $\Gamma^{(2)}$ is sufficient.

Superficially, our derivation of equation (13.49) differs from that used in Chapter 11 for the Landau–Ginzburg model. There we eliminated the bare parameters of the theory in favour of the values of vertex functions at a particular wavevector, and derived the renormalization-group equation by varying this wavevector. We didn't do this here because the bare parameters t and b are very interesting in themselves, being proportional to the temperature and the magnetic field. Instead, we derived (13.49) by varying the cutoff. Such details do not matter: the essential feature of a renormalizable theory is that small and large scales are divorced from each other, so one can be varied without affecting the other. The equation describing this variation can then be turned into something useful by dimensional analysis. This separation of short and large scales is the fundamental idea behind the renormalization group.

Let us continue with finding the critical exponents. At lowest order in T we know that Z_t and Z_s are independent of Λ, since $\Gamma^{(2)}$ does not depend on Λ to this order. We therefore expect that the Λ-dependence in Z_t and Z_s comes in at $O(T)$. Therefore let us write

$$Z_s = 1 + p \quad \text{and} \quad Z_t = 1 + q, \quad (13.54)$$

13.2 The non-linear σ-model

where p and q are $O(T)$, and so are small in the context of our low-temperature expansion. We didn't have to choose 1 for the constant terms, but it is the simplest choice. Now let us return to equations (13.34) and (13.35), and calculate the integral in $d = 2 + \epsilon$ dimensions. Note that ϵ here is $d - 2$ and not $4 - d$. We are interested in what happens close to two dimensions, not close to four dimensions. We have

$$\begin{aligned} I &= \int^\Lambda \frac{d^d\check{q}}{q^2 + h} \\ &= \int^\Lambda \frac{d^d\check{q}}{q^2} + \int^\Lambda d^d\check{q} \left(\frac{1}{q^2 + h} - \frac{1}{q^2} \right) \\ &= \Lambda^\epsilon \left(\frac{1}{2\pi} \log \left(\frac{\Lambda}{h^{1/2}} \right) + \frac{\gamma_e}{2\pi} + O(\epsilon) \right). \end{aligned} \tag{13.55}$$

Since we'll be letting $\epsilon \to 0$ we haven't kept the $O(\epsilon)$ terms. We have, however, kept the overall factor of Λ^ϵ even though it is 1 to $O(\epsilon)$, because our conscience will not allow us to set a dimensional quantity equal to one. On substituting this integral and the definitions (13.44) into (13.34), and keeping terms only to $O(T)$, we find that $\Gamma^{(2)}$ does not vary with Λ if p and q satisfy

$$\begin{aligned} p &= \frac{D-1}{2} \frac{\hat{t}}{2\pi} \log \frac{\Lambda}{\kappa} \\ q &= (2-D) \frac{\hat{t}}{2\pi} \log \frac{\Lambda}{\kappa}. \end{aligned} \tag{13.56}$$

(These are not the only possible values for p and q, since any Λ-independent expression could be added to them and they would still fulfil their rôle.) We thus obtain

$$\begin{aligned} \beta_t &= \epsilon t - \frac{\kappa^\epsilon}{2\pi}(D-2)t^2 + O(\epsilon^2 t, \epsilon t^2) \\ \gamma_s &= \frac{(D-1)\kappa^\epsilon t}{2\pi} + O(\epsilon t). \end{aligned} \tag{13.57}$$

We now have all the information we need about the behaviour of the system.

$t(b)/t$ is the ratio of the temperature as measured by someone whose length units have been stretched by a factor of b, to the true system temperature. If $\beta(t(b)) = 0$ then the first of equations (13.51) implies that $t(b)$ will not change with t. When $\epsilon > 0$ and $D > 2$ there are two such 'fixed points', one at $t = 0$ and the other at

$$t_c \equiv \frac{2\pi\epsilon}{D-2}\kappa^{-\epsilon}. \tag{13.58}$$

The $t = 0$ fixed point is attractive, while the $t = t_c$ fixed point is repulsive. The attractive fixed point is the usual low-temperature fixed point, while

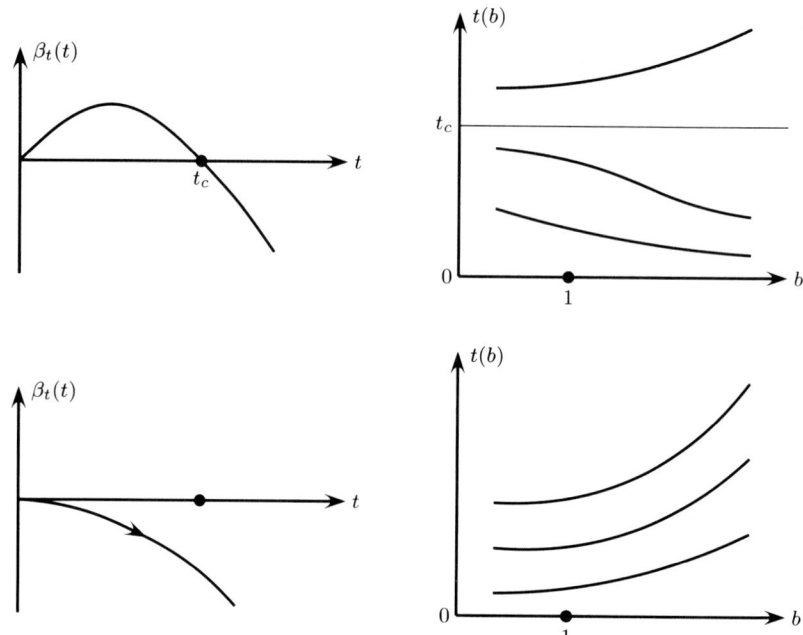

Figure 13.2 The upper two graphs show the situation in more than two dimensions. β_t has an attractive zero at $t = 0$, and a repulsive zero at $t = t_c$. The lower two graphs show the situation in less than two dimensions. β_t has only one zero, and it is repulsive.

the repulsive fixed point is associated with a phase transition. Since at zero temperature the system must be in its lowest-energy state with all the spins aligned, the phase transition is presumably associated with the onset of long-range order; this is established in the calculation of the exponent β in Problem 13.1. As $d \to 2^+$ it becomes harder for long-range order to take hold and t_c decreases. Notice that t_c also decreases as the number of spin components increases, which suggests that the presence of Goldstone modes really does discourage the onset of order. At $d = 2$ $(D > 2)$ there is no longer any transition. Below two dimensions there is only one fixed point at $t = 0$, and this is repulsive. There is no long-range order at any non-zero temperature. This behaviour is illustrated in Figure 13.2.

The case of two components is special. The $O(T^2)$ correction to β_t vanishes and we are left with just the lowest-order term. For $d > 2$ the low-temperature fixed point is repulsive, suggesting a phase transition at higher temperature, but we see no sign of it. It turns out that *all* higher-order corrections to β_t vanish in perturbation theory (see the footnote on page 341). However, this does not mean that there is no transition, as we

13.3 The Kosterlitz–Thouless transition

shall see in the next section.

It is now relatively straightforward to calculate five of the critical exponents; as usual, α requires some extra work. The calculations are left to the problems. The results are:

$$\beta = \frac{D-1}{2(D-2)}; \qquad \gamma = \frac{2}{\epsilon}; \qquad (13.59)$$

$$\delta = \frac{1}{\epsilon}\frac{4D-2}{D-1}; \qquad \eta = \frac{\epsilon}{D-2}; \qquad \nu = \frac{1}{\epsilon}.$$

The special nature of the two-component model stands out clearly in these results.

13.3 The Kosterlitz–Thouless transition

We have seen in the previous section that the non-linear σ-model for $d = 2$, $D = 2$ is rather special. We were unable to draw any definite conclusions regarding it from our ϵ-expansion about $d = 2$; indeed, historically it posed a great puzzle. In the late 1960s, it was known that the Mermin–Wagner theorem precluded the possibility of long-range order at low temperatures for a rotationally invariant model in two dimensions, and yet numerical studies of the system persisted in showing features reminiscent of a phase transition. These were not understood until the remarkable work of Berezinskii in 1970 and of Kosterlitz and Thouless in 1972. This demonstrated that the system *does* experience a phase transition, but the transition is *not* to a state of long-range order. The physics of this transition is fascinating, and it is instructive to understand why it lies beyond the scope of the methods we have used so far. Doing so involves backtracking to the underlying lattice model, namely the XY model introduced in §3.1.

For the XY model the angle $\theta_\mathbf{j}$ which gives the direction of the spin on site $\mathbf{j}$ is a convenient order parameter. To convert this into a continuous field we have to imagine that $\theta_\mathbf{j}$ is the value $\theta(\mathbf{x_j})$ of some smooth function of $\mathbf{x}$ at the position $\mathbf{x_j}$ of site $\mathbf{j}$; recall that when summing Boltzmann factors to form the partition function we include only order-parameter fields with terminating Fourier transforms, so $\theta(\mathbf{x})$ is smooth by construction. Unfortunately, this restriction is unacceptable. Figure 13.3(a) shows why: there the spins 'circulate' anti-clockwise—such a configuration of spins is called a **vortex**. Consequently, as one moves on a large circle around the centre of the figure, θ increases steadily from 0 to 2π. When we extend θ to a smooth function of $\mathbf{x}$, we find that contours of constant $\theta(\mathbf{x})$ run roughly radially. Following them inwards it becomes clear that they eventually collide, with the result that there are places at which neighbouring points have radically

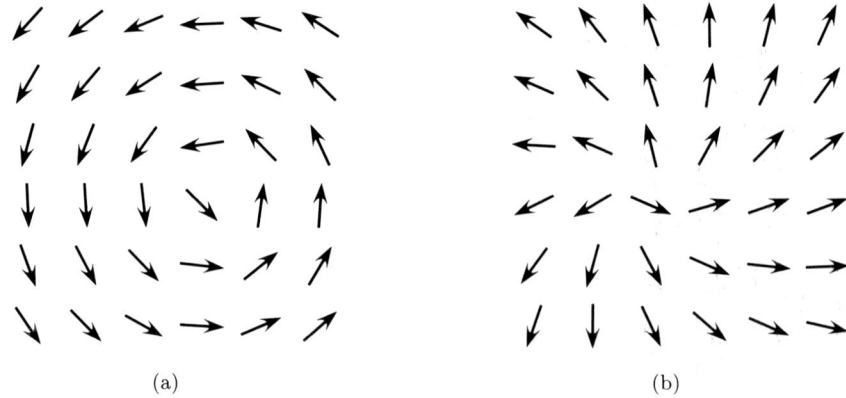

Figure 13.3 (a) A vortex with $q = 1$. The spins rotate once anticlockwise on travelling once anticlockwise around the centre of the figure. (b) This is also a $q = 1$ vortex. Each spin has been rotated (through 148°) relative to the vortex in (a), yet the spins still wind around once per circuit.

different values of θ. It follows that the original values of θ on the lattice cannot be derived from to a smooth function $\theta(\mathbf{x})$.

If we go ahead and evaluate Z by summing over smooth functions $\theta(\mathbf{x})$, we will be excluding from the sum all vortices of spins. But, as we shall soon see, the system's dynamics can be dominated by vortices. Hence conclusions derived from the non-linear σ-model cannot be safely applied to the XY model unless we can somehow extend the former to include the effects of vortices.

Fortunately, with a little cunning it *is* possible to adapt the non-linear σ-model to include the effects of vortices. We divide the field $\theta(\mathbf{x})$ into two components, a vortex field $\bar\theta(\mathbf{x})$ and a vortex-free field $\psi(\mathbf{x})$:

$$\theta(\mathbf{x}) = \bar\theta(\mathbf{x}) + \psi(\mathbf{x}) \tag{13.60}$$

Mathematically, the difference between $\bar\theta$ and ψ can be expressed by the line integrals

$$\oint_{\substack{\text{some}\\\text{path}}} \boldsymbol{\nabla}\bar\theta \cdot \mathrm{d}\mathbf{x} = 2\pi q \quad (\text{integer } q \neq 0) \tag{13.61a}$$

$$\oint_{\substack{\text{every}\\\text{path}}} \boldsymbol{\nabla}\psi \cdot \mathrm{d}\mathbf{x} = 0. \tag{13.61b}$$

$\psi(\mathbf{x})$ is everywhere smooth and can be expressed in terms of a rapidly convergent Fourier transform. $\bar\theta(\mathbf{x})$, by contrast, has singularities. We can always

13.3 The Kosterlitz–Thouless transition

choose ψ so that $\bar\theta$ is a local minimum of the energy: The difference between two fields with the same distribution of singularities is a smooth function. So amongst all the vortex fields $\bar\theta$ which, through (13.60), can represent any given field θ, we can select the one which minimizes the effective Hamiltonian.[6]

In its continuum formulation, the XY model's Hamiltonian is[7]

$$H = \int d^2x \, \tfrac{1}{2}\alpha^2 |\nabla\theta|^2. \tag{13.62}$$

It is straightforward to show (see Problem 13.2) that the field of minimum energy for which

$$\oint_S \nabla\theta \cdot d\mathbf{x} = 2\pi q, \tag{13.63}$$

where S is any path circling the origin once, is

$$\bar\theta = q\phi, \tag{13.64}$$

ϕ being the angle of plane polar coordinates. For this field,

$$|\nabla\theta| = \frac{q}{r}. \tag{13.65}$$

Using (13.62) and (13.65), we find that the energy of this vortex is

$$\int_{r_0}^{R} r\, dr \, \frac{\pi\alpha^2}{r^2} = \pi q^2 \alpha^2 \log(R/r_0), \tag{13.66}$$

where r_0 is a cutoff of the order of the lattice spacing and R is the linear size of the system. Notice that the vortex carries an infinite amount of energy as $(r_0/R) \to 0$, because as the lattice spacing shrinks, the field at the vortex centre has to vary more and more quickly. This is a reminder that vortices are possible only because our physical system really is made up of atoms, an inconvenient fact the non-linear σ-model would rather ignore. Two examples of vortices with $q \neq 1$ are shown in Figure 13.4.

It is easy to check that the sum of two or more vortices of the form (13.64), with different strengths and different centres, is also a local minimum

---

[6] The division of θ into $\bar\theta$ and ψ is still arbitrary to the extent that we can add a constant to $\bar\theta$ and subtract the same constant from ψ. This freedom can be removed by demanding that $\bar\theta$ be zero at the fourth site from the right in the second row (for example).

[7] Seeing the effective Hamiltonian in the form (13.62) makes it clear why for this model everything vanishes to all orders in perturbation theory. It is just the Hamiltonian of the Gaussian model—except for the crucial difference that θ is an angle and therefore values of θ that differ by multiples of 2π describe the same configuration. The significance of this cannot be brought out in perturbation theory, and requires a non-perturbative treatment.

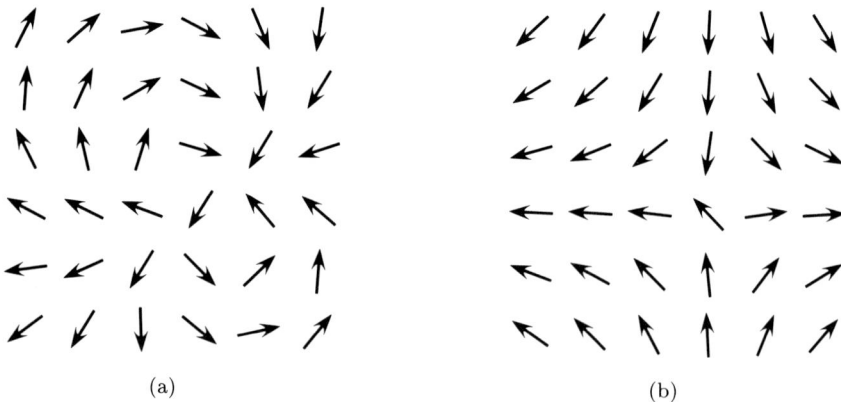

Figure 13.4 Vortices of strength (a) $q = 2$ (b) $q = -1$. Notice that the $q = -1$ vortex is *not* just a $q = 1$ vortex with the spins pointing around clockwise rather than anticlockwise. The spins in (b) are winding against the direction of rotation.

of the energy. Now that we know the structure of such fields, we are in a position to investigate the energy of the system when ψ is non-zero. This is

$$H = \int d^2\mathbf{x}\, \tfrac{1}{2}\alpha^2 |\boldsymbol{\nabla}\theta|^2 = \int d^2\mathbf{x}\, \tfrac{1}{2}\alpha^2 \left(|\boldsymbol{\nabla}\bar{\theta}|^2 + |\boldsymbol{\nabla}\psi|^2\right), \quad (13.67)$$

where we have used the fact that $\bar{\theta}$ minimizes the effective Hamiltonian to deduce that the cross-term proportional to $\boldsymbol{\nabla}\bar{\theta}\cdot\boldsymbol{\nabla}\psi$ vanishes. The vanishing of this cross-term has a very important consequence: it is possible to treat the ψ and $\bar{\theta}$ terms as two completely separate systems, since there is no interaction between them. The ψ system includes all the Goldstone modes which we have been discussing so far in this chapter; these modes prevent any long-range order developing in the ψ system and its thermodynamics is not particularly interesting. We shall therefore forget about it and concentrate instead on the $\bar{\theta}$ system, which describes the vortex structure.

13.3.1 The two-dimensional Coulomb gas

Equation (13.66) gives the energy of an isolated vortex; the next step is to find how vortices interact with each other. The easiest way to see this is to draw an analogy between a system of vortices and a system of electrical charges in two dimensions. If $\hat{\mathbf{k}}$ denotes the direction perpendicular to the plane of the system, then $\boldsymbol{\mathcal{E}} = \boldsymbol{\nabla}\bar{\theta} \times \hat{\mathbf{k}}$ (where $\bar{\theta} = q\phi$) is equal to the electric field produced by a charge of strength $2\pi\epsilon_0 q$ positioned at the centre of the vortex. Furthermore, the energy of an electric field is

$$E_{\text{electric}} = \int d^2\mathbf{x}\, \tfrac{1}{2}\epsilon_0 |\boldsymbol{\mathcal{E}}|^2, \quad (13.68)$$

13.3 The Kosterlitz–Thouless transition

whilst the energy of the system of vortices is

$$E_{\text{vortices}} = \int d^2\mathbf{x}\, \tfrac{1}{2}\alpha^2|\boldsymbol{\nabla}\overline{\theta}|^2 = \int d^2\mathbf{x}\, \tfrac{1}{2}\alpha^2|\boldsymbol{\nabla}\overline{\theta} \times \hat{\mathbf{k}}|^2 \qquad (13.69)$$
$$= \frac{\alpha^2}{\epsilon_0} E_{\text{electric}}.$$

From these two equations we deduce that the energy of a system of vortices of strengths $q_1, q_2, \ldots$ is α^2/ϵ_0 times the energy of a system of electric charges $2\pi\epsilon_0 q_1, 2\pi\epsilon_0 q_2, \ldots$ This means that there is almost a complete analogy between a system of vortices and the two-dimensional Coulomb gas. From now on, we shall discuss the system in electrical language, since this is more familiar.

As we've observed, a single vortex has an energy that diverges logarithmically with the size of the system, so at low temperatures very few isolated vortices are present. However, a pair of opposite charges—a dipole—has a finite total energy, because at large distances the fields from each charge almost cancel. The total energy of two charges $\pm q$ centred at $\mathbf{x}_1$ and $\mathbf{x}_2$ is

$$\Phi(|\mathbf{x}_1 - \mathbf{x}_2|) = 2\pi\alpha^2 q^2 \log\left(\frac{|\mathbf{x}_1 - \mathbf{x}_2|}{r_0}\right). \qquad (13.70)$$

Now although isolated vortices can be created and destroyed in a finite system (unlike electrical charges), in practice this is not an important process in a large sample. The appearance or disappearance of a new vortex requires a change in the orientations of the spins over the whole sample, and there is a large energy barrier preventing this. We shall therefore ignore the effects of isolated vortices on the system. But a dipole only affects the spins locally, and so it can be created or destroyed at will. At non-zero temperatures there will be a non-zero density of dipoles. Now (13.70) assumes that the two charges are in a vacuum. As the temperature rises and the density of dipoles rises with it, this will no longer be true, because there may be other dipoles nearby. Their presence will reduce the energy of the dipole, since they tend to align with and partially cancel its electric field. So the more dipoles there are, the lower each dipole's self-energy is. This in turn encourages the creation of more dipoles, until, as we shall see, there is a transition to a phase in which there are so many dipoles that it ceases to make sense to think of the charges as being bound in pairs to form dipoles, and we have rather to deal with a plasma of equal numbers of free, screened charges of each sign. This is the **Kosterlitz–Thouless** (K–T) transition.

To investigate this transition, we must calculate the dipole density n. So long as T is small and the dipoles are scarce and compact, this is easily done. We only consider dipoles composed of unit charges; the energy of a dipole goes as q^2, so dipoles composed of higher charges have significantly smaller Boltzmann factors. The states of the ith dipole in the system are labelled by the positions $\mathbf{x}_1^{(i)}$ and $\mathbf{x}_2^{(i)}$ of its two charges; we must also sum over the number of dipoles p. The partition function is

$$Z = \sum_{p=0}^{\infty} \frac{1}{p!} \int \frac{d\mathbf{x}_1^{(1)} d\mathbf{x}_2^{(1)}}{r_0^4} \cdots \frac{d\mathbf{x}_1^{(p)} d\mathbf{x}_2^{(p)}}{r_0^4} \exp\left[-\beta \left(\sum_{i=1}^{p} \Phi(|\mathbf{x}_1^{(i)} - \mathbf{x}_2^{(i)}|) - p\mu\right)\right]. \tag{13.71}$$

The factor $(p!)^{-1}$ prevents overcounting of states containing more than one dipole. We've treated the positions of the charges as continuous variables, and integrated over them; the factor of r_0^{-4} provides the correct conversion factor between the sum and the integral (compare equation (13.21)). μ, the chemical potential of a dipole, is there so that we can calculate $n \equiv \langle p \rangle$ by differentiating Z. Since we've ignored possible interactions between the dipoles, Z is quite easy to calculate. Its value is

$$Z = \exp\left[\frac{2\pi A}{r_0^4} \int_0^{\infty} r dr \, e^{-\beta(\Phi(r)-\mu)}\right], \tag{13.72}$$

where A is the system's area. Therefore

$$n = \frac{1}{\beta} \frac{\partial \log Z}{\partial \mu}\bigg|_{\mu=0} = \frac{2\pi}{r_0^4} \int_0^{\infty} r dr \, e^{-\beta \Phi(r)}. \tag{13.73}$$

But n becomes substantially harder to calculate once interactions between dipoles become important. To deal with this situation we'll use a mean-field approach, in which we model the dipoles' effect by a change in the permittivity of the medium. This permittivity will depend on the separation of the charges being considered. If we introduce a test charge and examine the electric field very close to it—on a scale small compared with the mean spacing of the dipoles—it will not be significantly modified from its value in a vacuum. On larger length scales, however, the polarizability of the system of dipoles intervening between the test charge and the point of measurement causes the electric field to be smaller than its value in a vacuum, and this is reflected in the permittivity on such scales exceeding unity.

In the spirit of mean-field theory, the density of pairs with separation r is approximately

$$dn(r) = \frac{2\pi}{r_0^4} e^{-\beta \Phi_{\text{eff}}(r)} r dr \tag{13.74}$$

(compare (13.73)), where $\Phi_{\text{eff}}(r)$ is the energy actually required to pull two opposite charges apart to separation r given that the surrounding medium has non-zero polarizability. Strictly, Φ_{eff} is related to the vacuum potential (13.70) by[8]

$$d\Phi_{\text{eff}}(r) = \frac{d\Phi(r)}{\epsilon(r)}, \tag{13.75}$$

---

[8] In standard electrostatics the force on a unit charge is the electric field $\mathcal{E} = \mathcal{E}_0/\epsilon$, where ϵ is the relative permittivity of the medium and $\mathcal{E}_0$ is what the electric field would be if the dielectric medium were replaced by vacuum.

13.3 The Kosterlitz–Thouless transition

where $\epsilon(r)$ is the permittivity seen by two charges a distance r apart, but we assume that $\epsilon(r)$ varies relatively slowly with r, and use instead the approximate expression

$$\Phi_{\text{eff}}(r) \approx \frac{\Phi(r)}{\epsilon(r)}. \tag{13.76}$$

When (13.76) is substituted into (13.74), we find (using (13.70))

$$dn(r) = \frac{2\pi}{r_0^4} \left(\frac{r}{r_0}\right)^{-\beta/(2\pi\epsilon_0\epsilon(r))} r\, dr. \tag{13.77}$$

This has expressed the dipole density in terms of the permittivity $\epsilon(r)$. As expected, the density increases with $\epsilon(r)$.

$\epsilon(r)$ differs from unity because of the polarizability of individual dipoles. Let us therefore concentrate on a single dipole of separation r and calculate its polarizability. The angle-dependent part of its total energy E in an applied field $\mathcal{E}$ is

$$E = -\mathcal{E} r \cos\vartheta, \tag{13.78}$$

where ϑ is the angle between the dipole and the applied field. The polarizability $\chi(r)$ of this system is

$$\chi(r) = \frac{\partial \langle r \cos\theta \rangle}{\partial \mathcal{E}}\bigg|_{\mathcal{E}=0}, \tag{13.79}$$

where the average is taken over all configurations with separation r, using the Boltzmann weight $e^{-\beta E}$. One finds

$$\chi(r) = \frac{\beta r^2 \int_0^{2\pi} d\vartheta \cos^2\vartheta}{\int_0^{2\pi} d\vartheta}$$
$$= \tfrac{1}{2}\beta r^2. \tag{13.80}$$

Now we can estimate $\epsilon(r)$. We suppose that only those dipoles with separation $r' \leq r$ contribute to it, so that[9]

$$\epsilon(r) = 1 + \int_{r'=0}^{r} \frac{\chi(r')}{\epsilon_0}\, dn(r'). \tag{13.81}$$

Differentiating with respect to r and using (13.77) gives:

$$d\epsilon(r) = \tfrac{1}{2}\beta\epsilon_0^{-1} 2\pi \left(\frac{r}{r_0}\right)^{[-\beta/(2\pi\epsilon_0\epsilon(r))]+4} d\log\left(\frac{r}{r_0}\right). \tag{13.82}$$

---

[9] In standard electrostatics we have

$$\epsilon = 1 + \frac{\chi}{\epsilon_0}.$$

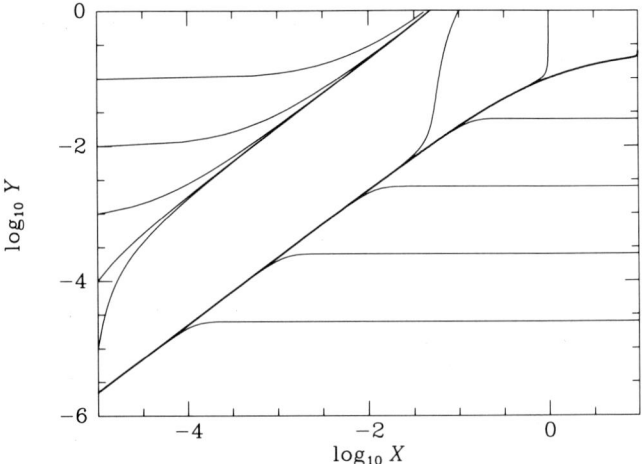

Figure 13.5 Solutions $Y(X)$ to the Kosterlitz–Thouless equation (13.84). For $X \ll 1$ the two linear solutions $Y = aX$, where $a = 0.2231$ or 18.71. can be seen clearly as the two nearly straight curves. Trajectories which start close to but just below $Y = 0.2231X$ veer away as X increases, yielding $Y = \text{constant}$ as $X \to \infty$. Solutions which start just above this line zoom to large Y at $X > 1$. Hence if we start from a point (X_0, Y_0) with $X_0 \ll 1$ which lies near the critical trajectory (the heavy curve), the value attained by Y at any given $X > 1$ is a rapidly-varying non-analytic function of Y_0. This non-analyticity gives rise to non-trivial critical exponents.

This is a non-linear first-order differential equation for $\epsilon(r)$. All we have to do now is to solve it—but, as with most non-linear differential equations, this is far from trivial. First, we write it more elegantly in terms of new, dimensionless variables

$$X \equiv \log\left(\frac{r}{r_0}\right) \qquad Y \equiv \frac{2\pi\epsilon_0}{\beta}\epsilon \qquad (13.83)$$

as

$$\frac{dY}{dX} = 2\pi^2 e^{-X(Y^{-1}-4)}. \qquad (13.84)$$

We would like the permittivity ϵ to become unity at short distances, so we would like to impose the following boundary condition on Y:

$$Y(X=0) = \frac{2\pi\epsilon_0}{\beta}, \qquad (13.85)$$

It is through this boundary condition that the temperature dependence of the solutions should enter.

We would expect the solutions to (13.84) to behave qualitatively as follows. Initially Y increases at some rate, determined by its value at $X = 0$.

13.3 The Kosterlitz–Thouless transition

If the initial value of Y is small (low temperature), the initial increase is not rapid and X grows large without $(Y^{-1} - 4)$ becoming negative. The exponential then ensures that the rate of increase diminishes rapidly, and Y asymptotes to some value. But if the initial value of Y is higher (higher temperature) the initial increase can be rapid enough that $(Y^{-1} - 4)$ becomes negative, and then when X does become large the exponential rapidly accelerates the increase, with the result that Y zooms off to infinity. There is a sharp division between these two different behaviours; this is the K–T phase transition.

Figure 13.5 shows a numerical integration of equation (13.84). This confirms that there is indeed a critical trajectory, drawn as the heavy curve in the figure, below which trajectories asymptote to small constant values of Y as $X \to \infty$, and above which trajectories zoom off to large Y as X becomes large. But analysis of our equation in the light of Figure 13.5 conclusively demonstrates that *all* trajectories starting from $X = 0$ and $Y > 0$ zoom off to infinity, rather than asymptoting to a constant value of Y.

If X and Y are both much less than 1, the 4 in the exponential of (13.84) can be neglected. The differential equation is then homogeneous, and a trial solution of the form $Y = aX$ yields

$$a = 2\pi^2 e^{-1/a}. \tag{13.86}$$

This equation has three real roots for a: 0, 0.2231, and 18.71. The first is of no physical interest, while the second and third correspond to the lower and upper diagonal lines in Figure 13.5, respectively. Thus the region within which solutions asymptote to finite Y does not reach to the line $X = 0$, $Y > 0$, but lies always below the line $Y = 0.2231X$. Consequently, every solution of (13.84) subject to the boundary condition (13.85) has divergent Y at large X, and we seem to have demonstrated that the macroscopic permittivity must diverge for any non-zero temperature. If this conclusion were correct there would be no phase transition.

Figure 13.6 shows two simulations of the $d = 2$ XY model, one below and one above T_c. A high degree of local order is apparent in the lower panel, which is entirely absent from the upper, hotter simulation. Thus these simulations confirm that something like the K–T transition *does* take place in the $d = 2$ XY model as equation (13.84) predicts when the boundary condition is imposed at $X > 0$.

However, equation (13.84) is a continuum approximation to a lattice problem, and it cannot be trusted for values of $X \ll 1$, which, by equation (13.83), correspond to dipole separations $r \simeq r_0$. Physically it would make quite as much sense to impose the condition $Y = 2\pi\epsilon_0/\beta$ at $X = 1$, i.e., at $r = er_0$, as at $X = 0$. Then the trajectories would fall into two classes according to their values of $Y(1)$. Those curves which began above the heavy curve would still go off to infinity as $X \to \infty$, but those which began below the heavy curve would asymptote to finite values of Y. Thus if we impose our boundary condition at any $X > 0$ there is a transition as $Y(1)$ crosses the full curve in Figure 13.5 (which corresponds to some T_c) from a phase with finite macroscopic ϵ to a phase in which the macroscopic value of ϵ is large. Of course ϵ does not really become infinite; when ϵ is large, the dipole density is also large and the approximations we made break down. But we can imagine that the low-temperature phase contains a relatively small number of tightly-bound dipoles, whereas the high-temperature phase is a vortex-antivortex plasma, with no clearly defined dipole pairs.

The sensitivity of the solutions of equation (13.84) to the precise location in X of the boundary condition arises because the energy of a dipole depends sensitively on the presence of other dipoles with separations comparable to r_0. Actually, very small dipoles do not occur because the system is really on a lattice rather than a continuum. So here as in the Landau–Ginzburg model we are faced with the problem that our results are sensitive to the smallest scales considered in the continuum model, and we have no *a priori* way of choosing the value of the short-scale cutoff. In the case of the Landau–Ginzburg model we have seen that it is possible to side-step this problem by expressing all quantities in macroscopic terms. No comparable procedure is available for the Kosterlitz–Thouless system, so we have had to resort to a crude exclusion of dipoles with $X < X_0$, which is what is physically implied by imposing the boundary condition $Y = 2\pi\epsilon_0/\beta$ at $X = X_0$ rather than at $X = 0$. In short we have made plausible the existence of a phase transition but we have not been able reliably to calculate the critical temperature. This is something that can be done numerically by means of simulations such as those shown in Figure 13.6.

A more sophisticated treatment of this problem has been given by Kosterlitz (1974). By performing a real-space renormalization calculation, he was able to show that the correlation length diverges above T_c as

$$\xi \sim e^{b/t^{1/2}}, \qquad \text{with} \qquad b \approx 1.5, \tag{13.87}$$

where $t \equiv (T - T_c)/T_c$. However below T_c the correlation length is always infinite. This is manifested in the calculation by the remarkable fact that even below the transition, the renormalization transformation does not take the system to a true low-temperature fixed point. Instead, it maps to one of a set of critical fixed points; which fixed point is reached, and hence which exponents characterize the large-distance behaviour of the system, depends on the temperature.

The 'singular' part of the free energy (see §5.7) behaves as

$$F_{\text{sing}} \sim \xi^{-2}. \tag{13.88}$$

The transition is therefore of infinite order in the Ehrenfest classification; all derivatives of the singular part of the free energy are continuous at $T = T_c$.

13.3 The Kosterlitz–Thouless transition

Figure 13.6 Two Monte-Carlo simulations of the $d = 2$ XY model. In the upper panel $T > T_c$ and the spins appear jumbled. In the lower panel $T < T_c$ and there is a degree of local order; numerous patches of aligned spins can be identified, and vortex–antivortex pairs can be made out if one looks carefully.

Many of the details of this picture have been confirmed by experiments on superconducting films and on thin films of superfluid helium. The thickness of these systems can be reduced so that they effectively have $d = 2$, and as we saw in §1.2, they have a complex order parameter with two real components, so $D = 2$. The superfluid density in helium can be inferred from the response of the film to torsions (Bishop and Reppy 1980, Maps and Hallock 1981, Agnolet et al. 1981) and the superconducting charge density in the superconductors from inductance measurements or from the resistance developed by the sample when carrying a non-zero current (Fiory et al. 1983). The hypothesis that the transition to the normal state is due to the appearance of isolated vortices can be checked by probing the dynamics of dissipation in the normal state; for example, Fiory et al. were able to ascribe the dissipation in their samples just above T_c to a single free vortex.

13.3.2 General Remarks

We conclude with a few general remarks on the K–T phase transition. We saw that the non-linear σ-model couldn't cope with it because of the singular nature of vortices. However, the Landau–Ginzburg model *is* able to deal with this transition. Consider a $d = 2$, $D = 2$ model, well below its transition temperature. The order parameter $\boldsymbol{\phi}$ spends most of its time near the minimum of the potential (which is shaped as in Figure 7.1), and so the system appears similar to the continuum XY-model. Now suppose we try to construct a vortex. Lines of constant $\bar{\theta}$ run radially inwards towards the origin. Close to the origin it becomes energetically advantageous for $\boldsymbol{\phi}$ to leave the minimum of the potential and decrease to zero. There is a large energy cost associated with doing this, but it allows the order parameter to be continuous everywhere and to have a terminating Fourier transform. So if we study the Landau–Ginzburg model, do we see a K–T transition without having to go to a lattice?

It depends on how carefully we look! The energy of such a vortex contains a contribution E of order μ^2/λ due to the fields near the origin. This energy E is also present in the expression for the energy of a vortex-antivortex pair, because the fields close to the centre of a vortex are not much altered when a dipole is formed. It follows that the Boltzmann factor $e^{-\beta E}$ is non-analytic in the coupling λ, and so we cannot hope to recover this when we expand Z in positive powers of λ. Hence vortices are not seen in perturbation theory. But if we tackle the Landau–Ginzburg model non-perturbatively (by a numerical simulation, for example), then we should see a K–T transition. One wonders whether any other non-perturbative phenomena are lost when a perturbative expansion is made.[10]

---

[10] Of course, having an indefinitely large source of unknown effects is very useful when a model does not agree with experiment. This excuse is not so useful in two dimensions, where numerical simulations are relatively straightforward, but in four (and more) dimensions it is used by particle physicists all the time.

13.3 The Kosterlitz–Thouless transition

The complete separation between the two systems $\bar{\theta}$ and ψ in (13.67) is very striking. However, it should be realized that on a lattice this separation is not exact, because line integrals such as those in equation (13.61) can only be defined approximately. These integrals measure how much θ rotates on going around a closed path. But on a lattice the angle between two adjacent spins is arbitrary to within $\pm 2\pi n$, and so the whole integral is arbitrary to within a sum of such factors. Fortunately, below T_c, when few vortices are present, most adjacent spins point in the same direction, and this allows us to define the line integrals; conversely, if conditions are so bad that the line integrals cannot be defined, many vortices will be present and we must be above T_c. So this problem is only severe enough to cast doubt on our treatment of the problem after the transition has taken place, when we know that the dipole model fails anyway.

Finally some remarks on what makes the $d = 2$, $D = 2$ system so special. There are two essential features of the K–T transition: vortices are local minima of the energy, and the energy of an isolated vortex is very large. The first feature is essential if the vortices are to decouple from the rest of the system. If they did not decouple it would not make much sense to talk about them and the physical picture would be much less clear (though it is not obvious that a non-perturbative phase transition could not still take place). Our vortices are local minima of the action because there is an integer-valued topological quantity (the strength, q) associated with them. In a small change, all quantities must change continuously, and therefore a quantity that can only take integral values must be constant. Because of this, a large change is needed to create or destroy a vortex. The existence of such an invariant depends on the dimension and on the symmetry group of the order parameter; it is not unique to $d = 2$.

However, the second feature is more special. If there is to be a transition from a phase of dipoles to a vortex plasma, it is necessary that at low temperatures charges should be predominantly bound in dipoles. This in turn requires that the energy of an isolated charge be very large, so as to compensate for the reduction in entropy when a bound vortex-antivortex pair is formed. In two dimensions, it is natural for isolated charges to have a large energy; any field obeying Poisson's equation will behave in this way. But in higher dimensions this 'confinement' is the exception rather than the rule. A field obeying Poisson's equation in three or more dimensions has a finite contribution to its energy from large distances, so there is no reason for charges to bind into dipoles, and there can be no transition at which the dipoles unbind. It is for this reason that two dimensions is special.

Problems

13.1 Calculate the critical exponents η, δ, β, γ and ν to lowest order in ϵ for the non-linear σ model. [Hint: considering the properties of the singular part of $\Gamma^{(0)}$ may help you with your calculation of δ; considering the function $S(b)$ may help with the calculation of β.]

13.2 By considering the $\mathcal{E}$-field analogue of the Hamiltonian of the XY model, show that the field (13.64) is a local minimum of the Hamiltonian.

13.3 In the vicinity of the K–T phase transition the macroscopic permittivity ϵ of a two-dimensional Coulomb gas is a sensitive function of temperature because the value attained by a solution $Y(X)$ to (13.84) at a given value of $X > 1$ can be a sensitive function of $Y(X_0)$, where $X_0 \ll 1$. One is tempted to characterize this sensitivity by linearizing the (13.84) about the critical trajectory in Figure 13.5. Show that this procedure cannot uncover non-analytic dependence of ϵ on T.

13.4 Show that dislocations in a two-dimensional crystal interact by a logarithmic potential. Hence argue that the Kosterlitz–Thouless model also applies to the melting of a two-dimensional crystal.

14
Universality

In Chapters 10 to 12 we have seen that the critical exponents of the Landau–Ginzburg model agree well with experiment (see Box 14.1 for a summary of results). This is good, but it is also slightly surprising. The Landau–Ginzburg Hamiltonian is one of the simplest possible non-Gaussian Hamiltonians, containing only one non-Gaussian term. Even the Ising model has an effective Hamiltonian which is much more complicated than this (see Appendix K). Yet not only do the critical exponents of the Landau–Ginzburg model agree with those of the Ising model, they agree with those of real physical systems too. This requires an explanation.

In Chapter 5 we saw that the critical behaviour of a system is determined by the fixed point in Hamiltonian space at which its Hamiltonian ends up after repeated rescalings. Since many Hamiltonians end up at the same fixed point, many systems share the same critical behaviour. It must be possible to describe the same phenomenon in the language of the Landau–Ginzburg model. Specifically, we should be able to show that adding other terms (such as $\int d^d\mathbf{x}\,\phi^6$) to the Landau–Ginzburg Hamiltonian leaves the model's critical behaviour unchanged. If this is true, it explains why we can learn about the critical behaviour of complicated systems by studying the Landau–Ginzburg model.

In the next section we shall see how the critical behaviour of the Gaussian model changes when we change its Hamiltonian. We begin with the Gaussian model because it is an especially simple form of the Landau–Ginzburg model, and the results we obtain will give us some idea of what

Box 14.1: Experimental data on critical exponents

Table 14.1. The critical exponents of the Ising model and of the liquid/vapour transition compared with those of the one-component Landau–Ginzburg model

	$\phi^4, d = 3$	Ising, $d = 3$	Fluid
α	0.110 ± 0.005	0.119 ± 0.006	$0.101–0.116$
β	0.325 ± 0.002	0.326 ± 0.004	$0.316–0.327$
γ	1.241 ± 0.002	1.239 ± 0.003	$1.23–1.25$
η	0.032 ± 0.005	0.024 ± 0.007	–
ν	0.630 ± 0.002	0.627 ± 0.002	0.625 ± 0.006

Table 14.1 compares the critical exponents in three dimensions of the one-component Landau–Ginzburg model calculated using the ϵ-expansion, with those of the Ising model calculated using the high-temperature series expansion. The agreement is excellent, despite the considerable differences between the Landau–Ginzburg and Ising models revealed in Appendix K. The table also contains the critical exponents for a real system, a fluid at the liquid/vapour critical point, and again the agreement is good.

Table 14.2. Comparison of the critical exponents α and ν for the helium superfluid transition with the exponents of the two-component Landau–Ginzburg model

	$\phi^4, D = 2, d = 3$	He superfluid
α	-0.007 ± 0.006	-0.014 ± 0.016
ν	0.670 ± 0.002	0.672 ± 0.001

Table 14.3. Comparison of some critical exponents for a ferromagnetic system with those of the three-component Landau–Ginzburg model

	$\phi^4, D = 3, d = 3$	Ni
α	-0.115 ± 0.009	0.04 ± 0.12
β	0.368 ± 0.004	0.358 ± 0.003
γ	1.390 ± 0.010	1.33 ± 0.02
ν	0.710 ± 0.007	0.64 ± 0.10

($D \equiv$ number of components)

Since the helium superfluid has a two-component order parameter, it should be compared with the two-component Landau–Ginzburg model. This is done in Table 14.2 for those critical exponents which can be measured (the order parameter itself is not measurable in this case). A ferromagnet has a three-component order parameter, and in Table 14.3 is compared with the $D = 3$ Landau–Ginzburg model. In each case the agreement is impressive.

might happen when we perturb the full model. We consider only one-component models in this chapter; similar calculations can be done for $D > 1$.

14.1 Perturbing the Gaussian Hamiltonian

We want to compare the critical behaviour of the Gaussian model with the critical behaviour of a slightly perturbed Gaussian model. Simply comparing the small-wavevector limits of the two sets of vertex functions is not directly helpful, because these limits are generally either zero or infinite (see equation (11.24)). To overcome this problem we must rescale the order parameter as we go along. Physically this corresponds to adjusting the sensitivity of our measuring instruments as we change the scale on which we observe the system; we did the same thing at the start of §5.4. To this end let us define the functions

$$F^{(n)}(\mathbf{k}_1,\ldots,\mathbf{k}_n;b) \equiv Z(b)^{-n/2}\Gamma^{(n)}(\mathbf{k}_1/b,\ldots,\mathbf{k}_n/b), \quad (14.1)$$

where $Z(b)$ is fixed by the condition that at some arbitrary fixed wavevectors $\mathbf{k}_{01}, \mathbf{k}_{02}$

$$F^{(2)}(\mathbf{k}_{01}, \mathbf{k}_{02}; b) = \Gamma^{(2)}(\mathbf{k}_{01}, \mathbf{k}_{02}). \quad (14.2)$$

Three things are going on here. First, we look at the $\Gamma^{(n)}$ on a scale larger by a factor of b, by dividing all of the wavevectors by this factor. Second, we change our units of length, so that the wavevectors are restored to their original values. Third, we turn up the sensitivity of our measuring instruments by a factor $Z^{1/2}(b)$. The $F^{(n)}$ are the vertex functions that result from these three processes. The object of the condition (14.2) is to ensure that $F^{(2)}$ has a finite limit as $b \to \infty$.

It is convenient to define 'reduced' versions of the $F^{(n)}$ with a δ-function factored out:

$$F^{(n)}(\mathbf{k}_1,\ldots,\mathbf{k}_n;b) \equiv F^{(n)}(\mathbf{k}_1,\ldots,\mathbf{k}_{n-1};b)(2\pi)^d\delta(\mathbf{k}_1+\cdots+\mathbf{k}_n), \quad (14.3)$$

just as we did for vertex functions in Chapter 8. In terms of the reduced $F^{(n)}$, equations (14.1) and (14.2) become

$$\begin{aligned}F^{(n)}(\mathbf{k}_1,\ldots,\mathbf{k}_{n-1};b) &= Z(b)^{-n/2}b^d\Gamma^{(n)}(\mathbf{k}_1/b,\ldots,\mathbf{k}_{n-1}/b)\\ F^{(2)}(\mathbf{k}_0;b) &= \Gamma^{(2)}(\mathbf{k}_0),\end{aligned} \quad (14.4)$$

where we have used the result that $\delta(\mathbf{k}/b) = b^d\delta(\mathbf{k})$. Let us suppose that $\lim_{b\to\infty} F^{(2)}(\mathbf{k}; b)$ exists. Then for large b we have

$$\Gamma^{(2)}(\mathbf{k}/b) \simeq b^{-d}Z(b)F^{(2)}(\mathbf{k};\infty). \quad (14.5)$$

Comparing this with equation (10.36) we see that the critical exponent η is

$$\eta = \lim_{b\to\infty}\frac{d\log Z(b)}{d\log b} + 2 - d. \quad (14.6)$$

The large-scale properties of the system are determined by the value of η and the $F^{(n)}(\mathbf{k}_i; \infty)$.

We compare the critical behaviour of two systems by calculating for each the functions $F^{(n)}(\mathbf{k}_1; \infty)$ and η. If they are equal, the two systems have the same critical behaviour. If η is the same for both systems but the $F^{(n)}(\mathbf{k}_i; \infty)$ differ, it is possible that the two sets of $F^{(n)}$ can be made equal to each other by rescaling the order parameter of one system. If they cannot be made equal in this way, the two systems have different critical behaviour.

Let us apply the above technique to the problem in hand. The vertex functions $\Gamma^{(n)}$ of the critical Gaussian Hamiltonian

$$H_G = \int d^d\mathbf{x}\, \tfrac{1}{2}\alpha^2|\nabla\phi|^2 \tag{14.7}$$

all vanish except for $\Gamma^{(2)}$, which is

$$\Gamma^{(2)}(\mathbf{k}) = \alpha^2 k^2. \tag{14.8}$$

Equations (14.4) give

$$Z(b)^{-1} b^d \frac{\alpha^2 k_0^2}{b^2} = \alpha^2 k_0^2, \tag{14.9}$$

and so

$$Z(b) = b^{d-2}. \tag{14.10}$$

From this and (14.6) it follows that for the Gaussian model

$$\eta = 0; \quad F^{(2)}(\mathbf{k}; \infty) = \alpha^2 k^2, \tag{14.11}$$

with all the other $F^{(n)}(\infty)$ vanishing.

Now let us add to H_G some even power of the order parameter as a small perturbation, giving the Hamiltonian

$$H = \int d^d\mathbf{x} \left[\tfrac{1}{2}\alpha^2|\nabla\phi|^2 + \frac{\lambda_n}{n!}\phi^n\right]. \tag{14.12}$$

We shall refer to pieces of the Hamiltonian as **operators**, although we imply nothing quantum-mechanical by this. The Feynman rules for Hamiltonians such as (14.12) are dealt with in Appendix N. There it is shown that the presence of the operator $\int d^d\mathbf{x}\, \phi^n$ adds to the Feynman rules an n-link vertex proportional to λ_n. To see whether this changes the critical behaviour we need only work to first order in λ_n, since, if they do not vanish, the first-order terms will dominate any terms of higher order for sufficiently small λ_n. (We are assuming that effects which cannot be expanded in a power series in λ_n can be neglected; see the end of this section.) At $O(\lambda_n)$ the new vertex contributes directly to $\Gamma^{(n)}$; by joining up m pairs of its links to form closed

14.1 The Gaussian Hamiltonian

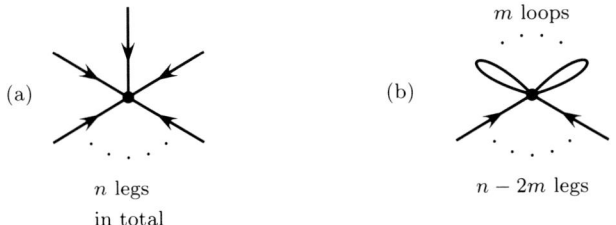

Figure 14.1 Contributions of an n-link vertex to (a) $\Gamma^{(n)}$ and (b) $\Gamma^{(n-2m)}$.

loops it also contributes to $\Gamma^{(n-2m)}$ for any $m < n/2$, but (again, to $O(\lambda_n)$) it doesn't contribute to $\Gamma^{(n')}$ for $n' > n$ (see Figure 14.1). The change in $\Gamma^{(n)}$ is independent of Λ; however, because of the loops, the changes to the $\Gamma^{(n-2m)}$ with $m > 0$ do depend on Λ. To $O(\lambda_n)$ the new vertex functions are

$$\Gamma^{(n-2m)}(\mathbf{k}_1, \ldots, \mathbf{k}_{n-2m-1}) = \lambda_n C_m \quad (n - 2m > 2),$$
$$\Gamma^{(2)}(\mathbf{k}) = \alpha^2 k^2 + \lambda_n C_{(n/2)-1}, \tag{14.13}$$

where

$$C_m \equiv \frac{1}{2^m m!} \left(\int^\Lambda \frac{d^d\check{\mathbf{q}}}{\alpha^2 q^2} \right)^m. \tag{14.14}$$

The integral appearing in (14.14) is

$$\int^\Lambda \frac{d^d\check{\mathbf{q}}}{\alpha^2 q^2} = \frac{\Omega_{d-1}}{(2\pi)^d(d-2)} \frac{\Lambda^{d-2}}{\alpha^2} \quad (d > 2). \tag{14.15}$$

When $d = 2$ there is a logarithmic divergence at small wavevector. To regulate this divergence let us put the system into a box of side αL, so that the integral over $\mathbf{q}$ is cut off for $q < L^{-1}$. This gives

$$\int^\Lambda \frac{d^d\check{\mathbf{q}}}{\alpha^2 q^2} = \frac{1}{2\pi\alpha^2} \log(\Lambda L) \quad (d = 2). \tag{14.16}$$

Equation (14.13) shows that to $O(\lambda_n)$ the addition of $\int d^d\mathbf{x}\, \phi^n$ has changed all the (even) vertex functions $\Gamma^{(n)}, \ldots, \Gamma^{(2)}$. However, this doesn't necessarily represent a change in the critical behaviour, because the system is no longer at its critical temperature:

$$\lim_{k \to 0} \Gamma^{(2)}(\mathbf{k}) = \lambda_n C_{(n/2)-1} \neq 0. \tag{14.17}$$

We can fix this by changing H to $H' \equiv H - \int d^d\mathbf{x}\, \tfrac{1}{2}\lambda_n C_{(n/2)-1}\phi^2$, which (to first order in λ_n) restores $\Gamma^{(2)}$ to the value it had in the critical Gaussian model. We could now proceed to study the critical behaviour of H', but it turns out that the smart thing to do is to add some more terms to H' to cancel out the changes in all the $\Gamma^{(n')}$ except for $\Gamma^{(n)}$. The reason for doing this will become clear later. As an example, let us find an operator which, when added to H_G, changes only $\Gamma^{(6)}$. We begin with

$$\delta_1 H = \int d^d\mathbf{x} \, \frac{\lambda_6}{6!} \phi^6. \tag{14.18}$$

This changes $\Gamma^{(6)}$, but by equation (14.13) it changes $\Gamma^{(4)}$ and $\Gamma^{(2)}$ as well. The change in $\Gamma^{(4)}$ is

$$\Delta\Gamma^{(4)} = \lambda_6 C_1 + O(\lambda_6^2), \tag{14.19}$$

where C_1 is defined by (14.14). To cancel this change we add a term

$$\delta_2 H = \int d^d\mathbf{x} \left(-\frac{\lambda_6 C_1}{4!} \phi^4 \right) \tag{14.20}$$

to the Hamiltonian. The sum of this and the ϕ^6 term that we started with causes no net change in $\Gamma^{(4)}$. However, $\Gamma^{(2)}$ is still changed, by an amount

$$\Delta\Gamma^{(2)} = \lambda_6 (C_2 - C_1^2) + O(\lambda_6^2). \tag{14.21}$$

So we have to add a term

$$\delta_3 H = \int d^d\mathbf{x} \, \frac{\lambda_6}{2!} (C_1^2 - C_2) \phi^2 \tag{14.22}$$

to cancel out the change in $\Gamma^{(2)}$. Now the only vertex function that changes when the operator

$$O_6 \equiv \int d^d\mathbf{x} \, \lambda_6 \left(\frac{1}{6!} \phi^6 - \frac{C_1}{4!} \phi^4 + \frac{C_1^2 - C_2}{2!} \phi^2 \right) \tag{14.23}$$

is added to H_G is $\Gamma^{(6)}$, which changes from zero to λ_6. In particular $\Gamma^{(2)}$ is unchanged, so the system is at its critical temperature and the Gaussian results

$$Z(b) = b^{d-2} \quad \text{and} \quad \eta = 0 \tag{14.24}$$

(which are determined by the behaviour of $\Gamma^{(2)}$) still apply. So whether the two systems look the same at large scales depends on the behaviour of $F^{(6)}(\mathbf{k}_i; b)$ as $b \to \infty$. From (14.24) and the definition (14.4) we have

$$F^{(6)}(\mathbf{k}_1, \ldots, \mathbf{k}_5; b) = b^{6-2d} \lambda_6. \tag{14.25}$$

For $d > 3$, $F^{(6)}(\mathbf{k}_i; \infty) = 0$ and the critical behaviour is that of the Gaussian model. The operator O_6 is then said to be **irrelevant**, because adding it to

14.1 The Gaussian Hamiltonian

H_G does not change the critical behaviour. For $d < 3$, $\Gamma^{(6)}(\mathbf{k}_i; \infty) = \infty$ (to first order in λ_6) and the critical behaviour is changed. Our $O(\lambda_6)$ results do not tell us what the new critical behaviour is, only that it is not that of the Gaussian model. In this case the operator O_6 is said to be **relevant**. In $d = 3$ we cannot tell whether O_6 is relevant without going to the next order in λ_6.

In a similar way we can construct the operator

$$O_n \equiv \int d^d\mathbf{x}\, \lambda_n \sum_{m=0}^{n/2-1} (-1)^m \frac{C_m}{(n-2m)!} \phi^{n-2m}, \qquad (14.26)$$

which to $O(\lambda_n)$ changes only $\Gamma^{(n)}$ when added to the Gaussian Hamiltonian. Specifically, the operator generates a reduced $F^{(n)}$ of the form

$$F^{(n)}(\mathbf{k}_1, \ldots, \mathbf{k}_{n-1}; b) = b^{n-d(n-2)/2} \lambda_n. \qquad (14.27)$$

If $n - d(n-2)/2 > 0$, the operator O_n is relevant; if $n - d(n-2)/2 < 0$, then O_n is irrelevant. Again the relevance of the operator depends on the dimension d. Note that if $d > 2$, then operators become more irrelevant as either d or n increases.

The operator $\int d^d\mathbf{x}\, \phi^n$ can be written as a linear combination of the O_m with $m \leq n$, and in this way we can find out what happens when it is added to H_G. Its effects are more complicated than those of the O_n, since by equations (14.13) it changes all of the $\Gamma^{(n-2m)}$ with $m \leq n/2$, and each of these changes goes as b to a different power. This is why we have studied the O_n rather than the $\int d^d\mathbf{x}\, \phi^n$.

There is a further class of operator to consider: those involving derivatives, such as $\int d^d\mathbf{x}\, |\nabla(\phi^2)|^2$. It is shown in Appendix N that an operator with n powers of ϕ and p derivatives gives rise to a vertex with n legs multiplied by p powers of wavevector. If we're interested only in models with reflection symmetry, p must be even. Again it is possible to construct a sum of a finite number of such operators that changes only $\Gamma^{(n)}$. The p powers of wavevector mean that the relevance of the operator is controlled by the sign of $n - p - d(n-2)/2$. Since p is positive, adding derivatives to an operator decreases its relevance. Problem 14.1 looks at such operators in more detail.

Table 14.4 shows the value of $n - d(n-2)/2$ for different values of n and d. For $d = 5$ (and in all higher dimensions) only $O_2 = \frac{1}{2}\int d^d\mathbf{x}\, \phi^2$ is relevant, and as the only effect of O_2 is to change the temperature it has no effect on critical behaviour. Of the operators containing derivatives, only $n = 2$, $p = 2$ has a non-negative table entry, and this operator is already present in the Gaussian model. So any model in five or more dimensions whose Hamiltonian can be written as an even polynomial in ϕ should have the critical exponents of the Gaussian model.

In four dimensions our first-order analysis is no longer sufficient to decide whether O_4 is relevant, and the critical behaviour need no longer be

Table 14.4. The scaling properties of the operator O_n in d dimensions

d	$n=2$	$n=4$	$n=6$	$n=8$
5	2	-1	-4	-7
4	2	0	-2	-4
3	2	1	0	-1
2	2	2	2	2

NOTES: $F^{(n)}(b)$ scales as $b^{\text{table entry}}$; if the table entry is positive/negative, O_n is relevant/irrelevant. To determine the relevance of an operator with n powers of ϕ and p derivatives, subtract p from the table entry.

that of the Gaussian model. But no powers of ϕ greater than the fourth are important since they can be written as a linear combination of O_2, O_4, and irrelevant operators. Their effect in the critical regime is merely to change the coefficients of the ϕ^2 and ϕ^4 terms already present in the Hamiltonian, and they do not give rise to any new physics themselves. Nothing new happens with operators involving derivatives either. To find out about critical behaviour in four dimensions it is therefore sufficient to study the Landau–Ginzburg model.

In three dimensions O_4 is relevant and O_6 may be relevant; all higher O_n are irrelevant. Adding a ϕ^4 term (which is a linear combination of the relevant operators O_2 and O_4) to the Hamiltonian moves us away from Gaussian critical behaviour. We cannot yet tell whether adding a ϕ^6 term involving the operator O_6 will change the critical behaviour still further; this is the subject of the next section. There are no new relevant operators involving derivatives.

In two dimensions all of the O_n are relevant. Adding any of them to the Gaussian Hamiltonian changes its critical behaviour. This means that we cannot make any claim for the unique importance of the Landau–Ginzburg model. Also, all the operators with $p=2$ become interesting. Their table entries are zero, and it is possible that at higher orders they could become relevant.

Since in three or more dimensions only finitely many operators change the critical behaviour of the Gaussian model, it should be possible to study the critical behaviour of complicated real systems using a very simple model Hamiltonian, and, as a corollary, many complicated real systems should share the same critical behaviour. This is what lies behind universality. In §14.2 we shall try to complete the picture by investigating the relevance of the operator O_6 in three dimensions; if it is irrelevant, this will explain why the Landau–Ginzburg model—which lacks a ϕ^6 term—is so successful. The success of this model suggests strongly that ϕ^6 is irrelevant, but it would be nice to see this coming from theory and not just from experiment.

14.1.1 The applicability of these results

In deriving the above results concerning the relevance of perturbations of the Gaussian model we have assumed that the effect of an operator can be expanded as a power series in its coefficient. In certain situations this is untrue. For example, adding $(\lambda_4/4!)\int d^d\mathbf{x}\,\phi^4$ to the Gaussian Hamiltonian with $\lambda_4 < 0$ causes the energy to become unbounded below, which is clearly a non-perturbative effect. Similar remarks apply to the addition of an odd power of ϕ to the Gaussian Hamiltonian. Our results concerning relevance do not apply to such cases.

There are also less dramatic non-perturbative effects. The critical exponents β and δ (and also α, γ, and ν below T_c) are not defined for the Gaussian model. However, adding a term $\lambda_n \phi^n$ with $n > 2$ to the Gaussian Hamiltonian causes them to take definite values, no matter how small λ_n is. Moreover, if the operator is 'irrelevant', these values are simply the values one obtains from Landau theory (see Problem 14.3), and are thus independent of the value of λ_n providing only it is non-zero. It follows that the difference between the model with $\lambda_n = 0$ and $\lambda_n > 0$ cannot be expanded in a power series about $\lambda_n = 0$. In this situation an operator's relevance or irrelevance determines not whether it affects Gaussian critical behaviour, but whether the actual critical behaviour differs from that predicted by Landau theory.

For these reasons it is best to regard our results for the Gaussian model as providing an ordering of operators with regard to their relevance, rather than an absolute division into relevant and irrelevant. Problem 7.3 demonstrates that the operator ϕ^6 generates its own set of critical exponents in Landau theory; however, it also shows that the operator ϕ^4 (which is more relevant than ϕ^6 for $d > 2$) overwhelms it. But note that at the level of Landau theory, neither of these operators changes the values of those critical exponents which are defined for the Gaussian model; in regard to these exponents our ideas of relevance and irrelevance apply in a straightforward way.

14.2 Perturbing the Landau–Ginzburg Hamiltonian

The results of the previous section apply to perturbations of the *Gaussian* model, but we are really interested in perturbing the Landau–Ginzburg model. The corrections to the Gaussian results in Table 14.4 are $O(\hat{g}_{4c})$, where $\hat{g}_{4c}$ is the critical value of the dimensionless four-vertex coupling defined by equation (10.64). From equation (10.67) we know that $\hat{g}_{4c} = 0.043$ in three dimensions, which is small, so we can hope that the results of the previous section for operators other than O_6 will not be affected. However, we were unable to classify O_6 as either relevant or irrelevant in the context of the $d = 3$ Gaussian model, so the non-Gaussian corrections are crucial.

We can only calculate these corrections approximately; the expression whose sign determines the relevance of a given operator is a power series in $\hat{g}_{4c}$, and we shall only calculate its first term. But this will give us an indication of the theory's predictions.

14.2.1 The case of three dimensions

As in Chapters 11 and 12 we use the renormalization group to investigate the behaviour of the vertex functions when their wavevector arguments are scaled. The theory with Hamiltonian

$$H = \int d^3x \left[\tfrac{1}{2}a^2|\nabla\phi|^2 + \tfrac{1}{2}\mu^2\phi^2 + \tfrac{1}{4!}\lambda_4\phi^4 + \tfrac{1}{6!}\lambda_6\phi^6\right] \tag{14.28}$$

is renormalizable (according to the power-counting arguments of §9.4), which means that the number of primitive Λ-sensitive integrals is equal to the number of parameters in H—four in this case. So the Λ-sensitivity of the vertex functions can be contained in four parameters a, m, $\hat{g}_4$, and $\hat{g}_6$, and by analogy with (9.31) we may write

$$\Gamma^{(n)}(\mathbf{k}_i; \mu^2, \lambda_4, \lambda_6, \Lambda) = a^n \Gamma_R^{(n)}(\mathbf{k}_i; m^2/a^2, \hat{g}_4, \hat{g}_6, \kappa'), \tag{14.29}$$

where

$$\begin{aligned}
a^2 &\equiv \left.\frac{d\Gamma^{(2)}}{dk^2}\right|_\kappa \\
m^2 &\equiv \Gamma^{(2)}(0) \\
\hat{g}_4 &\equiv \frac{1}{\kappa a (2\pi a)^3} \Gamma^{(4)}(\boldsymbol{\kappa}_1, \boldsymbol{\kappa}_2, \boldsymbol{\kappa}_3) \\
\hat{g}_6 &\equiv \frac{1}{a^6}\left(\Gamma^{(6)}(\boldsymbol{\kappa}_1,\ldots,\boldsymbol{\kappa}_5) - \Gamma^{(6)}(\boldsymbol{\kappa}_1,\ldots,\boldsymbol{\kappa}_5)\Big|_{\lambda_6=0}\right).
\end{aligned} \tag{14.30}$$

Apart from the presence of $\hat{g}_6$ and $\Gamma^{(6)}$, these are the normalization conditions for the ϕ^4 theory in three dimensions. $\hat{g}_4$ and $\hat{g}_6$ are dimensionless; the factor of $(2\pi)^3$ is included in the definition of $\hat{g}_4$ for consistency with equation (10.57), but we have not included a similar factor in the definition of $\hat{g}_6$. Notice that we have defined $\hat{g}_6$ as the difference between the actual value of $\Gamma^{(6)}$ and the value of $\Gamma^{(6)}$ in the absence of the ϕ^6 term, so that $\hat{g}_6$ vanishes when λ_6 does. At $T = T_c$ we have $\mu = \mu_c$ and $m^2 = 0$; from now on we won't show μ or m explicitly.

We use the same trick as in Chapter 11: we find how the $\Gamma_R^{(n)}$ scale with wavevector by deriving a renormalization group equation, and then use (14.29) to find out about the $\Gamma^{(n)}$. The invariance of the $\Gamma^{(n)}$ when κ is changed implies that

$$\left(\kappa\frac{\partial}{\partial\kappa} + \beta_4\frac{\partial}{\partial\hat{g}_4} + \beta_6\frac{\partial}{\partial\hat{g}_6} - \frac{n}{2}\gamma_\phi\right)\Gamma_R^{(n)}(\mathbf{k}_i; \hat{g}_4, \hat{g}_6, \kappa) = 0, \tag{14.31}$$

14.2 The Landau–Ginzburg Hamiltonian

where

$$\beta_4 \equiv \kappa \frac{\partial \hat{g}_4}{\partial \kappa}; \quad \beta_6 \equiv \kappa \frac{\partial \hat{g}_6}{\partial \kappa}; \quad \gamma_\phi \equiv -2\kappa \frac{\partial \log a}{\partial \kappa}. \tag{14.32}$$

(Here and subsequently, $\Gamma^{(n)}$ denotes a reduced vertex function.) The partial derivatives are all taken with the parameters of the bare theory held constant. Since β_4, β_6, and γ_ϕ are dimensionless, they are functions of $\hat{g}_4$ and $\hat{g}_6$ only (see §11.1).

By dimensional analysis $\Gamma_R^{(n)}$ must satisfy

$$\left(\kappa \frac{\partial}{\partial \kappa} - b \frac{\partial}{\partial b}\right) \Gamma_R^{(n)}(\mathbf{k}_i/b; \hat{g}_4, \hat{g}_6, \kappa) = (3 - \tfrac{1}{2}n) \Gamma_R^{(n)}(\mathbf{k}_i/b; \hat{g}_4, \hat{g}_6, \kappa). \tag{14.33}$$

Combining (14.33) and (14.31) gives an equation which tells us how the $\Gamma_R^{(n)}$ scale with wavevector:

$$\left(b \frac{\partial}{\partial b} + \beta_4 \frac{\partial}{\partial \hat{g}_4} + \beta_6 \frac{\partial}{\partial \hat{g}_6} - \frac{n}{2} \gamma_\phi + (3 - \tfrac{1}{2}n)\right) \Gamma_R^{(n)}(\mathbf{k}_i/b; \hat{g}_4, \hat{g}_6, \kappa) = 0. \tag{14.34}$$

This can be solved using the method of characteristics (§11.1). We define scale-dependent functions $\hat{g}_4(b)$, $\hat{g}_6(b)$, and $a(b)$ to be the solutions of the ordinary differential equations

$$b \frac{d\hat{g}_4(b)}{db} = -\beta_4\big(\hat{g}_4(b), \hat{g}_6(b)\big),$$

$$b \frac{d\hat{g}_6(b)}{db} = -\beta_6\big(\hat{g}_4(b), \hat{g}_6(b)\big), \tag{14.35}$$

$$b \frac{d \log a(b)}{db} = \tfrac{1}{2} \gamma_1\big(\hat{g}_4(b), \hat{g}_6(b)\big),$$

with the boundary conditions $\hat{g}_4(1) = \hat{g}_4$, $\hat{g}_6(1) = \hat{g}_6$, and $a(1) = a$. The solution of (14.34) may now be written

$$\Gamma_R^{(n)}(\mathbf{k}_i/b; \hat{g}_4, \hat{g}_6, \kappa) = b^{(n/2)-3} a^n(b) a^{-n} \Gamma_R^{(n)}(\mathbf{k}_i; \hat{g}_4(b), \hat{g}_6(b), \kappa), \tag{14.36}$$

and so from equation (14.29)

$$\Gamma^{(n)}(\mathbf{k}_i/b; \hat{g}_4, \hat{g}_6, \kappa) = b^{(n/2)-3} a^n(b) \Gamma^{(n)}(\mathbf{k}_i; \hat{g}_4(b), \hat{g}_6(b), \kappa). \tag{14.37}$$

This equation contains the information we need. From it we deduce that

$$\eta = 2 \lim_{b \to \infty} \frac{d \log a(b)}{d \log b} \quad \text{and} \quad F^{(n)}(\mathbf{k}_i; \infty) = \lim_{b \to \infty} \Gamma^{(n)}\big(\mathbf{k}_i; \hat{g}_4(b), \hat{g}_6(b), \kappa\big). \tag{14.38}$$

So, the critical properties of the system are determined by the limiting behaviour of $\hat{g}_4(b)$ and $\hat{g}_6(b)$ as $b \to \infty$.

If $\hat{g}_4$ and $\hat{g}_6$ are such that β_4 and β_6 vanish, equations (14.35) show that $\hat{g}_4(b)$, $\hat{g}_6(b)$, and $a(b)$ remain unchanged as b varies. We already know of one such fixed point—the non-Gaussian fixed point of ϕ^4 theory, where

$$\beta_4(\hat{g}_4, 0) = 0, \quad \hat{g}_4 \neq 0; \qquad \hat{g}_6 = 0. \tag{14.39}$$

That this is also a fixed point of the extended theory follows from the definitions of β_6 (equation (14.32)) and $\hat{g}_6$ (equation (14.30)), which show that β_6 is zero for any value of $\hat{g}_4$ so long as $\hat{g}_6 = 0$. If when $\hat{g}_6$ is made small but non-zero, $\hat{g}_4$ and $\hat{g}_6$ return to this fixed point as $b \to \infty$, the critical behaviour of the Hamiltonian (14.28) is that of the Landau–Ginzburg model. Let us write a general fixed point as $(\hat{g}_{4c}, \hat{g}_{6c})$, and expand $\beta_4(\hat{g}_4, \hat{g}_6)$ and $\beta_6(\hat{g}_4, \hat{g}_6)$ about this point:

$$\begin{aligned}\beta_4(\hat{g}_4, \hat{g}_6) &= p(\hat{g}_4 - \hat{g}_{4c}) + q(\hat{g}_6 - \hat{g}_{6c}) + \cdots, \\ \beta_6(\hat{g}_4, \hat{g}_6) &= r(\hat{g}_4 - \hat{g}_{4c}) + s(\hat{g}_6 - \hat{g}_{6c}) + \cdots.\end{aligned} \tag{14.40}$$

Substituting into the first two of equations (14.35) gives

$$\delta \begin{pmatrix} \hat{g}_4 \\ \hat{g}_6 \end{pmatrix} \approx - \begin{pmatrix} p & q \\ r & s \end{pmatrix} \begin{pmatrix} \hat{g}_4 - \hat{g}_{4c} \\ \hat{g}_6 - \hat{g}_{6c} \end{pmatrix} \frac{\delta b}{b}. \tag{14.41}$$

The fixed point is stable, therefore, if both the right eigenvalues of the matrix $\begin{pmatrix} p & q \\ r & s \end{pmatrix}$ are positive.

Let us investigate the stability of the fixed point $(\hat{g}_{4c}, 0)$. We already know enough to find the values of p and r. First, since β_4 is the usual Landau–Ginzburg β-function (see equation (10.63)),

$$p = \left(\frac{\partial \beta_4(\hat{g}_4, \hat{g}_6)}{\partial \hat{g}_4} \right)_{\hat{g}_6} \bigg|_{\substack{\hat{g}_4 = \hat{g}_{4c} \\ \hat{g}_6 = 0}} = 1. \tag{14.42}$$

Second, since β_6 vanishes for any value of $\hat{g}_4$ if $\hat{g}_6 = 0$,

$$r = \left(\frac{\partial \beta_6(\hat{g}_4, \hat{g}_6)}{\partial \hat{g}_4} \right)_{\hat{g}_6} \bigg|_{\substack{\hat{g}_4 = \hat{g}_{4c} \\ \hat{g}_6 = 0}} = 0. \tag{14.43}$$

The eigenvalues of the matrix are thus p and s. p is equal to one and therefore positive, so if the fixed point is to be stable with respect to a ϕ^6 perturbation then

$$s = \left(\frac{\partial \beta_6(\hat{g}_4, \hat{g}_6)}{\partial \hat{g}_6} \right)_{\hat{g}_4} \bigg|_{\substack{\hat{g}_4 = \hat{g}_{4c} \\ \hat{g}_6 = 0}} > 0. \tag{14.44}$$

14.2 The Landau–Ginzburg Hamiltonian

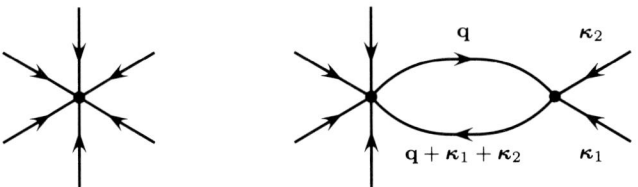

Figure 14.2 Two diagrams contributing to $\Gamma^{(6)}$ at one loop.

So we must calculate $\beta_6(\hat{g}_4, \hat{g}_6)$ and work out the derivative. This turns out to be quite straightforward.

To apply the definition (14.32) of β_6, we need an expression for $\hat{g}_6(\kappa)$. Since the derivative in (14.44) is taken at $\hat{g}_6 = 0$, we need only calculate β_6 to $O(\hat{g}_6)$. To find $\hat{g}_6$ in terms of κ we need the diagrams that contribute to $\Gamma^{(6)}$, which defines $\hat{g}_6$ through equations (14.30). The two diagrams that concern us to one loop are shown in Figure 14.2. The second is the only complicated one. The symmetry factor is $\frac{1}{2}$ since two lines join the same pair of vertices. However, there are actually 15 diagrams like this second one which differ only in the assignment of the six wavevectors to the external legs. Because we're working at a symmetry point all of these diagrams are equal, so

$$a^6 \hat{g}_6 = \Gamma^{(6)}(\kappa_1, \ldots, \kappa_5; \lambda_4, \lambda_6) - \Gamma^{(6)}(\kappa_1, \ldots, \kappa_5; \lambda_4, 0)$$
$$= \lambda_6 - \tfrac{15}{2}\lambda_6 \lambda_4 \int^\Lambda \frac{d^3 \mathbf{q}}{(2\pi)^3} \frac{1}{a^4 q^2 (\mathbf{q}+\boldsymbol{\kappa})^2}, \quad \boldsymbol{\kappa} \equiv \boldsymbol{\kappa}_1 + \boldsymbol{\kappa}_2. \tag{14.45}$$

The first term contributes nothing to the β-function. It is precisely what we had when studying the Gaussian model in the previous section. The second term is what we are interested in. The integral is not divergent at large wavevector so we can set $\Lambda \to \infty$. A similar integral has been done in Box 10.1; the result here is

$$\hat{g}_6 = \frac{\lambda_6}{a^6}\left(1 - \frac{15\pi^3}{4\kappa a}\frac{\lambda_4}{(2\pi a)^3}\right). \tag{14.46}$$

So the function β_6 is

$$\beta_6(\hat{g}_4, \hat{g}_6) = \kappa \left(\frac{\partial \hat{g}_6}{\partial \kappa}\right)_{\lambda_4, \lambda_6}$$
$$= \frac{15\pi^3}{4} \hat{g}_6 \hat{g}_4 + O(\hat{g}_6 \hat{g}_4^2), \tag{14.47}$$

Table 14.5. The critical exponents for the one-component Landau–Ginzburg model in two dimensions compared with those of the $d = 2$ Ising model

	$\phi^4, D = 1, d = 2$	Ising, $d = 2$
β	0.120 ± 0.015	0.125
γ	1.73 ± 0.06	1.75
η	0.26 ± 0.05	0.25
ν	0.99 ± 0.04	1

where we have used the definitions of $\hat{g}_4$ and $\hat{g}_6$ in equations (14.30) to express λ_4 and λ_6 in terms of $\hat{g}_4$ and $\hat{g}_6$.. Therefore

$$\left(\frac{\partial \beta_6}{\partial \hat{g}_6}\right)_{\hat{g}_4 = \hat{g}_{4c}} = \frac{15\pi^3}{4} \hat{g}_{4c} + O(\hat{g}_{4c}^2)$$
$$= 5 + O(\epsilon^2) \qquad (14.48)$$

(using the value of $\hat{g}_{4c}$ from equation (10.67)). The derivative is positive, so adding a ϕ^6 to the Landau–Ginzburg Hamiltonian has not changed its critical behaviour, at least to this order.

This result, while formally of order $\hat{g}_{4c}$, is numerically O(1). This suggests that higher-order corrections may be significant, and also that the Gaussian results of §14.1 may not be very reliable. But to the accuracy of our calculations we have shown that any three-dimensional physical system whose Hamiltonian can be written as an even functional of a one-component scalar field should have the same critical behaviour as the Landau–Ginzburg model. This astonishing claim is confirmed by the experimental results we have already seen. It is this that gives the Landau–Ginzburg model its great theoretical importance.

14.2.2 The case of two dimensions

This is all very satisfactory. The only potential difficulty is that our argument for the universal applicability of the Landau–Ginzburg model does not apply in dimension $d = 2$, whereas the model is actually able to reproduce the critical exponents of the $d = 2$ Ising model—see Table 14.5. We saw in §14.1 that *all* operators are relevant in two dimensions, so we wouldn't expect the Landau–Ginzburg model (whose Hamiltonian stops at ϕ^4) to have anything to do with the Ising model (whose Hamiltonian contains all even powers of ϕ; see Appendix K). But the agreement in Table 14.5 is excellent! This surely requires some further explanation.

What we actually showed in §14.1 is that in two dimensions all operators are relevant when added to the *Gaussian* Hamiltonian. It may be that, when

14.2 The Landau–Ginzburg Hamiltonian

added to the non-Gaussian Hamiltonian, the operators ϕ^6, $\phi^8\ldots$ are all irrelevant. As we have repeatedly emphasized, we cannot definitely answer this question one way or the other with existing mathematical techniques, because we cannot solve the theory exactly. However, we shall now show that a one-loop calculation, analogous to the one we've just done, does indeed overturn the Gaussian result in two dimensions for all operators ϕ^n with $n > 4$. This proves nothing; if the first-order corrections are so much larger than the zeroth-order result, they may themselves be overturned at second order. But it's still interesting to see it happening.

Let's add to the two-dimensional Landau–Ginzburg Hamiltonian an operator which contains ϕ^n as its highest power. As we saw in §14.1, adding in just ϕ^n changes not only $\Gamma^{(n)}$, but all the $\Gamma^{(m)}$ with $m < n$ as well. If we followed the same procedure as in the previous subsection we would find ourselves having to deal with $(n/2)-1$ β-functions all at once. Just as for the Gaussian model it is easier to deal with things one at a time, so we suppose that the actual operator that we add to the Hamiltonian is

$$O'_n = \int d^2\mathbf{x} \frac{\lambda_n}{n!} \phi^n + \{\text{lower powers of } \phi\}, \quad (14.49)$$

where the lower powers of ϕ are chosen so that O'_n changes only $\Gamma^{(n)}(\kappa)$, leaving all the $\Gamma^{(m)}(\kappa)$ with $m < n$ unaffected to $O(\lambda_n)$. O'_n here is analogous to O_n defined in (14.26), but the coefficients of the lower powers of ϕ are different here because of $O(\lambda_4)$ corrections. So we can write

$$\Gamma^{(n)}(\mathbf{k}_i; \mu^2, \lambda_4, \lambda_n, \Lambda) = a^n \Gamma^{(n)}_R(\mathbf{k}_i; m^2/a^2, \hat{g}_4, \hat{g}_n, \kappa), \quad (14.50)$$

where

$$\hat{g}_n \equiv \frac{1}{\kappa^2 a^n} \left(\Gamma^{(n)}(\boldsymbol{\kappa}_1, \ldots, \boldsymbol{\kappa}_{n-1}) - \Gamma^{(n)}(\boldsymbol{\kappa}_1, \ldots, \boldsymbol{\kappa}_{n-1}) \Big|_{\lambda_n=0} \right). \quad (14.51)$$

The analogously defined $\hat{g}_m$ with $4 < m < n$ all vanish. $\hat{g}_4$ is still defined as in (14.30).

Now we can proceed just as we did for the ϕ^6 theory in the previous section. By considering changes in the renormalization point κ we immediately get a renormalization group equation analogous to equation (14.31), with $\hat{g}_n$ and β_n replacing $\hat{g}_6$ and β_6. The β-function β_n is defined by

$$\beta_n = \kappa \frac{\partial \hat{g}_n}{\partial \kappa}, \quad (14.52)$$

with the parameters of the bare theory being held constant when taking the derivative. Everything else goes through as before, and we conclude that O_n is irrelevant if the following condition holds:

$$\left(\frac{\partial \beta_n(\hat{g}_4, \hat{g}_n)}{\partial \hat{g}_n} \right)_{\hat{g}_4} \Bigg|_{\substack{\hat{g}_4=\hat{g}_{4c} \\ \hat{g}_n=0}} > 0. \quad (14.53)$$

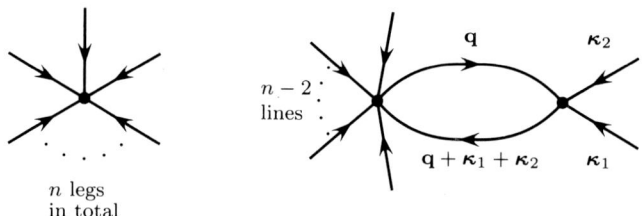

Figure 14.3 Two diagrams contributing to $\Gamma^{(n)}$ at one loop.

Again we work to first order in $\hat{g}_4$. The two diagrams to one loop that contribute are shown in Figure 14.3, where we recognize an n-line vertex from the presence of ϕ^n in the Hamiltonian. The multiplicity of the second diagram is $\frac{1}{2}n(n-1)$ (this is from the same source as the factor of 15 in (14.45)) so

$$g_n = \Gamma^{(n)}(\boldsymbol{\kappa}_1, \ldots, \boldsymbol{\kappa}_{n-1}; \lambda_4, \lambda_n) - \Gamma^{(n)}(\boldsymbol{\kappa}_1, \ldots, \boldsymbol{\kappa}_{n-1}; \lambda_4, 0)$$
$$= \lambda_n - \frac{n(n-1)}{4}\lambda_n\lambda_4 \int \frac{d^2\check{\mathbf{q}}}{a^4 q^2(\mathbf{q}+\boldsymbol{\kappa})^2}, \quad \boldsymbol{\kappa} \equiv \boldsymbol{\kappa}_1 + \boldsymbol{\kappa}_2. \quad (14.54)$$

This integral is infrared-divergent in two dimensions. The cure for this, as for similar integrals in Chapter 10, is to work in $4-\epsilon$ dimensions and to keep terms only to a given order in ϵ. The ϵ-expansion of the integral in (14.54) is

$$I = \frac{1}{(2\pi a)^2 a^2 \kappa^2}\left(\frac{1}{\epsilon} + \frac{1}{2} + O(\epsilon)\right). \quad (14.55)$$

Since we are working to first order in $\hat{g}_{4c}$, which is formally of order ϵ, we drop the $O(\epsilon)$ terms. The result is then

$$\hat{g}_n = \frac{\lambda_n}{a^n}\left(\frac{1}{\kappa^2} - \frac{n(n-1)}{4}\frac{\lambda_4}{(2\pi a)^2}\frac{1}{a^2\kappa^4}\right). \quad (14.56)$$

Now we can calculate β_n by differentiating this last equation. We obtain

$$\beta_n = \hat{g}_n\left(-2 + n(n-1)\hat{g}_{4c}\right)$$
$$= \hat{g}_n\left(-2 + 0.0675n(n-1)\right) \quad (14.57)$$
$$= 0.026\hat{g}_6 \quad \text{when} \quad n = 6.$$

The -2 is the contribution that we found in §14.1. We see that the $O(\hat{g}_{4c})$ correction is of the opposite sign, and grows rapidly as n rises. Using the $O(\epsilon)$ result $\hat{g}_{4c} = \epsilon/3\pi^2$ (equation (10.101)) we find that O_6 is *just* irrelevant. The higher O_n are still more irrelevant. The size of this first-order correction suggests that higher-order corrections are not likely to be small, and so we cannot come to a definite conclusion. However, this calculation has served to illustrate the unreliability of the Gaussian estimates, and so it is not at all surprising that the Gaussian result should be completely wrong in two dimensions.

14.3 Relevance and renormalizability

To a person familiar with particle physics the results of this chapter are likely to seem rather peculiar. Quantum field theory also involves functional integrals like the ones we have been doing, with an 'action' replacing the Hamiltonian. In particle physics it is usual to reject non-renormalizable theories,[1] because in such theories the vertex functions cannot be made Λ-insensitive and particle physics knows of no natural value for Λ.

The strange thing is that the operators we have classified as irrelevant, meaning that their inclusion in the Hamiltonian does not affect long-distance physics, are the same operators that a particle physicist would classify as non-renormalizable! These two viewpoints appear directly to contradict each other, and it is worth while looking more closely at what is really going on.

First let's outline the particle physicist's argument. Consider a theory with action (particle physics) or Hamiltonian (statistical physics)

$$H = \int d^d\mathbf{x} \left[\tfrac{1}{2}\alpha^2 (\boldsymbol{\nabla}\phi)^2 + \frac{\lambda_n}{n!} \phi^n \right]. \tag{14.58}$$

The vertex functions $\Gamma^{(n)}$ of this theory satisfy

$$\Gamma^{(n)}(\alpha, \lambda_n) = \alpha^n \Gamma^{(n)}(1, \lambda_{nR}), \quad \text{where} \quad \lambda_{nR} \equiv \frac{\lambda_n}{\alpha^n}, \tag{14.59}$$

by an argument analogous to that used in §9.5, and therefore $\Gamma^{(n)}$ takes the form of α^n times a power series in λ_{nR}. Now the dimension of λ_{nR} is $L^{D[\lambda_{nR}]}$, where

$$D[\lambda_{nR}] = d\left(\tfrac{1}{2}n - 1\right) - n. \tag{14.60}$$

If $D[\lambda_{nR}]$ is positive, then powers of λ_{nR} in the series for a particular vertex function have to be cancelled by powers of something with negative dimension. This 'something' could be the external wavevectors, but there is nothing to stop it being the cutoff-parameter Λ. So as we go to higher orders in λ_{nR} we expect to find more and more Λ-sensitivity. Taking enough powers of λ_{nR} will make any vertex function Λ-sensitive. But if $D[\lambda_{nR}]$ is zero or negative

---
[1] Recall that in Chapter 9 we defined a renormalizable theory to be one in which many values of the microscopic parameters correspond to each set of values of the macroscopic parameters.

this argument does not work, and there may be only a finite number of primitively Λ-sensitive diagrams. From the point of view of particle physics this latter case is acceptable, since a finite amount of Λ-sensitivity can be absorbed into physical parameters, but the former case is hopeless. Particle physicists certainly do not want their answers to depend on the undetermined value of Λ. And it is no good eliminating Λ in favour of the value of a vertex function at some renormalization point κ. This allows one formally to eliminate Λ, but this has the result that either the value of some vertex function at κ has to be $O(\Lambda)$ or Λ has to be $O(\kappa)$. So it looks as though the particle physicists are right: operators for which $D[\lambda_{nR}] > 0$ cause macroscopic physics to depend on the cutoff and so give rise to non-renormalizable theories. Yet, comparing (14.60) with (14.27) we see that we have classified such operators as irrelevant. What came over us?

Let's now look at the problem from the point of view of statistical physics. An important difference from particle physics is that here the cutoff has a physical meaning; it might be a few Ångströms. Let us write λ_{nR} as some dimensionless number ℓ_n times an appropriate power of the cutoff:

$$\lambda_{nR} = \ell_n \Lambda^{-D[\lambda_{nR}]}. \tag{14.61}$$

The number ℓ_n will probably be of order 1. For example, it is hard to imagine a distance of 10^9Å appearing in the microscopic description of an atomic system. Now let us consider once again what happens if $D[\lambda_{nR}]$ is positive. It is still true that each factor of λ_{nR} allows $D[\lambda_{nR}]$ extra powers of Λ to appear, but these are cancelled out by the negative powers of Λ in (14.61). Written in terms of ℓ_n, the vertex functions are not Λ-sensitive. In fact the zero-loop contribution to $\Gamma^{(n)}$ is $\ell_n \Lambda^{-D[\lambda_{nR}]}$, which is suppressed by powers of Λ. On the other hand, if $D[\lambda_{nR}]$ is negative, the positive powers of Λ that appear with ℓ_n are not cancelled by negative powers in the rest of the vertex function, because an integral over wavevector up to Λ will never go as a negative power of Λ at leading order. Therefore, taking enough powers of λ_{nR} will make any vertex function Λ-sensitive if $D[\lambda_{nR}] < 0$. This is precisely the opposite to what we found above! As we have argued throughout this chapter, it is those operators whose coupling has negative dimension in units of length that drastically change the large-scale behaviour. But how can two different groups of physicists come to opposite conclusions from the same starting point?

For statistical physicists Λ and the parameters appearing in the Hamiltonian are real; they define the theory, and from them one can calculate what will be measured. But for particle physicists neither Λ nor the parameters in the action (Hamiltonian) have any physical meaning; the theory is defined, not by them, but by the results of certain measurements carried out at 'macroscopic' vales of the wavevector. Suppose we know experimentally the value of $\Gamma^{(n)}$. Then, in particle physics, the value of the bare coupling λ_n has to be chosen to reproduce this measured value. To zero loops $\Gamma^{(n)} = \lambda_n$,

14.3 Relevance and renormalizability

so whatever the value chosen for the cutoff-parameter Λ we must keep the value of the bare coupling λ_n fixed, in order to reproduce the experimental result. Despite what dimensional analysis seems to imply, λ_n does not behave as some Λ-independent number times $\Lambda^{D[\lambda_n R]}$. Given this, the rest of the vertex functions then behave as the particle physicist would expect. There is no contradiction. Statistical physics says that the effects of the term ϕ^n disappear at sufficiently large scales; particle physics says that if λ_n is chosen large enough so that one of these effects does *not* disappear, but agrees with some measured value, then there will be many other non-vanishing effects as well. These are two distinct situations: in one λ_n has some natural value, while in the other it is forced to take an unnaturally large value in order to fit a measurement made at low wavevector, and the resulting theory is not acceptable. Hence the surprising result that irrelevant operators give rise to non-renormalizable theories, and *vice versa*. See Problem 14.6 for further discussion.

We are now in a position to make a full and clear statement about the nature of renormalizable theories in statistical physics. A renormalizable theory is one in which there are many different sets of microscopic parameters that generate the same macroscopic physics, to within corrections suppressed by powers of Λ. With this stipulation, the inclusion of an irrelevant operator in the Hamiltonian does not make the theory non-renormalizable, so long as the coefficient of the operator is O(1) when de-dimensionalized in powers of Λ. When approaching statistical physics from a particle physics background, one can get the idea that the reason for studying the Landau–Ginzburg model is that models containing higher powers of ϕ do not make sense, because they are 'non-renormalizable'. Then one worries whether the results obtained from the Landau–Ginzburg model are physically meaningful, since the Hamiltonians of real systems do contain higher powers of ϕ. There is no cause for such worry. The Landau–Ginzburg model is studied not because higher powers of ϕ crash the theory, but because they do not affect it.

To form a one-parameter family of macroscopically equivalent microscopic theories, it is generally necessary to include *all* relevant operators of a given symmetry. For example, if $\mu = 0$, there is no one-parameter set of values of $(\alpha, \lambda, \Lambda)$ whose members all give the same macroscopic physics. For the Landau–Ginzburg model to be renormalizable, the possibility of $\mu^2 \neq 0$ must be allowed. Note that this does not mean that a model with $\mu = 0$ is unphysical; rather, it means that the Hamiltonian of such a model does not maintain its form when Λ is varied. This is illustrated in Problem 14.5. Varying Λ is guaranteed not to introduce operators which break the symmetry of the Hamiltonian, but new relevant operators with the same symmetry can appear. In general, then, a renormalizable Hamiltonian must contain all relevant operators with a given symmetry.

Although the operators in renormalizable theories are always relevant, this does not mean that irrelevant operators never appear in particle physics.

Suppose that there really is some sort of cutoff in the physics of the vacuum—not necessarily due to space-time having an atomic structure, but some energy scale above and below which totally different degrees of freedom are important. If one writes an effective action for the degrees of freedom present at energies below the cutoff, there is no reason for it to be renormalizable. It will, in general, contain both relevant and irrelevant operators. Now suppose that we look at energies far below this 'cutoff' energy. The strongest interactions will be those governed by the relevant operators, and, as we have seen, these give rise to renormalizable theories. But in addition there will be irrelevant operators, generating weak non-renormalizable interactions suppressed by powers of Λ. These non-renormalizable interactions are not to have their strengths adjusted to fit measurements, but are genuinely weak. So at low energies we would expect the strongest interactions to be renormalizable, but with weak non-renormalizable corrections.

This is actually what happens in the electroweak theory. Above the electroweak scale the appropriate degrees of freedom for the weak interactions are fermions and the W and Z bosons, whose masses are about 100 GeV. This theory is renormalizable. At energies well below this these bosons are not seen, and the effective theory is one of weakly-interacting fermions. This effective low-energy theory is non-renormalizable, with a coupling constant

$$g_F = (200 \,\text{GeV})^{-2}. \tag{14.62}$$

This explains why the low-energy weak interactions are so weak. The bosons are too massive to be excited at low energies, and there is no renormalizable (relevant) operator in four dimensions that describes interacting fermions. Notice that the size of g_F indicates the energy at which the low-energy effective theory breaks down.

There is another extremely weak interaction in physics, and it is also non-renormalizable. This is gravity. The Newtonian gravitational constant in energy units is

$$G_N = (1.2 \times 10^{19} \,\text{GeV})^{-2} \equiv m_{\text{pl}}^{-2}, \tag{14.63}$$

where m_{pl} is referred to as the **Planck mass**. Traditionally, the non-renormalizability of gravity has been seen as a problem, and a particularly annoying one at that, since all the other interactions observed in nature can be renormalized. But from the point of view of statistical physics the non-renormalizability of gravity is completely natural. Suppose that at energies above m_{pl} physics is completely different from the physics that we know, and that the terms in which we describe the world are no longer appropriate. We can imagine taking the proper high-energy theory and rewriting it in terms of the fields that we observe in the low-energy world. This is like eliminating the W and Z bosons from the electroweak theory to obtain the low-energy version of the weak interactions, or eliminating the details of the structure of copper and zinc atoms from a description of β-brass to obtain the Ising model. There is no reason why the resulting theory should be renormalizable; the effective action will contain both relevant and irrelevant operators. But in the world of energies far below m_{pl}, only the relevant operators—which give rise to renormalizable theories—will survive. This explains why so much of the physics that we see is renormalizable. Non-renormalizable interactions will be suppressed by powers of m_{pl}, exactly like gravity. From this point of view, it is perfectly natural that the weakest of the known interactions—gravity—is non-renormalizable. Its weakness suggests that it is irrelevant, and therefore that it is non-renormalizable. In the same way that β-decay is the low-energy remnant of much richer physics above the electroweak scale, so gravity could be the low-energy remnant of much richer physics above the Planck scale. So non-renormalizable theories are actually the most interesting. They point to the existence of new physics and indicate the energy scale at which it might begin.

And so, on this exalted note, we end our study of critical phenomena.

Problems

14.1 Using the methods of §14.1, investigate the effect of adding an operator containing n powers of ϕ and p derivatives to the Gaussian Hamiltonian, and confirm the claim made in the text about the relevance of such operators.

14.2 Using the methods of §14.1, investigate the relevance of odd powers of ϕ. (Assume that a higher even power of ϕ is already present in the Hamiltonian so that the resulting model makes sense.) Does adding a term $\int d^d\mathbf{x}\,\phi^5$ affect the critical behaviour of the Landau–Ginzburg model in $d = 4$?

14.3 We've seen that the addition of a term $\lambda_n \int d^d\mathbf{x}\,\phi^n$ to the Gaussian Hamiltonian causes those exponents which are undefined in the Gaussian model to take their Landau-theory values, irrespective of how small λ_n is. To get an idea of the size of corrections to Landau theory, investigate the behaviour of the $O(\lambda_n^2)$ corrections to the Gaussian model as a function of the scaling parameter b, and compare them to the $O(\lambda_n)$ corrections.

14.4 To complete our discussion of universality for the $D = 1$ Landau–Ginzburg model, we should consider the scaling behaviour of the $\Gamma^{(n,l)}$ introduced in Chapter 12, since these are needed to calculate those exponents defined away from T_c. Develop a method for investigating their behaviour in the small-wavevector limit analogous to that used in §14.1 for the $\Gamma^{(n)}$, and so determine the effect on these operators of various small perturbations of the Gaussian Hamiltonian.

14.5 The closest analogue of blocking a lattice (in the sense of §5.1) in the context of the Landau–Ginzburg model is variation of Λ. This can be done by integrating out all the Fourier components of $\widetilde{\phi}(\mathbf{k})$ with $\Lambda_2 < |\mathbf{k}| < \Lambda_1$, where Λ_1 is the original, and Λ_2 is the new, value of the cutoff-parameter. The effective Hamiltonian H_2 after such a transformation is defined by

$$e^{-H_2} \equiv \int{}' \mathcal{D}\widetilde{\phi}\, e^{-H_1}.$$

The prime on the integral denotes that it is only over certain components of $\widetilde{\phi}$; H_1 is the original Hamiltonian of the system. If H_1 takes the form

$$H_1 = \int d^d x\, \tfrac{1}{2}\alpha^2|\nabla\phi|^2 + \tfrac{1}{4!}\lambda\phi^4,$$

show that H_2 contains a term proportional to ϕ^2, and find its coefficient to $O(\lambda)$.

14.6 We have seen that $O(\hat{g}_4)$ corrections make the operator $\int d^d x\, \phi^6$ irrelevant in three dimensions. Does this mean that the $\phi^4 + \phi^6$ model is non-renormalizable in three dimensions, despite its being renormalizable by power counting?

Appendices

Appendix A: **The magnetic scattering of neutrons**

The scattering of neutrons from condensed matter is a very complex business, involving many competing processes. A comprehensive study of the subject would fill volumes (indeed it *does* fill volumes—see, for example, the two-volume treatise by Lovesey 1984). For the purposes of this appendix however, we will only consider a very simple model in which neutrons are scattered from a system of fixed spins by the magnetic interaction of their dipole moments with those of the spins.

The cross section for the scattering of the neutron from an initial state $|i\rangle$ to a final state $|f\rangle$ is proportional to the modulus squared of the matrix element $\langle f|V|i\rangle$, where V is the interaction potential between the neutron and the spin system. V is given by

$$V = -\boldsymbol{\mu}_n \cdot \mathbf{B}, \tag{A.1}$$

where $\mathbf{B}$ is the magnetic field created by the dipole moments of the spins. The magnetic vector potential at $\mathbf{r}$ due to one such dipole at the origin is

$$\mathbf{A}(\mathbf{r}) = \frac{\mu_0}{4\pi} \frac{\boldsymbol{\mu} \times \mathbf{r}}{r^3}, \tag{A.2}$$

where $\boldsymbol{\mu}$ is the magnetic dipole moment of the spin (see any standard text on electromagnetism). If we have large numbers of dipoles with dipole moments $\boldsymbol{\mu}_l$ at positions $\mathbf{r}_l$ then the total vector potential is

$$\mathbf{A}(\mathbf{r}) = \frac{\mu_0}{4\pi} \sum_l \frac{\boldsymbol{\mu}_l \times (\mathbf{r} - \mathbf{r}_l)}{|\mathbf{r} - \mathbf{r}_l|^3}. \tag{A.3}$$

To get the magnetic field we take the curl of this potential. We will find it expedient to make use of the vector identity

$$\frac{\mathbf{r} - \mathbf{r}_l}{|\mathbf{r} - \mathbf{r}_l|^3} = -\nabla \left(\frac{1}{|\mathbf{r} - \mathbf{r}_l|}\right) \tag{A.4}$$

to write the magnetic field in the form

$$\mathbf{B}(\mathbf{r}) = -\frac{\mu_0}{4\pi} \sum_l \nabla \times \left[\boldsymbol{\mu}_l \times \nabla \left(\frac{1}{|\mathbf{r} - \mathbf{r}_l|}\right)\right]. \tag{A.5}$$

Assuming for the purposes of illustration that the scattering dipole moments are those of electrons, we can write the moments $\boldsymbol{\mu}_l$ in the form

$$\boldsymbol{\mu}_l = -\frac{e\hbar}{m_e}\mathbf{s}_l, \tag{A.6}$$

where the spin s_l is a dimensionless variable. Similarly, the moment of the neutron can be written

$$\boldsymbol{\mu}_n = \frac{g e \hbar}{m_n}\mathbf{s}_n, \quad \text{where } g_n = -1.91.$$

Equation (A.1) then becomes

$$V(\mathbf{r}, \mathbf{s}_n) = -\frac{g\mu_0 e^2 \hbar^2}{4\pi m_e m_n} \sum_l \mathbf{s}_n \cdot \nabla \times \left[\mathbf{s}_l \times \nabla \left(\frac{1}{|\mathbf{r} - \mathbf{r}_l|}\right)\right], \tag{A.7}$$

where $\mathbf{r}$ now represents the position vector of the neutron.

We write the initial state of the neutron as

$$|i\rangle = \frac{1}{\sqrt{v}} e^{i\mathbf{k}_i \cdot \mathbf{r}} |\mathbf{s}_n\rangle_i. \tag{A.8}$$

The i subscript here denotes the initial configuration. We have broken the neutron state into a spatial part, which we represent by a simple Schrödinger wavefunction, and a spin part. The factor of $v^{-1/2}$ ensures that the neutron wavefunction is normalized to one neutron per volume v. Similarly, we can write the final state of the neutron in the form

$$|f\rangle = \frac{1}{\sqrt{v}} e^{i\mathbf{k}_f \cdot \mathbf{r}} |\mathbf{s}_n\rangle_f. \tag{A.9}$$

Then the matrix element is

$$\langle f|V|i\rangle = \frac{1}{v}\langle \mathbf{s}_n|_f \int d^3\mathbf{r}\, e^{-i\mathbf{k}_f \cdot \mathbf{r}} V(\mathbf{r}, \mathbf{s}_n) e^{i\mathbf{k}_i \cdot \mathbf{r}} |\mathbf{s}_n\rangle_i. \tag{A.10}$$

We now substitute for $V(\mathbf{r}, \mathbf{s}_n)$ from equation (A.7) and use the identities

$$\int d^3r\, e^{i\mathbf{K}\cdot\mathbf{r}} \nabla \times \mathbf{F} \equiv -i \int d^3r\, e^{i\mathbf{K}\cdot\mathbf{r}} \mathbf{K} \times \mathbf{F},$$
$$\int d^3r\, e^{i\mathbf{K}\cdot\mathbf{r}} \nabla \phi \equiv -i \int d^3r\, e^{i\mathbf{K}\cdot\mathbf{r}} \mathbf{K} \phi. \quad (A.11)$$

These are strictly only true in the limit in which the confining volume v becomes infinite, but will be perfectly adequate for large v—the crucial point is that the fields $\mathbf{F}$ and ϕ should be small on the boundaries of v. Then

$$\langle f|V|i\rangle = \frac{g\mu_0 e^2 \hbar^2}{4\pi m_e m_n v} \langle s_n|_f \int d^3r\, e^{i\mathbf{K}\cdot\mathbf{r}} \sum_l \mathbf{s}_n \cdot \mathbf{K} \times (\mathbf{s}_l \times \mathbf{K}) \frac{1}{|\mathbf{r} - \mathbf{r}_l|} |s_n\rangle_i, \quad (A.12)$$

where

$$\mathbf{K} \equiv \mathbf{k}_i - \mathbf{k}_f. \quad (A.13)$$

The integral over $\mathbf{r}$ can now be performed, and we get

$$\langle f|V|i\rangle = \frac{g\mu_0 e^2 \hbar^2}{m_e m_n v} \langle s_n|_f \sum_l \mathbf{s}_n \cdot \frac{\mathbf{K} \times (\mathbf{s}_l \times \mathbf{K})}{K^2} e^{i\mathbf{K}\cdot\mathbf{R}_l} |s_n\rangle_i. \quad (A.14)$$

The vector quantity

$$\mathbf{p}_l \equiv \frac{\mathbf{K} \times (\mathbf{s}_l \times \mathbf{K})}{K^2} = \hat{\mathbf{K}} \times (\mathbf{s}_l \times \hat{\mathbf{K}}) \quad (A.15)$$

is the component of $\mathbf{s}_l$ in the plane perpendicular to $\hat{\mathbf{K}}$, which is the unit vector in the direction of $\mathbf{K}$. In terms of $\mathbf{p}_l$ the matrix element is

$$\langle f|V|i\rangle = \frac{g\mu_0 e^2 \hbar^2}{m_e m_n v} \sum_l e^{i\mathbf{K}\cdot\mathbf{R}_l} \langle s_n|_f\, \mathbf{s}_n \cdot \mathbf{p}_l\, |s_n\rangle_i. \quad (A.16)$$

Therefore

$$|\langle f|V|i\rangle|^2 = \left(\frac{g\mu_0 e^2 \hbar^2}{m_e m_n v}\right)^2 \sum_{l,m} e^{i\mathbf{K}\cdot(\mathbf{R}_l - \mathbf{R}_m)} \langle s_n|_i\, \mathbf{s}_n \cdot \mathbf{p}_m\, |s_n\rangle_f \langle s_n|_f\, \mathbf{s}_n \cdot \mathbf{p}_l\, |s_n\rangle_i. \quad (A.17)$$

Summing over the initial and final spin states of the neutron gives

$$\sum_{i,f} \langle s_n|_i\, \mathbf{s}_n \cdot \mathbf{p}_m\, |s_n\rangle_f \langle s_n|_f\, \mathbf{s}_n \cdot \mathbf{p}_l\, |s_n\rangle_i = \sum_i \sum_{a,b} p_l^a p_m^b \langle s_n|_i\, s_n^a s_n^b\, |s_n\rangle_i$$
$$= \tfrac{1}{4} \mathbf{p}_l \cdot \mathbf{p}_m, \quad (A.18)$$

where the superscripts a and b denote ordinary Cartesian components. In an experiment in which a large number of neutrons are scattered off the target over a long period of time, the value of this matrix element will get averaged over the thermal ensemble of states of the spin system. So the correct value is

$$|\langle f|V|i\rangle|^2 = \tfrac{1}{4}\left(\frac{g\mu_0 e^2\hbar^2}{m_e m_n v}\right)^2 \sum_{l,m} e^{i\mathbf{K}\cdot(\mathbf{R}_l-\mathbf{R}_m)}\langle \mathbf{p}_l\cdot\mathbf{p}_m\rangle, \qquad (A.19)$$

where the angle brackets around $\mathbf{p}_l\cdot\mathbf{p}_m$ denote a thermal average.

Equation (A.19) relates the scattering of neutrons by the target to the spatial Fourier transform of the correlation function $\langle \mathbf{p}_l\cdot\mathbf{p}_m\rangle$. Thus we can determine this correlation function by measuring the scattering as a function of $\mathbf{K}$. To take a simple example, imagine what would happen if we were to scatter unpolarized neutrons from an Ising magnet. We can make things particularly simple by deciding to fire our neutrons at the sample along a line perpendicular to the axis along which all the spins point. If we do this then it is simple to show using equation (A.15) that $\mathbf{p}_l\cdot\mathbf{p}_m = s_l s_m$. In this case measurement of the scattering gives the spin–spin correlation function directly.

Appendix B: The natural variables for thermodynamic potentials

It is often stated in texts on thermodynamics that each thermodynamic potential (U, H, F, and G) has associated with it a pair of 'natural' variables, such that if the potential is given as a function of these variables then all the thermodynamic properties of the system may be found.[1] The natural variables for the four potentials above are:

Thermodynamic potential	Natural variables
U	S, V
H	S, p
F	T, V
G	T, p

---

[1] For simplicity in this appendix we consider a simple p–V system with only two degrees of freedom.

Natural variables

As an example, suppose that the function $U(S,V)$ is known. Then, since

$$dU = TdS - pdV, \tag{B.1}$$

it follows that

$$T = \left(\frac{\partial U}{\partial S}\right)_V \quad \text{and} \quad p = -\left(\frac{\partial U}{\partial V}\right)_S. \tag{B.2}$$

Thus we have found the pressure and the entropy. This allows us to calculate the other three potentials H, F, and G. Also, the first of the previous two equations can be used to eliminate S in favour of V and T; the second then becomes the equation of state of the system, equating the pressure to a function of volume and temperature. By assiduously manipulating partial derivatives it is possible to find an expression for any quantity that occurs in classical thermodynamics—for example, $(\partial u/\partial g)_f$ is equal to

$$\frac{s\dfrac{\partial u}{\partial s}\dfrac{\partial^2 u}{\partial v \partial s} - \dfrac{\partial u}{\partial s}\dfrac{\partial u}{\partial v} - s\dfrac{\partial u}{\partial v}\dfrac{\partial^2 u}{\partial s^2}}{s\dfrac{\partial u}{\partial v}\dfrac{\partial^2 u}{\partial s^2} + v\dfrac{\partial u}{\partial v}\dfrac{\partial^2 u}{\partial v \partial s} - sv\left(\dfrac{\partial^2 u}{\partial v \partial s}\right)^2 + sv\dfrac{\partial^2 u}{\partial s^2}\dfrac{\partial^2 u}{\partial v^2}} \tag{B.3}$$

(the derivatives with respect to v are at constant s and vice versa). It is easy to check that the same could be done starting from any of the other three potentials, as long as they are expressed in terms of their natural variables.

But why must the potentials be expressed in terms of their natural variables if this is to be possible? Given $U(S,V)$, we have seen that S may be expressed in terms of V and T, thus allowing us to calculate $U(T,V)$. Why is it not then possible to *start* with $U(T,V)$ and reverse the process? What is it about natural variables that makes them so special?

The best way to understand what is going on is to start with $U(T,V)$ and to see what happens. Then we can write

$$dU = \left(\frac{\partial U}{\partial T}\right)_V dT + \left(\frac{\partial U}{\partial V}\right)_T dV \tag{B.4}$$
$$= C_V dT - p_T dV.$$

In this equation C_V is the heat capacity at constant volume, and p_T is the 'isothermal pressure'; that is, minus the rate of change of energy of the system with volume, at constant temperature. This is not the usual pressure. To calculate the usual pressure p (see equation (B.2)) we need an expression for the entropy of the system, so that the entropy may be held constant during the change in volume. There is no obvious way to find the entropy directly, but we know that

$$C_V = T\left(\frac{\partial S}{\partial T}\right)_V. \tag{B.5}$$

Since from (B.4) C_V is known as a function of V and T we might suppose that this equation could be integrated to give the entropy. However, this does not work. The preceding equation becomes on integration:

$$S(T,V) = \int^T dT' \, \frac{C_V(V,T')}{T'} + \Phi(V). \tag{B.6}$$

There is no way to determine the arbitrary function $\Phi(V)$ without further information. This function is not just an irrelevant constant of integration; it depends on V, and we need to know it in order to calculate quantities such as the pressure. We have been able to calculate the dependence of the entropy on temperature at constant volume, but we cannot find how the entropy changes with the volume. Without this knowledge the pressure cannot be calculated, and we can go no further.

A similar thing happens when we start from one of the other potentials expressed in terms of a pair of variables other than those in the table. Some quantities can be calculated, but there are always some that cannot. There is indeed something special about the natural variables. We are now in a position to understand what this is.

Experimentally, there are two different ways in which the energy of our system can be changed. We can do work on it, or we can supply heat to it. These two methods of energy transfer are quite distinct, and in a given change it is always experimentally possible to separate the energy supplied to the system into energy supplied as heat, and energy supplied as work. The expression for the differential dU in terms of its *natural* variables reflects this division into heat and work:

$$dU = \underbrace{TdS}_{\text{heat}} - \underbrace{pdV}_{\text{work}}. \tag{B.7}$$

The first term is the heat supplied to the system in a reversible change; the second is the work done on the system, also in a reversible change. It is therefore possible, given U as a function of V and S, to calculate the change of U in a process in which no work is done ($dV = 0$), or in a process in which no heat is supplied ($dS = 0$). Since heat and work are so fundamentally distinct from an experimental point of view, it is important to be able to handle these two situations. But let us suppose that we know $U(T,V)$ instead. In this case the two terms in the expression (B.4) for dU do not represent heat and work. We are still able to consider changes in which no work is done by putting $dV = 0$, as we know that no work is done if the volume does not change. But if the volume does change we do not have the information required to separate the change in energy into a part coming from work and a part coming from heat. This information cannot be obtained from $U(T,V)$ alone. The mathematics cannot know which part of dU is heat and which part is work unless we tell it. This explains our failure to calculate the entropy starting from $U(T, V)$. We were able to calculate $S(T)$ at constant V, since while V is constant no work is done and all of dU could be unambiguously labelled as heat. But once V changed, the division between work and heat could no longer be made and the entropy, defined through $dS = dQ/T$, became arbitrary.

So the reason why $U(S, V)$ is sufficient to find everything while $U(T, V)$ is not, is that the natural variables S and V are those that discriminate between work and heat, and we are vitally interested in making this distinction. The natural variables for all the other potentials follow from those for U by a series of Legendre transformations.

Appendix C: **Magnetic energy**

The problem of assigning an energy to a magnetic dipole in a magnetic field has been the cause of much confusion, because a dipole's energy depends on which components of the total system are included. Magnetic systems appear often in this book, so it is important to understand them thoroughly.

A magnetic dipole **m** in a magnetic field **B** feels a torque

$$\tau = \mathbf{m} \times \mathbf{B}. \tag{C.1}$$

This torque tries to line the dipole up with the field. As the angle between **m** and **B** changes, the torque does work. Let us take the zero of energy to be when the dipole and field are at right angles. Then the work done on the dipole in rotating it to a new orientation is

$$U = -\mathbf{m} \cdot \mathbf{B}, \tag{C.2}$$

as long as $|\mathbf{m}|$ and **B** remain unchanged during the rotation. U is a form of potential energy. It is lowest when **m** and **B** are parallel, as we would expect.

However, $|\mathbf{m}|$ and **B** may change when the dipole rotates unless arrangements are made to keep them fixed. Consider the dipole first. Let us model it as a perfectly conducting loop of wire of area A, carrying a current i: $|\mathbf{m}| = iA$. As the loop rotates, so the flux Φ linking it changes, and an e.m.f. $\mathcal{V}$ appears in the loop. This changes the current in the loop and so the value of $|\mathbf{m}|$. Now let us add to our model a battery to keep the current i in the loop constant. This battery will have to do work W, where

$$W = -\int dt\, \mathcal{V} i = -\int dt \left(-\frac{d\Phi}{dt}\right) i = i\Delta\Phi = +\mathbf{m} \cdot \mathbf{B}. \tag{C.3}$$

Compare this with equation (C.2). If we consider our system to be just the loop of wire and not the battery, then the work done on the loop by the battery is exactly equal to the work done by the loop on the external world. The two energies cancel, and the energy of the loop is independent of its orientation.

But it doesn't stop there! The part of the dipole's flux which links the solenoid that generates the external field **B** changes as the dipole rotates. Again we can use a battery connected to the solenoid to hold the current in the solenoid constant in spite of this. Similar arguments to those above give for the work done by this battery

$$W = +\mathbf{m} \cdot \mathbf{B}. \qquad (C.4)$$

If we consider our system now to be the loop of wire plus the solenoid, but not their associated batteries, the energy of this system is the sum of (C.2), (C.3), and (C.4), which is $+\mathbf{m} \cdot \mathbf{B}$. This system has its *maximum* energy when dipole and field are aligned, which is the exact opposite of what we started with!

Of course there is no inconsistency here. The three different energies apply to three different systems. However, statistical mechanics is formulated in terms of energies. We have to decide which energy is the correct one to use in the Boltzmann factor.

Imagine a dipole, free to rotate about a cooled lightly-damped bearing, in equilibrium with a low-pressure gas. The dipole is subject to the torque (C.1) from the magnetic field, and from collisions with the molecules of the gas. Left to its own devices the dipole will align itself with the magnetic field because of the torque (C.1). (The damping in the bearing ensures that it can dissipate its angular momentum; cooling the bearing ensures that the dipole is not excited by the bearing itself.) However, every so often a gas molecule will collide with the dipole and knock it away from its resting place. When this happens there is a restoring torque which any further collisions will have to overcome. At high temperatures the molecules will be moving quickly enough to manage this, and the dipole will swing around in all directions. But at low temperatures the torque exerted by the magnetic field will be large compared with the rate of angular momentum transfer by molecular collisions. The dipole will spend most of its time pointing almost in the direction of the field, as the molecules aren't going fast enough to overcome the restoring torque.

This intuitive argument suggests that the correct energy to use for a dipole in a magnetic field must be $-\mathbf{m} \cdot \mathbf{B}$. But is there a better reason why only the mechanical work done on the dipole contributes, and not the electrical work done on the dipole or on the solenoid? Yes. We know that

$$S = k_B \log \Omega, \qquad (C.5)$$

where S is the entropy of a system and Ω is the number of microstates compatible with the given values of the thermodynamic variables. Consider once more the dipole in the gas. The number of microstates of the gas with a given total energy increases very rapidly with that energy. When the dipole moves to line itself up with the field it knocks a few molecules about and increases the energy of the gas. By (C.5), this also increases the entropy of the gas. Now the crucial point is that it is *only* through this mechanical work that energy is exchanged between the gas and the dipole; it is true that in the alignment process more energy is extracted from the battery than goes into the gas, but since this energy does not come from the gas, it doesn't reduce the entropy of the gas. The mechanical work has increased the entropy of the gas, and this is what matters for statistical mechanics. The probability of finding the dipole in a given state is proportional to the number of microstates of the gas compatible with it (there is no Boltzmann factor, since the total energy of dipole plus gas does not change as the dipole moves), and so the dipole is most likely to be in that state which maximizes the entropy of the gas.

To understand this point a little more clearly, imagine a grain of sand undergoing Brownian motion in a beaker of water. The probability of finding it at height h will be proportional to $e^{-mgh/k_B T}$. Now we could look at the grain of sand with a magnifying glass and follow its movements. Every time it moves up, we could move a cannon-ball down through the same distance; every time it goes down, we could move the cannon-ball up. If we consider the system consisting of the cannon-ball and the beaker of water, but excluding the person lifting the cannon-ball, its energy is a minimum when the grain of sand jumps out of the beaker and the cannon-ball goes through the floor. But because the energy that moves the cannon-ball is not the thermal energy of the water, the grain of sand will continue to spend its time near the bottom of the beaker.

In the same way, whether or not we choose to use a battery to keep the current in the solenoid constant is up to us. It doesn't affect the statistical mechanics of the problem. Of course it does affect the behaviour of the system slightly, because the field **B** at the dipole due to the solenoid remains constant whereas without the aid of the battery it would change; similarly, the motion of the cannon-ball could gravitationally affect the beaker of water. But that is a different matter.

The dipole itself deserves some comment. We have modelled it as a loop of wire attached to a battery, which is not a very realistic model for the systems encountered in solid-state physics. The elementary dipoles in a ferromagnet do not have batteries attached to them! They keep their magnitude constant when they rotate without the need for an external source of energy. This doesn't affect any of the conclusions we have reached; it just means that for atomic dipoles the problem is simpler than we might fear.

Appendix D: **Connected correlation functions and $\log Z[J]$**

Consider a system whose microstates are labelled by the values of N variables $(s_1, \ldots, s_N)$. We shall refer to these variables as spins, but the discussion is perfectly general. In order to calculate correlation functions of these spins, we add a term

$$\delta H = -\frac{1}{\beta} \sum_i J_i s_i \tag{D.1}$$

to the Hamiltonian of the system and differentiate the partition function $Z(J_1, \ldots, J_n)$ with respect to the J_i:

$$\begin{aligned} G^{(n)}(i_1, \ldots, i_n) &\equiv \langle s_{i_1} \ldots s_{i_n} \rangle \\ &= \frac{1}{Z} \frac{\partial^n Z}{\partial J_{i_1} \ldots \partial J_{i_n}}. \end{aligned} \tag{D.2}$$

However, as we saw on page 43, the $G^{(n)}$ are not the most convenient way of describing the system, since each one contains a lot of information already present at lower orders. More useful are the connected correlation functions $G_c^{(n)}$ which do not duplicate information in this way. $G_c^{(n)}$ represents the difference between the actual value of $G^{(n)}$ and our 'best guess' at it, based on the $G_c^{(m)}$ with $m < n$. Equation (2.37) illustrates this idea for $G_c^{(3)}$. More generally our best guess at $G^{(n)}$ can be obtained by splitting up its n arguments in all possible ways, into at least two and at most n sub-groups. Each such **partition** of the n spins contributes a product of connected correlation functions to our best guess, the arguments of the correlation functions being the groups into which the spins have been divided. Then we sum over all possible partitions. For example, if $n = 7$, one way of partitioning the seven spins $(s_{i_1}, \ldots, s_{i_7})$ is

$$(s_{i_1}\, s_{i_2}\, s_{i_4})\ (s_{i_3}\, s_{i_5}\, s_{i_6})\ (s_{i_7}). \tag{D.3}$$

This partition contributes the term

$$G_c^{(3)}(i_1, i_2, i_4) G_c^{(3)}(i_3, i_5, i_6) G_c^{(1)}(i_7) \tag{D.4}$$

to our best guess at $G^{(7)}$. To get the other terms we have to split up the seven spins in all other possible ways and add the results. Each term involves only the correlation functions $G_c^{(1)}$ to $G_c^{(6)}$. The n-point connected correlation function $G_c^{(n)}$ is defined to be the difference between the actual value of $G^{(n)}$ and the sum of all these products:

$$G_c^{(n)} \equiv G^{(n)} - \sum_{\text{partitions}} \begin{bmatrix} \text{products of connected} \\ \text{correlation functions } G_c^{(m)} \\ \text{where } m < n \end{bmatrix}. \tag{D.5}$$

In the case $n = 7$ there are 786 terms in the sum of (D.5). Calculating $G_c^{(n)}$ by first finding $G^{(n)}$ using (D.2), and then using (D.5), is impractical for all but the smallest values of n.

We shall now prove the beautiful result that the connected correlation functions are obtained by differentiating $\log Z$:

$$G_c^{(n)}(i_1, \ldots, i_n) = \frac{\partial}{\partial J_{i_1}} \cdots \frac{\partial}{\partial J_{i_n}} \log Z. \tag{D.6}$$

This makes the $G_c^{(n)}$ as easy to calculate as the $G^{(n)}$. The proof of (D.6) will be by induction. First, notice that (D.6) implies that

$$\frac{\partial G_c^{(m)}(i_1, \ldots i_m)}{\partial J_{i_{m+1}}} = G_c^{(m+1)}(i_1, \ldots, i_{m+1}). \tag{D.7}$$

We shall show that this is true for $m = n$ if it is true for all $m < n$. Differentiate the expression (D.5) for $G_c^{(n)}$ with respect to $J_{i_{n+1}}$. All the terms in the sum on the right-hand side of (D.5) involve only those $G_c^{(m)}$ for which $m < n$, so (D.7) can be used to differentiate these terms. A term which is the product of p connected correlation functions will give rise to p new terms, in which each of the functions is differentiated in turn. From (D.7) it follows that these new terms will each be the product of $p - 1$ of the old connected correlation functions, plus one new connected correlation function with the point i_{n+1} as one argument. For example, differentiating the example term (D.4) above gives three new terms:

$$\frac{\partial}{\partial J_{i_8}} \left[G_c^{(3)}(i_1, i_2, i_4) G_c^{(3)}(i_3, i_5, i_6) G_c^{(1)}(i_7) \right] = \tag{D.8}$$
$$G_c^{(4)}(i_1, i_2, i_4, i_8) G_c^{(3)}(i_3, i_5, i_6) G_c^{(1)}(i_7)$$
$$+ G_c^{(3)}(i_1, i_2, i_4) G_c^{(4)}(i_3, i_5, i_6, i_8) G_c^{(1)}(i_7)$$
$$+ G_c^{(3)}(i_1, i_2, i_4) G_c^{(3)}(i_3, i_5, i_6) G_c^{(2)}(i_7, i_8).$$

The differentiation adds in the new point (here i_8) to each correlation function in turn. Every possible partition of the $n + 1$ spins is obtained by differentiating the right-hand side of (D.5) in this way, except for: (a) those partitions which involve $G^{(1)}(i_{n+1})$, and (b) the term $G_c^{(n+1)}(i_1, \ldots, i_{n+1})$.

Now let us differentiate $G^{(n)}$ on the right-hand side of (D.5). The definition (D.2) for $G^{(n)}$ gives

$$\frac{\partial G^{(n)}(i_1, \ldots, i_n)}{\partial J_{i_{n+1}}} = G^{(n+1)}(i_1, \ldots, i_{n+1}) - G^{(1)}(i_{n+1}) G^{(n)}(i_1, \ldots, i_n). \tag{D.9}$$

The second term on the right-hand side of (D.9), when combined with the terms from the previous paragraph, makes up a complete set of partitions of the $(n+1)$ spins into two or more groups. So we now have

$$\frac{\partial G_c^{(n)}}{\partial J_{i_{n+1}}} = G^{(n+1)} - \sum_{\text{partitions}} \left[\begin{array}{c} \text{products of connected} \\ \text{correlation functions } G_c^{(m)} \\ \text{where } m < n+1 \end{array}\right]. \quad (D.10)$$

Comparing this with (D.5) we see that

$$G_c^{(n+1)}(i_1, \ldots, i_{n+1}) = \frac{\partial G_c^{(n)}(i_1, \ldots, i_n)}{\partial J_{i_{n+1}}}. \quad (D.11)$$

So we have shown that if (D.7) holds for $m < n$, it also holds for $m = n$. It is easily checked that (D.7) works for $m = 1$, and therefore it is true for all m. Equation (D.6) now follows immediately since

$$G^{(1)}(i_1) = G_c^{(1)}(i_1) = \frac{\partial \log Z}{\partial J_{i_1}}. \quad (D.12)$$

Connected correlation functions are obtained by differentiating $\log Z[J]$.

Appendix E: The Gibbs free energy

In this appendix we are going to calculate the probability distribution for a quantity ϕ. It will turn out that for many systems this quantity is intimately related to the Gibbs free energy, so long as $T > T_c$.

The probability P that ϕ should lie within a particular range R is the sum of the probabilities for the system being in each of the microstates that are compatible with this condition:

$$P(\phi \text{ in } R) = \sum_{\substack{\text{states } \alpha: \\ \phi_\alpha \text{ in } R}} \frac{e^{-\beta E_\alpha}}{Z}, \quad (E.1)$$

where ϕ_α is the value taken by ϕ in the microstate α. The restriction on the microstates being summed over makes the sum in (E.1) harder than usual, but not impossible. Let us introduce a probability density $\rho(\varphi, \beta)$ and write:

$$P(\phi \text{ in } R) = \frac{1}{Z} \int_R d\varphi\, \rho(\varphi, \beta), \quad \text{where} \quad \rho(\varphi, \beta) \equiv \sum_{\text{all } \alpha} \delta(\phi_\alpha - \varphi) e^{-\beta E_\alpha}. \quad (E.2)$$

Gibbs free energy

The sum in the definition of $\rho(\varphi, \beta)$ is now over all microstates α. $\rho(\varphi, \beta)$ is the partition function of a system in which the quantity ϕ is constrained to take some particular value φ.

The δ-function in (E.2) may be written

$$\delta(\phi_\alpha - \varphi) = \frac{1}{2\pi i} \int_{-i\infty}^{i\infty} dJ\, e^{J(\phi_\alpha - \varphi)}. \quad (E.3)$$

Inserting this into the definition (E.2) of $\rho(\varphi, \beta)$ gives

$$\rho(\varphi, \beta) = \frac{1}{2\pi i} \int_{-i\infty}^{i\infty} dJ\, e^{-J\varphi} \sum_{\text{all } \alpha} e^{-\beta E_\alpha + J\phi_\alpha}. \quad (E.4)$$

Let us define

$$Z'(J, \beta) \equiv \sum_{\text{all } \alpha} e^{-\beta E_\alpha + J\phi_\alpha}. \quad (E.5)$$

Z' is the partition function of a system with Hamiltonian

$$H'(\alpha) = H(\alpha) - \frac{J\phi_\alpha}{\beta}, \quad (E.6)$$

where H is the Hamiltonian of the system we are studying; $H(\alpha) = E_\alpha$. If we can calculate Z', then equation (E.4) will give us $\rho(\varphi, \beta)$, from which $P(\phi \text{ in } R)$ will follow from (E.2).

To this end let us define the functions $F'(J, \beta)$ and $G'(J, \varphi, \beta)$ by

$$e^{-\beta F'(J,\beta)} \equiv Z'(J, \beta) \quad \text{and} \quad e^{-\beta G'(J,\varphi,\beta)} \equiv Z'(J, \beta) e^{-J\varphi}. \quad (E.7)$$

F' is the Helmholtz free energy of the modified system. The function G' satisfies

$$G'(J, \varphi, \beta) = F'(J, \beta) + \frac{J}{\beta}\varphi. \quad (E.8)$$

In this form $G'(J, \varphi, \beta)$ looks very much like the Gibbs free energy of the modified system. However, J and φ in (E.8) are independent variables, whereas for the true free energy φ would be the thermal average of the quantity ϕ in the field J. Consequently $G'(J, \varphi, \beta)$ is a function of three variables, whilst the true Gibbs free energy is a function of only two.

In terms of $G'(J, \varphi, \beta)$ the function $\rho(\varphi, \beta)$ becomes

$$\rho(\varphi, \beta) = \frac{1}{2\pi i} \int_{-i\infty}^{i\infty} dJ\, e^{-\beta G'(J,\varphi,\beta)}. \quad (E.9)$$

To calculate this integral let us expand $G'(J, \varphi, \beta)$ as a function of J about its stationary point,

$$\left(\frac{\partial G'(J, \varphi, \beta)}{\partial J}\right)_\varphi = 0. \quad (E.10)$$

This implies (using (E.8)) that

$$\frac{\partial F'(J,\beta)}{\partial J} + \frac{\varphi}{\beta} = 0. \tag{E.11}$$

$-\beta(\partial F'/\partial J)$ is the thermal average of ϕ in the presence of the field J. So (E.11) says that $G'(J,\phi,\beta)$ is stationary when J is such that $\langle\phi\rangle$ is equal to φ; let us call the value of J at this point J_0. $G'(J_0,\varphi,\beta)$ is equal to the Gibbs free energy $G'(\varphi,\beta)$ of the modified system. Expanding $G'(J,\varphi,\beta)$ about this point there is, of course, no term linear in J. The quadratic term is given by

$$\frac{\partial^2 G'(J,\varphi,\beta)}{\partial J^2} = \frac{\partial^2 F'(J,\beta)}{\partial J^2} \tag{E.12}$$

$$= -\tfrac{1}{\beta}\left(\langle\phi^2\rangle - \langle\phi\rangle^2\right).$$

This coefficient is always non-positive. It is proportional to the volume V of the system, because G' is extensive and J is intensive. Our expansion of $G'(J,\varphi,\beta)$ so far is

$$G'(J,\varphi,\beta) = G'(\varphi,\beta) - \frac{(J-J_0)^2}{2!}\left(\langle\phi^2\rangle - \langle\phi\rangle^2\right) + O(J^3). \tag{E.13}$$

This gives us

$$e^{-\beta G'(J,\varphi,\beta)} = e^{-\beta G'(\varphi,\beta)}\exp\left[\frac{(J+J_0)^2}{2!}\left(\langle\phi^2\rangle - \langle\phi\rangle^2\right) + \cdots\right]. \tag{E.14}$$

It is good to see that this expression tends to zero as $J \to \pm i\infty$; this means that we shall be able to do the integral in (E.9). For $|J| \gg (\langle\phi^2\rangle - \langle\phi\rangle^2)^{-1/2} \sim V^{-1/2}$ the integrand will be very small due to suppression by the exponential, so only values of $|J|$ up to $O(V^{-1/2})$ are important. Moreover, the coefficients of the terms in the expansion of $G'(J,\varphi,\beta)$ of $O(J^3)$ and higher are proportional to the volume, being derivatives of the Gibbs free energy, so for $|J| \lesssim O(V^{-1/2})$ all these higher terms are suppressed by positive powers of V. In the limit $V \to \infty$ they can therefore be neglected, and we need retain only the term quadratic in J.

Substituting (E.14) into (E.9), and defining $i\zeta \equiv J + J_0$, gives

$$\rho(\varphi,\beta) = \frac{1}{2\pi}\int_{-\infty}^{\infty} d\zeta\, e^{-\beta G'(\varphi,\beta)}\exp\left[-\tfrac{1}{2}\zeta^2\left(\langle\phi^2\rangle - \langle\phi\rangle^2\right)\right]$$

$$= e^{-\beta G'(\varphi,\beta)}\Big/\sqrt{2\pi\left(\langle\phi^2\rangle - \langle\phi\rangle^2\right)}. \tag{E.15}$$

In the thermodynamic limit the dominant ϕ-dependence comes from the exponential. Our result is then

$$\rho(\varphi,\beta) \propto e^{-\beta G'(\varphi,\beta)}. \tag{E.16}$$

Gibbs free energy

We have expressed ρ in terms of the Gibbs free energy of the modified system.

If the Hamiltonian of the system we are studying contains a term linear in ϕ, the Gibbs free energy G' of the modified system can be written in terms of the Gibbs free energy G of the real system. This is a special case, but it is an important one. It applies, for example, to magnetic systems in which ϕ is one component of the total spin and a term linear in ϕ represents the interaction of a magnetic field with the system. Let the Hamiltonian of the system we are studying be

$$H(\alpha) = H_0(\alpha) - B\phi_\alpha. \tag{E.17}$$

Then the Hamiltonian of the modified system is

$$H'(\alpha) = H_0(\alpha) - \left(B + \frac{J}{\beta}\right)\phi_\alpha. \tag{E.18}$$

The following argument is unfortunately rather subtle. Let us suppose that the system with Hamiltonian (E.17) has $\langle \phi \rangle = \varphi$ when B takes the value B_0. Now B in the Hamiltonian (E.18) is not B_0; rather, it is whatever magnetic field happens to be present. For this system, $\langle \phi \rangle = \varphi$ when J takes the value J_0, defined by equation (E.11). Therefore

$$B_0 = B + \frac{J_0}{\beta}. \tag{E.19}$$

In these circumstances, the two systems have identical Helmholtz free energies, because their Hamiltonians are identical. However, their Gibbs free energies differ, because $G(\varphi, \beta)$, the Gibbs free energy of the first system, is a Legendre transform with respect to B, while $G'(\varphi, \beta)$ is a Legendre transform with respect to J. Therefore

$$\begin{aligned} G(\varphi, \beta) &= F(B_0, \beta) + B_0\varphi \\ G'(\varphi, \beta) &= F'(J_0, \beta) + \frac{J_0\varphi}{\beta}. \end{aligned} \tag{E.20}$$

From the equality of F and F', and the last two equations, it follows that

$$G'(\varphi, \beta) = G(\varphi, \beta) - B\varphi. \tag{E.21}$$

Substituting (E.21) into equation (E.16) gives

$$\rho(\varphi, \beta) \propto e^{-\beta(G(\varphi,\beta) - B\varphi)}. \tag{E.22}$$

We have related the probability distribution for ϕ to the Gibbs free energy of the system: $\rho(\varphi, \beta)\,d\varphi$ is the probability that ϕ lies between φ and $\varphi + d\varphi$;

substituting into (E.2), we can find the probability that ϕ lies in any range R whatsoever.

On page 25 we defined the extended free energy F_3 to be the function $F_3(T, B, m)$ that generates m's probability distribution through $dP \propto e^{-\beta F_3} dm$. Comparing this with equation (E.22) we see that above T_c we have $F_3 = G - BM$, which is precisely equation (1.16).

There is, however, a flaw in the above reasoning. Consider a magnetic system with no applied field. If its symmetry is spontaneously broken, we have argued that $\rho(\varphi, \beta)$ should have a maximum away from $\varphi = 0$. Equation (E.22) then implies that $G(\varphi)$ has a minimum away from $\varphi = 0$ and therefore a maximum at $\varphi = 0$, assuming symmetry under $\varphi \leftrightarrow -\varphi$. But this is not possible, since we know from equation (2.47) that the Gibbs free energy is always convex. So equation (E.22) cannot be true in general.

The reason for this inconsistency is that in the presence of spontaneous symmetry breaking the terms of order J^3 and higher do not vanish when $V \to \infty$, and the neglect of these terms in (E.13) and the following equations has caused the error. When a symmetry is spontaneously broken a change in the external field of order V^{-1} causes a change in the thermal average of the total spin of order V as the spin swings around to follow the applied field. From this it follows that the $2n^{\text{th}}$ derivative of the Helmholtz free energy in zero applied field is of order V^{2n}, rather than of order V as we assumed above. The approximations that we made when deriving the result (E.15) break down, and $\rho(\phi, \beta)$ is no longer equal to the exponential of the Gibbs free energy.

This problem arises whenever $(\partial^2 G(\varphi, \beta)/\partial \varphi^2)$ is of order V^{-2}, so knowing G as a function of φ allows us to spot when the problem arises. In particular things go wrong for the 'flat' part of the graph of G against φ for a system with spontaneously broken symmetry (see Figure 2.6). This resolves the problem mentioned above.

Appendix F: Discrete Fourier transforms

If we have a set of N numbers, $\{x_j\}$ ($j = 0, \ldots, N-1$), its discrete Fourier transform is defined to be the set of numbers

$$\tilde{x}_q \equiv \sum_{j=0}^{N-1} x_j e^{-2\pi i j q / N} \quad (q = 0, \ldots, N-1). \tag{F.1}$$

Discrete Fourier transform theorem: *If the N numbers $\tilde{x}_q$ are defined by equation (F.1), then*

$$x_{j'} = \frac{1}{N} \sum_{q=0}^{N-1} \tilde{x}_q e^{2\pi i j' q / N}. \tag{F.2}$$

Discrete Fourier transforms

Proof: We multiply both sides of equation (F.1) by $(1/N)e^{2\pi i j' q/N}$ and sum over q, to obtain

$$\frac{1}{N} \sum_{q=0}^{N-1} \tilde{x}_q e^{2\pi i j' q/N} = \frac{1}{N} \sum_{q=0}^{N-1} \sum_{j=0}^{N-1} x_j e^{2\pi i (j'-j) q/N}$$

$$= \frac{1}{N} \sum_{j=0}^{N-1} x_j \sum_{q=0}^{N-1} e^{2\pi i (j'-j) q/N}.$$

(F.3)

The sum over q on the right side of equation (F.3) is a geometric progression, with sum N if $j' = j$, or (since j and j' are integers)

$$\frac{1 - e^{2\pi i (j'-j)}}{1 - e^{2\pi i (j'-j)/N}} = 0 \quad \text{if } j' \neq j.$$

(F.4)

Thus the only contributing term in the outer sum on the right of equation (F.3) is that for which $j = j'$. ◁

Discrete Fourier transforms share many of the properties of continuous Fourier transforms.[1] To demonstrate these properties, it is necessary to define the quantities x_j for j outside the range $[0, N-1]$ by the rule

$$x_j = x_{j+mN} \quad \text{for all integer } m,$$

(F.5)

i.e., to assume that x_j is periodic with period N. Note that $\tilde{x}_q$ is already periodic with period N (see equation (F.1)). We may now prove:

Discrete Fourier convolution theorem: *If the three sets of N numbers $\{x_j\}, \{y_j\}, \{z_j\}$ $(j = 0, \ldots, N-1)$ are related by*

$$z_j = \sum_{j'=0}^{N-1} y_{(j-j')} x_{j'},$$

(F.6)

then

$$\tilde{z}_q = \tilde{y}_q \tilde{x}_q.$$

(F.7)

Proof: Taking the discrete Fourier transform of both sides of equation (F.6) and rearranging the resulting double sum, we have

$$\tilde{z}_q = \sum_{j=0}^{N-1} e^{-2\pi i j q/N} \sum_{j'=0}^{N-1} y_{j-j'} x_{j'}$$

$$= \sum_{j'=0}^{N-1} x_{j'} e^{-2\pi i q j'/N} \sum_{j=0}^{N-1} y_{j-j'} e^{-2\pi i q (j-j')/N}.$$

(F.8)

---

[1] However, note that they are logically entirely distinct, and disaster will attend any attempt to invert a continuous Fourier transform as if it were a discrete transform, or *vice versa*.

If we now define $j'' \equiv (j - j')$, the inner sum in equation (F.8) becomes $\tilde{y}_q$. This is independent of j', so it may be taken out of the outer sum, which then yields $\tilde{x}_q$. ◁

Parseval's theorem:

$$\sum_j |x_j|^2 = \frac{1}{N} \sum_q |\tilde{x}_q|^2. \tag{F.9}$$

This result follows easily when the right side is evaluated directly from (F.1). ◁

The generalization of discrete Fourier transforms to more than one dimension is straightforward. Let $\mathbf{j} = (j_1, j_2)$ be a vector with integer components $0 \leq j_1 \leq L-1$ and $0 \leq j_2 \leq L-1$, and let $x_{\mathbf{j}}$ be a set of $N \equiv L^2$ numbers. Then the discrete transform of $x_{\mathbf{j}}$ is

$$\tilde{x}_{\mathbf{q}} \equiv \sum_{j_1, j_2 = 0}^{L-1} x_{\mathbf{j}} e^{-2\pi i \mathbf{j} \cdot \mathbf{q}/L}. \tag{F.10}$$

The inverse of this is

$$x_{\mathbf{j}'} = \frac{1}{N} \sum_{\mathbf{q}} \tilde{x}_{\mathbf{q}} e^{2\pi i \mathbf{j}' \cdot \mathbf{q}/L}. \tag{F.11}$$

The convolution theorem becomes the statement that the transform of

$$z_{\mathbf{j}} \equiv \sum_{\mathbf{j}'} y_{(\mathbf{j}-\mathbf{j}')} x_{\mathbf{j}'} \tag{F.12}$$

is

$$\tilde{z}_{\mathbf{q}} = \tilde{y}_{\mathbf{q}} \tilde{x}_{\mathbf{q}}. \tag{F.13}$$

Parseval's theorem becomes

$$\sum_{\mathbf{j}} |x_{\mathbf{j}}|^2 = \frac{1}{N} \sum_{\mathbf{q}} |\tilde{x}_{\mathbf{q}}|^2. \tag{F.14}$$

Appendix G: The method of steepest descent

Suppose one wishes to evaluate a contour integral of the form

$$I \equiv \int_\gamma dz\, e^g, \qquad (G.1)$$

where γ is some contour in the complex plane and $g(z)$ is an analytic function of z. Then one can obtain an approximate value for I as follows.

The modulus of the integrand is largest at the point z_s at which $u(z) \equiv \Re\mathrm{e}(g)$ peaks. At this point $\partial u/\partial x = \partial u/\partial y = 0$, where $z = x + iy$, so by the Cauchy-Riemann conditions, at z_s, $v(z) \equiv \Im\mathrm{m}(g)$ also has vanishing derivatives with respect to x and y. So at z_s, $dg/dz = 0$. Consequently, g's Taylor series about z_s reads:

$$g(z) = g(z_s) + \tfrac{1}{2!}g''(z_s)(z-z_s)^2 + \cdots . \qquad (G.2)$$

We write $g''(z_s) = -2ae^{i\phi}$ and $z - z_s = \epsilon e^{-i\phi'/2}$, where $a > 0$, ϵ, ϕ and ϕ' are all real numbers. Then

$$g(z) \simeq g(z_s) - ae^{i(\phi-\phi')}\epsilon^2 \qquad (G.3)$$

Now we deform the original contour of integration γ until it passes through z_s along the line $\phi' = \phi$. The contribution to the integral I from points near z_s can now be written

$$I' \equiv e^{-i\phi/2} e^{g(z_s)} \int_{\text{small } \epsilon} d\epsilon\, e^{-a\epsilon^2}. \qquad (G.4)$$

Although the Taylor expansion on which this expression is based is valid only for sufficiently small ϵ, the Gaussian nature of the integrand of (G.4) allows us to extend the limits of integration to $\pm\infty$ without appreciable error. Thus from the usual Gaussian integral we have

$$I' \simeq e^{-i\phi/2} e^{g(z_s)} \sqrt{\pi/a}. \qquad (G.5)$$

The method of steepest descent consists in equating I with the contribution I' from the neighbourhood of z_s:

$$\int_\gamma dz\, e^g \simeq e^{i\theta} e^{g(z_s)} \sqrt{\frac{2\pi}{|g''(z_s)|}} \quad \text{where} \quad \theta \equiv -\tfrac{1}{2}\arg[-g''(z_s)]. \qquad (G.6)$$

Appendix H: **Counting closed loops on a square lattice**

Equation (3.66) reduces the solution of the $d = 2$ Ising model to the problem of counting the number $g(l)$ of loops that can be drawn on a square lattice using l links. We must be careful to distinguish between loops and closed paths here. A **closed path** of l links is a route for getting from a site back to the same site in l steps; a **loop** is a particular pattern of links, with an even number of links meeting at each lattice point. For example, each ring of l links constitutes $2l$ closed paths, since there are l sites it could be considered to go from and to, and then it can be traversed in two different directions. However, it only represents one loop; this particular combination of l links occurs only once on the right-hand side of (3.64).

We call a closed path 'connected' if it consists of just one body of links, and 'disconnected' otherwise. Both connected and disconnected paths appear on the right-hand side of (3.64). Let $h(l)$ be the number of connected closed paths of length l, and define

$$D(l) \equiv \frac{1}{2l}h(l). \tag{H.1}$$

As we have seen on page 75, $h(l)$, and thus $D(l)$, are not difficult to calculate. We can find $g(l)$ (which includes both connected and disconnected loops) approximately in terms of $D(l)$ as follows. Given l_1 and l_2 such that $l_1 + l_2 = l$, we can find $D(l_1)$ loops of length l_1 and $D(l_2)$ loops of length l_2, and from these we can construct $D(l_1)D(l_2)$ disconnected loops containing l links. Generalizing this argument, we conclude that

$$g(l) = \sum_{n=1}^{l} \frac{1}{n!} \sum_{l_1 + \cdots + l_n = l} D(l_1) \ldots D(l_n), \tag{H.2}$$

where the sum is over all partitions of l into n pieces, and the factor $n!$ arises from the number of equivalent permutations of the l_i. However, as it stands, (H.2) is only an approximate expression for $g(l)$. The problem is that some loops are counted more than once, and some loops are counted that should not be counted at all.

Consider the loop shown in Figure H.1(a). This should contribute only once to the expression (H.2) for $g(8)$, but in fact it contributes three times. There are two different ways of traversing this loop, so it appears twice in $D(8)$; also it appears as a product of two four-link squares. The loop shown in Figure H.1(b) shouldn't appear at all, because the central link is traversed twice. Yet the loop contributes to $g(8)$, once as an eight-link path and once as a product of two four-link squares.

This difficulty can be overcome by modifying the definition of $D(l)$. We assign an 'amplitude' to each closed path, such that the total of the amplitudes for all the paths contributing to a loop gives the correct answer.

Closed loops on a square lattice

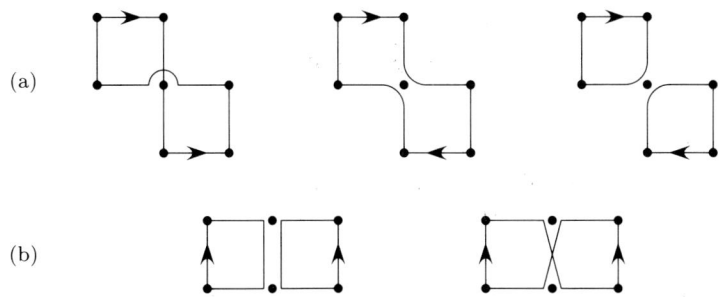

Figure H.1 (a) Three distinct contributions to the same (allowed) loop. (b) Two contributions to a forbidden loop.

For each path, we assign a factor of $e^{-i\pi/4}$ to a right turn and $e^{i\pi/4}$ to a left turn, and then we multiply together the factors for each turn on the path. The amplitude for a simple closed path is -1, while for a 'figure-of-eight' path like the first path in Figure H.1(a) or the second path in Figure H.1(b), the amplitude is $+1$. The three paths in Figure H.1(a) thus have amplitudes 1, 1, and -1 respectively, giving a total of 1; the two paths in Figure H.1(b) have amplitudes $+1$ and -1, giving a total of 0. Thus the allowed loop is counted once, while the forbidden loop is not counted. It can be proved (Sherman 1960, Burgoyne 1963) that this procedure gives the correct result for any loop; however, we do not know of a simple way of showing this.

Given that this procedure works, the next stage is to implement it. To achieve this goal we make each element of the matrix $\mathbf{M}$ introduced on page 75 into a 4×4 matrix $\mathbf{m_{ij}}$. Each element $m_{ij}^{\alpha\beta}$ of $\mathbf{m_{ij}}$ is associated with a direction of entry α to site $\mathbf{i}$ and a direction of entry β to site $\mathbf{j}$. For example, if $1, 2, 3, 4$ correspond to compass points E,N,W,S, respectively, then we shall call $m_{ij}^{3,4}$ the 'amplitude' associated with moving westwards into $\mathbf{i}$ and then southwards into $\mathbf{j}$. If it is impossible to get from $\mathbf{i}$ to $\mathbf{j}$ by such a path (doubling back is not allowed, else a link could be traversed twice) this entry is zero. If it is possible, the entry is $e^{i\theta/2}$, where θ is the angle the path turns through in passing $\mathbf{i}$.

When we calculate a single element of the sub-matrix $\mathbf{m}_{ij}^{(2)}$ of $\mathbf{M}^2$ we perform a double sum, first over all possible intermediate sites $\mathbf{k}$ and then over all possible directions in which we leave $\mathbf{i}$ (which, naturally, must coincide with the direction of arrival at $\mathbf{k}$). At the end $\mathbf{M}^2$ is a matrix of 4×4 matrices $\mathbf{m}_{ij}^{(2)}$, each element of which gives the amplitude for going from one given site to another in two steps and with specified directions on entering the first site and on entering the last. A diagonal element of a matrix $\mathbf{m}_{ii}^{(l)}$ gives the amplitude for entering site $\mathbf{i}$ twice from the same direction with l steps in between. Thus this element gives the amplitude for executing a closed path of l steps.

So let us redefine $D(l)$ to be

$$D(l) \equiv -\frac{1}{2l} \sum_{\text{sites } i} \text{Tr} \, \mathbf{m}_{ii}^{(l)}. \tag{H.3}$$

With $D(l)$ defined in this way, the expression (H.2) for $g(l)$ is correct. (The minus sign in (H.3) is to take account of the fact that the amplitude for a simple loop is -1.) Inserting (H.2) into (3.66) gives

$$\begin{aligned} Z_{\text{Ising}} &= \cosh^{2N}(\beta\epsilon) 2^N \left[1 + \sum_{l=1}^{\infty} v^l \sum_{n=1}^{l} \frac{1}{n!} \sum_{l_1+\cdots+l_n=l} D(l_1) \ldots D(l_n) \right] \\ &= \cosh^{2N}(\beta\epsilon) 2^N \left\{ 1 + \sum_{n=1}^{\infty} \frac{1}{n!} \left[\sum_{l=1}^{\infty} D(l) v^l \right]^n \right\} \\ &= \cosh^{2N}(\beta\epsilon) 2^N \exp \left[\sum_{l=0}^{\infty} D(l) v^l \right], \end{aligned} \tag{H.4}$$

where we have imposed the convention $D(0) = 1$.

It remains only to calculate $D(l)$. We do this by treating $\mathbf{M}$ as a $4N \times 4N$ matrix and finding its eigenvalues, λ_α, since

$$\begin{aligned} D(l) &= -\frac{1}{2l} \sum_{\text{sites } i} \text{Tr} \, \mathbf{m}_{ii}^{(l)} \\ &= -\frac{1}{2l} \sum_{\alpha=0}^{4N-1} \lambda_\alpha^l. \end{aligned} \tag{H.5}$$

Substituting this into equation (H.4) for Z and identifying the series for $\log(1-x)$ yields

$$\begin{aligned} Z_{\text{Ising}} &= \cosh^{2N}(\beta\epsilon) 2^N \exp \left[-\sum_{\alpha=0}^{4N-1} \sum_{l=0}^{\infty} \frac{\lambda_\alpha^l v^l}{2l} \right] \\ &= \cosh^{2N}(\beta\epsilon) 2^N \prod_{\alpha} (1 - v\lambda_\alpha)^{1/2}. \end{aligned} \tag{H.6}$$

Viewed as a matrix of matrices $\mathbf{m}_{ij}$, $\mathbf{M}$ is translationally invariant; $\mathbf{m}_{ij} = \mathbf{m}(\mathbf{i} - \mathbf{j})$. Hence its (matrix-valued) eigenvalues $\mathbf{M_q}$ are the components of its discrete Fourier transform (see Appendix F and above equation (3.44)):

$$\mathbf{M_q} = \sum_{\mathbf{j}} \mathbf{m}(\mathbf{j}) e^{-2\pi i \mathbf{q} \cdot \mathbf{j}/L}. \tag{H.7}$$

The only non-zero $\mathbf{m(j)}$ are for $\mathbf{j} = (\pm 1, 0)$ and $\mathbf{j} = (0, \pm 1)$ (see (3.45)). For example, when $\mathbf{j} = (-1, 0)$ the second site lies to the east of the first, so with our E,N,W,S ordering of directions $\mathbf{m}[(-1,0)]$ is non-zero only in the eastern column, 1. Thus

$$\mathbf{m}[(-1,0)] = \begin{pmatrix} 1 & 0 & 0 & 0 \\ e^{-i\pi/4} & 0 & 0 & 0 \\ 0 & 0 & 0 & 0 \\ e^{i\pi/4} & 0 & 0 & 0 \end{pmatrix}. \tag{H.8}$$

Similarly

$$\mathbf{m}[(1,0)] = \begin{pmatrix} 0 & 0 & 0 & 0 \\ 0 & 0 & e^{i\pi/4} & 0 \\ 0 & 0 & 1 & 0 \\ 0 & 0 & e^{-i\pi/4} & 0 \end{pmatrix}$$

$$\mathbf{m}[(0,-1)] = \begin{pmatrix} 0 & e^{i\pi/4} & 0 & 0 \\ 0 & 1 & 0 & 0 \\ 0 & e^{-i\pi/4} & 0 & 0 \\ 0 & 0 & 0 & 0 \end{pmatrix} \quad \mathbf{m}[(0,1)] = \begin{pmatrix} 0 & 0 & 0 & e^{-i\pi/4} \\ 0 & 0 & 0 & 0 \\ 0 & 0 & 0 & e^{i\pi/4} \\ 0 & 0 & 0 & 1 \end{pmatrix}. \tag{H.9}$$

Taking the Fourier transform amounts to multiplying $\mathbf{m(j)}$ by $Q_\mathbf{j} \equiv e^{-2\pi i \mathbf{q} \cdot \mathbf{j}/L}$ and adding the results together. We find

$$\widetilde{\mathbf{M}}(\mathbf{q}) = \begin{pmatrix} Q^{-1}_{(1,0)} & Q^{-1}_{(0,1)} e^{i\pi/4} & 0 & Q_{(0,1)} e^{-i\pi/4} \\ Q^{-1}_{(1,0)} e^{-i\pi/4} & Q^{-1}_{(0,1)} & Q_{(1,0)} e^{i\pi/4} & 0 \\ 0 & Q^{-1}_{(0,1)} e^{-i\pi/4} & Q_{(1,0)} & Q_{(0,1)} e^{i\pi/4} \\ Q^{-1}_{(1,0)} e^{i\pi/4} & 0 & Q_{(1,0)} e^{-i\pi/4} & Q_{(0,1)} \end{pmatrix}. \tag{H.10}$$

We could simply solve for the eigenvalues of this matrix and then form the product $\prod_\mathbf{q} \prod_{\alpha=1}^{4}(1 - v\lambda_\alpha)$ required in (H.6). But we reduce this to the evaluation of a determinant by observing that

$$\prod_{\alpha=1}^{4}(1 - v\lambda_\alpha) = |v\widetilde{\mathbf{M}}(\mathbf{q}) - \mathbf{I}| \tag{H.11}$$

$$= (1 + v^2)^2 - 2v(1 - v^2)\Re(Q_{(1,0)} + Q_{(0,1)})$$

Hence

$$Z_{\text{Ising}} = \cosh^{2N}(\beta\epsilon) 2^N$$
$$\times \prod_\mathbf{q} \{(1+v^2)^2 - 2v(1-v^2)[\cos(2\pi q_1/L) + \cos(2\pi q_2/L)]\}^{1/2}. \tag{H.12}$$

Appendix I: **Einstein's fluctuation theory**

Einstein argued that the probability P of a macroscopic configuration is proportional to the phase-space volume Ω compatible with it. In the microcanonical ensemble one can express the entropy S of any configuration as $S = k_B \log \Omega$ (see equation (2.12)). Hence

$$P[\delta\rho] \propto \Omega = e^{S/k_B}, \tag{I.1}$$

where S is the entropy of the density field $\delta\rho$. To find $S[\delta\rho]$ one imagines starting with the mean density distribution $\delta\rho = 0$ and changing the energy and specific volume of each fluid element until the desired form $\delta\rho(\mathbf{x})$ has been set up. Let lower-case letters u, v, etc., denote internal energy, volume, etc. per unit mass. Then from $TdS = dU + PdV$ we have that on setting up $\delta\rho(\mathbf{x})$ the net entropy changes by

$$\Delta S = \int d^3\mathbf{x}\, \rho(\mathbf{x}) \int_0^1 d\alpha\, \delta s(\mathbf{x}) = \int d^3\mathbf{x}\, \rho(\mathbf{x}) \int_0^1 \frac{d\alpha}{T} [\delta u + P\delta v]_\mathbf{x}, \tag{I.2}$$

where α parameterizes the various stages through which we deform the fluid to over-density $\delta\rho$. Since the whole system is to be regarded as a microcanonical ensemble, the total energy $U = \int d^3\mathbf{x}\, \rho u$ and the total volume are constant during the deformation parameterized by α. So if $T(\mathbf{x})$ and $P(\mathbf{x})$ were constant, the right-hand side of (I.2) would vanish. So we write

$$T(\mathbf{x},\alpha) \simeq T(\mathbf{x},0) + \alpha\delta T; \quad P(\mathbf{x},\alpha) \simeq P(\mathbf{x},0) + \alpha\delta P \tag{I.3}$$

and have to second order in δ

$$\begin{aligned}\Delta S &= \int d^3\mathbf{x}\, \rho(\mathbf{x})\tfrac{1}{2}\left[-\frac{\delta T(\delta u + P\delta v)}{T^2} + \frac{\delta P\delta v}{T}\right]_\mathbf{x} \\ &= \int d^3\mathbf{x}\, \rho(\mathbf{x})\tfrac{1}{2}\left[-\frac{\delta T \delta s}{T} + \frac{\delta P\delta v}{T}\right]_\mathbf{x}.\end{aligned} \tag{I.4}$$

When we now eliminate δs and δP through

$$\delta s = \left(\frac{\partial s}{\partial T}\right)_v \delta T + \left(\frac{\partial s}{\partial v}\right)_T \delta v \quad \text{and} \quad \delta P = \left(\frac{\partial P}{\partial T}\right)_v \delta T + \left(\frac{\partial P}{\partial v}\right)_T \delta v, \tag{I.5}$$

a Maxwell relation enables us to cancel the terms proportional to $\delta T \delta v$, and we have

$$\begin{aligned}\Delta S &= -\int d^3\mathbf{x}\, \frac{\rho}{2T}\left[\frac{C_v}{T}(\delta T)^2 + \frac{1}{\kappa_T v}(\delta v)^2\right] \\ &= -\int d^3\mathbf{x}\, \tfrac{1}{2}\left[\rho C_v\left(\frac{\delta T}{T}\right)^2 + \frac{1}{T\kappa_T}\left(\frac{\delta\rho}{\rho}\right)^2\right].\end{aligned} \tag{I.6}$$

This equation expresses the change in the overall entropy as a function of the magnitude of fluctuations in the local temperature and density. Substituting this into equation (I.1) we obtain a probability functional $P[\delta\rho, \delta T]$. To obtain the desired functional $P[\delta\rho]$ we must sum $P[\delta\rho, \delta T]$ over all configurations in which $\delta\rho[\mathbf{x}]$ has a particular form; in other words, we have to integrate $P[\delta\rho, \delta T]$ over all possible forms of $\delta T[\mathbf{x}]$. This integration merely generates a constant that is independent of $\delta\rho[\mathbf{x}]$. The functional (7.1) now follows.

Appendix J: The Gaussian transformation

It proves useful to consider the integral

$$I \equiv \int d^N \mathbf{x} \exp\left(-\tfrac{1}{4}\mathbf{x}\cdot\mathbf{L}^{-1}\cdot\mathbf{x} + \mathbf{s}\cdot\mathbf{x}\right), \tag{J.1}$$

where $\mathbf{x}$ and $\mathbf{s}$ are N-component vectors and $\mathbf{L}^{-1}$ is the inverse of a symmetric $N \times N$ matrix $\mathbf{L}$, all of whose eigenvalues λ_i exceed zero.

We evaluate the integral (J.1) by making an orthogonal transformation to the frame in which $\mathbf{L}$ is diagonal. Let x'_i and s'_i be the components of $\mathbf{x}$ and $\mathbf{s}$ in this frame. Then the argument of the exponential is

$$\sum_i \left(-\frac{x'^2_i}{4\lambda_i} + s'_i x'_i\right) = -\sum_i \left[\left(\frac{x'_i}{2\sqrt{\lambda_i}} - \sqrt{\lambda_i} s'_i\right)^2 - \lambda_i s'^2_i\right]$$
$$= -\sum_i \tfrac{1}{2} X^2_i + \sum_i \lambda_i s'^2_i, \tag{J.2}$$

where

$$X_i \equiv \sqrt{2}\left(\frac{x'_i}{2\sqrt{\lambda_i}} - \sqrt{\lambda_i} s'_i\right). \tag{J.3}$$

Since the transformation to the primed coordinates is orthogonal, we have

$$d^N\mathbf{x} = d^N\mathbf{x}' = 2^{N/2}\left(\prod_i \lambda_i\right)^{1/2} d^N\mathbf{X}$$
$$= 2^{N/2}\sqrt{\det \mathbf{L}}\, d^N\mathbf{X}. \tag{J.4}$$

Thus for any positive, symmetric matrix $\mathbf{L}$ we have

$$I = 2^{N/2}\sqrt{\det \mathbf{L}} \prod_i \int dX_i\, e^{-X^2_i/2} e^{\lambda_i s'^2_i}. \tag{J.5}$$

Since $\int dx\, e^{-x^2/2} = \sqrt{2\pi}$ and $\sum_i \lambda_i s'^2_i = \mathbf{s}\cdot\mathbf{L}\cdot\mathbf{s}$, we have finally

$$\int d^N\mathbf{x} \exp\left(-\tfrac{1}{4}\mathbf{x}\cdot\mathbf{L}^{-1}\cdot\mathbf{x} + \mathbf{s}\cdot\mathbf{x}\right) = (4\pi)^{N/2}\sqrt{\det \mathbf{L}}\, e^{\mathbf{s}\cdot\mathbf{L}\cdot\mathbf{s}}. \tag{J.6}$$

Appendix K: The Landau–Ginzburg model and the Ising model

Equation (3.3), generalized to an inhomogeneous field B that takes different values B_i on each lattice site, reads

$$Z_{\text{Ising}} = \sum_{\{s\}} \exp\left[\beta \sum_i s_i B_i - \tfrac{1}{2}\beta \sum_{ij} s_i J_{ij} s_j\right] \quad \text{(K.1)}$$
$$= \sum_{\{s\}} e^{\beta(\mathbf{s}\cdot\mathbf{B} - \mathbf{s}\cdot\mathbf{J}\cdot\mathbf{s})}.$$

We would like to identify the quadratic form in $\mathbf{s}$ of the exponent with the similar quadratic form on the right-hand side of equation (J.6). Unfortunately, $-\mathbf{J}$ is not a positive matrix. But since $s_i^2 = 1$ for all i,

$$-\mathbf{s}\cdot\mathbf{J}\cdot\mathbf{s} = \mathbf{s}\cdot\mathbf{L}\cdot\mathbf{s} - \alpha N, \quad \text{(K.2)}$$

where the matrix

$$\mathbf{L} \equiv \alpha\mathbf{I} - \mathbf{J} \quad \text{(K.3)}$$

can be made positive by taking α sufficiently large. Hence with (J.6) we may write

$$Z_{\text{Ising}} = e^{-\beta\alpha N} \sum_{\{s\}} e^{\beta(\mathbf{s}\cdot\mathbf{B} + \mathbf{s}\cdot\mathbf{L}\cdot\mathbf{s})}$$
$$= \frac{(4\pi)^{-N/2} e^{-\beta\alpha N}}{\sqrt{\det \mathbf{L}}}$$
$$\times \sum_{\{s\}} \int d^N\mathbf{x}\, \exp\left[-\tfrac{1}{4}\mathbf{x}\cdot(\beta\mathbf{L})^{-1}\cdot\mathbf{x} + \mathbf{s}\cdot(\mathbf{x} + \beta\mathbf{B})\right]$$
$$= \frac{(4\pi)^{-N/2} e^{-\beta\alpha N}}{\sqrt{\det \mathbf{L}}}$$
$$\times \sum_{\{s\}} \int d^N\mathbf{y}\, \exp\left[-\tfrac{1}{4}(\mathbf{y} - \beta\mathbf{B})\cdot(\beta\mathbf{L})^{-1}\cdot(\mathbf{y} - \beta\mathbf{B}) + \mathbf{s}\cdot\mathbf{y}\right],$$

$$\text{(K.4)}$$

where we have defined $\mathbf{y} \equiv \mathbf{x} + \beta\mathbf{B}$. The nice thing about (K.4) is that $\mathbf{s}$ appears only in the last term of the exponent. This enables us to do the sum over $\{s_i\}$. Since s_i takes only the values ± 1, we obtain

$$Z_{\text{Ising}} = \frac{(4\pi)^{-N/2} e^{-\beta\alpha N}}{\sqrt{\det \mathbf{L}}}$$
$$\times \int d^N\mathbf{y}\, \exp\left[-\tfrac{1}{4}(\mathbf{y} - \beta\mathbf{B})\cdot(\beta\mathbf{L})^{-1}\cdot(\mathbf{y} - \beta\mathbf{B})\right] \times 2^N \prod_i \cosh y_i.$$

$$\text{(K.5)}$$

The factors of 2^N now cancel and defining a new set of variables ϕ_i by[1]

$$\boldsymbol{\phi} \equiv (\beta \mathbf{L})^{-1} \cdot \mathbf{y}, \tag{K.6}$$

we can rewrite (K.5) in the form

$$Z_{\text{Ising}}(\beta; \mathbf{B}) = \frac{e^{-\beta \alpha N} e^{-\beta \mathbf{B} \cdot \mathbf{L}^{-1} \cdot \mathbf{B}/4}}{\pi^{N/2} \sqrt{\det \mathbf{L}}} \int d^N \mathbf{y}$$
$$\times \exp\left[-\tfrac{1}{4}\left(\mathbf{y} \cdot (\beta \mathbf{L})^{-1} \cdot \mathbf{y} - 2\mathbf{B} \cdot \mathbf{L}^{-1} \cdot \mathbf{y}\right) + \sum_i \log \cosh y_i \right]$$
$$= \pi^{-N/2} \beta^N \sqrt{\det \mathbf{L}} e^{-\beta \alpha N} e^{-\beta \mathbf{B} \cdot \mathbf{L}^{-1} \cdot \mathbf{B}/4} \int d^N \boldsymbol{\phi}$$
$$\times \exp\left[-\tfrac{1}{4}\beta \boldsymbol{\phi} \cdot \mathbf{L} \cdot \boldsymbol{\phi} + \tfrac{1}{2}\beta \mathbf{B} \cdot \boldsymbol{\phi} + \sum_i \log \cosh \left(\beta \sum_j L_{ij}\phi_j\right) \right]. \tag{K.7}$$

This demonstrates that the partition function of the Ising model is identical with the partition function of a model whose order parameter ϕ takes any real value rather than just the numbers ± 1. As the spherical model of §3.3 demonstrated, models whose order parameters can take any value can be easier to solve than the Ising model. Thus expressing Z_{Ising} in the form (K.7) is useful. However, the effective Hamiltonian of (K.7) is by no means simple, and some drastic approximations are necessary to make it tractable.

It is natural to ask at this stage what the physical interpretation of ϕ might be. At root ϕ is a mathematical abstraction, but if we imagine it to be a stochastic variable in its own right whose partition function is given by (K.7), then its expectation values are closely related to the expectation values of the order parameter s_i of the underlying Ising model. In fact, differentiating the logarithms of (K.4) and (K.7) with respect to B_i we have

$$\langle s_i \rangle = \frac{1}{\beta} \frac{\partial \log Z}{\partial B_i} = \frac{1}{2} \frac{\int d^N \boldsymbol{\phi}\, (\phi_i - L_{ij}^{-1} B_j) e^{-\mathcal{H}[\boldsymbol{\phi}]}}{\int d^N \boldsymbol{\phi}\, e^{-\mathcal{H}[\boldsymbol{\phi}]}}, \tag{K.8}$$

where the effective Hamiltonian $\mathcal{H}$ is

$$\mathcal{H}(\boldsymbol{\phi}) \equiv \tfrac{1}{4}\beta \boldsymbol{\phi} \cdot \mathbf{L} \cdot \boldsymbol{\phi} - \tfrac{1}{2}\beta \mathbf{B} \cdot \boldsymbol{\phi} - \sum_i \log \cosh y_i. \tag{K.9}$$

Thus the thermal average $\langle \tfrac{1}{2}(\phi_i - L_{ij}^{-1} B_j)\rangle$ is identical with the thermal average of s_i. Similarly

$$\langle s_i s_j \rangle = -\frac{1}{2\beta} L_{ij}^{-1} + \tfrac{1}{4}\langle \phi_i \phi_j \rangle. \tag{K.10}$$

---

[1] In this appendix each component of the vector $\boldsymbol{\phi}$ is the value of the *scalar* order parameter ϕ at a given site i, rather than a component of an D-dimensional vector as in much of the rest of the book. Similarly the field $\mathbf{B}$ is scalar-valued on each site.

For large α, $\mathbf{L}^{-1} \simeq \frac{1}{\alpha}\mathbf{I} + \frac{1}{\alpha^2}\mathbf{J}$ and the first term on the right of (K.10) is negligible, especially for $i \neq j$. Consequently, in this circumstance the moments of products of $\frac{1}{2}\phi_i$ at distant sites are identical with the corresponding moments of products of the s_i. As the critical point is approached, both $\sum_j \langle s_i s_j \rangle$ and $\sum_j \langle \phi_i \phi_j \rangle$ diverge while the sum over j of the remaining term in (K.10) does not. Thus the critical exponent γ associated with ϕ is the same as that for the physical order parameter s_i independently of the value of α. The identity of the other critical exponents can also be demonstrated.

We next show that the effective Hamiltonian (K.9) can be approximated by the effective Hamiltonian of the Landau–Ginzburg model. Recall that by (K.2) the diagonal elements of L_{ij} equal some positive number α while those corresponding to neighbouring sites equal a negative number, $-\epsilon$ say. In the case $\alpha = 2d\epsilon$, the matrix $\mathbf{L}$ is proportional to the crudest viable discrete representation on a lattice of the Laplacian operator. More generally

$$\sum_j L_{ij}\phi_j \simeq -\epsilon a^2 \nabla^2 \phi|_{\mathbf{x}_i} + (\alpha - 2d\epsilon)\phi_{\mathbf{x}_i}, \qquad (\text{K.11})$$

where a is the lattice spacing. Thus if we replace the set of numbers ϕ_i by a continuous function $\phi(\mathbf{x})$ and replace each $\sum_i$ by integrals $a^{-d} \int d^d\mathbf{x}$, (K.9) becomes

$$\mathcal{H}[\phi] = a^{-d} \int d^d\mathbf{x} \{ \tfrac{1}{4}\beta[-\epsilon a^2 \phi \nabla^2 \phi + (\alpha - 2d\epsilon)\phi^2] - \tfrac{1}{2}\beta B\phi - \log\cosh[y(\mathbf{x})] \}. \qquad (\text{K.12})$$

where $\phi = \phi(\mathbf{x})$, $B = B(\mathbf{x})$, and from (K.6) and (K.11)

$$y(\mathbf{x}) = \beta[-\epsilon a^2 \nabla^2 \phi|_{\mathbf{x}_i} + (\alpha - 2d\epsilon)\phi_{\mathbf{x}_i}]. \qquad (\text{K.13})$$

Integrating by parts the term in (K.12) containing ∇^2 and discarding the surface term, which involves only spins at the edge of the lattice, we have

$$\mathcal{H}[\phi] = a^{-d} \int d^d\mathbf{x} \{ \tfrac{1}{4}\beta[\epsilon a^2 (\nabla\phi)^2 + (\alpha - 2d\epsilon)\phi^2] - \tfrac{1}{2}\beta B\phi - \log\cosh y \}. \qquad (\text{K.14})$$

The square brackets now contain two of the three terms in the Landau–Ginzburg effective Hamiltonian $\mathcal{H}_{LG}$, (7.10). It remains only to persuade the log cosh term to yield the third term of $\mathcal{H}_{LG}$.

The effective Hamiltonian of (K.14) is a complicated functional of the field ϕ and its gradient. To get some understanding of the physical implications of this functional it is expedient to consider its behaviour in the neighbourhood of the field for which it is smallest—this is the field which makes the largest single contribution to the partition function. It is natural to assume that this field has vanishing gradient so that

$$\begin{aligned} y &\simeq \beta(\alpha - 2d\epsilon)\phi \\ &\equiv \phi/b, \end{aligned} \qquad (\text{K.15})$$

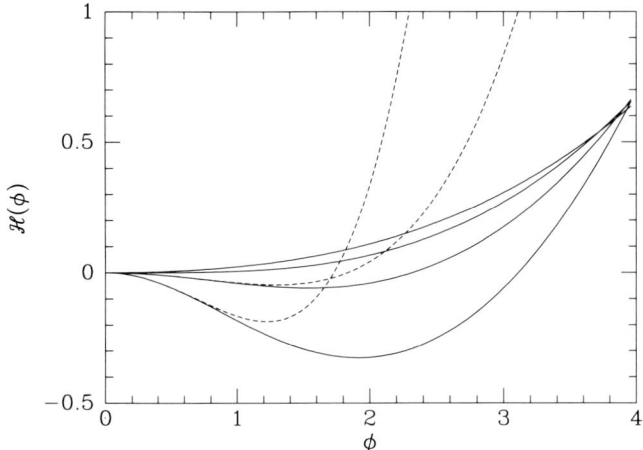

Figure K.1 The full curves show the function $\mathcal{H}(\phi)$ of equation (K.17) for various values of the parameter b defined by (K.16); from bottom to top $b = 1$, 1.5, 2, 2.5. The dashed curves show the Taylor series approximation (K.19) to $\mathcal{H}$ with $b = 1$ and 1.5.

where we have defined

$$b \equiv \frac{1}{\beta(\alpha - 2d\epsilon)}. \quad \text{(K.16)}$$

With these simplifications (K.14) becomes

$$\mathcal{H}(\phi(y))/V = \tfrac{1}{4}by^2 - \log\cosh y, \quad \text{(K.17)}$$

where V is the system's volume.

Now, $\log\cosh y$ is never negative and monotonically increases with y. So the last term in (K.17) always tends to diminish $\mathcal{H}$. On the other hand, for $y \gg 1$, $\log\cosh y$ grows only as y and hence as ϕ (see equation (K.13)), with the result that the term in $\mathcal{H}$ proportional to ϕ^2 eventually overwhelms it. This behaviour is shown by Figure K.1, in which $\mathcal{H}$ is plotted for several values of b. There is a small range of values of b near zero for which $\mathcal{H}(\phi)$ is negative for small ϕ before turning up towards $+\infty$. This is the range of values of b for which Landau theory predicts that $\langle\phi\rangle$ becomes non-zero. Negative values of b are of no physical interest since they cause $\mathcal{H}$ to diminish without limit for large ϕ.

The behaviour of $\mathcal{H}(\phi)$ shown in Figure K.1 can be understood by expanding $\mathcal{H}$ in a Taylor series in y. Since

$$\begin{aligned} 2\log\cosh y &= \log(1 + \sinh^2 y) \\ &\simeq \sinh^2 y - \tfrac{1}{2}\sinh^4 y \\ &= y^2 + (\tfrac{1}{3} - \tfrac{1}{2})y^4 + \cdots, \end{aligned} \quad \text{(K.18)}$$

we have
$$\mathcal{H}(\phi(y))/V = (\tfrac{1}{4}b - \tfrac{1}{2})y^2 + \tfrac{1}{12}y^4 + \cdots. \tag{K.19}$$

For $b < 2$ the coefficient of the quadratic term in this series is negative, which causes $\mathcal{H}$ to decrease before the quartic term drives it up towards positive values. Figure K.1 shows that the qualitative behaviour of $\mathcal{H}(\phi)$ near the minimum is reproduced by these two terms in the Taylor series alone, although as ϕ increases further the truncated Taylor series rapidly diverges from the exact expression; for large ϕ the first behaves as ϕ^4, while the second grows as ϕ^2. We now argue that the partition function is dominated by fields that lie sufficiently near the minimum in $\mathcal{H}$ for the difference between $\mathcal{H}$ and the Taylor series (K.19) to be unimportant. The validity of this approximation is unclear. But suppose we make this step into the dark. Then, when we use (K.15) to eliminate y from (K.19) in favour of ϕ and restore to $\mathcal{H}$ the term in $\nabla\phi$ from (K.14), we do, indeed, obtain the Landau–Ginzburg effective Hamiltonian. This then is the basis of the oft-made claim that near the critical temperature the Ising model's effective Hamiltonian can be expressed as $\mathcal{H}_{LG}$ plus additional terms. As we explain in Chapter 14, 'universality' assures us that these additional terms are of no consequence if all one seeks is critical exponents.

Appendix L: Functional differentiation and integration

The idea of a function of a finite number of variables is a familiar one. A function is a rule for assigning a number to each set of values which its arguments can take. For example, the functions

$$f(x) = x, \quad g(x_1, x_2) = x_1 + x_2, \quad h(x_1, \ldots, x_n) = \sum_{i=1}^{n} x_i \tag{L.1}$$

are functions of one, two, and n variables respectively. A **functional** is a generalization of this idea. Its argument is a *function*, and it assigns a number to each function for which it is defined. An example of a functional is

$$I[f(x)] = \int_{-1}^{+1} f(x)\,\mathrm{d}x. \tag{L.2}$$

The argument of a functional appears in square brackets; in this case the argument of the functional I is the function $f(x)$. Given a function $f(x)$, one integrates it from -1 to 1. The answer (if it exists) is the value of $I[f(x)]$ for that function. The functional $I[f(x)]$ assigns a number to each function $f(x)$, and is defined for all functions $f(x)$ for which the integral (L.2) exists. Note that $I[f(x)]$ is not in any sense a function of x. If $f(x) = x + 1$ and

L.1 Functional differentiation

$g(x) = 2x$, then $g(f(x))$ would mean $2x + 2$, which is a function of x. But $I[f(x)]$ is not a function of a function; it is a *functional* of a function. To avoid confusion we shall write $I[f]$ instead of $I[f(x)]$, unless we need to indicate the explicit form of $f(x)$. The square brackets make it clear that I is a functional and f a function.

Functionals are common in physics. The total electromagnetic field energy in a given volume is a functional of the electric and magnetic fields in that volume; the free energy of a magnetic material in a position-dependent magnetic field is a functional of the field, and a function of the temperature; the gravitational field at a given point in space is a function of that point and a functional of the matter distribution around it. Endless other examples are possible.

A functional can be thought of as the limiting case of a function as the number of its arguments becomes infinite. For example, the functional $I[f]$ above is the limit as $n \to \infty$ of the function $h(x_1, \ldots, x_n)$ of equation (L.1), if x_i is identified with $\frac{2}{n} f(-1 + 2i/n)$.

L.1 Functional differentiation

Many of the functions which appear in physics can be differentiated with respect to their arguments. It would be useful to extend the idea of differentiation to functionals as well. For example, the energy of a soap film is a functional of the shape of the film. The actual shape the film takes up is the one which minimizes the energy. If we could differentiate the energy functional with respect to the shape of the film, we should be able to find the shape that minimizes the energy by equating the derivative to zero.

The derivative of a functional $F[f]$ ought to tell us how much $F[f]$ changes for a given small change in f. There are many ways in which we could change $f(x)$, so let us consider what happens to F if we change $f(x)$ just in the neighbourhood of a particular value x_0 of its argument x. Let $\Delta(a; x)$ be some set of functions of x which satisfy

$$\int_{-\infty}^{+\infty} \Delta(a; x) \, dx = 1, \quad \text{and} \quad \lim_{a \to 0} \Delta(a; x) = 0, \ x \neq 0. \qquad (\text{L.3})$$

In other words, as $a \to 0$ the function $\Delta(a; x)$ becomes more and more like the Dirac delta function $\delta(x)$. The **functional derivative** of $F[f]$ can now be defined:

$$\frac{\delta F[f]}{\delta f(x_0)} = \lim_{a \to 0} \lim_{\epsilon \to 0} \frac{F[f(x) + \epsilon \Delta(a; x - x_0)] - F[f(x)]}{\epsilon}. \qquad (\text{L.4})$$

The notation on the left of the last equation means 'the functional derivative of F with respect to $f(x_0)$'. The expression on the right measures the rate of change of F as f is varied in the neighbourhood of x_0, and takes the limit of this rate of change as the change in f becomes completely localized and infinitesimal. The functional derivative is both a functional of f and a function of x_0. In practice the $a \to 0$ limit can be bypassed by simply using the Dirac δ-function straight away:

$$\frac{\delta F[f]}{\delta f(x_0)} = \lim_{\epsilon \to 0} \frac{F[f(x) + \epsilon\delta(x - x_0)] - F[f(x)]}{\epsilon}. \tag{L.5}$$

Only in the (for physicists) rare situations in which the meaning of $F[f(x) + \epsilon\delta(x - x_0)]$ is not immediately clear need the full limiting procedure be used.

The reader can check the following examples ($I[f]$ is the functional defined by (L.2)):

1. $\dfrac{\delta I[f]}{\delta f(x_0)} = \begin{cases} 1 & \text{if } -1 < x_0 < 1 \\ 0 & \text{otherwise.} \end{cases}$

2. $\dfrac{\delta I[f^2]}{\delta f(x_0)} = \begin{cases} 2f(x_0) & \text{if } -1 < x_0 < 1 \\ 0 & \text{otherwise.} \end{cases}$

3. $\dfrac{\delta f(2)}{\delta f(x_0)} = \delta(x_0 - 2).$

4. $\dfrac{\delta^2 I[f^3]}{\delta f(x_0)\delta f(x_1)} = \begin{cases} 6f(x_1)\delta(x_1 - x_0) & \text{if } -1 < \{x_0, x_1\} < 1 \\ 0 & \text{otherwise.} \end{cases}$

A functional derivative of great importance in this book is

$$\frac{\delta}{\delta J(\mathbf{x})} e^{\int d^d\mathbf{z}\, J(\mathbf{z})\phi(\mathbf{z})} = \phi(\mathbf{x}) e^{\int d^d\mathbf{z}\, J(\mathbf{z})\phi(\mathbf{z})}. \tag{L.6}$$

In Chapter 8 correlation functions of the order parameter ϕ are generated by applying this derivative to partition functions (e.g., equations (8.1) and (8.2)) which contain the above exponential as a factor.

Note that the dimensions of the functional derivative are *not* those of the functional divided by the function, as the notation (L.4) would suggest, but those of the functional divided by the product of the function and its argument. This arises because ϵ has the dimensions of f times x, and not those of f. Dimensional analysis is very useful in spotting mistakes, and this observation should help to prevent confusion.

Now we can write an expression for the change in F resulting from an arbitrary (but small) change in $f(x)$, not just one which is localized. If $f(x)$ changes to $f(x) + \eta(x)$, then the change in F is

$$\delta F = \int dx_0\, \frac{\delta F}{\delta f(x_0)} \eta(x_0) + O(\eta^2). \tag{L.7}$$

L.1 Functional differentiation

Compare this with the expression for the change in a function of several variables, familiar from ordinary calculus:

$$\delta h(x_1, \ldots, x_n) = \sum_{i=1}^{n} \frac{\partial h}{\partial x_i} \delta x_i + O(x_i^2). \tag{L.8}$$

For changes $\eta(x)$ which are not infinitesimal a Taylor series expansion, also analogous to that in ordinary calculus, can (sometimes) be written down:

$$F[f + \eta] = F[f] + \int dx_0 \frac{\delta F}{\delta f(x_0)} \eta(x_0) \tag{L.9}$$

$$+ \frac{1}{2!} \int dx_0 dx_1 \frac{\delta^2 F}{\delta f(x_0) \delta f(x_1)} \eta(x_0) \eta(x_1) + \cdots.$$

A very important rule, which we shall now derive, is the **chain rule** for functional differentiation. Suppose that F is a functional of a function g, and that this function g is itself a functional of a function f. (There is no reason why a function should not also be a functional; indeed, we have seen that the functional derivative $\delta F/\delta f(x_0)$ is both a functional of f and a function of x_0.) Now suppose that $f(x)$ changes, to $f(x) + \eta(x)$, where $\eta(x)$ is small. Then $g(y)$ will also change, let us say to $g(y) + \chi(y)$, and F will change to $F + \delta F$. From equation (L.7) these changes are related by

$$\delta F = \int dy \frac{\delta F}{\delta g(y)} \chi(y)$$
$$\chi(y) = \int dz \frac{\delta g(y)}{\delta f(z)} \eta(z). \tag{L.10}$$

Combining these two equations and reversing the order of the y- and z-integrations gives us

$$\delta F = \int dz \left[\int dy \frac{\delta F}{\delta g(y)} \frac{\delta g(y)}{\delta f(z)} \right] \eta(z). \tag{L.11}$$

Comparing (L.11) with (L.7), it follows that the functional derivative of F with respect to f is

$$\frac{\delta F}{\delta f(z)} = \int dy \frac{\delta F}{\delta g(y)} \frac{\delta g(y)}{\delta f(z)}. \tag{L.12}$$

This is the chain rule for functional differentiation.

L.2 The inverse of functional differentiation

One would expect 'functional integration' to be the inverse of functional differentiation, but it isn't. Before we consider the process which is called functional integration, let's try and see what the inverse of functional differentiation would be. First consider a function of two variables, $g(x_1, x_2)$. We can write

$$\mathrm{d}g = \frac{\partial g}{\partial x_1} \mathrm{d}x_1 + \frac{\partial g}{\partial x_2} \mathrm{d}x_2. \tag{L.13}$$

This may be integrated to give

$$\int_{x_1^0, x_2^0}^{x_1, x_2} \mathrm{d}g = g(x_1, x_2) - g(x_1^0, x_2^0)$$

$$= \int_{x_1^0, x_2^0}^{x_1, x_2} \frac{\partial g(x_1', x_2')}{\partial x_1'} \mathrm{d}x_1' + \int_{x_1^0, x_2^0}^{x_1, x_2} \frac{\partial g(x_1', x_2')}{\partial x_2'} \mathrm{d}x_2'. \tag{L.14}$$

The integrals are taken along some path in (x_1, x_2)-space, starting at (x_1^0, x_2^0) and finishing at (x_1, x_2). The actual path chosen is unimportant (as long as it is the same for each integral). In this way we recover the original function g to within a constant, depending on the choice of (x_1^0, x_2^0). Similarly for a function $h(\{x_i\})$ of n variables we can write

$$h(\{x_i\}) - h(\{x_i^0\}) = \sum_{j=1}^{n} \int_{\{x_i^0\}}^{\{x_i\}} \frac{\partial h}{\partial x_j'} \mathrm{d}x_j'. \tag{L.15}$$

This suggests the following as the inverse process to functional differentiation:

$$F[f] - F[f^{(0)}] = \int \mathrm{d}x \int_{f^{(0)}(x)}^{f(x)} \frac{\delta F[\zeta]}{\delta \zeta(x)} \mathrm{d}\zeta(x). \tag{L.16}$$

The '$\mathrm{d}\zeta(x)$' on the right-hand side of this equation is not an integral over all functions $\zeta(x)$ or anything like that; it is an ordinary one-dimensional integral over the value of the function ζ at the point x, from its (arbitrary) starting-value of $f^{(0)}(x)$ to its final value $f(x)$. The integrals are all taken over some path in function space from $f^{(0)}$ to f. (The function $\zeta(x)$ is analogous to the variables x_i' of (L.15). The natural notation, $f'(x)$, would be confusing.) With some thought this should become clear. Let's try an example. We saw above that the derivative of $F[f] = I[f^2]$ (where $I[f]$ is defined by (L.2)) is

$$\frac{\delta I[f^2]}{\delta f(x)} = \begin{cases} 2f(x) & \text{if } -1 < x < 1 \\ 0 & \text{otherwise,} \end{cases}$$

L.3 Functional integration

so let's put this into (L.16) and see if we get $I[f^2]$ back. Equation (L.16) becomes

$$F[f] - F[f^{(0)}] = \int_{-1}^{1} \mathrm{d}x \int_{f^{(0)}(x)}^{f(x)} 2\zeta(x)\mathrm{d}\zeta(x). \qquad \text{(L.17)}$$

The integral over x is only over the range $-1 < x < 1$ as the functional derivative vanishes outside this range. Inside this range the integral is simple, because the functional derivative at the point x depends only on the value of ζ at this point. It's as though the derivatives $(\partial h/\partial x_j)$ of the function $h(\{x_i\})$ above each depended only on x_j. So we don't have to worry about the path in function space that we take. The integral

$$\int_{f^{(0)}(x)}^{f(x)} 2\zeta(x)\mathrm{d}\zeta(x) \qquad \text{(L.18)}$$

appearing in (L.16) is $f^2(x) - \left(f^{(0)}(x)\right)^2$. (To make this clearer, put $u \equiv \zeta(x)$; this makes the integral look more familiar. Then (L.18) becomes

$$\int_{u=f^{(0)}(x)}^{u=f(x)} 2u\,\mathrm{d}u = \left[u^2\right]_{u=f^{(0)}(x)}^{u=f(x)} = f^2(x) - \left(f^{(0)}(x)\right)^2, \qquad \text{(L.19)}$$

as claimed.) Equation (L.17) then gives

$$F[f] - F[f^{(0)}] = \int_{x=-1}^{x=1} \mathrm{d}x \left(f^2(x) - \left(f^{(0)}(x)\right)^2\right), \qquad \text{(L.20)}$$

which is $I[f^2]$ to within an additive constant. The process defined in (L.16) is indeed the inverse of functional differentiation.

L.3 Functional integration

The process described in the last section is not the process which is known as functional integration. (Indeed, it doesn't seem to be used in physics at all.) The name 'functional integration' is reserved for the following. We consider a set S of functions (the elements of which may be required to satisfy boundary conditions or subject to some other restriction), work out the value of some functional F for every function in the set, and 'add up all the answers'. The result is called the **functional integral** of F over the set S. This idea is analogous to ordinary integration; the ordinary integral $\int f(x,y)\,\mathrm{d}x\mathrm{d}y$ is the sum of the values taken by f as its argument moves over a grid of points, times the size $\mathrm{d}x\mathrm{d}y$ of each cell of the grid; a functional integral is the sum of the values taken by a functional as its argument moves over some set of functions, times some measure of the 'volume' of function space associated with each function.

Why should we want to do this? One reason would be to calculate the partition function of a system whose state is described by some function $\phi(\mathbf{x})$, rather than by a finite set of variables. In calculating the partition function for such a system we would sum $e^{-\beta H[\phi]}$ over all functions $\phi(\mathbf{x})$: i.e., the partition function is the functional integral of $e^{-\beta H[\phi]}$ with respect to $\phi(\mathbf{x})$. Something similar happens in quantum mechanics, where the amplitude to get from one state to another is the sum over all classical paths $\mathcal{P}$ between the initial and final states weighted by a factor $e^{iS[\mathcal{P}]/\hbar}$, where $S[\mathcal{P}]$ is the action for the path $\mathcal{P}$. The action $S[\mathcal{P}]$ is a functional of the path $\mathcal{P}$, so this is another example of a functional integral. The reader will be able to provide other examples.

Now it is easy in principle to integrate over a single real variable, or a finite number of such variables. But integrating over a set of functions is another matter altogether. In general such a set will be infinite-dimensional, in that an infinite number of real numbers are needed to parameterize its elements. Then this infinite number of parameters must be integrated over, which will involve an infinite number of integrals. This is not obviously well defined, so some care is needed.

The way to go about it is to replace the functional we are integrating by a function of many variables, and then to do a multi-dimensional integral over these variables. By letting the number of variables tend to infinity, we should be able to get close to a 'true' functional integral. As an example, let's integrate the functional

$$A[f] = \exp\left[-k \int_{-L/2}^{L/2} \mathrm{d}x \, f^2(x)\right] \tag{L.21}$$

over all possible forms of the function $f(x)$ on the range $-L/2 < x < L/2$. In other words, let's evaluate the functional integral

$$K \equiv \int \mathcal{D}f \, \exp\left[-k \int_{-L/2}^{L/2} \mathrm{d}x \, f^2(x)\right]. \tag{L.22}$$

The notation '$\int \mathcal{D}f$' means 'integrate over all possible values of $f(x)$'. To give some meaning to this expression, we replace the functional $A[f]$ with the function

$$a(\{f_i\}) = \exp\left[-\frac{kL}{n} \sum_{i=1}^{n} f_i^2\right]. \tag{L.23}$$

In the limit $n \to \infty$, $a(\{f_i\})$ becomes the functional $A[f(x)]$ if we identify f_i with $f((\frac{i}{n} - \frac{1}{2})L)$. Then we replace the 'sum over all functions' $\int \mathcal{D}f$ with an integral over all the f_i. The resulting approximation to K, which we call K_n, is then

$$K_n = \int \mathrm{d}f_1 \ldots \mathrm{d}f_n \, \exp\left[-\frac{kL}{n} \sum_{i=1}^{n} f_i^2\right]. \tag{L.24}$$

L.3 Functional integration

We can work out K_n exactly, since the exponential factorizes into a product of exponentials, each of which depends on only one f_i. The result is

$$K_n = \prod_{i=1}^{n} \int \mathrm{d}f_i \, \exp\left[-\frac{kL}{n} f_i^2\right]$$
$$= \left(\frac{\pi n}{kL}\right)^{n/2}. \tag{L.25}$$

The functional integral K is supposed to be $\lim_{n \to \infty} K_n$. But K_n grows distressingly rapidly as n increases and this limit does not exist. We could define a normalizing factor for K_n to make the limit exist, but we would need to use a different factor for each value of k, since k appears raised to the power $-(n/2)$. Unlike ordinary integration, where the (Riemann) integral can be written as the limit of a sum normalized in a way which is independent of the particular function being integrated (for a wide class of functions), functional integration requires a different normalization for each functional if the result is to be finite.

Of what possible use can functional integration be, then? Let's think in physical terms. The smallest length scales which have been explored are of the order of 10^{-19} m. This is very short, but still a lot bigger than nothing at all. The idea of a continuous function of position can only ever be an approximation to what is known about the physical world. In condensed-matter physics, for example, things are made up of atoms. So an unrestricted functional integral can never arise in condensed-matter physics. If we were integrating over possible forms of the density $\rho(\mathbf{x})$ of a gas, we wouldn't mean all functions $\rho(\mathbf{x})$. We'd somehow have to exclude those functions which vary on very short distances, and restrict the set of functions over which we integrate. The particular problem should tell us what sort of restriction is required. Although integrating over all functions does not have any obvious physical meaning, and is not (in general) mathematically defined, the concept of integrating over a restricted set S of functions, compatible with the physics of the particular problem, is well-defined and useful.

L.3.1 The Gaussian functional integral

Virtually all the tractable functional integrals in physics are Gaussian or can be written in terms of Gaussian integrals. So let's consider the special case of Gaussian functional integrals. In particular, consider

$$G[J] \equiv \int_S \mathcal{D}\phi \, \exp\left[-\int \mathrm{d}^d\mathbf{x}\,\mathrm{d}^d\mathbf{y}\, \phi(\mathbf{x}) M(\mathbf{x},\mathbf{y}) \phi(\mathbf{y}) + \int \mathrm{d}^d\mathbf{z}\, \phi(\mathbf{z}) J(\mathbf{z})\right]. \tag{L.26}$$

We have written G as a functional of a function $J(\mathbf{z})$ because, although G is a functional of both $J(\mathbf{x})$ and $M(\mathbf{x},\mathbf{y})$, in the context of this book it is the dependence on J which is important. The set S of functions $\phi(\mathbf{x})$ which we are integrating over in (L.26) is as yet unspecified.

The function $M(\mathbf{x},\mathbf{y})$ is analogous to a matrix $\mathbf{M}$ connecting the two 'vectors' $\phi(\mathbf{x})$ and $\phi(\mathbf{y})$. In the example integral (L.22) done above (which was a Gaussian integral) this 'matrix' was $k\delta(\mathbf{x}-\mathbf{y})$, but in general $\mathbf{M}$ need not be diagonal. It can be chosen to be symmetric; that is, $M(\mathbf{x},\mathbf{y}) = M(\mathbf{y},\mathbf{x})$. We suppose in addition that its elements are real. In this case it will have a complete set of orthonormal eigenfunctions (eigenvectors) ψ_n and corresponding real eigenvalues λ_n:

$$\int d^d\mathbf{y}\, M(\mathbf{x},\mathbf{y})\psi_n(\mathbf{y}) = \lambda_n \psi_n(\mathbf{x}), \quad \text{or simply} \quad \mathbf{M}\psi_n = \lambda_n \psi_n. \tag{L.27}$$

We'll assume that these eigenfunctions form a discrete set; this will be true if the integrals in the exponent of (L.26) are only over a finite volume of space. If the integral (L.26) is to stand any chance of being well-defined, none of the eigenvalues λ_n can be negative, since otherwise the exponent in the functional integral would grow without limit with the magnitude of ψ_n.

Now let us expand $\phi(\mathbf{x})$ in terms of the $\psi_n(\mathbf{x})$:

$$\phi = \sum_n \phi_n \psi_n, \quad \text{where} \quad \phi_n = \int d^d\mathbf{x}\, \psi_n^*(\mathbf{x})\phi(\mathbf{x}) \equiv (\psi_n, \phi). \tag{L.28}$$

The function $J(\mathbf{x})$ in (L.26) can also be expanded in this way:

$$J(\mathbf{x}) = \sum_n J_n \psi_n(\mathbf{x}). \tag{L.29}$$

Now the exponent of (L.26) can be written

$$\int d^d\mathbf{x} d^d\mathbf{y}\, \phi(\mathbf{y}) M(\mathbf{x},\mathbf{y})\phi(\mathbf{x}) - \int d^d\mathbf{x}\, J(\mathbf{x})\phi(\mathbf{x})$$

$$= \sum_{n,m} (\phi_n \psi_n, \phi_m \mathbf{M}\psi_m) - \sum_{n,m} (J_n \psi_n, \phi_m \psi_m)$$

$$= \sum_{n,m} \phi_n \lambda_m \phi_m (\psi_n, \psi_m) - \sum_{n,m} J_n \phi_m (\psi_n, \psi_m)$$

$$= \sum_n (\lambda_n \phi_n^2 - J_n \phi_n). \tag{L.30}$$

So the integral (L.26) becomes

$$G[J] = \int \mathcal{D}\phi\, \exp\!\left[-\sum_n (\lambda_n \phi_n^2 - J_n \phi_n)\right]. \tag{L.31}$$

At this point we shall specify the set S of functions which we are integrating over. We define $G_N[J]$ to be the functional integral over the set of functions $\phi(\mathbf{x})$ which can be expanded as

$$\phi(\mathbf{x}) = \sum_{n=1}^{N} \phi_n \psi_n(\mathbf{x}), \tag{L.32}$$

L.3 Functional integration

just like (L.28), with the ϕ_n for $n > N$ zero; we will integrate over this set by integrating over the ϕ_n ($n \leq N$):

$$G_N[J] = \int d^d\phi_1 \ldots d^d\phi_N \, \exp\left[-\sum_{n=1}^{N}(\lambda_n \phi_n^2 - J_n\phi_n)\right]. \qquad (L.33)$$

This is our *definition* of the functional integral. Arranged in order of increasing eigenvalue, the functions $\psi_n(\mathbf{x})$ vary more and more rapidly with position as n increases. In considering only functions for which the ϕ_n ($n > N$) vanish, we are excluding functions which vary more rapidly with position than does $\psi_N(\mathbf{x})$. Seeing how $G_N[J]$ changes as N varies shows us how our results depend on the exclusion of the rapidly varying functions.

Let us calculate $G_N[J]$. The exponential in (L.33) factorizes into a product of one-dimensional Gaussian integrals and we obtain

$$\begin{aligned}G_N[J] &= \frac{\pi^{N/2}}{\left(\prod_{n=1}^{N}\lambda_n\right)^{1/2}} \exp\left[\sum_{n=1}^{N}\frac{J_n^2}{4\lambda_n}\right] \\ &= \frac{\pi^{N/2}}{\sqrt{\det' \mathbf{M}}} \exp\left[\sum_{n=1}^{N}\frac{J_n^2}{4\lambda_n}\right] \qquad (L.34) \\ &= G_N[0] \exp\left[\sum_{n=1}^{N}\frac{J_n^2}{4\lambda_n}\right].\end{aligned}$$

Continuing the analogy between functions and matrices which we began above, we have written $\det' \mathbf{M}$ for the product of the eigenvalues of the function $M(\mathbf{x}, \mathbf{y})$. The prime reminds us that we only multiply the first N eigenvalues. An interesting (and important) feature of (L.34) is that the J-dependence has factorized out. If we restrict ourselves to 'physical' forms of J—that is, functions $J(\mathbf{x})$ for which the J_n tend to zero as $n \to \infty$—the ratio $(G_N[J]/G_N[0])$ is perfectly well defined, even when $N \to \infty$. However, $G_N[J]$ itself will not be well defined in this limit, as the limit of the determinant does not exist.

So we have been able to 'do' the Gaussian integral (L.26). We have found a way of controlling the set of functions to that are integrated over, and we have an expression for the integral which tells us how the result depends on the set we choose. We have also succeeded in factoring out the J-dependence of the answer. The fact that we have succeeded in doing so much with the Gaussian integral perhaps suggests that the physics described by such integrals is not very interesting. Sadly, this is true; the statistical mechanics of a system whose partition function is a Gaussian functional integral is that of a set of non-interacting subsystems; similarly, Gaussian functional integrals in quantum field theory signal a theory of non-interacting particles. There can be some achievement in showing that a particular system which looks complicated can in fact be described in this way, but that is usually the end of interest in that system. Other *non-Gaussian* functional integrals are physically more interesting, but no-one knows how to do them. The best that can be done is to write them in terms of an infinite series of Gaussian functional integrals, and this is the reason for the importance of such integrals. This subject is dealt with in Chapter 8, where it is shown how the partition function of the Landau–Ginzburg model, which is a non-Gaussian functional integral, can be expressed in terms of the partition function of the Gaussian model, which is a Gaussian functional integral. We evaluate this integral below.

L.3.2 The partition function of the continuum Gaussian model

The partition function of the continuum Gaussian model may be written as the functional integral

$$Z_G[J] \equiv \int \mathcal{D}\phi\, e^{-H_G[J]}$$
$$= \int \mathcal{D}\phi\, \exp\left[-\int d^d x\, \left(\tfrac{1}{2}\alpha^2|\boldsymbol{\nabla}\phi|^2 + \tfrac{1}{2}\mu^2\phi^2 - J\phi\right)\right]; \qquad (L.35)$$

see §8.1. The set of functions $\phi(\mathbf{x})$ over which we are to integrate is not explicitly specified. The integral of (L.35) is of the form (L.26), as can more clearly be seen if the exponent is rewritten as

$$H_G = -\int d^d\mathbf{x}\, d^d\mathbf{y}\, \phi(\mathbf{y})\left[\frac{\delta(\mathbf{x}-\mathbf{y})}{2}\left(\alpha^2\nabla_\mathbf{x}^2 - \mu^2\right)\right]\phi(\mathbf{x}) - \int d^d\mathbf{x}\, J(\mathbf{x})\phi(\mathbf{x}), \qquad (L.36)$$

where the first term has been obtained by integrating by parts. To apply the method of the last subsection we first need to find the eigenfunctions of the operator in the square brackets. Then we shall replace the functional integral $Z_G[J]$ by a sequence of ordinary multiple integrals $Z_G[J; N]$ (analogous to the G_N of equation (L.33)) in which we integrate over all functions $\phi(\mathbf{x})$ that can be written as a linear combination of the first N eigenfunctions of that operator.

So let us find the eigenfunctions and eigenvalues of the operator

$$\mathbf{M}(\mathbf{x}, \mathbf{y}) = \delta(\mathbf{x}-\mathbf{y})\left(\alpha^2\nabla_\mathbf{x}^2 - \mu^2\right). \qquad (L.37)$$

Due to the simplicity of $\mathbf{M}$ this is straightforward. The eigenfunctions $\psi_\mathbf{k}$ and corresponding eigenvalues $\lambda_\mathbf{k}$ are

$$\psi_\mathbf{k}(\mathbf{x}) = e^{i\mathbf{k}\cdot\mathbf{x}}, \qquad \lambda_\mathbf{k} = -(\alpha^2 k^2 + \mu^2). \qquad (L.38)$$

To ensure that the eigenfunctions form a discrete set let us put the system in a cube of volume $V = L^d$, with periodic boundary conditions. Then the components of $\mathbf{k}$ are restricted to be integer multiples of $2\pi/L$.

L.3 Functional integration

Now that we know the eigenfunctions of $\mathbf{M}$, we can construct the integrals $Z_G[J; N]$. In fact it is more convenient to calculate $Z_G[J; \Lambda]$, which is the integral over those $\phi(\mathbf{x})$ that can be written as a linear combination of eigenfunctions $\psi_{\mathbf{k}}$ for which $|\mathbf{k}|$ is less than some particular value Λ, but the idea is the same. This restriction on $\phi(\mathbf{x})$ is precisely that which we imposed in §8.1 (equation (8.4)) on physical grounds. There we argued that only functions which did not vary too rapidly should be integrated over, and so we decided to consider only those functions having no Fourier components with wavevector greater than some cutoff Λ. Now we see the mathematical reason for doing this. The functional integral is not defined unless we restrict the set of functions that we are integrating over. Also, that particular form of cutoff is the one which arises naturally in working out the functional integral. Let us then write

$$\phi(\mathbf{x}) = \frac{1}{L^d} \sum_{\mathbf{k};\, k<\Lambda} \tilde{\phi}(\mathbf{k}) e^{i\mathbf{k}\cdot\mathbf{x}}, \qquad (\text{L.39})$$

where the normalization constant in front of the sum ensures that $\tilde{\phi}(\mathbf{k})$ is the Fourier transform of $\phi(\mathbf{x})$ in the limit $L \to \infty$. We also write

$$J(\mathbf{x}) = \frac{1}{L^d} \sum_{\mathbf{k};\, k<\Lambda} \tilde{J}(\mathbf{k}) e^{i\mathbf{k}\cdot\mathbf{x}}. \qquad (\text{L.40})$$

Substituting these expressions for $\tilde{\phi}$ and $\tilde{J}$ into the exponent in (L.35) gives

$$H_G = \frac{1}{L^d} \sum_{\mathbf{k};\, k<\Lambda} \tfrac{1}{2}(\alpha^2 k^2 + \mu^2)\tilde{\phi}(\mathbf{k})\tilde{\phi}(-\mathbf{k}) - \tilde{\phi}(\mathbf{k})\tilde{J}(-\mathbf{k}). \qquad (\text{L.41})$$

There is one complication, however. The quantities $\tilde{\phi}(\mathbf{k})$ and $\tilde{J}(\mathbf{k})$ are not all independent. If $\phi(\mathbf{x})$ and $J(\mathbf{x})$ are to be real, then their Fourier transforms must satisfy

$$\tilde{\phi}(-\mathbf{k}) = \tilde{\phi}^*(\mathbf{k}), \quad \text{and} \quad \tilde{J}(-\mathbf{k}) = \tilde{J}^*(\mathbf{k}). \qquad (\text{L.42})$$

So there are only half the number of independent variables that (L.41) would suggest. Let us denote the real and imaginary parts of $\tilde{\phi}$ and $\tilde{J}$ by subscripts 'R' and 'I'. Then we can write

$$H_G = \frac{1}{L^d} \sum_{\mathbf{k};\, k<\Lambda}{}' (\alpha^2 k^2 + \mu^2)\bigl(\tilde{\phi}_R^2(\mathbf{k}) + \tilde{\phi}_I^2(\mathbf{k})\bigr) + 2\bigl(\tilde{\phi}_R(\mathbf{k})\tilde{J}_R(\mathbf{k}) + \tilde{\phi}_I(\mathbf{k})\tilde{J}_I(\mathbf{k})\bigr).$$
$$(\text{L.43})$$

The prime on the $\sum$ indicates that the sum is over only half the possible values of $\mathbf{k}$—those for which $k_x \geq 0$, for example. Equation (L.43) shows

H_G in a form which makes the integration easy, since the $\widetilde{\phi}(\mathbf{k})$ for different $\mathbf{k}$ are decoupled from each other. The integral that we shall work out is

$$Z_G[J;\Lambda] = \prod_{\mathbf{k};\, k<\Lambda}' \int d\widetilde{\phi}_R(\mathbf{k})\, d\widetilde{\phi}_I(\mathbf{k})\, e^{-H_G}, \tag{L.44}$$

where H_G is the expression in (L.43). The prime on the $\prod$ indicates that again only half the possible values of $\mathbf{k}$ are being used. This integral is now explicitly the product of a large number of one-dimensional Gaussian integrals. Doing each of these integrals and multiplying the results together gives

$$Z_G[J;\Lambda] = \prod_{\mathbf{k};\, k<\Lambda}' \left(\frac{\pi L^d}{\alpha^2 k^2 + \mu^2} \exp\left[\frac{1}{L^d} \frac{\widetilde{J}_R^2(\mathbf{k}) + \widetilde{J}_I^2(\mathbf{k})}{\alpha^2 k^2 + \mu^2} \right] \right). \tag{L.45}$$

The factor of πL^d outside the exponential is unimportant, as the absolute normalization of the functional integral is not physically significant. Dropping these, we can rewrite $Z_G[J;\Lambda]$ as

$$Z_G[J;\Lambda] = \exp\left[-\sum_{\mathbf{k};\, k<\Lambda}' \log(\alpha^2 k^2 + \mu^2) \right] \times \exp\left[\frac{1}{L^d} \sum_{\mathbf{k};\, k<\Lambda}' \frac{\widetilde{J}_R^2(\mathbf{k}) + \widetilde{J}_I^2(\mathbf{k})}{\alpha^2 k^2 + \mu^2} \right]$$

$$= \exp\left[-\tfrac{1}{2} \sum_{\mathbf{k};\, k<\Lambda} \log(\alpha^2 k^2 + \mu^2) \right] \times \exp\left[\frac{1}{2L^d} \sum_{\mathbf{k};\, k<\Lambda} \frac{\widetilde{J}(\mathbf{k})\widetilde{J}(-\mathbf{k})}{\alpha^2 k^2 + \mu^2} \right], \tag{L.46}$$

where in the last line we have dropped the prime to indicate that now *all* values of k with $k < \Lambda$ should be summed over. Finally, let us take the limit $L \to \infty$, and write the sums in (L.46) as integrals over $\mathbf{k}$. This gives

$$Z_G[J;\Lambda] = \exp\left[\tfrac{1}{2} L^d \int_{k=0}^{\Lambda} d^d \check{\mathbf{k}}\, \log(\alpha^2 k^2 + \mu^2) \right] \tag{L.47}$$

$$\times \exp\left[\tfrac{1}{2} \int_{k=0}^{\Lambda} d^d \check{\mathbf{k}}\, \frac{\widetilde{J}(\mathbf{k})\widetilde{J}(-\mathbf{k})}{\alpha^2 k^2 + \mu^2} \right].$$

This is the result quoted in §8.1. The J-dependence has been factored out. Notice that the limit $\Lambda \to \infty$ does not exist.

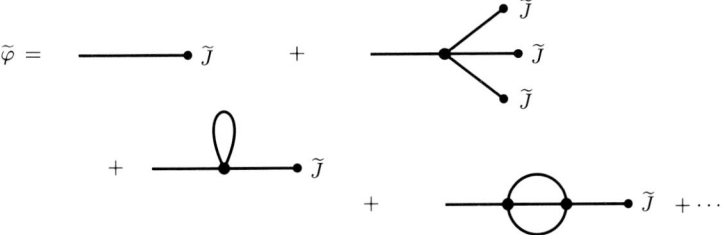

Figure M.1 The first few Feynman diagrams contributing to $\langle \tilde\varphi(\mathbf{k}) \rangle$.

Appendix M: The Feynman rules for the vertex functions

In this appendix we complete the proof, sketched in §8.4, that above T_c the $\Gamma^{(n)}$ given by the rules of Box 8.4 coincide with the functional derivatives of the dimensionless Gibbs free energy $\Gamma = \beta G$. That is,

- $\Gamma^{(0)}$ is $-\log Z_G[0]$ (where $Z_G[J]$ is the Gaussian partition function, given in equation (8.5)), minus the sum of all 1PI contributions to $\log Z[J]$.
- $\Gamma^{(1)}$ is minus the sum of all amputated 1PI one-point diagrams, *excluding* the one which involves the source J.
- $\Gamma^{(2)}$ is the inverse of the bare propagator, minus the sum of all amputated 1PI two-point diagrams.
- $\Gamma^{(n)}$, for $n > 2$, is minus the sum of all amputated 1PI n-point diagrams.

$\Gamma[\tilde\varphi]$ and the partition function $Z[J]$ are related by

$$\log Z[\tilde J] = \left[\int d^d\check{\mathbf{k}}\, \tilde J(\mathbf{k}) \tilde\varphi(-\mathbf{k})\right] - \Gamma[\tilde\varphi], \tag{M.1}$$

where $\tilde\varphi = \langle \tilde\phi \rangle$. Substituting equation (8.48) into (M.1) gives

$$\log Z[\tilde J] = -\Gamma^{(0)} + \left[\int d^d\check{\mathbf{k}}\, (\tilde J(-\mathbf{k}) - \Gamma^{(1)}(\mathbf{k})) \tilde\varphi(\mathbf{k})\right] \tag{M.2}$$

$$- \sum_{n=2}^{\infty} \frac{1}{n!} \int d^d\check{\mathbf{k}}_1 \ldots d^d\check{\mathbf{k}}_n\, \tilde\varphi(\mathbf{k}_1) \ldots \tilde\varphi(\mathbf{k}_n)\, \Gamma^{(n)}(\mathbf{k}_1, \ldots, \mathbf{k}_n).$$

$\tilde\varphi(\mathbf{k})$ can be calculated diagrammatically using the techniques of Chapter 8. The first few diagrams in its expansion are shown in Figure M.1.

We shall prove that, if the $\Gamma^{(n)}$ are calculated according to the rules of Box 8.4, then equation (M.2) is satisfied. Our proof is diagrammatic. From the rules of Box 8.3, we know that $\log Z[J]$ can be expanded as the sum of all diagrams with no external lines and with an arbitrary number of J-blobs.

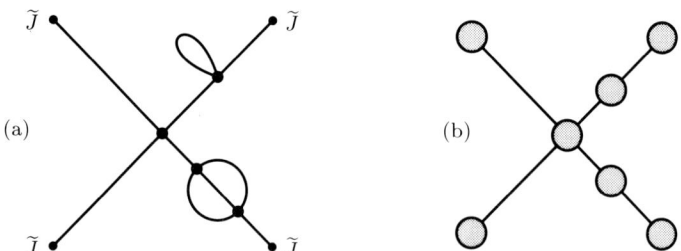

Figure M.2 (a) One diagram which contributes to $\log Z[J]$. (b) The 'tree representation' of the same diagram. Each 1PI subdiagram has been replaced by an 'island', these islands being joined together with bare propagators.

We begin by considering an arbitrary diagram in the expansion of $\log Z[J]$. This diagram is either itself 1PI or consists of two pieces joined by a bare propagator. In turn, these pieces are themselves either 1PI or .... Eventually no further subdivisions are possible and the diagram can be represented in the form shown in Figure M.2(b), where bare propagators join 1PI **islands** to form a tree structure (containing no loops). All the diagrams on the left-hand side of (M.2) can be written in this form. The idea behind the proof is that such diagrams will appear from the right-hand side of (M.2) when the $\widetilde{\varphi}$s appearing there are expressed in terms of $\widetilde{J}$ as in Figure M.1, so that the 1PI diagrams (from the $\Gamma^{(n)}$ on the right-hand side of (M.2)) 'grow legs'. We must show that when this is done each diagram in the expansion of $\log Z$ arises with the correct weight.

For definiteness, consider the diagram contributing to $\log Z$ shown in Figure M.2(a). From which terms on the right-hand side of (M.2) will it come? One place where it will arise is in the expansion of the term

$$-\frac{1}{4!}\int d^d\check{\mathbf{k}}_1\ldots d^d\check{\mathbf{k}}_4\, \widetilde{\varphi}(\mathbf{k}_1)\ldots\widetilde{\varphi}(\mathbf{k}_4)\,\Gamma^{(4)}(\mathbf{k}_1,\mathbf{k}_2,\mathbf{k}_3,\mathbf{k}_4). \qquad (\text{M.3})$$

The $\Gamma^{(4)}$ in (M.3) generates the four-point vertex at the centre of Figure M.2(a); the 'legs' of the diagram come from expanding the four $\widetilde{\varphi}$. The particular diagram in Figure M.2(a) turns up if we choose ' ✖ ' from $\Gamma^{(4)}$ (recall that an infinite number of diagrams contribute to $\Gamma^{(4)}$, of which ' ✖ ' is only one), and

in some permutation from the expansions of $\widetilde{\varphi}(\mathbf{k}_1)$ to $\widetilde{\varphi}(\mathbf{k}_4)$. It is not hard to see that the complete diagram of Figure M.2(a) arises with the correct symmetry factor. The symmetry factors of the 1PI 'islands' in Figure M.2(b)

Rules for the vertex functions

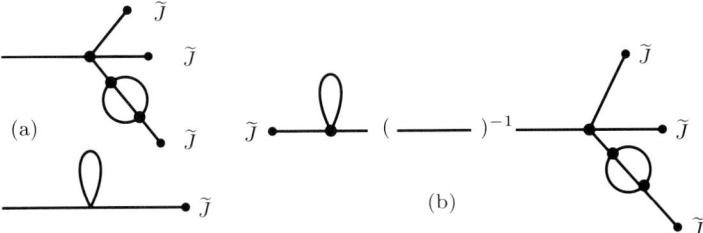

Figure M.3 (b) shows another way of obtaining the diagram of Figure M.2(a), this time by joining the two diagrams in (a) to the ends of the inverse of the bare propagator.

take care of themselves. The only rule relevant for determining any extra symmetry factor for the tree apart from those of the individual subdiagrams is the third rule of §8.2. For the diagram of Figure M.2(a) there is an extra factor of one-half, because the two legs in Figure M.2(a) marked with an arrow are the same and the places where they are attached are equivalent. This factor arises naturally when (M.3) is expanded, because there are only 4×3 ways of choosing the three different sorts of 'legs' present in Figure M.2(a) from the four $\tilde{\varphi}$s in (M.3), and not $4 \times 3 \times 2$ ways as there would be if all four legs were different. So the factor multiplying the diagram is $4 \times 3/4! = \frac{1}{2}$. One can quickly convince oneself that this works in general, and that expressions obtained by expanding terms on the right-hand side of equation (M.2) always have the correct symmetry factor.

This is looking promising. The difficulty is that the diagram of Figure M.2(a) also appears in the expansions of other terms on the right-hand side of (M.2). It comes from the terms

$$-\frac{1}{2!}\int d^d\check{k}_1 d^d\check{k}_2\, \tilde{\varphi}(k_1)\tilde{\varphi}(k_2\, \Gamma^{(2)}(k_1,k_2) \quad \text{and} \quad -\int d^d\check{k}\, \tilde{J}(k)\tilde{\phi}(-k). \qquad (M.4)$$

The point is that any of the 1PI islands of a particular diagram can be regarded as the $\Gamma^{(n)}$ on to which the legs of the diagram are joined. There are six 1PI islands in the diagram we are considering, so it appears six times (with the correct symmetry factor in each case), which is five times too many. If there are N vertex functions in the tree representation of a diagram then that diagram appears N times on the right-hand side of (M.2). There appears to be a problem.

Fortunately there is also a solution. It lies in the anomalous sign of the inverse of the bare propagator Δ^{-1} in the expression for $\Gamma^{(2)}$. Recall that $\Gamma^{(2)}$ is defined to be *minus* the sum of all two-point 1PI connected diagrams, *plus* the inverse of the bare propagator. We are about to see the reason for this peculiarity. It is not just the 1PI islands of Figure M.2(b) that can have legs joined to them to form the complete diagram; the bare propagators that join these islands can also be used as the 'body' of the diagram. The $n=2$ term in the sum in (M.2) contains the term

$$-\frac{1}{2!}\int d^d\check{\mathbf{k}}_1 d^d\check{\mathbf{k}}_2\, \widetilde{\varphi}(\mathbf{k}_1)\widetilde{\varphi}(\mathbf{k}_2)\, (2\pi)^d \delta(\mathbf{k}_1+\mathbf{k}_2)\Delta^{-1}(\mathbf{k}_1). \tag{M.5}$$

Therefore, if one chooses the two diagrams shown in Figure M.3(a) from the expansions of $\widetilde{\varphi}(\mathbf{k}_1)$ and $\widetilde{\varphi}(\mathbf{k}_2)$, and the inverse propagator from the expansion of $\Gamma^{(2)}$, the term (M.5) can give rise to the diagram of Figure M.2(a) in the way shown in Figure M.3(b). The symmetry factor is again correct. But because of the anomalous sign of Δ^{-1} in the definition of $\Gamma^{(2)}$, this diagram has the *opposite sign* to the six contributions considered so far. There are five bare propagators joining the six islands in Figure M.2(b), so five times the value of the diagram gets subtracted off. The total contribution is thus $6-5=1$, which is exactly correct. The anomalous sign of the bare propagator in the definition of $\Gamma^{(2)}$ has played a vital rôle.

Obviously, a tree with N islands has $N-1$ bare propagators joining them, so this mechanism is a general one; $N-(N-1)=1$. The only other points to clear up are the definitions of $\Gamma^{(1)}$ and $\Gamma^{(0)}$. The definition of $\Gamma^{(1)}$ excluded contributions containing the source $\widetilde{J}$, and from equation (M.2) we can see why. $\widetilde{J}$ is already present there, multiplying $\widetilde{\varphi}$. If $\widetilde{J}$ were also present in $\Gamma^{(1)}$ then connected diagrams containing a single factor of $\widetilde{J}$ would appear twice on the right-hand side of (M.2), once from the $\widetilde{J}\widetilde{\varphi}$ term and once from the $\Gamma^{(1)}\widetilde{\varphi}$ term. The exclusion of $\widetilde{J}$ from $\Gamma^{(1)}$ avoids this problem.

$\Gamma^{(0)}$ mops up the Gaussian contribution to the Gibbs free energy, along with all the diagrams that have not been considered so far—those diagrams in the expansion of $\log Z[J]$ which are themselves 1PI, and whose tree representation therefore consists of a single island. Such diagrams are not generated by any of the other terms on the right-hand side of (M.2), as these all contain factors of $\widetilde{\varphi}$ and so have 'legs'. If $\Gamma^{(0)}$ is defined to be minus the sum of all 1PI contributions to $\log Z$, all these diagrams are included. The Gaussian contribution to the Gibbs free energy does not appear as a diagram, and so it must be added in by hand.

We have thus established that if the $\Gamma^{(n)}$ are defined to be the objects from the rules of Box 8.4, every term in the expansion of $\log Z$ on the left-hand side of (M.2) appears on the right-hand side also, with the correct coefficient and sign. But clearly there are no terms on the right-hand side of (M.2) which do not appear somewhere in the expansion of $\log Z[J]$. Hence the two sides of the equation must be equal, and $\Gamma^{(n)}$ may be equivalently defined by the rules of Box 8.4 or as the nth functional derivative of the Gibbs free energy evaluated at $\widetilde{\varphi}=0$.

Appendix N: **Feynman rules for generalized Landau–Ginzburg models**

There are three main ways to generalize the Landau–Ginzburg Hamiltonian. First, powers of ϕ other than ϕ^2 or ϕ^4 may be included; second, there may be terms other than $|\boldsymbol{\nabla}\phi|^2$ involving derivatives of ϕ; finally, the order parameter may have more than one component. In Chapter 8 we showed how to calculate the free energy and correlation functions of the Landau–Ginzburg model by drawing Feynman diagrams. There are Feynman rules for the more general models too, which we shall explain in this appendix.

The good news is that almost all of what we did in Chapter 8 is completely general. The partition functions of generalized models can be written as sums of Feynman diagrams made of links and vertices, and their logarithms are given by sums of the connected diagrams only. The functional derivatives of the Gibbs free energy are still the vertex functions, the building blocks of the connected correlation functions. The only things that change for more general Hamiltonians are the diagrams themselves, and a few numerological details.

N.1 Higher powers and derivatives of the order parameter

Consider a model with Hamiltonian

$$H[J] = \int \mathrm{d}^d \mathbf{x} \left(\tfrac{1}{2}\left(\alpha^2 |\boldsymbol{\nabla}\phi|^2 + \mu^2 \phi^2 \right) + \frac{\lambda_6}{6!}\phi^6 - J\phi \right). \tag{N.1}$$

This is the Landau–Ginzburg Hamiltonian but with ϕ^6 replacing ϕ^4. The partition function $Z[J]$ is

$$Z[J] = \int \mathcal{D}\phi\, \mathrm{e}^{-H[J]}. \tag{N.2}$$

To derive the Feynman rules for this model we shall proceed as we did in §8.2. The first step is to move the non-quadratic term in H outside the functional integral (compare equation (8.12)):

$$Z[J] = \exp\left[-\frac{\lambda_6}{6!} \int \mathrm{d}^d\mathbf{z}\, \frac{\delta^6}{\delta J^6(\mathbf{z})} \right] Z_G[J]. \tag{N.3}$$

Z_G, the Gaussian partition function, is given in equation (8.5). From (N.3) follow the Feynman rules, just as in §8.2, equation (8.15) and below. The four functional derivatives in that equation produced vertices with four legs; the six derivatives in (N.3) produce vertices with six legs, since they give zero unless there are six J-blobs all at the same position for them to act upon. None of the arguments about the form of Feynman diagrams in §8.2 depend on the number of legs per vertex, so diagrams for the modified Hamiltonian (N.1) can be drawn just as for the Landau–Ginzburg model, except that vertices must have six legs instead of four. What about symmetry factors? There is only one argument in §8.2.2 which changes, and that is the first one, about the cancellation of the 1/4! that appears in the Landau–Ginzburg Hamiltonian. That cancels because of the 4! ways of applying the four derivatives to the four Js at each vertex. Here there are six derivatives, and six Js per vertex, which generates a factor of 6!. But we have written the coefficient of ϕ^6 in the Hamiltonian (N.1) as $\lambda_6/6!$ for just this reason. The factors of 6! cancel, and the symmetry factor rules go through *exactly* as before. Everything else—Feynman rules in k-space, the arguments that motivate the introduction of vertex functions, and the result that vertex functions are functional derivatives of the Gibbs free energy—is also model-independent. To sum up, the only changes are:

- Diagrams must be drawn with six-link vertices, rather than with four-link vertices, and
- With each such vertex is associated a factor $-\lambda_6$, instead of $-\lambda$.

If both ϕ^4 and ϕ^6 terms are present in the Hamiltonian, we can write the partition function as

$$Z[J] = \exp\left[-\frac{\lambda_4}{4!}\int d^d\mathbf{z}\,\frac{\delta^4}{\delta J^4(\mathbf{z})} - \frac{\lambda_6}{6!}\int d^d\mathbf{z}\,\frac{\delta^6}{\delta J^6(\mathbf{z})}\right] Z_G[J]. \quad (N.4)$$

The two terms in the exponential mean that both sorts of vertex are present, with a factor $-\lambda_4$ for each four-link vertex and a factor $-\lambda_6$ for each six-link vertex. The diagrammatic expansion is now a double power series in λ_4 and λ_6.

In general, if there is a term $(\lambda_n/n!)\int d^d\mathbf{x}\,\phi^n$ in the Hamiltonian then n-link vertices, each with an associated $-\lambda_n$, appear in the Feynman rules. But when diagrams contain vertices with many legs there is one point that can cause confusion. The diagram in Figure N.1 contributes to the vertex function $\Gamma^{(6)}$. The corresponding expression C, omitting the symmetry factor, is

$$C(\mathbf{k}_1, \mathbf{k}_2, \mathbf{k}_3, \mathbf{k}_4, \mathbf{k}_5, \mathbf{k}_6) = \quad (N.5)$$

$$\lambda_4 \lambda_6 \int \frac{d^d\check{\mathbf{q}}}{(\alpha^2 q^2 + \mu^2)(\alpha^2(\mathbf{q}+\mathbf{k}_5+\mathbf{k}_6)^2 + \mu^2)} \delta(\mathbf{k}_1 + \cdots + \mathbf{k}_6).$$

What is the symmetry factor? The only rule that is relevant is the one about more than one link joining the same two vertices. In Figure N.1 two links join the two vertices, so the symmetry factor is $\frac{1}{2}$. This is correct. What can easily be forgotten is that there are a further *fourteen* diagrams that look just like the one in Figure N.1, except that the wavevectors $\mathbf{k}_1, \ldots, \mathbf{k}_6$ are assigned to the external legs differently.

Simply changing around the wavevectors at one vertex does *not* make a different diagram. The different permutations of the links at each vertex

N.1 Feynman rules

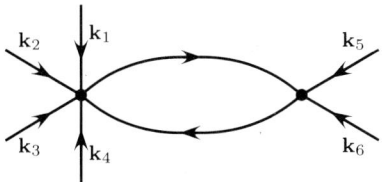

Figure N.1 A graph contributing to $\Gamma^{(6)}$, discussed in the text. Distinct graphs are produced when $\mathbf{k}_5$ or $\mathbf{k}_6$ is exchanged with any of $\mathbf{k}_1$ to $\mathbf{k}_4$.

have already been taken into account when deriving the symmetry factor rules. So swapping $\mathbf{k}_5$ and $\mathbf{k}_6$ in Figure N.1 gives nothing new. But swapping two wavevectors at *different* vertices does change things. The result is a completely new diagram, usually with a different algebraic expression corresponding to it. Swapping $\mathbf{k}_1$ and $\mathbf{k}_5$ in Figure N.1 produces a new diagram; the $(\mathbf{q} + \mathbf{k}_5 + \mathbf{k}_6)^2$ in the denominator of (N.5) is replaced by $(\mathbf{q} + \mathbf{k}_1 + \mathbf{k}_6)^2$. The total number of distinct diagrams in the present case is the number of ways of choosing a pair from six objects, without regard to order: $(6 \times 5)/2 = 15$. This factor of 15 is logically different to a symmetry factor. Symmetry factors multiply single diagrams, but here there are fifteen different diagrams, each with symmetry factor $\frac{1}{2}$.

It may be that all the external wavevectors are the same; for an example of this, see §8.4.3. In this case all the different diagrams have the same value, and the number corresponding to the '15' above looks more like a symmetry factor. Do not forget that it is a completely distinct idea.

There is a very close correspondence between the Feynman rules and the terms in the Hamiltonian. We have just seen that if $\int d^d\mathbf{x}\, \phi^n$ appears in the Hamiltonian, an n-link vertex appears in the Feynman rules. The correspondence between term and rule is even closer than 'one leg per vertex per power of ϕ'. Each of the factors of $\phi(\mathbf{x})$ appearing in the above term is at the same position $\mathbf{x}$, and that position is integrated over; similarly, the n links of the n-point vertex all meet at one point, whose position is then integrated over. This also works in k-space. Written in terms of the Fourier transform of the order parameter, the above term is

$$\int d^d\check{\mathbf{q}}_1 \ldots d^d\check{\mathbf{q}}_n\, \widetilde{\phi}(\mathbf{q}_1) \ldots \widetilde{\phi}(\mathbf{q}_n)\, \delta(\mathbf{q}_1 + \cdots + \mathbf{q}_n). \qquad (\text{N.6})$$

Each of the $\widetilde{\phi}$ has a different wavevector for its argument, and these are all integrated over, subject only to the restriction that the total of the wavevectors must be zero. The corresponding Feynman rule is that there is an n-link vertex, with a different wavevector for each link, and these are all integrated over subject to the restriction that the sum of the wavevectors vanishes.

This close correspondence is entirely general, and it enables Feynman rules to be 'read off' from any Hamiltonian with very little effort. The reason why it works is simple. When non-quadratic terms in the Hamiltonian are moved outside the functional integral, each factor of $\phi(\mathbf{x})$ (or $\tilde\phi(\mathbf{q})$) is replaced by a functional derivative with respect to $J(\mathbf{x})$ (or $\tilde J(-\mathbf{q})$). So there is a one-to-one correspondence between factors of the order parameter in the Hamiltonian and derivatives with respect to the source. Then these derivatives act on the Gaussian partition function. There is only a non-zero answer when there are J-blobs at the positions (or with the wavevectors) appearing in the functional derivatives. In this way the form of the Hamiltonian feeds directly through to the Feynman rules. We can now use this to find the Feynman rules for other sorts of terms in the Hamiltonian.

The Hamiltonian may contain derivatives of ϕ other than $|\nabla\phi|^2$; for example

$$H[J] = \int d^d\mathbf{x}\left(\tfrac{1}{2}\left(\alpha^2|\nabla\phi|^2 + \mu^2\phi^2\right) + \frac{\lambda_4}{4!}\phi^4 + \frac{\lambda_4'}{4!}\phi^3\nabla^2\phi - J\phi\right). \quad (\text{N.7})$$

The new term introduces a new vertex into the Feynman rules. It is easier to see what this is if we work in k-space. The integral over the two quartic terms in (N.7) can be written as

$$\frac{\lambda_4}{4!}\int d^d\check{\mathbf{q}}_1\ldots d^d\check{\mathbf{q}}_4\,\tilde\phi(\mathbf{q}_1)\tilde\phi(\mathbf{q}_2)\tilde\phi(\mathbf{q}_3)\tilde\phi(\mathbf{q}_4)\,\delta(\mathbf{q}_1+\cdots+\mathbf{q}_4) \quad (\text{N.8})$$

$$+\frac{\lambda_4'}{4!}\int d^d\check{\mathbf{q}}_1'\ldots d^d\check{\mathbf{q}}_4'\,\tilde\phi(\mathbf{q}_1')\tilde\phi(\mathbf{q}_2')\tilde\phi(\mathbf{q}_3')\tilde\phi(\mathbf{q}_4')\left(-(q_4')^2\right)\delta(\mathbf{q}_1'+\cdots+\mathbf{q}_4').$$

We know that in k-space the ϕ^4 term gives rise to a four-leg vertex, multiplied by a factor $-\lambda_4$, with the wavevectors on the four legs adding up to zero. By analogy, we might guess that the term $\phi^3\nabla^2\phi$ gives rise to a vertex with four legs, multiplied by a factor of $-\lambda_4'(-q_4^2)$ where $\mathbf{q}_4$ is the wavevector on one of the legs, with the four wavevectors again adding up to zero. But this isn't quite correct. The $(1/4!)$ that multiplies the ordinary ϕ^4 term in the Hamiltonian is cancelled out by the 4! ways in which the vertex can be connected to its four links. For the ϕ^4 term each of these 24 ways gives the same answer, so the $(1/4!)$ is completely cancelled. But for the new vertex these 24 ways are not all the same. Six of them are multiplied by $-q_1^2$, six by $-q_2^2$, six by $-q_3^2$, and six more by $-q_4^2$. So the actual factor for the vertex is

$$-\lambda_4'\frac{-(q_1^2+q_2^2+q_3^2+q_4^2)}{4}. \quad (\text{N.9})$$

Notice that this is symmetric in the four wavevectors.

The correct rules for other terms containing derivatives can be worked out in a similar way. If there are n powers of ϕ in the term, its coefficient

N.2 Feynman rules

should be written in the form $\lambda_n/n!$. The form of the vertex can be deduced from the Fourier transform of the term, as we did in the example above. The powers of wavevector that appear in the Fourier transform should be symmetrized. The factor associated with the vertex is then $-\lambda_n$ multiplied by the symmetrized sum, divided by the number of terms in the sum.

N.2 A D-component order parameter

The simplest Hamiltonian involving an D-component order parameter is the natural generalization of the Gaussian model:

$$H_G[J_1,\ldots,J_n] = \int d^d\mathbf{x} \sum_{i=1}^{D} \left(\tfrac{1}{2}\alpha^2|\boldsymbol{\nabla}\phi_i|^2 + \tfrac{1}{2}\mu_i^2\phi_i^2 - J_i\phi_i\right). \quad (\text{N.10})$$

This Hamiltonian is a sum of n one-component Gaussian Hamiltonians, with each component of the order parameter coupled to a different source field J_i and having a different mass μ_i. The partition function for this model is

$$\begin{aligned}
Z_G[J_1,\ldots,J_n] &= \int \mathcal{D}\phi_1\ldots\mathcal{D}\phi_n \, e^{-H_G[\{J_i\}]} \\
&= \prod_{i=1}^{D} \int \mathcal{D}\phi_i \exp\left[-\int d^d\mathbf{x}\,\left(\tfrac{1}{2}\alpha^2|\boldsymbol{\nabla}\phi_i|^2 + \tfrac{1}{2}\mu_i^2\phi_i^2 - J_i\phi_i\right)\right] \\
&= \prod_{i=1}^{D} Z_G[\mu_i; J_i].
\end{aligned}$$
(N.11)

It is the product of the D partition functions for the ϕ_i. The D components of the order parameter behave independently of each other. If terms of the form $\phi_a\phi_b$ or $\boldsymbol{\nabla}\phi_a \cdot \boldsymbol{\nabla}\phi_b$ with $a \neq b$ are present, they can be removed by rescaling and rotating the ϕ_i. We shall assume that this has been done.

The correlation functions of this model must carry labels to specify which components of the order parameter they refer to, for example

$$G^{(3)}_{112}(\mathbf{x}_1,\mathbf{x}_2,\mathbf{x}_3) \equiv \langle \phi_1(\mathbf{x}_1)\phi_1(\mathbf{x}_2)\phi_2(\mathbf{x}_3)\rangle. \quad (\text{N.12})$$

We can represent them diagrammatically by having D different sorts of J-blobs, links and external points, each carrying a number from 1 to D as a **label**. In practice it is sufficient to label the links, since a link must always connect two points with the same label.[1] A link with label i has the value $(\alpha^2 k^2 + \mu_i^2)^{-1}$. A diagram that contributes to the above correlation function is shown in Figure N.2(a).

---

[1] If there were non-diagonal terms in the Hamiltonian, such as $\phi_1\phi_2$, this would not be true. We do not consider such cases.

Non-Gaussian Hamiltonians involving order parameters with several components are more complicated than their one-component analogues because there are many non-Gaussian terms of a given order: if $D = 6$ there are 56 distinct quartic terms. Fortunately the coefficients of such terms are often related by symmetries of the model being considered. If a $D = 6$ model is symmetric under the exchange of any two components of the order parameter, the term $\phi_1\phi_4\phi_5\phi_6$ in the Hamiltonian must have the same coefficient as the term $\phi_2\phi_3\phi_4\phi_5$, and so on. Instead of 56 independent coefficients there will be just 4. Real physical systems often have symmetries, and this makes the task of modelling them easier than it might otherwise be.

A model involving a D-component order parameter is $O(D)$ **symmetric** if it is invariant under 'rotations' of the D components into each other. To be precise, the model is $O(D)$ symmetric if the Hamiltonian is invariant under the transformation

$$\begin{pmatrix} \phi_1 \\ \vdots \\ \phi_D \end{pmatrix} \mapsto \mathbf{R} \begin{pmatrix} \phi_1 \\ \vdots \\ \phi_D \end{pmatrix} \qquad (\text{N.13})$$

where $\mathbf{R}$ is an orthogonal $D \times D$ matrix with unit determinant. $O(D)$ symmetry severely limits the form of a Hamiltonian. The Gaussian Hamiltonian (N.10) is not $O(D)$ symmetric unless all of the μ_i are equal and the J_i vanish. There is only one $O(D)$ symmetric quartic term:

$$\frac{\lambda}{4!} \int d^d\mathbf{x} \left(|\boldsymbol{\phi}|^2 \right)^2. \qquad (\text{N.14})$$

Models with $O(D)$ symmetry often arise in practice. The rotation group in three dimensions is $O(3)$.

The effect of non-quadratic terms in the Hamiltonian is to produce additional vertices in the Feynman rules, just as in the one-component case. For example, the quartic term in

$$H = H_G + \frac{\lambda}{4!} \int d^d\mathbf{x}\, \phi_1\phi_3\phi_4^2, \qquad (\text{N.15})$$

produces a vertex where four links meet, one with label 1, one with label 3, and two with label 4. Figure N.2(b) shows the $O(\lambda)$ contribution to the four-point correlation function $G^{(4)}_{1134}$ for this Hamiltonian.

If there are several non-quadratic terms in the Hamiltonian, each has its own vertex. Labels of links and J-blobs that are not fixed by the labels on the external points of a diagram—**internal** labels—must be summed over all of their possible values, in the same way that we integrate over the internal wavevectors of a diagram.

The detailed rules for symmetry factors can be derived as was done in Chapter 8. We omit the derivation and just state the results.

N.2 Feynman rules

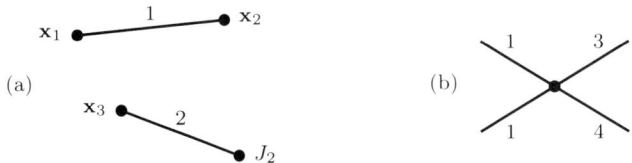

Figure N.2 Examples of Feynman diagrams for the multicomponent Landau–Ginzburg model. Each link has a label, representing one of the field components. (a) shows a disconnected diagram contributing to $G^{(3)}{}_{112}$; (b) shows a diagram contributing to $G_c^{(4)}{}_{1134}$.

- If the term $\int d^d x \, (\prod_{i=1}^m \phi_i^{n_i})$ appears in the Hamiltonian, write its coefficient as $\lambda / \prod(n_i!)$. Then there is a factor of $(-\lambda)$ associated with each appearance of this vertex in a diagram.
- Multiply by $\frac{1}{2}$ for each link whose two ends are joined to the same vertex.
- For each set of l identically-labelled links joining the same two vertices, multiply by $(l!)^{-1}$. So if three links labelled 1, 1, and 3 join the same two vertices, there is a factor of $1/2!$.
- If the internal points of a diagram and their associated labels can be rearranged in r ways and yet leave the diagram looking exactly the same (with regard to points, links, and labels), divide by r.

Although straightforward in principle, drawing all the appropriate diagrams is in practice much harder than in the one-component case. A Hamiltonian with only two or three non-quadratic terms would be quite easy to deal with but usually there are very many such terms, each with its own vertex. The O(D) symmetric quartic term $(|\boldsymbol{\phi}|^2)^2$ gives $D(D+1)/2$ different terms when expanded. But fortunately these terms are of two types only; ϕ_i^4, and $\phi_i^2 \phi_j^2$. This makes calculations much easier than they would be for a completely general quartic Hamiltonian, although they can still be tricky, particularly at two or more loops.

Let's work through an example to illustrate these rules: the two-loop contribution to the two-point vertex function $\Gamma_{aa}^{(2)}(\mathbf{k})$ shown in Figure N.3, calculated for the model with the O(D) symmetric Hamiltonian

$$\int d^d \mathbf{x} \, \tfrac{1}{2} \alpha^2 |\nabla \boldsymbol{\phi}|^2 + \tfrac{1}{2} \mu^2 |\boldsymbol{\phi}|^2 + \frac{\lambda}{4!} \int d^d \mathbf{x} \, (|\boldsymbol{\phi}|^2)^2. \tag{N.16}$$

This model is known as the D-component Landau–Ginzburg model. Expanding the quartic term gives

$$(|\boldsymbol{\phi}|^2)^2 = \sum_{i=1}^D \phi_i^4 + \sum_{i<j=1}^D 2\phi_i^2 \phi_j^2. \tag{N.17}$$

428 Appendix N: Feynman rules for generalized Landau–Ginzburg models

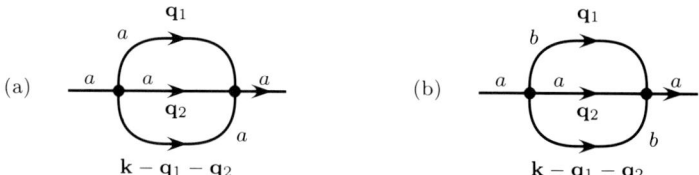

Figure N.3 The figure shows two two-loop diagrams that contribute to $\Gamma^{(2)}$ of the D-component Landau–Ginzburg model. (a) Here both vertices couple together four 'a' labels. (b) Here both vertices couple together two 'a' and two 'b' labels. There are $D-1$ diagrams like this, one for each label b different from a.

There are two types of four-link vertices. Type 1 couples together four links with the same labels; type 2 couples together two pairs of links with different labels.

The two vertices in our diagram must be both type 1 or both type 2. Both cases have identical wavevector dependence; only their coefficients differ. Consider the former case first (Figure N.3(a)). The term ϕ_a^4 has coefficient $\lambda/4!$ in the Hamiltonian, and so by the first of the above rules each vertex contributes $-\lambda$. The symmetry factor is $1/3!$, by the third rule above. So the coefficient of the diagram with two type 1 vertices is $\lambda^2/6$.

The term $\phi_a^2 \phi_b^2$ has coefficient $\lambda/12$ in the Hamiltonian, so a type 2 vertex gives a factor of $-\lambda/3$. The symmetry factor this time is $1/2!$. There are $D-1$ such diagrams, one for each value of the label b, so the total contribution of the diagrams with two type 2 vertices is $(D-1)\lambda^2/18$. The total of both sorts of diagram is

$$\lambda^2 \frac{D+2}{18} \int \frac{d^d\check{q}_1 d^d\check{q}_2}{(\alpha^2 q_1^2 + \mu^2)(\alpha^2 q_2^2 + \mu^2)(\alpha^2(\mathbf{k} - \mathbf{q}_1 - \mathbf{q}_2)^2 + \mu^2)}. \tag{N.18}$$

This calculation is quite a lot trickier than it would have been for a one-component field, but by no means impossible.

Answers

Answers for Chapter 1

1.1

Exponent	Definition
α	$c_v \sim \alpha^{-1}\big((|T-T_c|/T_c)^{-\alpha}-1\big),\ T\to T_c,\ \bar{\rho}=\rho_c$
β	$\rho_{\text{liquid}}-\rho_{\text{gas}}\sim(T_c-T)^\beta,\ T\to T_c\text{ from below},\ \bar{\rho}=\rho_c$
γ	$\kappa_T\sim|T-T_c|^{-\gamma},\ T\to T_c,\ \bar{\rho}=\rho_c$
δ	$\rho-\rho_{\text{gas}}\sim(p-p_c)^{1/\delta},\ p\to p_c,\ T=T_c$
η	$G^{(2)}(r)\sim 1/r^{d-2+\eta},\ T=T_c,\ \bar{\rho}=\rho_c$
ν	$\xi\sim|T-T_c|^{-\nu},\ \bar{\rho}=\rho_c$

1.2 From Tables 1.1 and 1.3 we have that these critical exponents describe non-analytic dependence of f's derivatives on its natural variables, (T, B). In the cases of certain exponents (e.g. γ and α when they exceed zero) this nonanalyticity leads directly to an observable diverging at T_c. In other cases it causes sufficiently high derivatives of an observable to diverge at T_c. Hence any exponent can in principle be determined from f_s by studying a sufficiently high derivative of the latter in the limit $T\to T_c$, where the corresponding derivative of f_a will be negligible.

1.3 Equation (1.28) may be written

$$G_c^{(2)}(r,t)=\phi(rt^{(2-\alpha)/d})/r^{d-2+\eta},$$

where $\phi(x)\equiv\psi(x^d)$. The characteristic length scale is clearly $l\equiv t^{(\alpha-2)/d}\propto\xi\propto t^{-\nu}$. Hence $\nu d=2-\alpha$ as required.

With (2.28) we have

$$\chi\sim t^{-\gamma}\sim\int_0^\infty dr\,r^{d-1}\frac{\psi(r^d t^{2-\alpha})}{r^{d-2+\eta}}\int d\Omega_{d-1}$$

$$=\Omega_{d-1}t^{-(2-\eta)\nu}\int dy\,y^{1-\eta}\phi(y),$$

where Ω_d is the area of the unit d-sphere (see Problem 3.4) and Josephson's law has been used. Fisher's law follows by comparing the exponents of t.

1.4 Write $\boldsymbol{\xi}=\sum_{\mathbf{k}}\boldsymbol{\Xi}_{\mathbf{k}}e^{i(\mathbf{k}\cdot\mathbf{x})}e^{-i\omega_{\mathbf{k}}t}$, where the $\boldsymbol{\Xi}_{\mathbf{k}}$ have random phases, and, by the reality of $\boldsymbol{\xi}$, $\boldsymbol{\Xi}_{-\mathbf{k}}=\boldsymbol{\Xi}_{\mathbf{k}}^*$ and $\omega_{-\mathbf{k}}=-\omega_{\mathbf{k}}$. Then

$$\langle\dot{\xi}^2\rangle=-\sum_{\mathbf{kk}'}\langle\boldsymbol{\Xi}_{\mathbf{k}}\cdot\boldsymbol{\Xi}_{\mathbf{k}'}\rangle\omega_{\mathbf{k}}\omega_{\mathbf{k}'}$$

$$\times e^{i(\mathbf{k}+\mathbf{k}')\cdot\mathbf{x}}e^{-i(\omega_{\mathbf{k}}+\omega_{\mathbf{k}'})t}$$

$$=\sum_{\mathbf{k}}\langle\Xi_{\mathbf{k}}^2\rangle\omega_{\mathbf{k}}^2$$

Hence

$$\text{K.E.}=L^3\tfrac{1}{2}\rho\sum_{\mathbf{k}}\langle\Xi_{\mathbf{k}}^2\rangle\omega_{\mathbf{k}}^2$$

$$=\sum_{\mathbf{k}}k_BT$$

which implies that

$$\langle\Xi_{\mathbf{k}}^2\rangle=\frac{2k_BT}{\rho\omega_{\mathbf{k}}^2 L^3}.$$

Now

$$\langle \boldsymbol{\xi}(\mathbf{x}) \cdot \boldsymbol{\xi}(\mathbf{x}') \rangle = \sum_{\mathbf{k}\mathbf{k}'} \langle \boldsymbol{\Xi}_{\mathbf{k}} \cdot \boldsymbol{\Xi}_{\mathbf{k}'} \rangle e^{i(\mathbf{k} \cdot \mathbf{x} - \omega_{\mathbf{k}} t)}$$
$$\times e^{i(\mathbf{k}' \cdot \mathbf{x}' - \omega_{\mathbf{k}'} t)}$$
$$= \frac{2k_B T}{\rho L^3} \sum_{\mathbf{k}} \frac{e^{i\mathbf{k} \cdot (\mathbf{x}-\mathbf{x}')}}{\omega_{\mathbf{k}}^2}$$

Approximating $\sum_{\mathbf{k}} \simeq (L/2\pi)^3 \int d^3\mathbf{k}$ and taking $\omega_{\mathbf{k}} = kc$, we have

$$\langle \boldsymbol{\xi}(\mathbf{x}) \cdot \boldsymbol{\xi}(\mathbf{x}') \rangle = \frac{2k_B T}{\rho c^2 (2\pi)^2} \int dk$$
$$\int d\theta \times \sin\theta e^{ikr\cos\theta}$$
$$= \frac{4k_B T}{\rho c^2 (2\pi)^2} \int dk \frac{\sin kr}{kr} \sim \frac{1}{r}. \quad (\text{a1.1})$$

Hence $\eta = 0$.

The integral in (a1.1) is dominated by $k < \pi/r$ because waves with wavelength greater than the separation of $\mathbf{x}$ and $\mathbf{x}'$ provide correlated displacements at the two points, while shorter-wavelength waves provide decorrelating displacements. In this model the correlation length is of order L. To obtain a more realistic picture one has to include viscosity and thermal conduction—see, e.g., Appendix D of Stanley (1967).

1.5 From the equation of continuity, $\partial_t \delta\rho + \rho_0 \nabla \cdot \dot{\boldsymbol{\xi}} = 0$, it follows that the displacements $\boldsymbol{\xi}(x,t)$ of particles in an acoustic mode are related to the over-density $\delta\rho$ by $\delta\rho = -\rho_0 i\mathbf{k} \cdot \boldsymbol{\xi}$. Then proceeding in close analogy with the previous problem, one finds

$$\langle \delta\rho(\mathbf{x})\delta\rho(\mathbf{x}') \rangle = \rho_0^2 \sum_{\mathbf{k}} k^2 \langle \Xi_{\mathbf{k}}^2 \rangle e^{i\mathbf{k} \cdot (\mathbf{x}-\mathbf{x}')}.$$

Since the K.E. is still $\propto \dot{\xi}^2$, $\langle \Xi_{\mathbf{k}}^2 \rangle$ is again $\propto \omega_{\mathbf{k}}^{-2} \propto k^{-2}$, so when one approximates $\sum_{\mathbf{k}}$ by an integral and carries out the angular integration, one finds

$$\langle \delta\rho(\mathbf{x})\delta\rho(\mathbf{x}') \rangle \sim \int dk\, k^2 \frac{\sin kr}{kr}$$

from which the quoted result follows.

1.6 Let $\mathcal{G}(x,p) = xp - g(x)$ and $x_p \equiv x(p)$. Then $\overline{g}(p) = \mathcal{G}(x_p, p)$ is the value of $\mathcal{G}$ at its extremal point (x_p, p):

$$\left.\frac{\partial \mathcal{G}}{\partial x}\right|_{x_p} = (p - g'(x_p)) = 0.$$

Analogously defining $\overline{\mathcal{G}} \equiv pq - \overline{g}(p)$, we have

$$\frac{\partial \overline{\mathcal{G}}}{\partial p} = q - \overline{g}'(p)$$
$$= q - \left(\frac{\partial \mathcal{G}(x_p, p)}{\partial x_p}\frac{dx_p}{dp} + \frac{\partial \mathcal{G}}{\partial p}\right)$$
$$= q - \left[(p - g'(x_p))\frac{dx_p}{dp} + x_p\right]$$
$$= q - x_p,$$

so $\overline{\mathcal{G}}$ has an extremum when $x_p = q$. $\overline{\overline{g}}$ is the value of this extremum:

$$\overline{\overline{g}}(q) = \overline{\mathcal{G}}(p_q, q)$$
$$= qp_q - \overline{g}(p_q)$$
$$= qp_q - [x_{p_q} p_q - g(x_{p_q})].$$

Since we know that $x_{p_q} = q$ the result now follows.

1.7 If f is to have a maximum at $(T, B) = (T_c, 0)$, the ratio

$$\mathcal{R} \equiv \frac{\partial^2 f}{\partial T^2}\frac{\partial^2 f}{\partial B^2} \Big/ \left(\frac{\partial^2 f}{\partial T \partial B}\right)^2$$

must tend to a number greater than unity as $(T_c, 0)$ is approached along any path. Consider approaching $(T_c, 0)$ along the line $T < T_c$, $B = 0$. By Table 1.3, we then have $\mathcal{R} = c_b \chi/(\partial m/\partial T)^2$, and given that along this line $c_B \sim t^{-\alpha}$, $\chi \sim t^{-\gamma}$ and $m \sim t^\beta$, it follows that Rushbrooke's inequality $\alpha + \gamma \geq 2(1 - \beta)$ must hold if $\mathcal{R}$ is not to tend to zero.

1.8 We consider the ratio $\mathcal{R}$ defined in the solution to the previous problem. As before we identify $-(\partial^2 f/\partial T^2) = c_B \sim t^{-\alpha}$ and $-(\partial^2 f/\partial T \partial B) = \partial m/\partial T \sim t^{\beta-1}$, but now we recall that at $T = T_c$, $B \sim m^\delta$, so

$$-\frac{\partial^2 f}{\partial B^2} = 1 \Big/ \left(\frac{\partial B}{\partial m}\right)_T \sim m^{1-\delta} \quad (\text{a1.2})$$

for $T = T_c$. But as T approaches T_c from below, the $B(m)$ curves tend smoothly to $B \sim m^\delta$ (see Figure 1.5). So we can use (a1.2) on $B = 0$, $T < T_c$ with m replaced by $m_0 \sim t^\beta$. We then have $-(\partial^2 f/\partial B^2) \sim t^{\beta(1-\delta)}$ and Griffiths' inequality $\alpha + \beta(1 + \delta) \geq 2$ follows from the requirement that $\mathcal{R}$ should not tend to zero as $(T_c, 0)$ is approached.

Answers for Chapter 2

2.1 False; the probability of a system having energy E is proportional to $e^{-E/k_B T}$ times the number of microstates with this energy. A system at a given temperature minimizes its Helmholtz free energy, which is a cross between energy and density of states (equations (2.13) and (2.12)).

Answers for Chapter 2

2.2 The partition function is

$$Z = \int d^d\mathbf{p}_1 \ldots d^d\mathbf{p}_N \, d^d\mathbf{q}_1 \ldots d^d\mathbf{q}_N \, e^{-H/k_BT}$$

$$= \int d^d\mathbf{p}_1 \ldots d^d\mathbf{p}_N \, \exp[\sum \frac{p_i^2}{2mk_BT}]$$

$$\times \int d^d\mathbf{q}_1 \ldots d^d\mathbf{q}_N \, \exp[\frac{U(\mathbf{q}_1,\ldots,\mathbf{q}_N)}{k_BT}], \quad \text{(a2.1)}$$

which has factorized as claimed. The mean kinetic energy $\bar{E}$ of particle 1 is given by

$$\bar{E} = \frac{1}{Z} \int \{d^d\mathbf{p}_i\} \{d^d\mathbf{q}_i\} \, (p_1^2/2m) e^{-H/k_BT}. \quad \text{(a2.2)}$$

The integrals over the $\mathbf{q}_i$ and over $\mathbf{p}_2$ to $\mathbf{p}_N$ cancel, leaving a Gaussian integral: the result is $\bar{E} = k_BTd/2$ (d is the number of dimensions).

Treating the marbles as being in thermal equilibrium gives $T \sim 10^{20}$ K. However, only the degrees of freedom associated with the motions of the centres of mass of the marbles are at this temperature. The remaining $\sim N_A$ degrees of freedom per marble are only at room temperature, so when the shaking stops and the energy of motion gets shared out evenly between the modes, the actual temperature rise is only a small fraction of a degree.

The factorization of Z for a classical system means that the kinetic energy term holds no interesting information; in the models we build we omit such a term. Quantum-mechanically Z does not factorize, because the $\mathbf{q}_i$ and $\mathbf{p}_i$ do not commute; however, we are concerned with long-distance phenomena, for which classical physics is presumably a good approximation, and so our omission of kinetic terms should not affect our evaluation of critical exponents.

2.3 Using equation (2.9) gives

$$S = k_B \left(\log n_0 + \frac{n_1}{n_0} e^{-\beta \Delta E} (1 - \beta \Delta E) \right.$$
$$\left. + \mathrm{O}(e^{-2\beta \Delta E}) \right) + \Phi(\{V\}), \quad \text{(a2.3)}$$

where $\Delta E \equiv E_2 - E_1$. When $\beta \to \infty$ this becomes

$$S = k_B \log n_0 + \Phi(\{V\}). \quad \text{(a2.4)}$$

Unless Φ is a constant, the entropy at absolute zero depends on the constraints, which contradicts the third law. The third law is not being used here as additional experimental input; rather, it is being used to determine which quantity corresponds to classical entropy.

2.4 δ_{r_1,r_2} is 1 when the particles are in the same box, and 0 otherwise, so its thermal average is the required probability p. This average can be got by differentiating Z, since δ_{r_1,r_2} appears in H:

$$Z = \sum_{r_1,r_2=1}^{n} \exp[\beta\kappa\delta_{r_1,r_2}]$$
$$= n(n-1+e^{\beta\kappa}); \quad \text{(a2.5)}$$
$$p = \frac{1}{\beta} \frac{\partial \log Z}{\partial \kappa},$$

which gives the required answer. $-\kappa p$ is the mean energy; since the change in n takes place at constant energy, equating the mean energies for $n=2$ and $n=6$ gives

$$T' = \frac{1}{(k_B/\kappa)\log 5 + T^{-1}}; \quad \text{(a2.6)}$$

this shows a decrease in temperature.

2.5 The partition function for one spin is $Z_1 = 2\cosh(\mu\beta B)$; for N spins, $Z_N = Z_1^N$. F, G, and M are all directly proportional to N, so f, g, and m are independent of N and therefore show no non-analytic behaviour as $N \to \infty$. For example, $m = \tanh(\mu\beta B)$; this has the form of Figure 2.5(a) rather than Figure 2.6(a).

The next part can be treated as a counting problem: if the total spin is zero, $(N/2)$ spins must be up and $(N/2)$ must be down. This can happen in $N!/[(N/2)!]^2$ ways, each with Boltzmann factor 1, so the probability p is

$$p = \frac{N!}{[(N/2)!]^2 Z} \sim (\cosh(\mu\beta B))^{-N} \quad \text{(a2.7)}$$

in the limit of large N. Equation (2.49) can also be used, but the algebra is harder.

2.6 The partition function of this system in a magnetic field B is

$$Z_N = e^{-\beta N\mu B} + e^{\beta N\mu B} = 2\cosh(N\beta\mu B); \quad \text{(a2.8)}$$

we've taken the zero of the spin interaction energy to be when all spins are aligned. The magnetization per spin is $m = \tanh(N\mu B)$. In the limit $N \to \infty$ this has the form of Figure 2.6(a) rather than Figure 2.5(a).

We have $(\partial f/\partial B) = -\mu\tanh(N\mu B)$; this is discontinuous at $B = 0$ when $N \to \infty$. The expression for g is more complicated, but it is of order N^{-1} for all values of $|m| < 1$.

2.7 A magnetic system with Hamiltonian H, total spin S and subject to a magnetic field B has a partition function

$$Z(B) = \sum_\alpha e^{-\beta(H_\alpha - BS)}. \quad \text{(a2.9)}$$

δM, the RMS fluctuation in S about its mean value M, is small for a macroscopic system, and so

$$Z(B) \approx e^{\beta BM} \sum_{\alpha'} e^{-\beta H}, \quad \text{(a2.10)}$$

where we've used the fact that S is nearly constant to take a factor outside the sum. The sum is now only over those states α' for which S is within δM of M. Comparing (a2.10) with (2.56) shows that

$$\rho(M)\mathrm{d}M = \frac{\mathrm{d}M}{\delta M}\frac{Z(B)}{Z(0)}\mathrm{e}^{-\beta BM}. \quad (\text{a2.11})$$

Using equations (2.13) and (2.16), and ignoring the M-dependence of δM, gives (2.49).

Answers for Chapter 3

3.1 Reverse the sign of every spin and of J: this transformation (i) leaves invariant the energy of each configuration, and (ii) turns an antiferromagnet into a ferromagnet, and *vice versa*.

3.2 We use (3.31) with **F** the matrix which equals the usual Pauli spin matrix σ_z in the frame in which **T** is given by (3.18). The eigenvectors of **T** are $2^{-1/2}(1,-1)$ and $2^{-1/2}(1,1)$, and when we take these for our coordinate directions **F** becomes σ_y/i. Moreover, **T**'s eigenvalues are $2\cosh(\beta\epsilon)$ and $2\sinh(\beta\epsilon)$, so (3.31) becomes

$$\langle s_0 s_n \rangle = 0 + (-1)^2 \tanh^n(\beta\epsilon).$$

3.3 The quoted result follows immediately on differentiation of (3.25). By (2.28) and the last problem,

$$\chi/\beta = 1 + 2\tanh(\beta\epsilon) + 2\tanh^2(\beta\epsilon) + \cdots$$
$$= \frac{2}{1-\tanh(\beta\epsilon)} - 1$$
$$= \mathrm{e}^{2\beta\epsilon},$$

which agrees with the quoted result evaluated at $B=0$.

3.4

$$I = \pi^{(n+1)/2} = \int_0^\infty r^n \mathrm{d}r\, \mathrm{e}^{-r^2} \int \mathrm{d}\Omega_n$$
$$= \tfrac{1}{2}\Omega_n \int_0^\infty \mathrm{d}\rho\, \rho^{(n-1)/2}\mathrm{e}^{-\rho}$$
$$= \tfrac{1}{2}\Omega_n[\tfrac{1}{2}(n-1)]!.$$

$\Omega_1 = 2\pi$ and $\Omega_3 = 2\pi^2$ follow immediately. $\Omega_2 = 4\pi$ follows from $\tfrac{1}{2}! = \tfrac{1}{2}\sqrt{\pi}$.

3.5 The spin vector $(s_1, s_2, \ldots, s_N)$ is confined to an $(N-1)$-dimensional sphere of radius $\sqrt{N}$, so

$$\lim_{T\to\infty}\langle s_i^p \rangle = \frac{1}{N^{(N-1)/2}\Omega_{N-1}}$$
$$\times \int \mathrm{d}\theta\, \sqrt{N}\sin\theta N^{p/2}\cos^p\theta$$
$$\times \int N^{(N-2)/2} \sin^{N-2}\theta \mathrm{d}\Omega_{N-2}$$
$$= \frac{\Omega_{N-2}}{\Omega_{N-1}} N^{p/2} I \simeq \frac{N^{(p+1)/2}}{\sqrt{\pi}} I,$$

where $I \equiv \int \mathrm{d}\theta \cos^p\theta \sin^{N-1}\theta$ and Stirling's approximation $\log N! \sim N\log N - N$ has been used. In the limit $N\to\infty$ I becomes more and more dominated by the region $\theta \simeq \pi/2$, where $\sin\theta \simeq 1$. With $\phi \equiv \frac{\pi}{2} - \theta$,

$$I = \int \mathrm{d}\phi\, \sin^p\phi\cos^{N-1}\phi$$
$$\simeq \int_{-1}^1 \mathrm{d}\phi\, \phi^p(1-\tfrac{1}{2}\phi^2)^{N-1}.$$

If p is odd I clearly vanishes. For even p define x by $x/(N-1) = \tfrac{1}{2}\phi^2$ to find that

$$I = 2\int_0^{(N-1)/2} \frac{\mathrm{d}x}{N-1}\left(\frac{2x}{N-1}\right)^{(p-1)/2}$$
$$\times \left(1 - \frac{x}{N-1}\right)^{N-1}$$
$$\simeq \left(\frac{2}{N}\right)^{(p+1)/2}[\tfrac{1}{2}(p-1)]!,$$

where use has been made of

$$\lim_{n\to\infty}(1-x/n)^n = \mathrm{e}^{-x}.$$

3.6 Adding a term $-\beta B \sum_i s_i = -\beta B \tilde{s}_0$ to the Hamiltonian modifies (3.37) to

$$Z = \frac{\mathrm{e}^{N\alpha}}{2\pi\mathrm{i}} \int_{\alpha-\mathrm{i}\infty}^{\alpha+\mathrm{i}\infty}\mathrm{d}p\, \mathrm{e}^{pN}\int \mathrm{d}s_1\ldots\mathrm{d}s_N$$
$$\times \exp\left[\beta B\tilde{s}_0 - \sum_{ij}\left(p\delta_{ij} + \tfrac{1}{2}\beta J_{ij}\right)s_i s_j\right]$$

On transforming to the new integration variables $y_i \equiv \tilde{s}_i/\sqrt{N}$, only the integral over y_0 changes from the field-free case. Hence (3.44) becomes

$$Z = \frac{\pi^{(N-1)/2}\mathrm{e}^{N\alpha}}{2\pi\mathrm{i}}\int_{\alpha-\mathrm{i}\infty}^{\alpha+\mathrm{i}\infty}\mathrm{d}p$$
$$\times \exp\left[pN - \tfrac{1}{2}\sum_{\mathbf{q}\neq 0}(p+\tfrac{1}{2}\beta\tilde{J}_{\mathbf{q}})\right]$$
$$\times \int_{-\infty}^\infty \mathrm{d}y_0\, \exp[\beta B\sqrt{N}y_0 - (p+\beta\tilde{J}_0/2)y_0^2]. \quad (\text{a3.1})$$

The integral over y_0 has value

$$\exp[(\tfrac{1}{2}\beta B)^2 N/(p+\tfrac{1}{2}\beta\tilde{J}_0)]\sqrt{\pi/(p+\tfrac{1}{2}\beta\tilde{J}_0)}.$$

Hence the sum over **q** in (a3.1) can be extended to include $\mathbf{q}=0$ and again converted to an integral. It then follows that (3.51) becomes

$$Z = (\beta\epsilon)^{1-N/2}\frac{\pi^{N/2}\mathrm{e}^{N\alpha}}{2\pi\mathrm{i}}\int_{\alpha'-\mathrm{i}\infty}^{\alpha'+\mathrm{i}\infty}\mathrm{d}\zeta\,\mathrm{e}^{G(\zeta)},$$

where

$$G(\zeta) \equiv g(\zeta) + \frac{(\tfrac{1}{2}\beta B)^2 N}{\beta\epsilon(\zeta - d)}.$$

Equations (3.56) and (3.57) are now obtained by replacing g with G in (3.55) and determining ζ_s from the requirement that G', rather than g', vanish.

Answers for Chapter 4

3.7 Differentiating (3.58) we have
$$\chi = \frac{1}{2\epsilon(\zeta_s - 3)} - \frac{B}{2\epsilon(\zeta_s - 3)^2}\frac{d\zeta_s}{dB}$$

The first term agrees with Box 3.3 and the second vanishes with B provided $\zeta_s > 3$.

3.8 Differentiating the integral $I(\zeta_s)$ of (3.54) we have
$$\frac{dI}{d\zeta_s} = -\int_0^{2\pi} \frac{d\omega_1 d\omega_2 d\omega_3}{(\zeta_s - \sum_k \cos\omega_k)^2}$$
$$\simeq -4\pi \int \frac{\omega^2 d\omega}{((\zeta_s - 3) + \tfrac{1}{2}\omega^2)^2}.$$

Here we've used the fact that the integral is dominated by contributions near the origin of ω-space, where $\cos\omega_k \simeq 1 - \tfrac{1}{2}\omega_k^2$, and have extended the range of integration to $-\infty < \omega_k < \infty$. Evaluating the integral, we find $dI/d\zeta_s = -2\sqrt{2}\pi^2/(\zeta_s - 3)^{1/2}$. Integrating this up it follows that
$$I(\zeta_s) \simeq I(3) - 4\sqrt{2}\pi^2(\zeta_s - 3)^{1/2}.$$

Substituting this approximation in (3.54), the required result follows.

3.9 With the approximation to the integral of (3.57) derived in the last problem, this equation becomes at $\beta = \beta_c$
$$\frac{(\zeta_s - 3)^{1/2}}{\sqrt{2\pi}} = \frac{2(\tfrac{1}{2}\beta_c B)^2}{\beta_c \epsilon(\zeta_s - 3)^2}.$$

It follows that on the critical isotherm $(\zeta_s - 3) \sim B^{4/5}$. Plugging this result into (3.56) and recalling that ζ_s is a stationary point of g, we conclude that $f \sim B^{6/5}$, and thus that $m \sim B^{1/5}$ as required.

3.10
$$u = \frac{\sum_\alpha E_\alpha e^{-\beta E_\alpha}}{\sum_\alpha e^{-\beta E_\alpha}}$$
$$= \frac{\sum_\alpha E_\alpha(1 - \beta E_\alpha + \cdots)}{\sum_\alpha(1 - \beta E_\alpha + \cdots)}$$
$$\simeq u_{\max} - \beta\Big(\frac{1}{N}\sum_\alpha E_\alpha^2 - u_{\max}^2\Big)$$
$$+ O((\beta E_\alpha)^2).$$

Answers for Chapter 4

4.1 The denominator in the expression for $P(\alpha \to \alpha')$ is unaltered if α and α' are exchanged. We therefore have
$$\frac{P(\alpha \to \alpha')}{P(\alpha' \to \alpha)} = \exp[-\beta(E_{\alpha'} - E_\alpha)].$$

This satisfies the microreversibility criterion. P is never zero, so it satisfies the accessibility criterion provided that the algorithm for choosing α' allows any configuration to be reached from an initial α in a finite number of steps.

As (a) $E_{\alpha'} - E_\alpha$ becomes large and positive, the denominator of P tends to $\exp[\tfrac{1}{2}\beta(E_{\alpha'} - E_\alpha)]$. We therefore find
$$P(\alpha \to \alpha') \sim \exp[-\beta(E_{\alpha'} - E_\alpha)].$$

As (b) $E_{\alpha'} - E_\alpha$ becomes large and negative, the denominator of P tends to $\exp[-\tfrac{1}{2}\beta(E_{\alpha'} - E_\alpha)]$, and we obtain
$$P(\alpha \to \alpha') \sim \text{constant}.$$

In both these limits P therefore reduces to the Metropolis algorithm. The Metropolis algorithm is usually preferred because computing the probability of accepting a move requires many fewer floating point operations.

4.2 On integrating by parts, we find that
$$\int_{-\infty}^\infty ds\, g\frac{dg'}{ds} = -\int_{-\infty}^\infty ds\, g'\frac{dg}{ds}$$

for any two differentiable functions g and g' that vanish sufficiently quickly at infinity. The operator d/ds is therefore not self-adjoint. But
$$\int_{-\infty}^\infty ds\, g\frac{d^2 g'}{ds^2} = +\int_{-\infty}^\infty ds\, g'\frac{d^2 g}{ds^2},$$

so the operator d^2/ds^2 is self-adjoint. On expanding $\hat{O}_{\text{FP}}$ we find
$$\hat{O}_{\text{FP}} = \sum_i \Big[\Gamma\Big(\frac{\partial H}{\partial s_i}\frac{\partial}{\partial s_i} + \frac{\partial^2 H}{\partial s_i^2}\Big) + \frac{\sigma^2}{2}\frac{\partial^2}{\partial s_i^2}\Big].$$

The first term on the right contains a single derivative and is therefore not self-adjoint.

The transformed operator $\hat{O}'_{\text{FP}}$ expands to
$$\hat{O}'_{\text{FP}} = \frac{\sigma^2}{2}\sum_i\Big[\frac{\partial^2}{\partial s_i^2} + \frac{\beta}{2}\frac{\partial^2}{\partial s_i^2} - \frac{\beta^2}{4}\Big(\frac{\partial H}{\partial s_i}\Big)^2\Big],$$

in which all terms contain either zero or two derivative operators. This operator is therefore self-adjoint. Now suppose that a function f_n is an eigenfunction of $\hat{O}'_{\text{FP}}$ with eigenvalue λ_n. Then
$$\lambda_n = -\frac{\sigma^2}{2}\prod_i \int_{-\infty}^\infty ds_i\, f_n^* \hat{O}'_{\text{FP}} f_n$$
$$= -\frac{\sigma^2}{2}\prod_i \int_{-\infty}^\infty ds_i \Big|\Big(\frac{\partial}{\partial s_i} + \tfrac{1}{2}\beta\frac{\partial H}{\partial s_i}\Big)f_n\Big|^2$$
$$\leq 0.$$

This establishes that the eigenvalues are non-positive. It is easy to see that the operator in brackets in the last equation gives zero only when it acts on $f_0 \propto \exp(-\beta H/2)$. The function f_0 is therefore the eigenfunction of $\hat{O}'_{\text{FP}}$ with the largest eigenvalue.

It is a standard result that the eigenfunctions of a self-adjoint operator form a complete orthonormal set. We therefore expand an arbitrary initial probability density W in the form

$$W(\mathbf{s}, t=0) = \exp(-\beta H/2) \sum_n a_n f_n(\mathbf{s}).$$

The solution to the Fokker-Planck equation at a later time is then

$$W(\mathbf{s}, t) = \exp(-\beta H/2)$$
$$\times \sum_n a_n \exp(\lambda_n t) f_n(\mathbf{s}).$$

After a sufficiently long time the sum is dominated by the largest value of λ_n. So, unless we are unlucky enough to have chosen an initial W for which $a_0 = 0$, we have for t large

$$W(\mathbf{s}, t) \propto a_0 \exp(-\beta H).$$

The distribution W therefore relaxes towards the Gibbs distribution, as was to be shown.

4.3 The probability that the Langevin equation evolves the variables s_i into the variables $s_i + \Delta s_i$ after a timestep Δt is simply the probability for the random variables f_i to take the values

$$f_i = \frac{\Delta s_i}{\Delta t} + \Gamma \frac{\partial H}{\partial s_i}.$$

Given that the f_i are Gaussianly distributed with mean zero and variance $\sigma^2/\Delta t$, we at once obtain the desired result. This is a Markov process because the f_i are chosen independently at each timestep.

The accessibility criterion is satisfied because there is a non-zero probability for any value of Δs_i. The ratio of the forward and backward transition probabilities is

$$\exp\left[\frac{-\Delta t}{2\sigma^2}\left\{\sum_i \left(\frac{\Delta s_i}{\Delta t} + \Gamma\frac{\partial H}{\partial s_i}\right)^2 - \left(\frac{-\Delta s_i}{\Delta t} + \Gamma\frac{\partial H}{\partial s_i}\right)^2\right\}\right]$$

$$= \exp\left(-\beta \Delta s_i \frac{\partial H}{\partial s_i} + O(\Delta t)\right)$$

$$= \exp[-\beta \Delta H + O(\Delta t)].$$

In the limit $\Delta t \to 0$ this is just what we require for the microreversibility criterion.

4.4 The accessibility criterion is satisfied because, by a suitable choice of the unit vector $\hat{\mathbf{n}}$, a spin $\mathbf{s}$ can be transformed to anywhere on the D-dimensional unit sphere. The proof of microreversibility follows the lines given for the Ising model in the text: on computing the ratio of backward and forward probabilities, only the boundary terms of the flipped spin clusters fail to cancel. The interaction energy between any pair of spins can be written as a sum of two parts, one (involving their components parallel to $\hat{\mathbf{n}}$) which changes sign when one of the spins is reflected in the plane perpendicular to $\mathbf{n}$, and another (involving the components perpendicular to $\mathbf{n}$) which is unchanged. The boundary terms in the ratio of the probabilities give exactly the exponential of β times the energy difference arising from the components parallel to $\hat{\mathbf{n}}$.

Answers for Chapter 5

5.1 The probabilities P_p and P_a for adjacent spins to be parallel and anti-parallel are proportional to e^K and e^{-K} respectively, so the properly normalized probabilities are

$$P_p = \frac{e^K}{e^K + e^{-K}}, \qquad P_a = \frac{e^{-K}}{e^K + e^{-K}}. \tag{a5.1}$$

The ratio P_p/P_a is e^{2K}, so if we can calculate P'_p and P'_a, the probabilities for the renormalized model, we can take their ratio to find K'.

The probability of two adjacent sites being parallel on the decimated lattice is the sum of the probabilities of the configurations ↑↑↑, ↓↓↓, ↑↓↑ and ↓↑↓ on the original lattice, which is

$$P'_p = \frac{e^{2K}}{(e^K + e^{-K})^2} + \frac{e^{-2K}}{(e^K + e^{-K})^2}$$
$$= \frac{\cosh 2K}{2\cosh^2 K}. \tag{a5.2}$$

Similarly, P'_a is the sum of the probabilities of the configurations ↑↑↓, ↑↓↓, ↓↑↑ and ↓↓↑ on the original lattice. All of these have the same probability in fact, and we get

$$P'_a = \frac{1}{2\cosh^2 K}. \tag{a5.3}$$

Then the renormalized coupling constant comes from

$$e^{2K'} = \frac{P'_p}{P'_a} = \frac{1}{\cosh 2K}. \tag{a5.4}$$

So

$$K' = \tfrac{1}{2} \log \cosh 2K. \tag{a5.5}$$

The only fixed point solutions $K = K' = K^*$ for this equation are at $K^* = 0$ and $K^* = \infty$, which correspond to the high- and low-temperature fixed points $T = \infty$ and $T = 0$ respectively.

5.2 Referring to the decimated sites as 'odd' and the retained sites as 'even', we have from equation (5.112)

$$H'_n = -2^{N'} \sum_{\text{even spins}} t'_n \log \sum_{\text{odd spins}} e^{-\mathcal{H}}. \tag{a5.6}$$

The effective Hamiltonian can be written

$$\mathcal{H} = -K \sum_{\langle ij \rangle} \sigma_i \sigma_j - L \sum_{\langle ij \rangle_e} \sigma_i \sigma_j - L \sum_{\langle ij \rangle_o} \sigma_i \sigma_j, \tag{a5.7}$$

where $\langle ij \rangle$ indicates that sites i and j are nearest neighbours and $\langle ij \rangle_e$ and $\langle ij \rangle_o$ indicate nearest neighbours on the sub-lattices of even and odd sites respectively. Expanding the

Answers for Chapter 5

Boltzmann factor $e^{-\mathcal{H}}$ of equation (a5.6) in powers of $\mathcal{H}$, we can then write

$$e^{-\mathcal{H}} = 1 - \mathcal{H} + \tfrac{1}{2}\mathcal{H}^2 + \cdots$$

$$= 1 + K \sum_{\langle ij \rangle} \sigma_i \sigma_j + L \sum_{\langle ij \rangle_e} \sigma_i \sigma_j$$

$$+ L \sum_{\langle ij \rangle_o} \sigma_i \sigma_j + K^2 \sum_{\substack{\langle ij \rangle \\ \langle kl \rangle}} \sigma_i \sigma_j \sigma_k \sigma_l,$$

(a5.8)

where we have kept terms up to $O(K^2)$ and $O(L)$. Now we perform the sum over odd sites in equation (a5.6) and the second and fourth sums in this expression vanish, because they contain exactly as many terms for which the summand is -1 as they do for which it is $+1$. The same is almost true of the last sum. Each term in this sum necessarily includes two odd and two even spins. All the terms in this sum will cancel out when we sum over the odd spins, except those for which the two odd spins are the same spin. Considering the different ways in which two even spins can be adjacent to the same odd spin, we arrive at

$$\sum_{\text{odd spins}} e^{-\mathcal{H}} = 2^{N'} + 2^{4N'} K^2$$

$$+ (L + 2K^2) \sum_{\langle ij \rangle'} \sigma_i \sigma_j$$

$$+ K^2 \sum_{\langle ij \rangle'_{nn}} \sigma_i \sigma_j,$$

(a5.9)

where $\langle ij \rangle'$ and $\langle ij \rangle'_{nn}$ indicate that i and j are nearest and next-nearest neighbours respectively on the renormalized lattice. We now take the logarithm, and then the second sum in (a5.6), over the even spins, picks out the two terms and gives us our renormalized parameters K' and L' as in the problem.

To find the fixed points, we set $K' = K = K^*$ and $L' = L = L^*$, to get

$$K^* = 2K^{*2} + L^*,$$
$$L^* = K^{*2}.$$

(a5.10)

One solution is $K^* = L^* = 0$, which is the high-temperature fixed point. The other we can find by substituting the second equation into the first. The critical fixed point turns out to be at

$$K^* = \tfrac{1}{3}, \quad L^* = \tfrac{1}{9}.$$

(a5.11)

To linearize the renormalization equations about this point we define $\delta K \equiv K - K^*$ and $\delta L \equiv L - L^*$, and write

$$\delta K' = \frac{\partial K'}{\partial K}\delta K + \frac{\partial K'}{\partial L}\delta L = 4K^* \delta K + \delta L,$$

$$\delta L' = \frac{\partial L'}{\partial K}\delta K + \frac{\partial L'}{\partial L}\delta L = 2K^* \delta K,$$

(a5.12)

where all the derivatives are evaluated at the critical fixed point (K^*, L^*). Substituting in the values for K^* and L^*, we find that

$$\begin{pmatrix} K' \\ L' \end{pmatrix} = \mathbf{M} \begin{pmatrix} K \\ L \end{pmatrix},$$

(a5.13)

where the matrix $\mathbf{M}$ is

$$\mathbf{M} \equiv \begin{pmatrix} 4/3 & 1 \\ 2/3 & 0 \end{pmatrix}.$$

(a5.14)

The eigenvalues are $\tfrac{1}{3}(2 \pm \sqrt{10})$. One is negative, as mentioned in §5.9, the other is $\lambda_R = 1.720\ldots$. This is the one which gives us the exponent ν. Using equation (5.26) with $b = \sqrt{2}$, we get

$$\nu = 0.638\ldots$$

(a5.15)

5.3 There are, of course, many different ways to solve this problem on a computer. As an example, we give the central function of the FORTRAN program we used to perform the calculation. This function takes as its argument the interaction parameter K and returns the corresponding value for the renormalized parameter K'. The function would be used as part of a larger program to plot the graph of K' against K, or to find the stationary point and the critical exponent ν.

```
      function kprime(k)
C     Subroutine to calculate the
C     value of the renormalized
C     coupling constant.

      integer b1,b2,block1,block2
      integer s1,s2,s3,s4,s5,s6,s7,s8
      real k,kprime,sum1,sum2
      real factor1,factor2

C     In the outermost loop, we run
C     through the values of the two
C     block spins.  One of them does
C     not actually need to change,
C     since the system is up-down
C     symmetric.

      b2 = 1
      sum1 = 0.
      do b1 = -1,1,2

         sum2 = 0.

C        Now go through the values of the
C        sub-spins compatible with those
C        of the blocks.

         do s1 = -1,1,2
         do s2 = -1,1,2
         do s3 = -1,1,2
         do s4 = -1,1,2
         do s5 = -1,1,2
         do s6 = -1,1,2
         do s7 = -1,1,2
         do s8 = -1,1,2
            block1 = s1+s2+s3+s4
```

```
                block2 = s5+s6+s7+s8
                if (block1.eq.0) then
                    factor1 = 0.5
                elseif (block1*b1.gt.0) then
                    factor1 = 1.
                else
                    factor1 = 0.
                endif
                if (block2.eq.0) then
                    factor2 = 0.5
                elseif (block2*b2.gt.0) then
                    factor2 = 1.
                else
                    factor2 = 0.
                endif
                h = s1*s2+s1*s3+s2*s1+s2*s4+
X                   s3*s4+s3*s5+s4*s3+s4*s6+
X                   s5*s6+s5*s7+s6*s5+s6*s8+
X                   s7*s8+s7*s1+s8*s7+s8*s2
                sum2 = sum2 + factor1*
X                       factor2*exp(k*h)
              enddo
              enddo
              enddo
              enddo
              enddo
              enddo
              enddo
              enddo
              sum1 = sum1 + b1*b2*log(sum2)
            enddo
            kprime = sum1/2.
            return
            end
```

5.4 After n successive renormalizations, we have from equations (5.35) and (5.44):

$$\xi^{(n)} = \xi^{(0)} \prod_{i=1}^{n} b^{-1},$$

$$\chi^{(n)} = \chi^{(0)} \prod_{i=1}^{n} b^{d(1-2\omega^{(i)})}. \quad \text{(a5.16)}$$

If the susceptibility $\chi^{(n)}$ diverges with $\xi^{(n)}$ according to equation (5.64):

$$\chi^{(n)} \sim \xi^{(n)\gamma/\nu}, \quad \text{(a5.17)}$$

then, on substituting from (a5.16) we find that

$$\chi^{(0)} \sim \xi^{(0)\gamma/\nu}, \quad \text{(a5.18)}$$

where γ/ν takes the same value as in equation (a5.17), regardless of the value of n. However, equation (5.67) tells us that $\gamma/\nu = 2 - \eta$, so the value of η is also independent of n. Along with the result of §5.4.4, we now have proof that two of the exponents in this model are independent of n. The scaling laws (5.80) then tell us that the same is true of all the other exponents.

5.5 If the sets $\mathcal{T}_n$ and $\mathcal{T}_m$ are different, then there must be at least one spin which is in one set but not in the other. Summing over the two values of such a spin gives zero. If the two sets are the same, then the product $t_m t_n$ takes the value $+1$ for each of the 2^N terms in the sum over the values of the N spins. Hence result.

Answers for Chapter 6

6.1 We generalize our variational Hamiltonian for the Ising model to allow a separate variational parameter for each site:

$$H_0 = -\sum_i \lambda^{(i)} s_i.$$

Therefore

$$F_0(\{\lambda^{(i)}\}) = -\frac{1}{\beta} \sum_i \log[2\cosh(\beta\lambda^{(i)})].$$

The remaining part of the Hamiltonian is

$$H_1 = -\epsilon \sum_{\text{pairs } ij} s_i s_j + \sum_i (\lambda^{(i)} - B_i) s_i.$$

Adding to F_0 the thermal average of H_1 in the Gibbs distribution generated by H_0, we obtain the variational free energy

$$F_{\text{var}}(\{\lambda^{(i)}\}) = \sum_i \left[-\frac{1}{\beta} \log[2\cosh(\beta\lambda^{(i)})] \right.$$

$$\left. + (\lambda^{(i)} - B_i)\tanh(\beta\lambda^{(i)}) \right]$$

$$- \epsilon \sum_{\text{pairs } ij} \tanh(\beta\lambda^{(i)})\tanh(\beta\lambda^{(j)}).$$

Equating the partial derivative with respect to $\lambda^{(i)}$ to zero, we obtain

$$\lambda_{\min}^{(i)} - B_i = \epsilon \sum_j \tanh(\beta\lambda_{\min}^{(j)}),$$

where the sum goes over sites j neighbouring i. Using this to substitute for the sum over j above, and remembering that a sum over all i and j counts each pair twice, we find

$$F_{\text{var}}(\{\lambda_{\min}^{(k)}\}) = \sum_i \left[-\frac{1}{\beta} \log[2\cosh(\beta\lambda^{(i)})] \right.$$

$$\left. + (\lambda^{(i)} - B_i)\tanh(\beta\lambda^{(i)}) \right].$$

This is equation (6.43) of the text.

6.2 We construct a translationally invariant trial Hamiltonian H_0 in the form

$$H_0 = \frac{1}{2\beta} \sum_i s_i^2 + \sum_i \lambda s_i.$$

The corresponding free energy is then easily shown to be

$$F_0 = -N \tfrac{1}{2} \beta \lambda^2 + \text{constant independent of } \lambda.$$

The average value of the spin s_i in the probability distribution corresponding to H_0 is

$$\langle s_i \rangle_0 = -\beta \lambda.$$

The remaining part of the Hamiltonian is

$$H_1 = -\epsilon \sum_{\text{pairs } ij} s_i s_j - \sum_i \lambda s_i,$$

so the variational free energy is

$$\begin{aligned}F_{\text{var}} &= F_0 + \langle H_1 \rangle_0 \\ &= N\left[-\tfrac{1}{2}\beta\lambda^2 - \tfrac{1}{2}z\epsilon(\beta\lambda)^2 + \lambda^2\beta\right] \\ &= \tfrac{1}{2}N\lambda^2(\beta - z\epsilon\beta^2).\end{aligned}$$

For $\beta < \beta_c = (\epsilon z)^{-1}$, F_{var} is minimized by $\lambda = 0$ and therefore $\langle s_i \rangle = 0$. However, for $\beta < \beta_c$ there is no minimum; the variational free energy can be made arbitrarily negative by a sufficiently large λ, and therefore a sufficiently large (in magnitude) $\langle s_i \rangle$.

6.3 In an inhomogeneous magnetic field, the above analysis changes because the variational parameter must be site-dependent and there is an additional term in the Hamiltonian. If we choose to put the extra term in H_1 (we could equally well have chosen to place it in H_0) we find that the expectation of a spin is

$$\langle s_i \rangle_0 = -\beta \lambda^{(i)},$$

and the variational free energy becomes

$$\begin{aligned}F_{\text{var}} = &-\tfrac{1}{2}\beta \sum_i \lambda^{(i)2} - \beta^2 \epsilon \sum_{\text{pairs } ij} \lambda^{(i)2} \\ &+ \beta \sum_i \lambda^{(i)}(\lambda^{(i)} + B_i).\end{aligned}$$

Making this stationary with respect to $\lambda^{(i)}$ we get

$$\lambda^{(i)} + B_i - \beta\epsilon \sum_{\text{neighbours } j} \lambda^{(j)} = 0.$$

When we eliminate the variational parameters in favour of the average values of the spins, we find

$$\langle s_i \rangle = \beta\left(B_i + \epsilon \sum_{\text{neighbours } j} \langle s_j \rangle\right).$$

This is the same as the first-order expansion, equation (6.45), of the corresponding result for the Ising model. The correlation functions that are deduced from the two expressions are therefore also the same.

6.4 In terms of the discrete Fourier transform of the spins, the Hamiltonian of the Gaussian model with nearest-neighbour interactions can be written

$$\begin{aligned}H = &\frac{1}{2N\beta} \sum_{\mathbf{q}} |\tilde{s}_{\mathbf{q}}|^2 [1 - 2\beta\epsilon \sum_{l=1}^d \cos(2\pi q_l/L)] \\ &+ \frac{1}{N} \sum_{\mathbf{q}} \tilde{s}_{\mathbf{q}} \tilde{B}_{-\mathbf{q}}.\end{aligned}$$

In zero magnetic field, H is a positive definite quadratic form so long as $\beta < \beta_c = (\epsilon z)^{-1}$. We identify $\beta = \beta_c$ as the point at which the model becomes ill-defined, in agreement with mean-field theory.

Performing the Gaussian integral over the real (or imaginary) part of $\tilde{s}_{\mathbf{q}}$, we find for the partition function in a magnetic field

$$Z = (\text{factor independent of } B)$$
$$\times \exp\left[\frac{1}{2N} \sum_{\mathbf{q}} \frac{\tilde{B}_{\mathbf{q}} \tilde{B}_{-\mathbf{q}}}{1 - 2\beta\epsilon \sum_{l=1}^d \cos(2\pi q_l/L)}\right].$$

Taking the second derivative with respect to the magnetic field we obtain by the linear response theorem

$$\tilde{G}_c^{(2)}(\mathbf{q}) = \frac{1}{1 - 2\beta\epsilon \sum_{l=1}^d \cos(2\pi q_l/L)}.$$

This agrees with the mean-field result (equation (6.48)) for the Ising model, and therefore (see the previous problem) with the mean-field result for the Gaussian model.

Answers for Chapter 7

7.1 The Landau approximation to f is

$$f = \beta^{-1}(\text{constant} + \tfrac{1}{2}\phi^2 + \tfrac{1}{3}a\phi^3 + \tfrac{1}{4}b\phi^4).$$

Differentiating w.r.t. ϕ and solving the resulting cubic equation we find that possible values of the mean field are

$$\phi_0 = \begin{cases} 0, \\ \dfrac{-a \pm \sqrt{a^2 - 4b}}{2b}, & a^2 > 4b. \end{cases}$$

At $\phi = 0$, $\partial^2 f/\partial\phi^2 = \beta^{-1}$, so $\phi = 0$ is always a local minimum $f = f(0)$ of f. If $a^2 \leq 4b$ it is the only minimum. But if $a^2 > 4b$, there is a second local minimum. It first appears at

$$(\phi, f) = \left(-\frac{2}{a}, f(0) + \frac{1}{3a^2\beta}\right).$$

As b increases, the value of this local minimum in f decreases, and eventually drops below $f(0)$ at $|\phi_0| > 0$.

7.2 The mean field ϕ_0 in the presence of a non-zero field B is determined by (7.23). A plot of $\mu^2\phi + \tfrac{1}{3!}\lambda\phi^3$ versus ϕ demonstrates that ϕ_0 changes continuously as μ decreases through zero; μ^2 is the slope of the graph at $\phi = 0$:

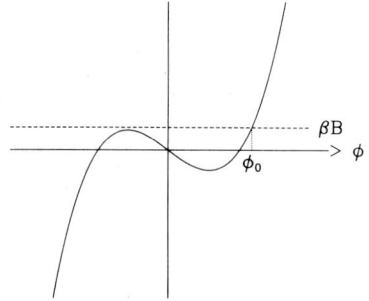

Differentiating (7.17) to find u yields

$$u = \tfrac{1}{2}\phi_0^2 \frac{\mathrm{d}\mu^2}{\mathrm{d}\beta} + \tfrac{1}{4!}\phi_0^4 \frac{\mathrm{d}\lambda}{\mathrm{d}\beta} - B\phi_0$$
$$+ \left(\mu^2 \phi_0 + \tfrac{1}{3!}\lambda\phi_0^3 - \beta B\right)\frac{\mathrm{d}\phi_0}{\mathrm{d}\beta}$$

and we see that u is continuous since it is a sum of continuous and non-singular terms.

7.3 Extremizing $f = \mathcal{H}/\beta$ w.r.t. ϕ yields $\mu^2 \phi_0 + \lambda \phi_0^5 = 0$, implying

$$\phi_0 = \begin{cases} 0, & T > T_c; \\ (|\mu^2|/\lambda)^{1/4}, & T < T_c. \end{cases} \quad (\text{a7.1})$$

Hence $\beta = \tfrac{1}{4}$. With this value of ϕ_0, f is given by

$$\beta f = \text{const} + \begin{cases} 0, & T > T_c; \\ -\tfrac{1}{3}|\mu^2|^{3/2}/\lambda^{1/2}, & T < T_c. \end{cases}$$

Differentiating to get $u = \mathrm{d}(\beta f)/\mathrm{d}\beta$ we find for $T < T_c$

$$u = -\tfrac{1}{2}(|\mu^2|/\lambda)^{1/2}\frac{\mathrm{d}|\mu^2|}{\mathrm{d}\beta} + \tfrac{1}{6}(|\mu^2|/\lambda)^{3/2}\frac{\mathrm{d}\lambda}{\mathrm{d}\beta},$$

which is continuous at $T = T_c$. Differentiating again to get $k_B c_B = -\beta^2 (\mathrm{d}u/\mathrm{d}\beta)$ we find that c_B contains a term $\propto |\mu^2|^{-1/2}$, implying that $\alpha = \tfrac{1}{2}$.

Adding to $\mathcal{H}$ a term $-\beta B$ and differentiating w.r.t. ϕ to find the mean field, we have

$$\mu^2 \phi_0 + \lambda \phi_0^5 = \beta B. \quad (\text{a7.2})$$

At $T = T_c$, $\mu^2 = 0$ and $\phi_0 \propto B^{1/5}$, implying $\delta = 5$. Differentiating (a7.2) w.r.t. B, we have

$$(\mu^2 + 5\lambda\phi_0^4)\chi = \beta,$$

so with (a7.1), $\chi \propto 1/|\mu^2|$ on both sides of T_c, implying that $\gamma = 1$.

When a term $\tfrac{1}{4}\lambda_4 \phi^4$ is present, ϕ_0 is the root of

$$\mu^2 \phi + \lambda_4 \phi^3 + \lambda\phi^5 = \beta B. \quad (\text{a7.3})$$

At $B = 0$ the transition still occurs as μ^2 passes through zero and $\phi_0 = 0$. Hence, near the transition, the third term on the left of (a7.3) is negligible compared to the second term, and for $\lambda_4 \neq 0$, the value taken by $\mathrm{d}\phi_0/\mathrm{d}\beta$ at the transition is as in Landau–Ginzburg theory, and the critical exponents take their mean-field values. For example, given that $\lambda\mu^2 \ll \lambda_4^2$, when $B = 0$ (a7.3) implies that $\phi_0^2 = -\mu^2/\lambda_4^2$.

Answers for Chapter 8

8.1 From the definition,

$$\langle \Phi^2 \rangle = \int \mathrm{d}^d \mathbf{x}\, \mathrm{d}^d \mathbf{y}\, \mathrm{e}^{-(x^2+y^2)/a^2} \langle \phi(\mathbf{x})\phi(\mathbf{y})\rangle. \quad (\text{a8.1})$$

Writing

$$\langle \phi(\mathbf{x})\phi(\mathbf{y})\rangle = G_c^{(2)}(\mathbf{x}, \mathbf{y})$$
$$= \int \mathrm{d}^d \tilde{k}\, \mathrm{e}^{\mathrm{i}\mathbf{k}\cdot(\mathbf{x}-\mathbf{y})} \widetilde{G}_c^{(2)}(\mathbf{k}), \quad (\text{a8.2})$$

and doing the $\mathbf{x}$ and $\mathbf{y}$ integrals, gives the required answer. In the Gaussian model, $\widetilde{G}_c^{(2)}(\mathbf{k}) = (\alpha^2 k^2 + \mu^2)^{-1}$. The condition $a\mu/\alpha \gg 1$ means that $\alpha^2 k^2$ in the denominator can be neglected in the integral, giving

$$\langle \Phi^2 \rangle = \left(\frac{\pi}{2}\right)^{3/2} \frac{a^3}{\mu^2}. \quad (\text{a8.3})$$

The absence of α from the answer reflects the fact that fluctuations in ϕ on scales much greater than the correlation length are determined by the ϕ^2 term in the Hamiltonian, rather than by the gradient term.

8.2 Expanding $\mathrm{e}^{-\lambda x^4}$ and exchanging the sum and integral gives

$$I(\lambda) = \sum_{n=0}^{\infty} \frac{(-\lambda)^n}{n!} \int_{-\infty}^{\infty} \mathrm{d}x\, x^{4n} \mathrm{e}^{-\mu^2 x^2}. \quad (\text{a8.4})$$

Evaluating the integrals gives

$$I(\lambda) = \frac{\pi^{1/2}}{\mu} \sum_{n=0}^{\infty} \left(\frac{-\lambda}{4\mu^4}\right)^n \frac{(4n-1)!!}{n!}. \quad (\text{a8.5})$$

The ratio of successive terms in this series is proportional to n for large n, and so the series always diverges. However, if λ/μ^4 is small, the first few terms can give a fair approximation to I. For example, if $\mu = 1$ and $\lambda = 0.1$, then $I = 1.674\ldots$, while the first three terms of the series give $I = 1.697\ldots$.

8.3

(i)

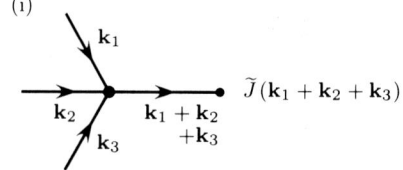

Answers for Chapter 8

The expression for this diagram is

$$\frac{-\lambda \tilde{J}(\mathbf{k}_1 + \mathbf{k}_2 + \mathbf{k}_3)}{(\alpha^2 k_1^2 + \mu^2)(\alpha^2 k_2^2 + \mu^2)(\alpha^2 k_3^2 + \mu^2)}$$
$$\times \frac{1}{(\alpha^2(\mathbf{k}_1 + \mathbf{k}_2 + \mathbf{k}_3)^2 + \mu^2)}.$$

Since it contributes to $G_c^{(3)}$, it has propagators for its external legs. It vanishes if $J = 0$, as we would expect by symmetry. The symmetry factor is 1.

(ii)

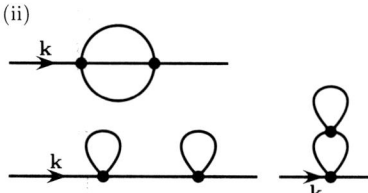

The expression for the top diagram is

$$\frac{\lambda^2}{6} \frac{1}{(\alpha^2 k^2 + \mu^2)^2} \int^\Lambda \frac{d^d\check{q}_1 d^d\check{q}_2}{(\alpha^2 q_1^2 + \mu^2)(\alpha^2 q_2^2 + \mu^2)}$$
$$\times \frac{1}{(\alpha^2(\mathbf{k} - \mathbf{q}_1 - \mathbf{q}_2)^2 + \mu^2)}.$$

The symmetry factor of $\frac{1}{6}$ comes from the three links joining the same two points.

The bottom left diagram is

$$\frac{\lambda^2}{4} \frac{1}{(\alpha^2 k^2 + \mu^2)^3} \left(\int^\Lambda \frac{d^d\check{q}}{\alpha^2 q^2 + \mu^2} \right)^2.$$

There is a symmetry factor of $\frac{1}{2}$ from each link with its two ends joined to the same vertex.

The bottom right diagram is

$$\frac{\lambda^2}{4} \frac{1}{(\alpha^2 k^2 + \mu^2)^2} \int^\Lambda \frac{d^d\check{q}_1 d^d\check{q}_2}{(\alpha^2 q_1^2 + \mu^2)^2(\alpha^2 q_2^2 + \mu^2)}.$$

Wavevector $\mathbf{q}_2$ flows around the top link, and $\mathbf{q}_1$ flows in the two lower links. A symmetry factor of $\frac{1}{2}$ comes from the top link, whose ends are both joined to the same vertex; there is another factor of $\frac{1}{2}$, because the two links below it join the same pair of vertices.

(iii)

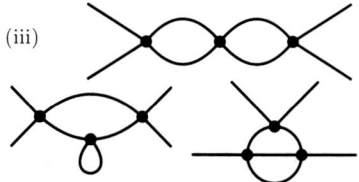

Only 1PI diagrams contribute to $\Gamma^{(4)}$. There is an extra minus sign, and there are no propagators for external links. The wavevectors flowing into the diagram are $\mathbf{k}_1$, $\mathbf{k}_2$, $\mathbf{k}_3$, and $\mathbf{k}_4$, starting north-west of vertical and going around anticlockwise.

The expression for the top diagram is

$$\frac{\lambda^3}{4} \int^\Lambda \frac{d^d\check{q}_1 d^d\check{q}_2}{(\alpha^2 q_1^2 + \mu^2)(\alpha^2(\mathbf{q}_1 + \mathbf{k}_1 + \mathbf{k}_2)^2 + \mu^2)}$$
$$\times \frac{1}{(\alpha^2 q_2^2 + \mu^2)(\alpha^2(\mathbf{q}_2 + \mathbf{k}_1 + \mathbf{k}_2)^2 + \mu^2)}.$$

Notice that this can be written in several different ways; $-\mathbf{k}_3 - \mathbf{k}_4$ could replace $\mathbf{k}_1 + \mathbf{k}_2$; $\mathbf{q}_1$ and $\mathbf{q}_2$ could have constants added. Any correct answer can be written in the above form by a suitable change of variables. The symmetry factor is $\frac{1}{2}$ for each pair of links joining the same two vertices.

The expression for the bottom left diagram is

$$\frac{\lambda^3}{2} \int^\Lambda \frac{d^d\check{q}_1 d^d\check{q}_2}{(\alpha^2 q_1^2 + \mu^2)^2(\alpha^2 q_2^2 + \mu^2)}$$
$$\times \frac{1}{(\alpha^2(\mathbf{q}_1 + \mathbf{k}_1 + \mathbf{k}_2)^2 + \mu^2)}.$$

$\mathbf{q}_1$ flows around the main loop; $\mathbf{q}_2$ flows around the small wart. Note that the bottom link of the main loop contributes two propagators, not one. The only contribution to the symmetry factor is $\frac{1}{2}$ from the wart. If there were a wart on the top link of the main loop, the symmetry factor would be $\frac{1}{8}$; two factors of $\frac{1}{2}$ from the warts, and a factor of $\frac{1}{2}$ because the bottom and top links would be identical.

The expression for the bottom right diagram is

$$\frac{\lambda^3}{2} \int^\Lambda \frac{d^d\check{q}_1 d^d\check{q}_2}{(\alpha^2 q_1^2 + \mu^2)(\alpha^2(\mathbf{k}_1 + \mathbf{q}_1 - \mathbf{q}_2)^2 + \mu^2)}$$
$$\times \frac{1}{(\alpha^2 q_2^2 + \mu^2)(\alpha^2(\mathbf{k}_1 + \mathbf{k}_4 + \mathbf{q}_2)^2 + \mu^2)}.$$

The symmetry factor is $\frac{1}{2}$ from the two links that join the same two vertices.

However, the above three expressions are not the complete two-loop contribution to $\Gamma^{(4)}$. The wavevectors can be assigned to the external legs of these diagrams in a number of topologically distinct ways, and this generates new contributions. Suppose that, in the first diagram, $\mathbf{k}_1$ remains joined to the top left. To the bottom left can be joined $\mathbf{k}_2$, $\mathbf{k}_3$, or $\mathbf{k}_4$. Once one has been chosen, the remaining two must be joined to the links on the right. The order of the links in each pair is immaterial. So there are three diagrams like the first one. Exactly the same argument applies to the bottom left diagram. The bottom right diagram is different. Six distinct pairs of wavevectors can appear on the top pair of links; the other two must then go to the links at the side. Which one goes on which side does not matter; the two possible arrangements can be deformed into each other without cutting links. So in all, there are six such diagrams.

8.4 Functionally differentiating (8.48) three times gives

$$\Gamma^{(3)}(\mathbf{k}_1,\mathbf{k}_2,\mathbf{k}_3;\tilde{\varphi}_0) = \sum_{n=0}^{\infty} \frac{1}{n!} \int d^d\check{\mathbf{q}}_1 \ldots d^d\check{\mathbf{q}}_n$$

$$\times \tilde{\varphi}(-\mathbf{q}_1)\ldots\tilde{\varphi}(-\mathbf{q}_n)\Gamma^{(n+3)}(\mathbf{k}_1,\mathbf{k}_2,\mathbf{q}_1,\ldots,\mathbf{q}_n)$$

(note that $\Gamma^{(n)}$ is symmetric under the exchange of any two of its wavevector arguments, so the order of the arguments in $\Gamma^{(n+3)}$ in (a8.5) does not matter).

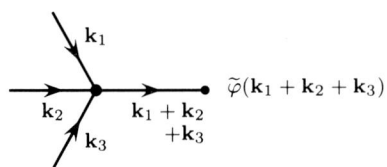

The only term in the expansion of $\Gamma^{(3)}$ that contributes at $O(\lambda)$ is the term involving $\Gamma^{(4)}$; the diagram is shown above. The corresponding expression is

$$\lambda\tilde{\varphi}(\mathbf{k}_1+\mathbf{k}_2+\mathbf{k}_3).$$

This should be compared with a8.3(i). The expressions differ by a minus sign and the inclusions of propagators for the external links, but otherwise they are analogous. The reason is that $\log Z[J]$ can be written

$$\log Z[J] = \sum_{n=0}^{\infty} \frac{1}{n!} \int d^d\check{\mathbf{q}}_1 \ldots d^d\check{\mathbf{q}}_n$$

$$\times \tilde{J}(-\mathbf{q}_1)\ldots\tilde{J}(-\mathbf{q}_n)\tilde{G}_c^{(n)}(\mathbf{q}_1,\ldots,\mathbf{q}_n)$$

(because of equation (8.21)), which is just like equation (8.48) for Γ. In particular, differentiating this gives

$$\tilde{G}_c^{(3)}(\mathbf{k}_1,\mathbf{k}_2,\mathbf{k}_3;J) = \sum_{n=0}^{\infty} \frac{1}{n!} \int d^d\check{\mathbf{q}}_1\ldots d^d\check{\mathbf{q}}_n$$

$$\times \tilde{J}(-\mathbf{q}_1)\ldots\tilde{J}(-\mathbf{q}_n)\tilde{G}_c^{(n+3)}(\mathbf{k}_1,\mathbf{k}_2,\mathbf{q}_1,\ldots,\mathbf{q}_n)$$

which has the same structure as the series for $\Gamma^{(3)}$.

8.5 Draw the Feynman diagram contributing to $\Gamma^{(n)}$ without its external legs (which in any case do not contribute to the corresponding expression), and imagine attaching a surface all the way around the edge of the diagram, to make a closed surface in 3-space. This closed surface can be viewed as a polyhedron, whose edges are the internal links of the Feynman diagram, whose vertices are the vertices of the diagram, and whose faces are the loops of the diagram plus the extra surface used to make the closed surface. Euler's theorem for the polyhedron (vertices + faces = edges +2) then gives equation (8.67).

8.6 For small J, the maxima of the exponent are at

$$\phi = \pm\sqrt{\frac{6m^2}{\lambda}} + \frac{J}{2m^2} \quad (m^2 = -\mu^2); \quad \text{(a8.6)}$$

at these points the exponent has the values

$$-V\frac{3m^4}{2\lambda} \pm V\sqrt{\frac{6m^2}{\lambda}}J. \quad \text{(a8.7)}$$

The value of $\langle\phi\rangle$, averaged over these two points, is

$$\langle\phi\rangle = \sqrt{\frac{6m^2}{\lambda}}\tanh\left(VJ\sqrt{\frac{6m^2}{\lambda}}\right). \quad \text{(a8.8)}$$

In the limit $V \to \infty$ the gradient of $\langle\phi\rangle$ vs. J becomes infinite at $J = 0$, and $\langle\phi(J)\rangle$ is like Figure 2.6(a). It follows from this (or from direct calculation) that f and g have the forms of the other graphs in this figure.

Since $\langle\phi\rangle$, f, and g are only non-analytic for infinite V, we might hope that perturbation theory in a finite volume would reproduce their correct analytic forms. This doesn't work: the above results, expanded in powers of V, contain only positive powers of V. Perturbation theory doesn't produce positive powers of V, because the $V \to \infty$ limit of a finite number of such terms does not exist, and perturbation theory generates only a finite number of terms at one time. Moreover, the results cannot be expanded in a series of non-negative powers of λ.

8.7 The required probability is the sum of the probabilities of finding the system in each microstate for which $(\int d^d\mathbf{x}\,\phi)/V$ lies between ϕ_0 and $\phi_0 + d\phi_0$. Now

$$\frac{1}{V}\int d^d\mathbf{x}\,\phi = \frac{\tilde{\phi}(0)}{V}. \quad \text{(a8.9)}$$

So the probability is the functional integral of e^{-H}/Z over all of the $\mathbf{k} \neq 0$ components of $\tilde{\phi}$, plus an integral over $\tilde{\phi}(0)$ between $V\phi_0$ and $V(\phi_0 + \delta\phi_0)$; this gives the claimed expression for $\rho(\phi_0)$.

To proceed, write $\phi(\mathbf{x}) = \phi_0 + \psi(\mathbf{x})$, where ϕ_0 is independent of position and $\int d^d\mathbf{x}\,\psi(\mathbf{x}) = 0$. Then the Landau–Ginzburg Hamiltonian becomes:

$$H = V\left(\tfrac{1}{2}\mu^2\phi_0^2 + \tfrac{1}{4!}\phi_0^4\right)$$

$$+ \int d^d\mathbf{x}\,\tfrac{1}{2}\alpha^2|\nabla\psi|^2 + \tfrac{1}{2}\mu^2\psi^2 \quad \text{(a8.10)}$$

$$+ \tfrac{1}{4!}\lambda(\psi^4 + 4\psi^3\phi_0 + 6\psi^2\phi_0^2).$$

Then

$$-\log\rho(\phi_0) = V(\tfrac{1}{2}\mu^2\phi_0^2 + \tfrac{1}{4!}\lambda\phi_0^4) \quad \text{(a8.11)}$$

$$-\log Z_\psi(\phi_0) + \log Z_\psi(0),$$

where

$$Z_\psi(\phi_0) \equiv \int \mathcal{D}\psi\,e^{-H_\psi}, \quad \text{(a8.12)}$$

H_ψ being all the ψ-dependent terms from H. The first term in (a8.11) is the zero-loop contribution to Γ (equation (8.63)). The remaining two terms generate the higher-loop contributions to Γ. The Feynman rules for Z_ψ are: a propagator $(\alpha^2 k^2 + \mu^2)^{-1}$ when $\mathbf{k} \neq 0$, and zero when $\mathbf{k} = 0$ (since ψ has no $\mathbf{k} = 0$ component); a four-leg vertex $(-\lambda)$; a three-leg vertex $(-\lambda)\phi_0$; a two-leg vertex $(-\lambda)\phi_0^2/2$. Each logarithm is the sum of connected diagrams with no external legs; only terms involving ϕ_0 survive in the difference of the two logarithms. It's easy to see that all these diagrams are 1PI: because the diagrams have no external legs, conservation of wavevector demands that the wavevector flowing along the link joining two 1PI 'islands' of a diagram must vanish. But the propagator vanishes at zero wavevector, so there are no 1PR diagrams. The $\psi^2 \phi_0^2/2$ vertex in the rules for $-\log \rho(\phi_0)$ joins two ψ-legs and provides a factor $\phi_0^2/2$. It corresponds to the vertices in diagrams such as Figure 8.11 contributing to $\Gamma(\phi_0)$, where two internal links join two external links carrying zero wavevector; remember, from (8.69), that a factor of ϕ_0 goes with each of these external links. Similarly, the $\psi^3 \phi_0$ vertex corresponds to a vertex in a diagram contributing to $\Gamma(\phi_0)$ where three internal links join one external link carrying zero wavevector. Even the symmetry factors work.

So Γ does have a physical significance, even though below the critical temperature it is no longer the Gibbs free energy.

Answers for Chapter 9

9.1 Equation (6.81) states that at T_c,
$$\widetilde{G}_c^{(2)}(k) = 1/\Gamma^{(2)}(k) \sim k^{-2+\eta}.$$
Differentiating we find $d\Gamma^{(2)}/dk^2 \sim k^{-\eta}$.

9.2 Differentiating (9.13) we have
$$\frac{d\Gamma^{(2)}}{dk^2} = a^2 - \tfrac{1}{6}\lambda^2\left(B + k^2 \frac{\partial B}{\partial k^2}\right).$$
Hence one needs to show that $\kappa^2(\partial B/\partial k^2)_\kappa = -B(\kappa, \kappa)$. But differentiating (9.14) yields
$$\frac{\partial B}{\partial k^2} = -\frac{\Delta A}{k^4} + \frac{1}{k^2}\frac{\partial \Delta A}{\partial k^2},$$
and from the first line of (9.15) we have
$$B = \frac{\Delta A}{k^2} - \frac{dA}{dk^2}\bigg|_\kappa.$$
Since $d\Delta A/dk^2 = dA/dk^2$, the result now follows.

9.3 Since the value of $\Gamma^{(2)}(q)$ must be the same in each case,
$$\alpha^2 q^2 + \mu^2 + \sum_i D_i(q) = a^2 q^2 + m^2 + \sum_i D_i'(q).$$
(a9.1)

From the definitions of m^2 and a^2,
$$m^2 = \mu^2 + \sum_i D_i(0);$$
$$a^2 = \alpha^2 + \frac{d}{dq^2}\sum_i D_i(q^2)\bigg|_{q=\kappa}.$$
Substituting the definitions of m^2 and a^2 into (a9.1) gives
$$D_i'(q) = D_i(q) - D_i(0) - q^2 \frac{d}{dq^2}D_i(q^2)\bigg|_{q=\kappa}$$
as the modified contribution of the ith diagram to $\Gamma^{(2)}$.

9.4 It suffices to show that each of the three integrands of (9.19) separately gives rise to a Λ-insensitive integral. So consider
$$I(K) \equiv \int_0^\Lambda \frac{d^d \tilde{q}}{(a^2 q^2 + m^2)^2} \frac{2a^2 \mathbf{q} \cdot \mathbf{K} - a^2 K^2}{a^2(\mathbf{K} - \mathbf{q})^2 + m^2}$$
(a9.2)
As in Box 9.1 we average the integrand over all directions of $\mathbf{K}$ before integrating over $\mathbf{q}$. The angle-dependent part of this integral is identical with that of equation (1) of Box 9.1. So after angle-averaging it is
$$\left\langle \frac{2a^2 \mathbf{q} \cdot \mathbf{K} - a^2 K^2}{a^2(\mathbf{K} - \mathbf{q})^2 + m^2} \right\rangle_\mathbf{K} = \frac{K^2}{\mathcal{Q}^2}\left(2\frac{q^2}{\mathcal{Q}^2} - 1\right) + $$
$$+ O((k^2/\mathcal{Q}^2)^2),$$
where $\mathcal{Q} \equiv K^2 + q^2 + m^2/a^2$. Thus at large q the angle-average of the integrand of (a9.2) scales as q^6 and the required result follows.

9.5 From equation (9.21) we have
$$\Gamma^{(4)}(\mathbf{k}_1, \mathbf{k}_2, \mathbf{k}_3) - \Gamma^{(4)}(\boldsymbol{\kappa}_1, \boldsymbol{\kappa}_2, \boldsymbol{\kappa}_3) + g$$
$$= g - \frac{g^2}{2}\int \frac{d^d \tilde{q}}{(a^2 q^2 + m^2)} \times$$
$$\left(\frac{2a^2 \mathbf{q} \cdot (\mathbf{k}_{12} - \boldsymbol{\kappa}_{12}) - a^2(k_{12}^2 - \kappa_{12}^2)}{[a^2(\mathbf{k}_{12} - \mathbf{q})^2 + m^2][a^2(\boldsymbol{\kappa}_{12} - \mathbf{q})^2 + m^2]}\right.$$
$$\left. + 2 \text{ permutations}\right) + O(g^3),$$
where $\mathbf{k}_{12} \equiv \mathbf{k}_1 + \mathbf{k}_2$ and $\boldsymbol{\kappa}_{12} \equiv \boldsymbol{\kappa}_1 + \boldsymbol{\kappa}_2$.

Answers for Chapter 10

10.1 On making the given substitution, the integral in (10.64) can be evaluated and comes to π^3. This should be the same as $J(1)$ if we set $\epsilon = 1$, $d = 3$ in equation (10.94). Doing this, we get
$$J(1) = \pi^{3/2}\frac{[(-\tfrac{1}{2})!]^3}{0!}.$$
(a10.1)
Since $0! = 1$ and $(-\tfrac{1}{2})! = \sqrt{\pi}$ (see Box 10.2), we have the required agreement.

10.2 Taking the logarithm of equation (9.13) and setting $m = 0$, we have

$$\log \Gamma^{(2)}(k) = \log a^2 k^2$$
$$+ \log\left[1 - \frac{\lambda^2}{6a^2} B(k, \kappa)\right]$$
$$= 2\log ak - \frac{\lambda^2}{6a^2} B(k, \kappa) + O(\lambda^3).$$
(a10.2)

Equation (9.15) gives us $B(k, \kappa)$, but we can ignore the second term in the integrand where differentiation with respect to k is concerned, since it has no dependence on k. The numerator of the first term can then be re-written as $\alpha^2[Q^2 - (\mathbf{k} - \mathbf{Q})^2]$. Replacing α with a as described in §9.2 and setting $m = 0$, we have

$$B(k, \kappa) = \frac{1}{k^2} \int_0^\Lambda \frac{d^d \check{\mathbf{q}}_1 d^d \check{\mathbf{q}}_2}{a^6 q_1^2 q_2^2}$$
$$\times \left[\frac{1}{(\mathbf{k} - \mathbf{Q})^2} - \frac{1}{Q^2}\right]$$
$$+ \text{ terms independent of } k.$$
(a10.3)

Comparing this and (a10.2) with (10.43) and (10.41), one can convince oneself that we should still arrive at equation (10.51).

10.3 Proof of the generalized Feynman formula follows exactly the lines of the proof for two parameters. We have

$$\prod_{i=1}^n \frac{1}{a_i^{\alpha_i}} = \frac{1}{\prod_i (\alpha_i - 1)!} \int_0^\infty dt_1 \ldots dt_n$$
$$\times e^{-\Sigma_i a_i t_i} t_1^{\alpha_1 - 1} \ldots t_n^{\alpha_n - 1}.$$
(a10.4)

Defining Feynman parameters as suggested, with the constraint that $\sum_i u_i = 1$, we find that the Jacobian is again simply s and as before the integral over s can be performed using equation (10.83) to give the desired result, equation (10.89).

10.4 We have

$$\sum_{n=0}^\infty \sum_{m=1}^\infty a_{nm} \epsilon^n \hat{g}_c^m = 0.$$

Differentiating w.r.t. a_{pq} gives

$$\epsilon^p \hat{g}_c^q + \left(\sum_{n,m} a_{nm} \epsilon^n m \hat{g}_c^{m-1}\right) \sum_r \epsilon^r \frac{\partial b_r}{\partial a_{pq}}.$$

Since $\hat{g}_c$ is not larger than $O(\epsilon)$, the first term is at least $O(\epsilon^{p+q})$. Hence the coefficient of any lower power of ϵ in the second term must vanish. The $n = 0$, $m = 1$ term of the quantity in brackets is $O(1)$. However a_{01} vanishes in the Landau–Ginzburg model, and so the bracket is in fact $O(\epsilon)$. It follows that the coefficient of ϵ^2 in the second term is $\partial b_1/\partial a_{pq}$,

which must therefore vanish if $p + q > 2$. Repeatedly applying this argument we find that

$$\frac{\partial b_r}{\partial a_{pq}} = 0 \quad \text{if} \quad p + q > r + 1.$$

So the $O(\epsilon)$ value of $\hat{g}_c$ calculated in (10.101) does not change when the β-function is calculated to higher orders.

Answers for Chapter 11

11.1 The partial derivative $\partial H/\partial x_i$, when ψ and $\mathbf{p}$ take the values that solve the differential equation, can be expressed as

$$\frac{\partial H}{\partial x_i} = \left(\frac{\partial H}{\partial x_i}\right)_{\mathbf{p}, \psi} + \left(\frac{\partial H}{\partial p_j}\right)_{\mathbf{x}, \psi} \frac{\partial p_j}{\partial x_i}$$
$$+ \left(\frac{\partial H}{\partial \psi}\right)_{\mathbf{x}, \mathbf{p}} \frac{\partial \psi}{\partial x_i}$$
$$= \left(\frac{\partial H}{\partial x_i}\right)_{\mathbf{p}, \psi} + \frac{dx_j}{d\lambda} \frac{\partial p_j}{\partial x_i} + \left(\frac{\partial H}{\partial \psi}\right)_{\mathbf{x}, \mathbf{p}} p_i$$
$$= 0.$$

The summation convention is used in the second term on the right and we have used the equations satisfied by the curve $\mathbf{x}(\lambda)$ and the definition of p_i to go from the first line to the second. We can equate the whole expression to zero because, when the differential equation is satisfied, $H = 0$ for all $\mathbf{x}$.

Because $p_i \equiv \partial \psi/\partial x_i$, we know that $\partial p_i/\partial x_j = \partial p_j/\partial x_i$. Using this, we see that the second term on the right above is simply $dp_i/d\lambda$. Therefore,

$$\frac{dp_i}{d\lambda} = -p_i \left(\frac{\partial H}{\partial \psi}\right)_{\mathbf{x}, \mathbf{p}} - \left(\frac{\partial H}{\partial x_i}\right)_{\mathbf{x}, \psi}.$$

To get a differential equation for the function ψ itself, we write

$$\frac{d\psi}{d\lambda} = \frac{\partial \psi}{\partial x_i} \frac{dx_i}{d\lambda} = p_i \left(\frac{\partial H}{\partial p_i}\right)_{\mathbf{x}, \lambda}.$$

11.2 (a) With this form for β, equation (11.21) becomes a separable first-order differential equation which can be integrated immediately to give

$$\log\left|\frac{\hat{g}(b) - \hat{g}_c}{\hat{g}(b=1) - \hat{g}_c}\right| = -A \log b,$$

or

$$\hat{g}(b) - \hat{g}_c \sim b^{-A}.$$

The running coupling constant therefore approaches a value $\hat{g}_c$ at which the β-function has a simple zero in a power-law manner for large b if $A > 0$.

(b) The integration here goes the same way as in (a), but this time the result (assuming $\hat{g}(1) > \hat{g}_c$) is

$$\frac{1}{p-1}[(\hat{g}(1) - \hat{g}_c)^{1-p} - (\hat{g}(b) - \hat{g}_c)^{1-p}]$$
$$= -B \log b.$$

In the limit of large b we therefore have

$$\hat{g}(b) - \hat{g}_c \sim [(p-1)B \log b]^{1/(1-p)}.$$

In this case the convergence of $\hat{g}(b)$ towards $\hat{g}_c$ goes only logarithmically. For larger and larger p, the convergence goes more and more slowly.

(c) The integration of the differential equation now gives

$$\tfrac{1}{2}[(\hat{g}(b) - \hat{g}_c)^2 - (\hat{g}(1) - \hat{g}_c)^2] = -C \log b.$$

Solving for $\hat{g}(b)$, we find

$$\hat{g}(b) = \hat{g}_c \pm \sqrt{-2C \log b - (\hat{g}(1) - \hat{g}_c)^2}.$$

No solution is possible when b is bigger than a critical value

$$b_{\text{crit}} = \exp[-2C(\hat{g}(1) - \hat{g}_c)^2].$$

Assume $C < 0$ so $b_{\text{crit}} \geq 1$. The trajectories $\hat{g}(b)$ then have rather a curious appearance: pairs of trajectories, given by the positive and negative square roots in the solution, meet and 'annihilate' at $\hat{g} = \hat{g}_c$. The trajectories that start with $\hat{g}(1)$ closest to $\hat{g}_c$ have the smallest values of b_{crit} and are the first to disappear. It is not possible to integrate the differential equation to arbitrarily large b from any given initial value $\hat{g}(1)$, so the methods used in this chapter to find the vertex functions at large b are not valid with a β-function of this form. Fortunately this type of singularity in the β-function is thought not to occur in the physical systems with which we are concerned.

11.3 Putting $\eta = \epsilon^2/54$ into (11.40) gives

$$\delta = 3 + \epsilon + \tfrac{13}{27}\epsilon^2 + O(\epsilon^3)$$

$$= 4.48 + O(\epsilon^3).$$

This differs substantially from $\delta = 4.89\ldots$ found directly from the scaling relation. In fact, expanding δ in powers of ϵ has thrown away accuracy unnecessarily. There are two contributions to the $O(\epsilon^3)$ terms in the above equation: the uncertainty in the value of η at this order, and the $O(\epsilon^3)$ terms neglected in the expansion of the scaling relation. Only the former represents true uncertainty, since we believe the scaling relation to be exact. Even if η were wrong by a factor of 2 (which it is, nearly) the value of δ calculated from the scaling relation would only change by 0.1.

Answers for Chapter 12

12.1 We write

$$\Gamma(t_R, \phi) = \sum_{n=0}^{\infty} \frac{\phi^n}{n!} \Gamma^{(n)}(t_R),$$

where we have only shown the dependence of Γ on t_R and on ϕ. The heat capacity is proportional to the second derivative of Γ with temperature. Below T_c this derivative should be evaluated with $\phi = \langle \phi \rangle$. However, if we evaluate the derivative at $T = T_c$ all ϕs are to be put equal to zero after the differentiation. The remaining terms are

$$\left.\frac{d^2\Gamma}{dt_R^2}\right|_{t_R=0} = \Gamma^{(2)}\left(\frac{d\phi}{dt_R}\right)^2$$

$$+\Gamma^{(1)}\left(\frac{d^2\phi}{dt_R^2}\right) + 2\Gamma^{(1,1)} + \Gamma^{(0,2)}.$$

But both $\Gamma^{(1)}$ and $\Gamma^{(1,1)}$ vanish by symmetry at T_c, and $\Gamma^{(2)}(0)$ also vanishes at T_c (it is the inverse susceptibility). The only remaining term is $\Gamma^{(0,2)}$, which is the second derivative of Γ w.r.t. T at $\phi = 0$. So at T_c these two derivatives are equal.

12.2 Suppose we are given the entropy as a function of B and T. Then we can compute

$$\left(\frac{\partial S}{\partial T}\right)_M = \left(\frac{\partial S}{\partial T}\right)_B + \left(\frac{\partial S}{\partial B}\right)_T \left(\frac{\partial B}{\partial T}\right)_M.$$

The difference between the two heat capacities is therefore

$$C_B - C_M = T\left[\left(\frac{\partial S}{\partial T}\right)_B - \left(\frac{\partial S}{\partial T}\right)_M\right]$$

$$= -T\left(\frac{\partial S}{\partial B}\right)_T \left(\frac{\partial B}{\partial T}\right)_M.$$

Now a Maxwell relation (derived by differentiating the Helmholtz free energy) gives us

$$\left(\frac{\partial S}{\partial B}\right)_T = \left(\frac{\partial M}{\partial T}\right)_B.$$

Another standard manipulation of derivatives gives

$$\left(\frac{\partial B}{\partial T}\right)_M = -\left(\frac{\partial B}{\partial M}\right)_T \left(\frac{\partial M}{\partial T}\right)_B.$$

Therefore

$$C_B - C_M = T\left(\frac{\partial M}{\partial T}\right)_B^2 \left(\frac{\partial B}{\partial M}\right)_T = T\frac{\alpha_B^2}{\chi_T}.$$

Above T_c, the magnetization in zero field is zero whatever the temperature. α_B is therefore zero and the two heat capacities are identical. (This can also be seen by noting that the Gibbs free energy at zero magnetization and the Helmholtz free energy at zero external field are numerically equal.) So, we can compute α from C_M.

As $T \to T_c$ from below, the difference between C_B and C_M is singular because α_B is. However, the difference behaves with temperature like

$$\frac{\alpha_B^2}{\chi_T} \sim |T_c - T|^{2\beta - 2 - \gamma'}.$$

The Rushbrooke scaling relation, which we verify in this chapter, shows that the power of $T_c - T$ on the right is exactly $-\alpha$. Therefore, the difference $C_B - C_M$ diverges at the same rate as C_M itself, and C_B and C_M diverge with the same power of temperature as T_c is approached from below.

12.3 The left-hand side of equation (12.24) is a physical vertex function which must be independent of the renormalization wavevector κ. Putting $n = 0$ and $l = 2$ we can therefore write

$$\kappa \frac{\partial}{\partial \kappa}[Z_2^{-2}\Gamma_R^{(0,2)}(\mathbf{k}_1, \ldots, \mathbf{p}_{l-1}; \hat{g}, \kappa) + A] = 0,$$

where the derivative is to be taken at constant α, λ and Λ. Writing out the terms that arise from the differentiation of the product, and from the dependence of $\Gamma_R^{(0,2)}$ on the arguments $\hat{g}$ and κ, we find

$$\left[\kappa\frac{\partial}{\partial \kappa} + \beta\frac{\partial}{\partial \hat{g}} + 2\gamma_2\right]\Gamma_R^{(0,2)}(\mathbf{k}_1, \ldots, \mathbf{p}_{l-1}; \hat{g}, \kappa)$$
$$= B,$$

where β, γ_2 and B are defined by (11.8), (12.29) and (12.52).

Answers for Chapter 13

13.1 The starting point is

$$\Gamma^{(n)}(\mathbf{k}_i/b; \Lambda, S, t, h) \qquad (a13.1)$$
$$= b^{-d}S^{-n/2}(b)\Gamma_R^{(n)}(\mathbf{k}_i; \Lambda, t(b), h(b)),$$

which follows from (13.43) and (13.52). First we must find the functions $S(b)$, $t(b)$ and $h(b)$. For temperatures at or near t_c (13.51), (13.53), (13.57) and (13.58) yield

$$\log S(b) \sim -\tfrac{1}{2}\epsilon\frac{D-1}{D-2}\log b$$

$$\log h(b) \sim \left(d - \tfrac{1}{2}\epsilon\frac{D-1}{D-2}\right)\log b$$

$$\log dt(b) \sim \epsilon \log b,$$

where $dt(b) \equiv |t_c - t(b)|$.

The exponent η is defined in terms of $\Gamma^{(2)}$ at t_c, for $h = 0$. In this case (a13.1) becomes

$$\Gamma^{(2)}(\mathbf{k}/b; t_c, 0) = b^{-d}S^{-1}(b)\Gamma_R^{(n)}(\mathbf{k}; t_c, 0).$$

We expect that $\Gamma^{(2)}(\mathbf{k}/b) \sim b^{\eta-2}$. Using the expression for $S(b)$ gives at once

$$\eta = \frac{\epsilon}{D-2}.$$

The exponent δ is defined in terms of the response of the system to a magnetic field at t_c. None of the $\Gamma^{(n)}$ correspond directly to the magnetization m. However,

$$m \sim \frac{\partial \Gamma^{(0)}}{\partial h}.$$

Recall (see footnote on page 331) that the $\Gamma^{(n)}$ for this model are explicit functions of the magnetic field, which we regard as a parameter of the model. Hence $\Gamma^{(0)}$ is the Helmholtz free energy, so far as h and b are concerned, and the previous equation holds. Integrating we obtain

$$\Gamma^{(0)} \sim h^{(\delta+1)/\delta}.$$

At t_c,

$$\Gamma^{(0)}(; t_c, h) = b^{-d}\Gamma_R^{(0)}(; t_c, h(b)).$$

The LHS is independent of b. Demanding that the RHS be independent of b also, and writing the dependence of $\Gamma_R^{(0)}$ on $h(b)$ in terms of δ, gives

$$\delta = \frac{4}{\epsilon}\frac{D-2}{D-1}.$$

Notice that δ and η satisfy the usual scaling relation.

The exponent β is defined in terms of the behaviour of m below t_c. The trick used when finding δ doesn't work here (try it), so we need a new one. It is clear that the magnetization is directly proportional to the spin length S. Now the relation

$$\Gamma^{(n)}(0; t_c - dt, 0)$$
$$= b^{-d}S^{-n/2}(b)\Gamma_R^{(n)}(0; t_c - dt(b), 0)$$

tells us that if dt is changed to $dt(b)$, then multiplying S by $S(b)$ keeps the model the same. If $\epsilon > 0$, $S(b)$ shrinks as $dt(b)$ grows, which is reasonable; as the temperature is lowered, the magnetization grows and so S must be reduced to counteract this. This argument shows that

$$m(dt(b)) \sim S^{-1}(b).$$

Since $m(dt) \sim dt^\beta$, this gives

$$\beta = \tfrac{1}{2}\frac{D-1}{D-2}.$$

The exponent γ is defined in terms of the susceptibility $\chi \sim 1/\Gamma^{(2)}(0)$. For $\mathbf{k} = 0$, $h = 0$ we have

$$\Gamma^{(2)}(0; t_c - dt, 0)$$
$$= b^{-d}S^{-1}(b)\Gamma_R^{(2)}(0; t_c - dt(b), 0).$$

The LHS is independent of b; since $\Gamma^{(2)} \sim dt^\gamma$, the RHS yields

$$\gamma = \frac{2}{\epsilon}.$$

The exponent ν is defined in terms of $\Gamma^{(2)}$ below t_c, at non-zero $\mathbf{k}$. We have

$$\Gamma^{(2)}(\mathbf{k}/b; t_c - dt, 0)$$
$$= b^{-d}S^{-1}(b)\Gamma_R^{(2)}(\mathbf{k}; t_c - dt(b), 0) \sim b^X f(b/\xi),$$

by definition. The equality shows that $\Gamma^{(2)}(\mathbf{k})$ is of the form of a power of b times a function of $dt(b) \sim dt(1)b^\epsilon$, or equivalently a function of $b\,dt(1)^{1/\epsilon}$. Comparing this with the second line, we see that

$$\xi \sim dt(1)^{-1/\epsilon}.$$

Hence, from the definition of β,

$$\nu = \frac{1}{\epsilon}.$$

13.2 Since the energy of the XY model can be written as the energy of an electric field (see (13.69)), the result follows immediately if one recalls that the solutions to Poisson's equation generate electric fields of minimum energy for given boundary conditions. We have observed in the text that the electric field from a point charge $2\pi\epsilon_0 q$ corresponds to $\bar{\theta} = q\phi$, a vortex of strength q.

To show that the solutions of $\nabla \cdot \mathcal{E} = \rho/\epsilon_0$ have minimum energy, we must consider

$$F[\mathcal{E}] \equiv \int d^2\mathbf{x}\, \tfrac{1}{2}\epsilon_0 |\mathcal{E}|^2$$
$$+ \int d^2\mathbf{x}\, \lambda(\mathbf{x})(\nabla\cdot\mathcal{E} - \rho/\epsilon_0).$$

$\lambda(\mathbf{x})$ is an infinite set of Lagrange undetermined multipliers, which impose the bracketed constraint. We must equate the functional derivative of F w.r.t. $\mathcal{E}$ to zero. This gives

$$\epsilon_0 \mathcal{E} = \nabla\lambda$$

(we have integrated by parts the $\lambda\nabla\cdot\mathcal{E}$ term), and so

$$\nabla^2\lambda = \rho.$$

In words, the energy takes an extreme value if $\mathcal{E}$ is the gradient of a solution to Poisson's equation. This is what we sought to prove.

13.3 The solution of the linearized equation will be of the form $\delta Y(X) = f(X)\delta Y(X_0)$, where $F(X)$ is some rapidly growing function of X. So $\delta Y(X_1)$ is inevitably proportional to $\delta Y(X_0)$ and thus to the temperature T.

13.4 Consider an edge dislocation formed by removing all atoms on the x-axis at $x > 0$. This stresses the lattice by stretching bonds that are roughly tangential to a circle about the origin, and compressing radially directed bonds: a circle about the origin of diameter r has one fewer atoms on it than it would in a perfect lattice. Moreover the strain around such a circle is proportional to the fraction of the circle's atoms that are missing: $1/r$. Hence the elastic energy density distance r from the dislocation is proportional to $1/r^2$, just like the electrostatic energy density around a line of charge. Let another dislocation be located on the x-axis at $x = x_1 > 0$. Then the total stress energy will be a decreasing function of x_1 if the second dislocation is associated with an extra line of atoms on the axis at $x > x_1$, i.e., if this second dislocation has opposite 'charge' to the first. That the attractive force between the dislocations scales as $1/x_1$ follows from the dependence on x_1 of the total stress energy. For the application of this picture to melting see Peierls 1979.

Answers for Chapter 14

14.1 Since the operator will be represented by a vertex with n legs, it will affect only those $\Gamma^{(m)}$ with $m \leq n$. We therefore imagine constructing an operator $O_{n,p}$ which changes only $\Gamma^{(n)}$; we do not need to know the detailed form of this operator.

The value of the reduced $\Gamma^{(n)}$ in the presence of this operator is

$$\Gamma^{(n)}(\{\mathbf{k}_i\}) \sim \lambda_{n,p} A_p(\{\mathbf{k}_i\}),$$

where A_p is a homogeneous function of degree p. This follows from the Feynman rules for such operators (see Appendix N). From this, (14.4) and (14.10) it follows that

$$F^{(n)}(\{\mathbf{k}_i\}; b) = Z(b)^{-n/2} b^d \Gamma^{(n)}(\{\mathbf{k}_i/b\})$$
$$= b^{n-p-d(n-2)/2} \Gamma^{(n)}(\{\mathbf{k}_i\}).$$

Hence the relevance of the operator is controlled by the sign of $n - p - d(n-2)/2$, as claimed.

14.2 The analysis goes through exactly as in §14.1. The only purpose of the even power of ϕ is to reassure us; since we work to first order in couplings only, its presence never affects the relevance of the odd power of ϕ. If we construct an operator O_n whose only effect is to change $\Gamma^{(n)}$, we conclude that O_n is relevant/irrelevant as $n - d(n-2)/2$ is positive/negative.

When $d = 4$, O_5 is irrelevant. However, this does *not* make ϕ^5 irrelevant. ϕ^5 can be written as some linear combination of O_5 and the relevant operators O_3 and O_1. These latter operators involve ϕ and ϕ^3, which are not already present in the Landau–Ginzburg Hamiltonian. This is different from the case of ϕ^6, whose effect can be absorbed by changes in μ and λ.

14.3 To $O(\lambda_n^2)$, $\lambda_n \int d^d\mathbf{x}\, \phi^n$ contributes to the vertex functions $\Gamma^{(2(n-m))}$ with $1 \leq m \leq (n-1)$. The diagrams divide into two classes: those containing loops that join two legs of the same vertex (first diagram), and those without such loops (second diagram).

Diagrams of the first sort can be eliminated by adding to H_G the operator O_n of (14.26) instead of $\lambda_n \int d^d\mathbf{x}\, \phi^n$; we consider them no further. The remaining diagrams can be labelled by the number m of links joining the two vertices; these links give rise to $(m-1)$ loops. The expression I_m for such a diagram is of the form

$$I_m(\{\mathbf{k}_i\}; b) \sim \int \frac{d^d\tilde{\mathbf{q}}_1 \ldots d^d\tilde{\mathbf{q}}_{m-1}}{(q_i + k)^{2m}},$$

where k is a function of the external wavevectors. We are only interested in the b-dependence of such a diagram. If I_m diverges more than logarithmically at large wavevector, it will not depend on b as $b \to \infty$. Such diagrams are

not important. Otherwise, power counting implies that
$$I_m(\{\mathbf{k}_i\}; b) \sim b^{2m-(m-1)d}.$$
Using (14.4) with the Gaussian value for $Z(b)$, we see that the contribution δf of this diagram to $F^{(2(n-m))}$ goes like
$$\delta f \sim b^{-(d-2)(2n-m)/2+d+(2m-(m-1)d)}$$
$$\sim b^{2n-d(n-2)}.$$

The exponent is precisely twice the quantity which in §14.1 we showed determines the relevance of O_n. Hence, if an operator is relevant, higher-order corrections are likely to grow with b; if an operator is irrelevant, the higher-order corrections are likely to vanish as $b \to \infty$. In particular, this makes plausible the assertion made in §14.1.1 that irrelevant operators have Landau theory critical exponents.

14.4 Let us define (reduced) functions $F^{(n,l)}(\mathbf{k}_i; \mathbf{p}_i; b)$ by
$$F^{(n,l)}(\mathbf{k}_1, \ldots, \mathbf{k}_n; \mathbf{p}_1, \ldots, \mathbf{p}_{l-1}; b)$$
$$= Z(b)^{-n/2} Z_2^l(b) b^d \Gamma^{(n,l)}(\mathbf{k}_i/b; \mathbf{p}_i/b),$$
where
$$F^{(2,0)}(\mathbf{k}_0; b) = \Gamma^{(2,0)}(\mathbf{k}_0)$$
$$F^{(2,1)}(\mathbf{k}_1; \mathbf{p}_1; b) = \Gamma^{(2,1)}(\mathbf{k}_1; \mathbf{p}_1)$$
(compare (14.4)). The $F^{(n,l)}(b)$ characterize the theory in the limit $b \to \infty$. In addition to rescaling the field, the factor of $Z_2^l(b)$ in the first equation allows for the possibility that we may have to rescale the temperature in this limit in order that the $F^{(n,l)}(b)$ have a finite limit as $b \to \infty$. (This is *not* the Z_2 that appeared in Chapter 12.) We have used $F^{(2,1)}$ to define Z_2 so as to avoid the complications associated with $\Gamma^{(0,1)}$ (see (12.24)).

It follows that $Z(b) = b^{d-2}$; since $\Gamma^{(2,1)} = 1$ in the Gaussian model, the above imply that $Z_2(b) = b^{-2}$. Now let O_n (equation (14.26)) be added to the Hamiltonian. Sketches will confirm that O_n changes only those $\Gamma^{(m,l)}$ for which $m + 2l = n$. The figure shows a typical diagram.

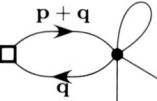

This shows a contribution to $\Gamma^{(2,1)}$ caused by ϕ^6. The integral for the loop with $\mathbf{p}/b$ flowing in at the square block is
$$I = \int^\Lambda \frac{d^d \check{q}}{\alpha^4 q^2 (\mathbf{q} + (\mathbf{p}/b))^2}$$
$$\sim b^{4-d} \text{ if } d < 4;$$
$$\sim \text{const. if } d \geq 4$$

(we require only its b-dependence). Thus,
$$\Gamma^{(m,l)} \sim b^{(4-d)l} \quad \text{or} \quad \text{const.},$$
depending on the dimension. Putting this all together gives us
$$F^{(m,l)}(b) \sim b^{(m+2l)-d(m+2l-2)/2},$$
if $d < 4$. The sign of the exponent determines the relevance of O_n. Since $m + 2l = n$, this is exactly the condition we found in §14.1. Those operators we classified as irrelevant for not affecting the $\Gamma^{(m)}$ do not affect the $\Gamma^{(m,l)}$ either. When $d \geq 4$ the details are different but the conclusion is the same.

14.5 Let us write $\phi = \phi_0 + \varphi$, where $\tilde{\phi}_0(\mathbf{k})$ is only non-zero for $0 \leq k \leq \Lambda_2$, while $\tilde{\varphi}(\mathbf{k})$ is only non-zero for $\Lambda_2 \leq k \leq \Lambda_1$. Then
$$e^{-H_2[\phi_0]} = \int \mathcal{D}\varphi \, e^{-H_1[\phi_0 + \varphi]},$$
with
$$H_1 = \int d^d \mathbf{x} \left(\tfrac{1}{2}\alpha^2 (|\boldsymbol{\nabla}\phi_0|^2 + \boldsymbol{\nabla}\varphi|^2) + \tfrac{\lambda}{4!}(\phi_0^4 + \varphi^4) \right.$$
$$\left. + \tfrac{1}{4!}\lambda(\phi_0^4 + 4\phi_0^3\varphi + 6\phi_0^2\varphi^2 + 4\phi_0\varphi^3 + \varphi^4) \right).$$
The terms not involving φ factor out of the functional integral, leaving us with a Hamiltonian for φ with terms whose coefficients involve ϕ_0. In this way ϕ_0 enters the Feynman rules for this functional integral, and the resulting diagrams are functions of ϕ_0. H_2 is given by minus the sum of the connected diagrams with no external legs.

To answer the question we must identify Feynman diagrams with coefficients proportional to ϕ_0^2. Two such diagrams are shown below:

$\phi_0^2 \bullet\!\!\bigcirc \qquad \phi_0 \bullet\!\!=\!\!\bullet \phi_0$

The first diagram contains one $\tilde{\varphi}^2$ vertex, with a coefficient proportional to $\lambda \phi_0^2$; the second diagram contains two three-leg φ^3 vertices, each with a coefficient proportional to $\lambda \phi_0$. The first diagram is the only $O(\lambda)$ diagram proportional to ϕ_0^2, so it's the one we want. Rewriting the $\phi_0^2 \varphi^2$ term in H_1 as
$$\tfrac{1}{2}\lambda \int d^d\check{q}_1 d^d\check{q}_2 \, \tilde{\phi}_0(\mathbf{q}_1)\tilde{\phi}_0(\mathbf{q}_2)$$
$$\times \left(\tfrac{1}{2}\int d^d\check{q}_3 \, \varphi(\mathbf{q}_3)\varphi(-\mathbf{q}_1 - \mathbf{q}_2 - \mathbf{q}_3)\right),$$
the piece outside the bracket is (minus) the coefficient of the φ^2 vertex. This vertex is a bit nastier even than the general vertices considered in Appendix N, because (as far as the φ field is concerned) wavevector is not conserved. To avoid confusion it's best to write down the bare propagator for φ in full, including its δ-function:
$$\tilde{G}_c^{(2)}(\mathbf{k}_1, \mathbf{k}_2) = \frac{(2\pi)^d \delta(\mathbf{k}_1 + \mathbf{k}_2)}{\alpha^2 k_1^2}.$$

Answers for Chapter 14

Then the value of the diagram is

$$-\tfrac{1}{2}\lambda \int d^d\check{q}_1 d^d\check{q}_2\, \widetilde{\phi}_0(\mathbf{q}_1)\widetilde{\phi}_0(\mathbf{q}_2)$$

$$\times \tfrac{1}{2}\int_{\Lambda_2}^{\Lambda_1} d^d\check{q}_3\, \frac{(2\pi)^d \delta(\mathbf{q}_3 - \mathbf{q}_1 - \mathbf{q}_2 - \mathbf{q}_3)}{\alpha^2 q_3^2},$$

where the factor of $\tfrac{1}{2}$ on the second line remains because the two ends of a link are joined to the same vertex. The integral over $\mathbf{q}_3$ is only for $\Lambda_2 \leq q_3 \leq \Lambda_1$. Notice that the δ-function is providing a restriction on $\mathbf{q}_1$ and $\mathbf{q}_2$; this ensures that the resulting term in H_2 conserves wavevector. The final contribution to H_2 is a term

$$\tfrac{1}{2}\frac{\lambda \Lambda_1^{d-2}\Omega_{d-1}}{\alpha^2(d-2)}\left(1 - \left(\frac{\Lambda_2}{\Lambda_1}\right)^{d-2}\right)$$

$$\times \tfrac{1}{2}\int d^d\check{q}\, \widetilde{\phi}_0(\mathbf{q})\widetilde{\phi}_0(-\mathbf{q}).$$

As claimed, there is a mass term in the new Hamiltonian. At higher orders in λ many irrelevant terms will be present too, but as we have seen they do not affect the long-distance physics.

14.6 Yes, it does. It means that if you want the ϕ^6 term to have an effect of a particular size at a particular distance, you must choose a value for λ_6 that grows as some power of Λ, faster than dimensional analysis would indicate. Look at things from the point of view of the bare theory. As Λ is increased one of the terms in its Hamiltonian becomes quite unreasonably big compared to all of the others when expressed in dimensionless units. Such a theory must give the correct value of $\Gamma^{(6)}$ at the renormalization point by construction, but it is most unlikely to give sensible macroscopic physics for any other vertex function. The dependence on Λ cannot be absorbed into a small number of parameters, unless these parameters include Λ.

This problem is believed to afflict the Landau–Ginzburg model in four dimensions; perturbatively, $\hat{g}_c = 0$ is a UV-unstable fixed point when $d = 4$. This is important for physics in that the Standard model involves Higgs particles with an action similar to the Landau–Ginzburg Hamiltonian. If the Higgs particle exists and is described by such an action, this would point to new physics at a scale of only a few TeV.

References

Agnolet, G., Teitel, S. L. and Reppy, J. D. 1981. *Phys. Rev. Lett.*, **47**, 1537. Experimental investigation of superfluid helium films.

Ahlers, G. 1980. *Rev. Mod. Phys.*, **52**, 489. Critical phenomena at low temperatures.

Amit, D. J. 1978. *Field Theory, the Renormalization Group, and Critical Phenomena*, World Scientific, Singapore.

Anders, R. and Stierstadt, K. 1981. *Solid State Commun.*, **39**, 185. Measurement of η for Ni.

Andersen, H. C. 1980. *J. Chem. Phys.*, **72**, 2384. On molecular-dynamics simulations with a random element.

Bak, P. Tang, C. and Wiesenfeld, K. 1988. *Phys. Rev.*, **38**, 36. Introduction of the idea that earthquakes may be a manifestation of self-organizing criticality. See also the same authors in *Phys. Rev. Lett.*, **59**, 381. (1987) for similar arguments applied to sand piles.

Baker, G. A. 1972. *Phys. Rev. B*, **5**, 2622.

Ball, R. C., Blunt, M. J. and Rath Spivack, O. 1989. *Proc. Roy. Soc.*, **A423**, 123. On fractals and scale invariance in diffusion-limited growth phenomena.

Bally, D., Grabchev, B., Popovici, M., Totia, M. and Lungu, A. M. 1968. *J. Appl. Phys.*, **39**, 459. Measurement of γ for Fe.

Bardeen, J., Cooper, L. N. & Schrieffer, J. R. 1957. *Phys. Rev.*, **108**, 1175. The classic explanation of superconductivity.

Baxter, R. J. 1984. *Exactly Soluble Models in Statistical Mechanics*, Academic Press, London.

Baxter, R. J. and Wu, F. Y. 1973. *Phys. Rev. Lett.*, **31**, 1294. Introduction of the triple-spin interaction model.

Bell, T. L. and Wilson, K. G. 1975. *Phys. Rev. B*, **11**, 3431. On the difficulty of calculating the renormalization exponent ω_c.

Berezinskii, V. L. 1970. *Sov. Phys. JETP*, **32**, 493. Early work on the phase transition of the $d = 2$ XY model.

References

Berlin, T. H. & Kac, M. 1952. *Phys. Rev.*, **86**, 821.   Introduction of the Gaussian and Spherical models.

Bishop, D. J. and Reppy, J. D. 1980. *Phys. Rev. B*, **22**, 5171.   Experimental investigation of superfluid helium films.

Blöte, H. W. J. and Swendsen, R. H. 1979. *Phys. Rev. B*, **20**, 2077.   Calculation of the critical exponents of the $d = 3$ Ising model by Monte-Carlo renormalization.

Burgoyne, P. N. 1963. *J. Math. Phys.*, **4**, 1320.   Improved proof of Sherman's (1960) result, togther with a clear account of the history of the loop-counting method.

Cabannes, H. 1975. *The Padé Approximants Method and its Application to Mechanics*, Springer, Berlin.

Carlson, J., M. and Langer, J., S. 1989. *Phys. Rev. Lett.*, **62**, 2632.   Earthquakes and self-organizing criticality.

Chang, R. F., Barstyn, H. and Sengers, J. V. 1979. *Phys. Rev. A*, **19**, 866.   Measurements of γ, η and ν for trimethylpentane–nitroethane.

Cohen, J. D. and Carver, T. R. 1977. *Phys. Rev. B*, **15**, 5350.   Measurement of β for Ni.

Collet, P. and Eckmann, J.-P. 1978. *Lecture Notes in Physics*, **74**, Springer, Berlin. Study of the hierarchical model.

Collins, M. F., Minkiewicz, V. J., Nathans, R., Passell, L. and Shirane, G. 1969. *Phys. Rev.*, **179**, 417.   Measurement of γ for Fe and Ni.

de Gennes, P. G. 1972. *Phys. Lett.*, **38A**, 339.   On the application of the renormalization group to studies of polymers.

Dunn, A. G., Essam, J. W. and Richie, D. S. 1975. *J. Phys. C*, **8**, 4219.   Evaluation of series expansions for the percolation model.

Dyson, F. J. 1969. *Commun. Math. Phys.*, **12**, 91.   The hierachical model.

Dyson, F. J. 1971. *Commun. Math. Phys.*, **21**, 269.   More on the hierachical model.

Eckmann, J.-P. Magnen, J. and Seneor, R. 1975. *Commun. Math. Phys.*, **39**, 251. Borel resummation of the ϵ-expansion.

Essam J. W. 1972. In *Phase Transitions and Critical Phenomena*, C. Domb and M. S. Green (eds.), Vol. 2, Academic Press, London.

Ferrenberg, A. M. and Landau, D. P. 1991. *Phys. Rev. B*, **44**, 5081.   Monte-Carlo results for the $d = 3$ Ising model.

Feynman, R. P. 1955. *Phys. Rev.*, **97**, 660.   Gives a variational principle for the lowest energy of a system described by a path integral.

Feynman, R. P. 1972. *Statistical Mechanics*, Benjamin, Reading.   In chapter 5 Z is obtained for the $d = 2$ Ising model by exact summation of the high-T expansion.

Feynman, R. P. 1985. *Surely you're joking, Mr. Feynman!*, Norton, New York.

Fiory, A. T., Hebard, A. F. and Glaberson, W. I. 1983. *Phys. Rev. B*, **28**, 5075.   Determination of the superconducting charge density in superconducting films.

Fisher, M. E. 1967. *Rep. Prog. Phys.*, **30**, 615.   A classic review of the theory of critical phenomena.

Fisher, M. E. 1969. *Phys. Rev.*, **180**, 594.   Derivation of the eponymous inequality.

Fisher, M. E. 1974. *Rev. Mod. Phys.*, **46**, 597.   A standard review of the theory of critical phenomena.

Gaunt, D. S. & Guttmann, A. J. 1974. *Phase Transitions and Critical Phenomena*, eds C. Domb & M. S. Green, vol. 3, p. 181. Academic Press, London.   Asymptotic analysis of power-series coefficients.

Golstone, J. 1961. *Nuovo Cimento*, **19**, 154.

Greer, S. C. and Moldover, M. R. 1981. *Ann. Rev. Phys. Chem.*, **32**, 233. Evidence for the equivalence of α and α' in fluid systems and a review of various connected measurements.

Griffiths, R. B. 1965. *Phys. Rev. Lett.*, **14**, 623 and J. Chem. Phys., **43**, 1958. Derivation of the eponymous inequality.

Guggenheim, E. A. 1945. *J. Chem. Phys.*, **13**, 253. Classic demonstration of the law of corresponding states.

Heller, P. 1967. *Rep. Prog. Phys.*, **30**, 731. A review of experimental investigations of critical phenomena.

Hiroyoshi, H. 1980. *J. Phys. Soc. Japan.*, **48**, 830. Measurement of δ for Ni.

Hoover, W. G. 1985. *Phys. Rev. A*, **31**, 1695. On a technique for fixing the temperature of a molecular-dynamics simulation.

Ising, E. 1925. *Z. der Physik*, **31**, 253. The first solution of the $d=1$ Ising model.

Josephson, B. D. 1967. *Proc. Phys. Soc.*, **92**, 269, 276. Derivation of the eponymous inequality.

Kac, M. & Ward, J. C. 1952. *Phys. Rev.*, **88**, 1332. Introduction of the idea that calculating Z_{Ising} is a matter of counting closed loops.

Kadanoff, L. P. 1966. *Physics*, **2**, 263. The introduction of the idea that exponents could be derived from real-space scaling arguments.

Kaufman, B. & Onsager, L. 1949. *Phys. Rev.*, **76**, 1244. First calculation of $G_c^{(2)}$ for the $d=2$ Ising model.

Kobeissi, M. A. 1981. *Phys. Rev. B*, **24**, 2380. Measurement of β for Fe.

Kogut, J. B. 1986. *J. Stat. Phys.*, **43**, 771. Discusses methods for overcoming critical slowing down.

Kolb, E. W. and Turner, M. S. 1989. *The Early Universe*, Addison-Wesley, Redwood City. Introduction to the importance of phase transitions for the origin of the observable Universe.

Kosterlitz J. M. 1974. *J. Phys. C*, **7**, 1046. Application of real-space renormalization to the $d=2$ XY model.

Kosterlitz J. M. and Thouless D. J. 1973. *J. Phys. C*, **6**, 1181. Classic paper on the Kosterlitz–Thouless transition.

Kumar, A., Krishnamurthy, H. R. and Gopal, E. S. R. 1983. *Phys. Rep.*, **98**, 57. A comprehensive review of critical phenomena in fluid systems.

Landau, L. D. 1937. *Phys. Z. Sowjetunion*, **11**, 26. (Translated in *J.E.T.P.* **7**, 19. (1937)). Introduces Landau theory.

Landau, L. D. and Lifshitz, E. M. 1969. *Statistical Physics*, Pergamon Press, Oxford.

Lenz, W. 1920. *Phys. Zeitschrift*, **21**, 613. Introduction of the Ising model.

Ma, S.-K. 1973. *Rev. Mod. Phys.*, **45**, 589. A useful general review of critical phenomena.

Ma, S.-K. 1976. *The Modern Theory of Critical Phenomena*, Benjamin, Reading.

Mandlebrot, B. B. 1974. *J. Fluid Mech.*, **62**, 331. A discussion of scale-freedom in turbulence.

Maps, J. and Hallock, R. B. 1981. *Phys. Rev. Lett.*, **47**, 1533. Experimental investigation of superfluid helium films.

Maris, H. J. and Kadanoff, L. P. 1978. *Am. J. Phys.*, **46**, 652. A useful general review of critical phenomena.

Meakin, P. and Tolman, S. 1989. *Proc. Roy. Soc.*, **A423**, 133. On fractals and scale invariance in diffusion-limited growth phenomena.

Mermin, N. D. and Wagner, H. 1966. *Phys. Rev. Lett.*, **17**, 1133. Contains a proof that long range order cannot occur in with $D>1$ and $d \leq 2$.

References

Metropolis, N., Rosenbluth, A., Rosenbluth, M. Teller, A. and Teller, E. 1953. *J. Chem. Phys.*, **21**, 1087. Introduction of the Metropolis algorithm for molecular-dynamics simulations.

Moldover, M. R., Sengers, J. V., Gamon, R. W., Hocken, R. J. 1979. *Rev. Mod. Phys.*, **51**, 79. The effects of gravity on the measurement of exponents for fluids.

Montroll, E. W., Potts, R. B. and Ward, J. C. 1963. *J. Math. Phys.*, **4**, 308. Gives history of Onsager's formula for $\langle s \rangle$ in the $d=2$ Ising model. Calculates general two-point correlation functions for that model.

Nauenberg, M. and Nienhuis, B. 1974. *Phys. Rev. Lett.*, **33**, 344. Application of the renormalization group to the Ising model.

Niemeijer, Th. and van Leeuwen, J. M. J. 1976. In *Phase Transitions and Critical Phenomena*, C. Domb and M. S. Green (eds.), Vol. 6, Academic Press, London. The general relation between the eigenvalues of the renormalization matrix and the exponent η.

Nose, S. 1984. *J. Chem. Phys.*, **81**, 511. On a technique for fixing the temperature of a molecular-dynamics simulation.

Onsager, L. 1944. *Phys. Rev.*, **65**, 117. The first solution of the $d=2$ Ising model.

Parisi, G. 1988. *Statistical Field Theory*, Addison Wesley, Redwood City.

Pawley, G. S., Swendsen, R. H., Wallace, D. J. and Wilson, K. G. 1984. *Phys. Rev. B*, **29**, 4030.

Peebles, P. J. E., *The Large Scale Structure of the Universe*, Princeton University Press, Princeton. The standard introduction to the study of clustering by galaxies.

Peierls, R. 1934. *Phys. Rev.*, **54**, 918. Proof of an inequality between the true value of Z and approximations to it.

Peierls, R. 1979. *Surprises in Theoretical Physics*, Princeton University Press, Princeton. An elegant discussion of the application of the Kosterlitz-Thouless transition to melting.

Polyakov, A. M. 1987. *Gauge Fields and Strings*, Contemporary Concepts in Physics, vol. 3, Harwood Academic Publishers. A proposal for the analytic solution of the $d=3$ Ising model.

Rastelli, E. and Realto, L. 1969 *Physics Lett. A*, **30**, 172. Evidence that $\gamma \simeq \gamma'$.

Reichl, L. E. 1980. *A Modern Course in Statistical Physics*, Arnold.

Rocker, W., Kohlhaus, R. and Schöpgens, A.W. 1971. *Z. Agnew. Phys.*, **32**, 164. Measurement of γ, γ' and δ for Fe.

Rowlinson, J. S., Winton, F. L. S. 1982. *Liquids and Liquid Mixtures*, Butterworth, London. Discusses the definition of the order parameter for binary fluids and gives some experimental results for critical exponents.

Rushbrooke, G. S. 1963 *J. Chem. Phys.*, **39**, 842. Derivation of the eponymous inequality.

Sahni, P. S. and Banavar, J. R. 1981. *Phys. Lett. A*, **85**, 56. Monte-Carlo renormalization for the $d=3$ Ising model on a face-centred cubic lattice.

Salomaa, M. M., Volovik, G. E. and Landau, L. D. 1987. *Rev. Mod. Phys*, **59**, 533. A review of experiments on superfluidity in ^{3}He.

Sherman, S. 1960. *J. Math. Phys.*, **1**, 202. The first proof that the 'fermionic' weighting of loops summed to obtain Z for the $d=2$ Ising model gives correct count. See also *J. Math. Phys.*, **4**, 1213 (1963).

Soeffge, F. 1980. *Phil. Mag. B*, **42**, 47. Measurement of ν for Ni.

Stanley, H. E. 1969. *J. Phys. Soc. Japan*, **26S**, 102. Demonstrates that the spherical model corresponds to the $D \to \infty$ limit of the Heisenberg model.

Stanley, H. E. 1971. *Introduction to Phase Transitions and Critical Phenomena*, Oxford University Press, Oxford.

Stephenson, R. L. 1964. *J. Math. Phys.*, **5**, 1009.   A calculation of the correlation functions of the $d = 2$ Ising model.

Suter, R. M. and Hohenemser, C. 1978. *Phys. Rev. Lett.*, **41**, 705.   Measurement of ν for Fe.

Swendsen, R. H. 1979a. *Phys. Rev. Lett.*, **42**, 859.   The development of Monte-Carlo renormalization.

Swendsen, R. H. 1979b. *Phys. Rev. B*, **20**, 2080.   Comparison of two renormalization schemes for the $d = 2$ Ising model.

Swendsen, R. H. 1982. In *Topics in Current Physics 30*, T. W. Burkhardt and J. M. J. van Leeuwen (eds.), Springer, Berlin.   Further development of Monte-Carlo renormalization.

Swendsen, R. H. and Wang, J.-S. 1987. *Phys. Rev. Lett.*, **58**, 86.   Introduces the eponymous algorithm for relaxing Monte-Carlo simulations.

Sykes, M. F., Essam, J. W. and Gaunt, D. S. 1965. *J. Math. Phys.*, **6**, 283.   Low temperature expansions for the $d = 2$ and $d = 3$ Ising models.

't Hooft, G. and Veltman, M. 1972. *Nucl. Phys. B*, **44**, 189.   Introduces dimensional regularization.

Vygovskiy, V. P. and Yergin, Yu. V. 1972. *Phys. Met. Metallogr.*, **34.3**, 40.   Measurement of β, γ and δ for Ni.

Wallace, D. J. 1976. In *Phase Transitions and Critical Phenomena*, C. Domb and M. S. Green (eds.), Vol. 6, Academic Press, London.   Calculation of η to $O(\epsilon^4)$.

Widom, B. 1965. *J. Chem. Phys.*, **43**, 3892, 3898.   Introduction of the eponymous scaling hypothesis.

Wilson, K. G. and Kogut, J. 1974. *Phys. Rep.*, **12C**, 75.   An early exposition of the renormalization group in quantitative form.

Wolff, U. 1989. *Phys. Rev. Lett.*, **62**, 361.   Introduces the eponymous algorithm for relaxing Monte-Carlo simulations.

Yang, C. N. 1952. *Phys. Rev.*, **85**, 808.   Derivation of the mean magnetization of the $d = 2$ Ising model in closed form.

Young, A. P. and Stinchcombe, R. B. 1975. *J. Phys. C*, **8**, L535.   Application of the renormalization group to the percolation model.

Zinn-Justin, J. 1989. *Quantum Field Theory and Critical Phenomena*, Oxford University Press, Oxford.

Index

Page numbers in boldface denote definitions or main references, and should be referred to first.

1PI (one-particle irreducible) 212
1PR (one-particle reducible) 212

A

a, as replacement for α 237
Abelian group 284
absolutely convergent integrals 66
accessibility assumption **93**, 94–95, 107, 109, 112
α (characteristic length) 181
    replacement by a 237
α (critical exponent) 5
    expression in terms of other exponents 29, 140, 314
    in mean-field theory 174
    in real-space renormalization 137–139
    of two-dimensional Ising model 78
    of Landau–Ginzburg model 312–314, 317–318
    of spherical model 70
    relation to flow function γ_2 314
α' (critical exponent) 5
    in real-space renormalization 139
    of Landau–Ginzburg model 314
amputated diagrams 212
    and vertex functions involving ϕ^2 305
analytic continuation,
    and validity of renormalization group equations 309
    in number of dimensions 270
anti-ferromagnetic/paramagnetic transition 13
approximation
    in real-space renormalization 119
    in Ising model 149
    in percolation 146
    of one model by another 321
asymptotic series 196
    example 226
attractive fixed points 120
autocorrelation function 104

B

bare coupling constant, divergence of 256, 260, 264
bare dimensionless coupling constant
    in massive theory 253
    in massless theory 263
bare mass, relation to renormalized mass 254
bare propagator **192**, 204
Berezinskii, V. L. 339
Berlin, T. H. 58
β (critical exponent) 8
    expression in terms of other exponents 29, 140, 316
    in mean-field theory 173
    in real-space renormalization 136
    of two-dimensional Ising model 79
    of Landau–Ginzburg model 315–318
    of non-linear σ-model 339, 351
    of spherical model 71
    relation to flow functions γ_1 and γ_2 314
β-brass 57
β-function 265, **284**, 363, 367
    and fixed points 290–291
    and running coupling constant 289
    for massive theory 257ff
    for massless theory 265ff
    of non-linear σ-model 335
    zeros of 258, 266, 278
        and critical phenomena 291–292
        and universality 364–369
        Gaussian 259
        infrared stable 291
        of non-linear σ model 337–338
        ultraviolet stable 290
binary fluids 11
blob, in Feynman diagrams 197
block variables **115**, 182, 229, 242
blocking transformations 114
    connection to Landau–Ginzburg model 230
Bogoliubov inequality 165
Boltzmann factor 34

Boltzmann's constant k_B 34
bond percolation, 60
    critical exponent for 145
    duality argument for 144
    series expansions for 146
    treatment by real-space renormalization 143
        approximations in 146
Brownian motion 383
bubble graph **230**, 243
    adding to links 234

C

calculation of flow functions
    β and γ_1 296
    γ_2 316
cannon-ball 383
canonical ensemble, sampling in molecular dynamics 98
chain rule for functional differentiation 407
characteristics, method of **287–289**, 295, 311, 315, 336, 363
chemical potential 23, **26**
classical thermodynamics 35–38, 42, 378
compressibility, isothermal 7, 179
confinement of charge in two dimensions, and Kosterlitz–Thouless transition 351
connected correlation functions 18, **43–44**, 384
    analogy with dishes 44
    and Feynman integrals 205ff
    and vertex functions 213, 218
    as derivatives of $\log Z$ 384ff
    connected diagrams 204
    diagrammatic representation 208, 209
    for multi-component model 425
    higher-order 44
    in Gaussian model 192
    in mean-field theory 168ff
    in wavevector space 204ff
    of powers of ϕ^2 300
    renormalization of 131
    rules for finding 208, 209
    two-point 18
connected diagrams 198
    and connected correlation functions 204
    and $\log Z$ 204
    building blocks for 211
constants of motion, and molecular dynamics 97
constraints 35
    and thermal averages 41
continuum Gaussian model 189
contradictory viewpoints 369
convex function 24
convexity inequality 165
convolution theorem for DFTs 391
Cooper pairs 14
coordination number 56
correlation functions 16ff, **42**, 64, 384
    analogy with menu 44
    best guess at 384
    connected: see connected correlation functions
    diagrammatic representation 199ff
    higher-order 44
    in mean-field theory 168
    measurement by neutron scattering 378
    of Gaussian model 192
    of Landau–Ginzburg model at low temperatures 322
    of Ising model 78
    of ϕ^2 300
        in wavevector space 302
    two-point 17
correlation length **18**, 65
    and critical slowing down 105
    divergence of 125
        in Landau–Ginzburg model 311
    in Gaussian model 192
    renormalization of 122, 131
correlations
    between configurations in simulations 104
    between spins 43–44
corresponding states, law of 23
Coulomb gas, two-dimensional, and vortices 342
coupling constant 181
    bare dimensionless
        in massive theory 253
        in massless theory 263
    divergence of 256, 260, 264
    in massless theory 264
    renormalization of 238, 250, 256
    renormalized dimensionless
        critical value for 259, 266
        in massive theory 257
        in massless theory 265
    renormalized
        in massive theory 257
        in massless theory 264
critical behaviour 1–447
    and small-wavevector limits 210
    comparing different systems 356
critical dimension 255
    lower 255, 320ff
        for rotationally symmetric models 322–324
        investigating with non-linear σ-model 325, 337–338
    upper 172, 255, 292
critical dynamics 124
critical exponents 20
    and divergences 5
    calculating from simulations 111
    dynamical 105, 108–109
    for percolation 145
    for renormalized system 132
    in Landau theory 184
    in mean-field theory 173–176
    numerical values 22
    for $T \neq T_c$ 135, 311–316
    for $T = T_c$ 133, 293–295
    of non-linear σ-model 339, 351
    primed equal to unprimed 127, 137, 139, 314
    α 5

Index

expression in terms of other exponents 140
in mean-field theory 174
in real-space renormalization 137–139
of two-dimensional Ising model 78
of Landau–Ginzburg model 312–314, 317–318
of spherical model 70
relation to flow function γ_2 314
α' 5
in real-space renormalization 139
of Landau–Ginzburg model 314
β 8
expression in terms of other exponents 140
in mean-field theory 173
in real-space renormalization 136
of two-dimensional Ising model 79
of Landau–Ginzburg model 315–318
of non-linear σ-model 339, 351
of spherical model 71
relation to flow functions γ_1 and γ_2 314
δ 8
behaviour under renormalization 132
expression in terms of other exponents 140, 295
in mean-field theory 174
in real-space renormalization 134
of Landau–Ginzburg model 294–295, 297
of non-linear σ model 339, 351
of spherical model 70, 83
relation to flow function γ_1 295
η **18**, 32, 355
at $d = 4$ and above 263
below $d = 4$ 264
in Landau–Ginzburg model 261, 293, 297
by dimensional regularization 273
using ϵ-expansion 278
using Feynman parameters 275
using renormalization group 297
in mean-field theory 176
in real-space renormalization 130, 133
of non-linear σ-model 339, 351
relation to flow function γ_1 293
γ 7
of spherical model 70
above $d = 4$ 255
below $d = 4$ 256
expression in terms of other exponents 140, 312
in Landau–Ginzburg model 254
by dimensional regularization 271
using ϵ-expansion 280
using renormalization group 317–318
in mean-field theory 174
in real-space renormalization 137
of non-linear σ-model 339, 351

relation to flow functions γ_1 and γ_2 312
γ' (critical exponent) 7
equality to γ 7, 137, 314
in real-space renormalization 137
ν 20
in mean-field theory 174
in real-space renormalization 124, 127
of Landau–Ginzburg model 311, 317–318
of non-linear σ-model 339, 351
relation to flow function γ_2 311
ν' 127, 314
critical fixed point 124
behaviour near 135
linearization about 141
in percolation calculation 145
location in Monte Carlo renormalization 155
critical opalescence **8–9**, 10, 18, 178, 246
and thermodynamic potentials 27
critical point **4**, 41
critical region 247
critical slowing down 104
algorithms avoiding 106–111
critical surface 122
critical temperature 51
Curie temperature **2**, 6
cutoff parameter Λ 181, **190–192**, 207
dependence of physical quantities on 191, 194, 305, 369
in non-linear σ-model 328
in physics of vacuum 372
physical meaning 370

D

data collapse 21
$\Delta(\mathbf{x})$ (two-point function in Gaussian model) 191–193
decimation 116
for one-dimensional Ising model 156
for two-dimensional Ising model 156
decorrelation time 105
δ (critical exponent) 8
behaviour under renormalization 132
expression in terms of other exponents 140, 295
in mean-field theory 174
in real-space renormalization 134
of Landau–Ginzburg model 294–295, 297
of non-linear σ-model 339, 351
of spherical model 70, 83
relation to flow function γ_1 295
δ function
and wavevector conservation 206
relating full and reduced correlation functions 207, 213, 305
with zero argument 223, 322
detailed balance 93
DFT: see discrete Fourier transform
diagrams
amputated 212
disconnected 198

456 Index

connected 198, 204
Feynman rules for 195
  for ϕ^2 correlation functions 301, 303
  similar but distinct 224, 365, 422
dimensional analysis 363, 371
  in derivation of renormalization group
    equations **285**, 310, 313, 335–336
dimensional regularization **269**, 316, 328
  in calculation of exponent η 273
  in calculation of exponent γ 271
dimensionality, variation of 269
dimensionless coupling constant
  bare
    in massive theory 253
    in massless theory 263
    divergence of 264
  renormalized
    critical value for 259, 266
    in massive theory 257
    in massless theory 265
dimensionless Gibbs free energy Γ **216**, 304, 318
  below T_c 220-221
  connection with $\Gamma^{(n)}$ above T_c 216ff
dimensionless variables in Feynman integrals 251
dimensions of quantities in Landau–Ginzburg model 181, 215
  of $\Gamma_R^{(n)}$ 285
  of quantities in non-linear σ-model 331, 334
dipoles, and Kosterlitz–Thouless transition 343–347
disconnected diagrams 198
discrete Fourier transforms, convolution theorem 391
discrete scaling symmetry 114
divergence
  infrared 249–253
  of correlation length 125
    in Landau–Ginzburg model 311
  of coupling constant 256, 260, 264
  of transverse fluctuations in $d \leq 2$ 323
  ultraviolet 249–253
divergences and critical exponents 5
domain, magnetic 3
domain of attraction of fixed point 291
dressed parameters, see renormalized parameters
duality argument for percolation 144
dynamical critical exponent 105, 108–109
dynamical universality 105

E

effective action 372
effective Hamiltonian 117
Ehrenfest, P. 26
eigenvalue, magnetic 129, 142
eigenvalue, thermal 130
Einstein fluctuation theory 179, 182, 398

electroweak theory; weakness of low-energy interactions 372
entropy 36, 52, 379, 383
ϵ-expansion **268–270**, 278–280
  around $d = 2$ 337
  around $d = 4$ **268–270**, 278–280, 316, 325, 368
    in calculation of η 278–280
    in calculation of flow function γ_1 296–297
    in calculation of flow function γ_2 316
    in calculation of γ 280
equations of state 37
ergodic hypothesis, and molecular dynamics 97
estimators 87
  fluctuations in 88
η **18**, 32, 355
  at $d = 4$ and above 263
  below $d = 4$ 264
  in Landau–Ginzburg model 261, 293, 297
    by dimensional regularization 273
    using ϵ-expansion 278
    using Feynman parameters 275
    using renormalization group 297
  in mean-field theory 176
  in real-space renormalization 130, 133
  of non-linear σ-model 339, 351
  relation to flow function γ_1 293
Euler equation 285–286
Euler's Γ-function 272
even parameters in real-space renormalization 130
excitation energies, of ordered Landau–Ginzburg model 321
excluded volume, in non-ideal gas 163
experimental data 1–16, 22
  and evidence for universality 354
  and Kosterlitz–Thouless transition 350
extended free energy 25
extensive quantities 2, **22**
external field, renormalization of 129
external points in Feynman graphs 199

F

factorial function, extension to non-integers 272
ferromagnet
  correlation functions 43
  spontaneous symmetry breaking 48
  transition to paramagnet 2, 12
ferromagnetic/paramagnetic transition 2, 12
Feynman diagrams 195ff
  internal and external points in 199
Feynman integrals 195ff
  dimensionless variables in 251
  power counting in 252
Feynman parameters 273–276
Feynman, R. P. 164, 273
Feynman rules 195ff
  correspondence with terms in Hamiltonian 423
  derivative terms 424
  for $G_c^{(n)}$ 208

Index

for $\widetilde{G}_c^{(n)}$ 209
for $\Gamma^{(n)}$ 214
for $G_c^{(n,l)}$ 301
for $\widetilde{G}_c^{(n,l)}$ 303
for $\Gamma^{(n,l)}$ 304–305
for Gibbs free energy 216
for non-linear σ-model 329–330
in wavevector space 204ff, 303ff
ϕ^6 term 421
ϕ^n term 422
modification of 238, 240
multi-component model 427
non-quadratic multi-component terms 426
$F^{(n)}$ 355
field renormalization **235–238**, 245–246
field source 181
fields, smoothness of 229
finite systems 51, 150, 227
first-order phase changes 23
fish graph **239**, 244
Fisher, M. E. 28
Fisher scaling law 29, 32, 140, 312
fixed points 120, **289–292**, 295, 296
    attractive 120
    critical 124
        behaviour near 135
        linearization about 141
            in percolation calculation 145
            location in Monte Carlo renormalization 155
    Gaussian, stability of 292, 355ff
    high-temperature 121, 128
    in context of Landau–Ginzburg model 353
    infrared stable **291**, 314
    linearization about **121**, 141
    low-temperature **121–122**, 128, 337
    in real-space renormalization 120
    marginal 121
    mixed 121
    repulsive 120
    non-Gaussian, stability of 292
        in ϕ^6 model 364ff
    in non-linear σ-model 337
    ultraviolet stable 290
fizzy drinks 1
flow functions 285
    β 257ff, 265ff, **284ff**, 363, 367
    γ_1 284–284
    γ_2 310
    for non-linear σ-model 335
flows in Hamiltonian space 120
fluctuations 18, 36, 40–44, 191, 192, 321
    and correlation functions 40
    and critical opalescence 8
    and Ginzburg criterion 247
    Einstein's theory for 399
    in energy 40–41
    in statistical estimators 87–88, 90
    in a general quantity 41
    in Landau–Ginzburg model at different lengths 194
    neglect of in mean-field theory 177
Fokker–Planck equation 102
food 44
Fourier convolution theorem, discrete 391
Fourier transformation
    of Landau–Ginzburg order parameter ϕ 190
    of ϕ^2 302
    discrete 391
        convolution theorem for 391
        Parseval's theorem for 392
four-point vertex function 238, 241
free energy
    additive renormalization of 138
    extended 25
    Gibbs 21, 37, 50
    Helmholtz 21, 37
    in real-space renormalization 137, 141
    of Ising ring 62
    of two-dimensional Ising model 76
    variational, and Bogoliubov inequality 165
frustrated systems 13
functional derivative 189, 199, 300, 303-4, **405**
    chain rule 407
    dimensions 406
    examples 406
    inverse of 408
    w.r.t. source field 189, 406
functional integral 189, 195, **409**
    as limit of multiple integral 410
    example 410
    Gaussian 411ff
    in particle physics 369
functional measure 245
functional Taylor series 216, 309, 407
    expansion around critical temperature 301
functionals 404ff
    as limit of functions 405
    examples of 405
functions, convex 24

G

$\Gamma^{(n)}$: see vertex functions
$G^{(n)}$: see correlation functions
$\Gamma^{(0)}, \Gamma^{(1)}, \Gamma^{(2)}$, anomalous definitions of 420
$\Gamma[\tilde{\varphi}]$: see dimensionless Gibbs free energy
γ (critical exponent) 7
    of spherical model 70
    above $d = 4$ 255
    below $d = 4$ 256
    expression in terms of other exponents 140, 312
    in Landau–Ginzburg model 254
        by dimensional regularization 271
        using ϵ-expansion 280
        using renormalization group 317–318
    in mean-field theory 174
    in real-space renormalization 137
    of non-linear σ-model 339, 351
    relation to flow functions γ_1 and γ_2 312
γ' (critical exponent) 7

equality to γ 7, 137, 314
    in real-space renormalization 137
γ_1 (flow function) 285
    calculation to order ϵ^2 296
γ_2 (flow function) 310
    calculation to order ϵ^2 316–317
Γ-function of Euler 272
gap parameter, for superconductors 14
Gaussian functional integral 411ff
Gaussian model **58–59**, 95, 98, 322, 331
    and molecular dynamics 98
    continuum version 189
        correlation length 192
        diagrammatic representation of 198
        evaluation of partition function 414
        $G^{(n)}$ and $G_c^{(n)}$ 189
        relevance of perturbations 355ff, 359, 361, 373
    correlation functions 192
        use in Landau–Ginzburg model 196
    in mean-field theory 177
    in real-space renormalization 117
    value of ω for 129
Gaussian transformation 399
    and Ising model 400
Gaussian white noise 101
generalized Ising models 147
generalized Landau–Ginzburg models 421ff
$G_c^{(n)}$: see connected correlation functions
$\hat{g}_6$ 362
Gibbs probability distribution **34–35**, 47
    sampling numerically 85, 90
Gibbs free energy 21, 37, 50–51, 215ff, 331, 386ff
    analyticity 49–52, 220–221
    connection with 1PI diagrams 219–221
    convexity 51, 221, 390
    dimensionless version 216
    of Landau–Ginzburg model 215ff, 223ff
Ginzburg criterion 246–247
Goldstone bosons 324
Goldstone, J. 324
Goldstone modes **324**, 328
graphs
    bubble **230**, 243
        adding to links 234
    fish, **239**, 244
    Saturn **231–232**, 235, 238, 243, 296–297
gravity, non-renormalizability of 372
Griffiths, R. B. 29
Griffiths scaling law **29**, 32, 140

H

Hamiltonian 33, **38**
    density 179
    effective 117
    for box model 38
    for continuum Gaussian model 189
    for Gaussian model 58
    for Heisenberg model 58
    for Ising model 55
    for non-linear σ-model 329
    space 119
        fixed points in 120
        flows in 120
He I and He II (normal and superfluid helium) **13**, 348
heat 36, 380
heat capacity 36, 39, 318
heat bath 34
Heisenberg model **58**, 95, 109, 179, 325
helium three 15
Helmholtz free energy 21, 37, 45, 48, 50, 62, 331
    additive renormalization of 138
    cutoff dependence 191
    in real-space renormalization 137, 141
    of Gaussian model 191
    of Landau–Ginzburg model 203ff, 222
    of hydrogen/oxygen mixture 47
hidden symmetry (see also spontaneous symmetry breaking) 324
hierarchical model 119
high-temperature expansions 72
    numerical evaluation of 80
    loops in 394
high-temperature fixed point 121, 128
homogeneity hypothesis **27–28**, 142
homogeneous functions 285

I

importance sampling 89
independence, statistical 91
    of configurations in simulations 103
independent loops, in Feynman diagrams 222
infinite fluctuations, in Gaussian model 194
infinite-range interactions 171
infinitesimal transformations, of renormalization group 284
infrared behaviour, of non-linear σ-model 331
infrared divergences **249–253**, 357, 368
infrared stable fixed point **291**, 314
inhomogeneous term in renormalization group equation 313
integrable systems, and molecular dynamics 97
integrals
    absolutely convergent 66
    Feynman 195ff
        dimensionless variables in 251
        power counting in 252
    in non-integer dimensions 271
intensive quantities 2
interaction of vortices 342
internal labels 426
internal points in Feynman graphs 199
invariants, and Kosterlitz–Thouless transition 351
inverse of functional differentiation 408
inverse temperature 21
irrelevant operators 358–359
    and non-renormalizable theories in particle physics 371
Ising, E. 55
Ising model **55**, 182

Index 459

agreement of critical exponents with
    Landau–Ginzburg model 353
  in two dimensions 366ff
and Landau–Ginzburg model 400
and mean-field theory 159, 166–168
and universality 404
anti-ferromagnetic 82
approximations in real-space renormalization treatment of 149
calculation of renormalized parameters in 148
correlation functions 78
  within mean-field theory 168–171
defining block variables for 116
Gaussian transformation 400
generalized 147
high-temperature expansions for 72
in real-space renormalization 116, 117, 147ff, 156
majority rule for 149
Monte Carlo renormalization studies of 155–156
normalization condition for 118
one-dimensional 61
  absence of phase transition in 62, 160, 176, 322
  decimation treatment 156
  real-space renormalization treatment 150
  solution by transfer matrices 62
simulation of 85, 88–89, 95, 105–109
three-dimensional 160
  Monte Carlo renormalization studies 156
  critical exponents for 318
two-dimensional 85, 160
  decimation treatment 156
  exact partition function of 401
  Monte Carlo renormalization studies 155
  real-space renormalization treatment 150
with infinite-range interactions 171–172
islands 418
isothermal compressibility 7, 179
isotherms, of water 4
isotropic lattice 18

J

J-blob **197**, 208, 209
Josephson, B. D. 28
Josephson scaling law **29**, 31, 140, 314

K

Kac, M. 58
Kadanoff, L. P. 29, 113
Kadanoff scaling hypothesis 29
κ-independence of physical quantities 310
κ_T (isothermal compressibility) 7, 179
k_B (Boltzmann's constant) 34
k-space 205
  correlation functions 204ff

Feynman rules 205, **209**
versus position space 113, 210
Kosterlitz, J. M. 339
Kosterlitz–Thouless transition 339–351
confinement of charge in 351

L

labels in multi-component models 425
Λ-sensitivity (cutoff-sensitivity) 191, 194, 369
  at high orders 369
  in correlation functions of ϕ^2 305
  in non-linear σ-model 332
  of internal energy in Gaussian model 307
  of specific heat in Gaussian model 308
Landau, L. D. 158
Landau theory 183, 222, 255, 262–263, 403
  critical exponents for 184
  Ginzburg criterion for validity of 246
  internal energy in 185
  latent heat in 185
  specific heat in 185
  susceptibility in 186
Landau–Ginzburg model 181
  absence of absolute temperature in 325, 329
  and blocking transformations 230
  and Ising model 400
  and universality 404
  at low T 321
  critical exponents 271ff, 293ff, 311ff
  dimensions of quantities involved in 181
  Feynman graph expansion for partition function 195ff
  formulation 178
  generalizations 421ff
Langevin equation **100–103**
  and Fokker-Planck equation 102
  integration of compared to molecular dynamics 103
Laplacian operator, discrete representation of 402
large dimensionality, limit of 172
large-scale properties, characterization of 356
latent heat 1, 7
  and Landau theory 185
lattice
  blocking of 114
  isotropic 18
  renormalization of 114
lattice gas 56
law of corresponding states 23
Legendre transformation **24**, 26, 32, 215, 304, 331, 381
Lenz, W. 55
line integrals, and vortices 340
linear renormalization scheme 116, 127
  normalization constant in 127–128
  ω parameter in 128
  problems with 129–131
linear response theorem **42**, 83, 168, 210
linearization about fixed points 121, 141
  in percolation calculation 145
link, in Feynman diagrams 197
liquid–gas transition 11

long-range interactions, neglect of in real-space renormalization 151
long-range order, at low temperatures 321–324
   absence in one dimension 65
loops, on lattice and high-temperature expansions 394
loops, in Feynman diagrams
   expansion in 221–223
   and renormalizability 243, 305
   independent 222
lower critical dimension 255
   for rotationally symmetric models 322–324
   investigating with non-linear σ-model 325, 337–338
low-temperature expansion 321
low-temperature fixed point **121–122**, 128, 337

M

m^2, dependence on t 246
m, as replacement for μ 234
magnetic dipole, energy of 381
magnetic eigenvalue 129, 142
magnetic energy 381ff
magnetic field, in non-linear σ-model 331
magnetic moment 2
magnetization
   renormalization of 131
   spontaneous 48–52, 216, 221, 390
majority rule 116
   for Ising model 149
marginal fixed points 121
Markov chain 92
Markov process 92
mass 231
   bare 181
   relation between bare and renormalized 254
mass renormalization 230–231
mean energy 35
mean-field theory **158–177**, 186, 255, 263
   and infinite-range interactions 171
   correlation functions in 168
   critical exponents in 173–176
   of Ising model 159
   of non-ideal gas 162
   of percolation 161
   variational derivation of 164
measure, in non-linear σ-model 326–328
Mermin, N. D. 324
Mermin–Wagner theorem **324**, 339
metastability and spontaneous symmetry breaking 47
methanol 11
method of characteristics **287–289**, 295, 311, 315, 336, 363
method of steepest descent 172, 393
Metropolis algorithm 94ff
microcanonical ensemble **98**, 398
microreversibility **93–94**, 106–108

microstates 33
mixed fixed points 121, 124
m_{pl} (Planck mass) 372
molecular dynamics **95–100**
   and ergodic hypothesis 97
   and integrable systems 97
   and sampling of configuration space 97
Monte Carlo methods **92–95**, 106–110, 153–155
Monte Carlo renormalization 153
   for Ising model 155
   location of critical fixed point in 155
   problems with 153
μ (bare mass) 181
   expression in terms of m 254
   replacement by m 234
multi-component systems 322
multi-component order parameter 425
   in Gaussian model 425

N

natural variables 21–22, 37, 378
neutrons, scattering of 17, 375
Newton's equations 95
n-hexane 11
n-point correlation function $G^{(n)}$ 42
non-Gaussian fixed point 364
non-integer dimensions, definition of integrals in 271
non-interacting spins 43
non-linear σ-model **325–339**
   absence of long-range order below $d = 2$ 332
   absence of vortices in 349–350
   $beta$-function for 335
   critical exponents for 339, 351
   cutoff parameter for 328
   $d = 2$ and $D = 2$ as a special case 338
   dimensions of quantities in 331, 334
   Feynman rules for 329–330
   fixed points for 337
   flow functions for 335
   Hamiltonian for 329
   infrared behaviour of 331
   investigation of critical dimension using 325, 337–338
   Λ-sensitivity of 332
   magnetic field in 331
   $O(D)$ symmetry in 325–328
   renormalizability of 333, 336
   renormalization group equation for 335
   two-point vertex function in 329
non-renormalizable theories 369
non-zero external field 129
normal modes 97
normalization conditions
   in real-space renormalization 118
      for Ising model 118
   for Landau–Ginzburg model 283
   for ϕ^6 model 362
n-sphere, surface area of 83
ν (critical exponent) 20
   in mean-field theory 174
   in real-space renormalization 124, 127

of Landau–Ginzburg model 311, 317–318
of non-linear σ-model 339, 351
relation to flow function γ_2 311
ν' (critical exponent) 127, 314
numerical algorithms, structure of 91

O

odd parameters, in real-space renormalization 129
ω (parameter in linear renormalization scheme) 127–129
   critical value of 129, 130
one-loop Gibbs free energy 225–226
one-particle irreducible (1PI) diagrams 212
   and $\Gamma^{(n,l)}$ 304
one-particle reducible (1PR) diagrams 212
$O(D)$ symmetry 426
   in non-linear σ-model 325–328
Onnes, K. 14
opalescence, critical **8–9**, 10, 18, 178, 246
   and thermodynamic potentials 27
operators 356
   relevant and irrelevant 358–359
   O_n 359
   O'_n 367
order below T_c 321
order parameter 9
   fluctuations in 194
   Fourier transform 190
   multi-component 425
   of XY model 339
   rescaling of 355

P

Padé approximants 82
paramagnetic/anti-ferromagnetic transition 13
paramagnetic/ferromagnetic transition 2, 12
parameters
   bare **231**, 241–242
   renormalized **231**, 241–244, 245, 283, 305
Parseval's theorem 392
particle physics, contrasted with statistical physics 369
particle-in-boxes model 38, 43, 45
   correlation functions 45
partition function **35**, 47
   derivatives of 40ff, 42, 45, 384
   for continuum Gaussian model 189–190
   for Landau–Ginzburg model 195ff
      analyticity 196
   for one-dimensional Ising model 62
   for particle-in-boxes model 39, 45
   for spherical model 66–72, 83
   for two-dimensional Ising model 73–78
paths on lattice, and high-temperature expansions 394
Peierls R. E. 164
penetration depth 15
percolation 60, 161
   bond 60
   critical exponent for 145
   duality argument for 144
   correlation length of 60
   series expansions for 146
   site 60
   treatment by real-space renormalization 143
      approximations in 146
perturbation of Gaussian model 355ff, 359, 361, 373
perturbation of Landau–Ginzburg model 361ff
perturbation theory
   for Landau–Ginzburg partition function 195ff
   convergence 196
   going to all orders in 261
phase changes, continuous 3
   implications for free energy 23
   critical points 2
physical trajectory 124
Planck length 31
Planck mass m_{pl} 372
position space
   Feynman rules 208
   versus k-space 210
   renormalization 113
      approximations in 119
position-dependent temperature 300
potentials, thermodynamic 21, 35ff, 378
potential, chemical 23, **26**
Potts model 58
power counting in Feynman integrals 242, 252
power laws 30
   and failure of Taylor expansions 309
ϕ^n as perturbation of Gaussian model 356ff
pressure 37
primitively divergent diagrams 243
   for $\Gamma^{(n,l)}$ 307
probability distribution
   for spin 51
   relation to Gibbs free energy 221, 227, 386ff
propagator 232
   bare **192**, 204
   as function of dimension 193
$\tilde{\phi}(\mathbf{k})$ 190

Q

quantum field theory 30, 188, 192, 212, 369ff
   and perverse notation 307

R

radius of convergence 196
   of expansion around T_c 309
random numbers 92
real-space renormalization **113**, 158
   approximations in 119, 146, 149
   critical exponents from 133ff
   even and odd parameters 130
   fixed points 120ff
   neglect of long-range interactions in 151
   treatment of bond percolation 143ff
   treatment of Ising model 147ff, 156

rectangular lattice, blocking of 115
reduced correlation functions 207
reduced vertex functions **213**, 305
reference field, for functional Taylor expansion 301
refractive index 8
regular part of thermodynamic potentials 25
regularity of renormalization transformation 141
relevance and renormalizability 369ff
relevance of operators 359
    in Gaussian model 355ff, 359, 373
    dependence on dimension 359–360
    in two dimensions 366ff
    of O_6 in three dimensions 362ff
relevant perturbation: see relevance of operators
renormalizability 242
    and relevance 369ff
    of Landau–Ginzburg model 241ff, 305
    of non-linear σ-model 333, 336
renormalization 31
    at higher orders 241
    at $m \neq 0$ 299
    field renormalization **235–238**, 245–246
        and coupling constant 238
    of connected correction function G_c 132
    of correlation length 122, 131
    of coupling constant 238, 250, 256
    of external field 129
    of $\Gamma^{(n,l)}$ 308
    of Hamiltonian 117
    of lattice 114
    of magnetization 131
    of ϕ^6 model in $d = 3$ 362
    of susceptibility 132
    point **240**, 245, 284, 286
    real-space 113
    schemes
        decimation 116
        linear 116, 127
        majority rule 116
            for Ising model 149
        restrictions 116
    the truth about 371
    transformation, regularity of 141
renormalization group 113, **284**
    derived by varying cutoff 336
    derived by varying renormalization point 283–285
    equation **284–285**, 294–295, 300, 310–311, 315–316, 319, 329
        for $\Gamma^{(0,2)}$ 313
    for non-linear σ-model 335
    for ϕ^n model in $d = 2$ 367
    for ϕ^6 model 362
    for $T \neq T_c$ 310
    solution of **287–289**, 295, 311, 314, 315, 336
    position-space 113
    real-space 113
renormalized coupling constant
    dimensionless
        critical value for 259, 266

    in massive theory 257
    in massless theory 265
    in massive theory 257
    in massless theory 264
renormalized mass, relation to bare mass 254
renormalized parameters **231**, 241–244, 245, 283, 305
renormalized vertex functions 283, 307, 362
    in non-linear σ-model 334
repulsive fixed points 120
    Gaussian, in $d < 4$ 292
    of non-linear σ-model 338
rotational invariance 320, 323
running coupling constant 289
running value of t_R 311
Rushbrooke G. S. 29
Rushbrooke scaling law **29**, 32, 140

S

sampling of configurations 87
sand 383
Saturn graph **231–232**, 235, 238, 243, 296–297
scaling 29
scaling laws 27, 140, 282
    Fisher law **29**, 32, 140
    Griffiths law **29**, 32, 140
    Josephson law **29**, 31, 140, 314
    Rushbrooke law **29**, 32, 140
scattering of neutrons 375
screening by vortices 343
series expansions for percolation 146
Sherman theorem for path amplitudes 395
similar but distinct diagrams 224, 365, 422
simulations, numerical 84ff
    and calculation of critical exponents 111, 153
singular portions of potentials 25
site percolation 60
small-wavevector limit **210**, 355
Smoluchowski equation 102
solution to renormalization group equation **287–289**, 295, 311, 314, 315, 336
    at $T = T_c$ 287–289
    at $T \neq T_c$ 311
source field J **181**, 189, 192
    Fourier transform $\tilde{J}$ 190
source-blob **197**, 208, 209
specific heat
    and Landau theory 185
    constant field versus constant magnetization 312, 318
sphere, n-dimensional, surface area of 83
spherical model **58–59**, 66–72, 95
spin, probability distribution for 51
    relation to Gibbs free energy 221, 227, 386ff
spins, non-interacting 43
spontaneous magnetization 48–52, 216, 221, 390
spontaneous symmetry breaking 48–52, 184, 216, 221, 324, 390
    and quantum field theory 31

finite systems 51, 227
  tests for 51
square block, in diagrams for ϕ^2 correlation functions 301, 304
square lattice, blocking of 114
statistical physicist's argument 370
steepest descent, method of 172, 393
step function $\Theta(x)$ 206
stiffness 324
structure charts, describing algorithms **91**, 95, 100, 103, 104, 108, 110
superconductors 14
  thin-film 348
superfluids 13, 15
suppression of fluctuations in Gaussian model 194
susceptibility 7, 42, 209
  and landau theory 186
  and mean-field theory 173
  in Landau–Ginzburg model 254
  renormalization of 132
Swendsen, R. H. 106, 153
Swendsen–Wang algorithm 106–108
symmetry, discrete scaling 114
symmetry, broken: see spontaneous symmetry breaking
symmetry factor **199**, 305
  examples 203, 226
  for ϕ^2 correlation functions 302
  for multi-component models 426
  rules for 200–203, 208, 209, 214
symmetry, hidden 324
symmetry point 264
symmetry, unbroken 50

T

theories, unrenormalizable 244
thermal averages **11**, 34, 41, 300
  of microscopic quantities 42ff
thermal eigenvalue 130
thermal equilibrium 33–35, 47
thermodynamic limit **41**, 51, 227, 388
thermodynamic potentials 21, 35ff, 378
  extended 25
thermodynamics 35–38, 42, 378
thin films, of superfluid helium 348
third law of thermodynamics 36, 52
't Hooft, G. 270
Thouless, D. J. 339
Toeplitz determinants 79
trajectory 124
transfer matrices, and one-dimensional Ising model 61
transformation, Gaussian 399
transforms Legendre **24**, 26, 32
transition temperature 1
  and thermodynamic potentials 26
translational invariance 16, 166, 207
tree structure of connected diagrams 212, 418
triangular lattice, blocking of 115
$\Theta(x)$ (step function) 206

turning point, in renormalization-group flow 135
two-point correlation function 16
two-point vertex function 230, 241
  in non-linear σ-model 329
type I and II superconductors 15

U

ultraviolet divergences 249–253
ultraviolet stable fixed point 290
unbiased estimator 87
unit matrix 331
universality 21, 156, 259, 353ff, 360
  class **21**, 182
  dynamical 105
universality, and connection of Landau–Ginzburg and Ising models 404
unreliability of Gaussian estimates of relevance 369
unrenormalizable theories 244
upper critical dimension 255

V

van der Waals theory of non-ideal gas 23, 163
variance, of estimator 88
Veltman, M. 270
Verlet algorithm 96
vertex **198**, 421
vertex functions 211ff, **212**, 216ff, 228, 300, 304
  anomalous definitions of $\Gamma^{(0)}$, $\Gamma^{(1)}$, $\Gamma^{(2)}$ 215, **420**
  as building blocks of connected diagrams 212
  definition below T_c 220–221
  dependence on a 245
  field-dependent 226
  four-point 238, 241
  reduced 213
  relation to Gibbs free energy 215, 417ff
  renormalized 246
  rules for finding 214
  small wavevector limit 355
  two-point 230, 241
    in non-linear σ-model 329
vortices 321, **339**
  dipoles of 343
  energy of 341
  interactions 342
    analogy with electrical charges 343
  in Landau–Ginzburg model 350
  separation from Goldstone modes 342
  unobtainable from non-linear σ-model 340

W

Wagner, H. 324
Wang, J-S. 106
water
  energy fluctuations 41
  isotherms of 4
  metastability 47
  transition to ice 2

transition to steam 3
wavevector conservation **206**, 304–305
wavevector space 205
  correlation functions 204ff
  Feynman rules 205, **209**
  versus position space 210
wavevectors in Landau–Ginzburg model 190
  integration over 304
weak non-renormalizable interactions 372
  gravity as example 372
Weiss P. E. 158
Weiss theory of ferromagnetism 158
Widom homogeneity hypothesis **27–28**, 142
Wilson, K. G. 113
Wolff, U. 108
Wolff algorithm 108–110
work 35, 36, 380

considerable amount of 332

X

XY model 57–58
  and Kosterlitz-Thouless transition 339
  in two dimensions 320, 339–348
  vortices in 339

Z

Z_2, and renormalization of temperature 306–307
$Z(b)$, scaling factor for order parameter 355
$Z[J]$ for Landau–Ginzburg model, analyticity 196
  rules for finding 195ff